PROCEEDINGS OF SYMPOSIA
IN PURE MATHEMATICS
Volume 48

The Mathematical Heritage of Hermann Weyl

R. O. Wells, Jr., Editor

AMERICAN MATHEMATICAL SOCIETY
PROVIDENCE, RHODE ISLAND

PROCEEDINGS OF THE SYMPOSIUM
ON THE MATHEMATICAL HERITAGE OF HERMANN WEYL
HELD AT DUKE UNIVERSITY
DURHAM, NORTH CAROLINA
MAY 12-16, 1987

with support from the National Science Foundation, Grant DMS-8611533

1980 *Mathematics Subject Classification* (1985 *Revision*). Primary 00, 01.

Library of Congress Cataloging-in-Publication Data

Symposium on the Mathematical Heritage of Hermann Weyl
(1987: Duke University, Durham, North Carolina)
The mathematical heritage of Hermann Weyl/R. O. Wells, Jr., editor.
p. cm.--(Proceedings of symposia in pure mathematics, ISSN 0082-0717;
v. 48)
ISBN 0-8218-1482-6
1. Groups, Theory of--Congresses. 2. Weyl, Hermann, 1885-1955--Congresses. I. Wells, R. O. (Raymond O'Neil), 1940- . II. Title. III. Series.
QA171.S96 1987
512'.22--dc19   88-19367

Table of Contents

Preface

This volume celebrates the rich legacy we have inherited from Hermann Weyl, one of the great mathematicians of this century. It represents the proceedings of a conference held in his honor at Duke University, May 12-16, 1987, two years after the 1985 centennial of his birth. This conference lasted 5 days and featured 23 speakers, almost all of whom have contributed articles to this volume.

The organizing committee consisted of Michael Atiyah (Oxford), Lipman Bers (Columbia), Felix Browder (Chicago), S. S. Chern (Berkeley), B. D. Mostow (Yale), and myself as chairman. We decided to have a wide spectrum of speakers representing many of the diverse areas in which Weyl made significant contributions. We intended from the beginning to have a conference of sufficient size so that mathematicians, graduate students, and others interested in Hermann Weyl's mathematics would be able to attend such a conference from all over the country and from abroad. There were other celebrations of Hermann Weyl's 100th birthday, most notably the lectures by Armand Borel, Roger Penrose, and C. N. Yang in Zürich in October of 1985. The conference at Duke allowed the North American mathematical community to participate in the celebration, noting that the last part of Weyl's career was spent at the Institute for Advanced Study in Princeton.

The speakers at the conference and the titles of their talks were:

Raoul Bott (Harvard), Induced representations

Felix E. Browder (Rutgers), Hermann Weyl as a philosopher, and the difference it made to his mathematics and physics

Dennis P. Sullivan (CUNY and IHES), Riemann surfaces applied to one-dimensional dynamical systems

Robert P. Langlands (IAS), Representation theory and arithmetic

David A. Vogan, Jr. (MIT), Non-commutative algebras and unitary representations

Roger E. Howe (Yale), The oscillator semigroup

James G. Arthur (Toronto), Harmonic analysis and the trace formula

James I. Lepowsky (Rutgers), Vertex operators and the Monster

I. M. Singer (MIT), Some mathematical aspects of string theory

Louis Nirenberg (NYU), Nonlinear elliptic equations

Phillip A. Griffiths (Duke), Value distribution theory

Robert L. Bryant (Rice), Surfaces in conformal geometry

H. Blaine Lawson, Jr. (SUNY, Stony Brook), Algebraic cycles and homotopy

S.-T. Yau (California, San Diego), Yang-Mills theory over Kähler manifolds

Ronald G. Douglas (SUNY, Stony Brook), Invariants for elliptic operators

Michael F. Atiyah (Oxford), New invariants for manifolds of dimensions 3 and 4

Clifford Taubes (Harvard), The stable topology of self-moduli spaces: A nonlinear Hodge theory

Roger Penrose (Oxford), Fundamental asymmetry in physical laws

Edward Witten (Princeton), Some mathematical applications of quantum field theory

There were some 300 participants who attended the conference, and the high level of exposition and inspiring lectures had a magnetic effect on the entire conference. Both C. N. Yang (SUNY, Stony Brook) and Harry Furstenberg (Hebrew University) were scheduled to speak, but were unable to attend for personal reasons, and their presence was missed.

Hermann Weyl was one of the most successful mathematicians of all time to tackle successfully serious problems in physics as well as philosophy. Currently we are witnessing a tremendous interaction between mathematics and physics, on a scale and in a fashion not seen in recent times since the Göttingen days of the invention of quantum mechanics in the early decades of this century. A significant number of the papers in this volume bear vivid testimony to this time-honored tradition of the mutual influence of these two, in principle quite different, intellectual activities.

The eighteen papers in this proceedings are presented in the order given, which was partly topical. The opening contribution is by Raoul Bott who spent some time with Hermann Weyl at the Institute for Advanced Study, and the closing two papers by Roger Penrose and Edward Witten represent contributions in the area of physics, a subject which was very important to Weyl. Included in the collection is a second paper by Roger Howe (which was not presented at the conference) on the classical groups, a subject in which Weyl made major contributions, and for which Howe's survey is most appropriate.

This conference was supported financially by the National Science Foundation, whose generous help is much appreciated. It was sponsored by the American Mathematical Society whose centennial is in 1988. I'm happy that the volume can help celebrate both the centennial of Hermann Weyl and make a contribution to the celebration of the mathematical heritage of the 100-year-old American Mathematical Society at the same time. The dedicated staff of the AMS, in particular Dottie Smith and John Balletto, took care of numerous logistical details and organized a splendid conference setting for the participants. I want to thank them for doing a splendid job. Duke University was very generous in its support, and I want to thank, in particular, Michael C. Reed, Chairman of the Mathematics Department, and Ann Tustall, Administrative Assistant, for their generous help and assistance after the Duke site was chosen. Phillip Griffiths, Provost of

Duke University, as well as a mathematical protégé of the heritage of Hermann Weyl, helped insure that everything went very smoothly. I'd like to thank him for both his assistance and for his fine lecture at the conference. Finally, I want to thank all of the other speakers for contributing beautiful lectures, and for writing up their ideas in a fashion which gives the contemporary reader some idea of the breadth and beauty of the ideas of Hermann Weyl.

R. O. Wells, Jr.
Rice University
May 1988

Proceedings of Symposia in Pure Mathematics
Volume 48 (1988)

On Induced Representations

RAOUL BOTT

It is with a great sense of reverence and gratitude that I open these proceedings in honour of Hermann Weyl. Reverence and gratitude for the breadth and beauty of his work I believe we all share, but by a stroke of quite exceptional good fortune I not only knew Hermann Weyl to a certain extent, but am even beholden to him for an intercession which proved to be pivotal in my life. For this reason I would like to preface my more technical remarks by a few personal words of recollection and appreciation.

At a crucial point of a lecture, Hermann Weyl had the habit of lifting his shoulders and then letting them fall again. This motion seemed to convey the inevitability of that particular turn of thought, the God-given nature of our subject, and the minor role that he himself might have had in its development. Hermann Weyl's seminar lectures were not particularly easy to understand, but they always had about them this air of ease, of a natural motion in the inevitable stream of the subject itself. I forget what he talked about in 1948—or was it '49—when he visited the Carnegie Institute of Technology, where I was at the time just finishing my metamorphosis from engineering to applied mathematics, but I remember the light feel of it quite distinctly and it was later reinforced often.

After that lecture we went to dinner—ah, the blessing of a small graduate school where the handful of graduate students are a natural part of all colloquium dinners—and I got to know Hermann Weyl personally. He professed an interest in my thesis—in network theory—and the next morning I presented some aspects of it to him, and in particular the existence and uniqueness proof for the distribution of currents in a network consisting of resistors alone. Let me present this argument to you here so that you may see how amusing it must have been for Hermann Weyl to hear my recital.

Let us deal with the network problem in its simplest guise, as indicated in Figure 1.

We force external currents $\{I_j\}$ to flow into the network, at the vertexes $\{j\}$, and ask what voltage distributions $\{E_j\}$ will be measured at these same vertexes

1980 *Mathematics Subject Classification* (1985 *Revision*). Primary 14F12, 14F40, 14F20, 58D15, 22E46, 58A14, 53C20.

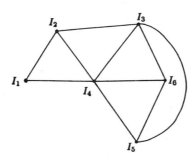

FIGURE 1

after a steady state has been achieved. The edges of the network are made of resistors with impedance g_α, that is, resistance $r_\alpha = 1/g_\alpha$.

If we denote by $C^0(N)$ and $C^1(N)$ the **R**-valued cochains on N and write

$$(1) \qquad\qquad C^0(N) \xrightarrow{\delta} C^1(N)$$

for the usual coboundary, then—so I told him, of course in a quite different notation—the g_α should be interpreted as an inner product on $C^1(N)$. Thus a 1-cochain $f \in C^1(N)$ should be given the norm

$$(2) \qquad\qquad (f, f) = \sum_\alpha f^2(\alpha) g_\alpha, \qquad \alpha\text{—oriented edges,}$$

while 0-cochains in $C^0(N)$ should be given the norm

$$(3) \qquad\qquad (f, f) = \sum_p f^2(p), \qquad p\text{—vertices.}$$

With this agreed, δ now has a well-defined adjoint

$$\delta^* : C^1(N) \to C^0(N)$$

and in terms of it the Kirchoff Laws take the form:

$$(4) \qquad\qquad \delta^* \delta E = I.$$

Here we have of course interpreted both E and I as 0-cochains, that is, elements of $C^0(N)$. From the transparent positivity of our inner products it now follows that for (1)–(3) to have a solution, I must satisfy the compatibility equation:

$$(5) \qquad\qquad I \in \operatorname{Im} \delta^* \leftrightarrow I \text{ perpendicular to } \operatorname{Ker} \delta \ (= H^0(N))$$

and if I satisfies this condition then E is unique, again up to $\operatorname{Ker} \delta = H^0(N)$. All this then translates to the fact that

(a) *the sum of the currents into each component of N must be zero,* and that

(b) *once a single vertex in each component of N is grounded* (i.e., E is set equal to zero at p) *then* (1)–(3) *uniquely determines E.*

Hermann Weyl chuckled after hearing this story and referred me to a mini-paper in the Rendiconti di Palermo where he had given the identical argument 25 years earlier. Whether at that time he mentioned his far-reaching "method of orthogonal projection" in the harmonic theory of N-manifolds I do not recall, and I became aware of it and the connection to this example only a year later, when—thanks in part to his intervention—I was invited to the Princeton Institute and spent the years 1949–1951 there.

I do not want to repeat here what I have already written on two other occasions about the excitement of these years at Princeton. Let me just add a few words concerning Hermann Weyl. Of all the great figures at Princeton at the time, he seemed to me the only one who felt truly at ease in the place. Just as mathematics was second nature to him, so this was a natural milieu for him. He took the position that his "hardworking" days were behind him and that he would now enjoy mathematics—grosso modo—at second hand. I often had lunch at his table in the little cafeteria above Fuld Hall, and recall remarks such as: "You look as if you hadn't slept too much, young man—*now I* sleep wonderfully!" Or again—"I don't see how you young people can do mathematics now; it has become so complicated. In my day we had such simple problems to solve."

It is of course not true that Weyl had at this time—his age must have been close to my own now—retired from mathematics.

In 1949 he very actively ran a seminar on the Hodge theory where finally two water-tight arguments for its validity were being presented, one by de Rham and one by Kodaira. And later in 1954 he gives a masterly account of sheaf theory, line bundles, and such in his address to the International Congress, awarding the Fields Medal to Serre and Kodaira.

This account also beautifully displays his warmth and joy at new mathematics, and his gift for enthusiastic praise. To Kodaira he says, "Your work has more than one connection with what I tried to do in my younger years; but you have reached heights of which I never dreamt. Since you came to Princeton in 1949 it has been one of the greatest joys of my life to watch your mathematical development."

This is then the Hermann Weyl I remember: generous, warm, encouraging, friendly—in no way overbearing. If anything, there was a genuine humility about his lunchtime reminiscences, his references to the mathematical genius of C. L. Siegel, D. Hilbert, von Neumann and others. I remember an occasion when he reminisced about Peter—the Peter of the Peter-Weyl Theorem. He spoke of him with affection and great good humor. He even extolled Peter's wisdom for us. It seems that after a year of research, during which time he and Hermann Weyl discovered the beautiful theorem which now bears their name, Peter decided against pursuing a research career. He preferred his schoolmaster's hut, high in the mountains. "There is a wise man," Weyl would remark, his eyes dancing with mirth.

I mention this aspect of Hermann Weyl's nature because it is not apparent in his writing. Indeed there, as well as on more formal occasions and lectures, it

was hard to miss Weyl's Olympian quality. On these occasions he could not—
and would not—hide his poetic nature, his immense cultural overview and his
gift for words.

Some people found this offputting. A few weeks ago I met a physicist at Cern
who told me that he had to put down *Classical groups* because he felt himself
enveloped by a Wagnerian drama. For all those let me record here just one last
anecdote from the years 1949–50. Once while standing in line at the cafeteria I
brought up his early book, *Die Idee der Riemannschen Fläche* in which so many
of the concepts we now take as a matter of course were first clearly enunciated.
Weyl made a face. "I can't bear to look at it now," he said. "How pompous one
can be at 26!" And I remember that we then—quite characteristically—drifted
from this topic to discuss the notion of "identity" and the tenuous thread that
links a 60-year-old to his 26-year-old incarnation.

But let me turn now to my subject proper, and very briefly outline the Hermann Weyl character formula and some of the developments it initiated.

Hermann Weyl brought a geometric and global point of view to the whole
subject of semisimple Lie group theory which has stood as a model for many of
us, and still points the way in the most recent developments. Let me recall the
first such link, which, A. Borel tells us in his excellent account of Weyl's work
on groups [**B**], was already singled out by I. Shur for special praise.

Let then K be a compact connected semisimple Lie group and $T \subset K$ be its
maximal torus. For $K = \mathrm{U}(n)$, T should be thought of as the group of diagonal
matrices:

$$(6) \qquad T = D(n) = \left\{ \begin{matrix} z_1 & & \\ & \ddots & \\ & & z_n \end{matrix} \right\}, \qquad z_i \in S^1,$$

and the first fundamental fact of the subject is the following extension of the
spectral theorem.

$$(7) \qquad \textit{Every } k \in K \textit{ is conjugate to an element of } T.$$

It follows that class-functions f on K are determined by their values $\iota_T^* f$ on
T alone, and it therefore stands to reason that if dk denotes a left-right invariant
volume on K then there must be a measure du on T with the property

$$(8) \qquad \int_K f \, dk = \int_T \iota_T^* f \cdot du$$

for all integrable class-functions f on K.

Hermann Weyl now finds two beautiful formulas for du, in terms of the "positive roots $\alpha > 0$ of K on T," and the "Weyl group W of K." Recall here that
W is the group of automorphisms of T induced by *inner automorphisms of K*,
or equivalently

$$(9) \qquad W = \text{Normalizer of } T/\text{Centralizer of } T,$$

while the "roots" $\Sigma = \Sigma(K)$ of K are by definition the characters of the irreducible representation into which the tangent space of K/T at the coset T decomposes under the left action of T.

In Lie algebra terms we have

$$(\mathfrak{k}/\mathfrak{t}) \otimes \mathbf{C} \simeq \sum_{\alpha} E_{\alpha}, \qquad \alpha \in \hat{T}, \tag{10}$$

with $\hat{T} = \text{Hom}(T, \mathbf{C}^*)$ the character group of T, while $\mathfrak{k}$ and $\mathfrak{t}$ are the Lie algebra of K and T respectively.

Because the vector space $\mathfrak{k}/\mathfrak{t}$ is real, every root α occurs with the inverse α^{-1} so that it is natural to partition Σ into a "positive set" of roots Σ_+, and their inverses, the negative roots:

$\Sigma = \Sigma_+ \amalg \Sigma_-$. Of course this choice is to be made with some compatibility relative to the Lie structure of $k \otimes \mathbf{C}$; that is, one would like the relations:

$$[E_{\alpha}, E_{\beta}] \subset E_{\alpha\beta} \tag{11}$$

to hold whenever α, β and $\alpha\beta$ are in Σ_+. Weyl shows that such choices of Σ_+ do exist and in fact that they are in 1-1 correspondence with the fundamental chambers into which the action of W partitions t.

It took twenty more years (and the Nirenberg-Newlander theorem!) for this compatibility condition to be interpreted in its most geometric form, namely as an integrability condition for a homogeneous complex structure on K/T. Indeed a choice of Σ_+ induces an almost complex structure on K/T by declaring that the E_{α}, $\alpha > 0$, generate the $(1,0)$ part of the tangent space to K/T at T and then translating this subspace by the group action. This done, it then becomes clear that (11) is precisely the integrability condition for this structure.

But I am getting ahead of my story, namely Hermann Weyl's formula for du, in the *semisimple* case. It is the following one:

$$d\mu = |A|^2 \cdot dt \cdot \frac{1}{|W|}, \qquad |W| = \text{order } W, \tag{12}$$

with

$$A = \prod_{\alpha > 0} (\alpha^{1/2} - \alpha^{-1/2}). \tag{13}$$

Furthermore this A is not only well defined, but is antisymmetric as regards the action of W on $\hat{T}$, and can also be described by the expression:

$$A = \Sigma(-1)^{|w|} \cdot \rho^w \tag{14}$$

where $\rho \in \hat{T}$ is the positive square root of the product of all the $\alpha \in \Sigma_+$:

$$\rho^2 = \prod_{\alpha > 0} \alpha. \tag{15}$$

This formula is delicate—in the sense, for example, that it does not hold for U_2 but only for SU_2—which *is* semisimple. Indeed for U_2 there is only one positive root: $z_1 z_2^{-1}$, which clearly has no square root in $\hat{T} \approx \mathbf{Z} \oplus \mathbf{Z}$. This root restricts to z_1^2 on SU_2 and there clearly has one. Thus for SU_2

$$(16) \qquad A = z_1 - z_1^{-1}.$$

In terms of these concepts the Weyl character formula is now given by the following

THEOREM 1 (1925). *Let V be an irreducible K-module (i.e., representation)*

and χ_V its character. Then

$$(17) \qquad \chi_V | T = A^{-1} \cdot \left(\sum_{w \in W} (-1)^w (\lambda \rho)^w \right)$$

where $\lambda \rho \in \hat{T}$ is a unique positive character on T.

I have always loved this formula, and regret that I learned it only years after I had left the Institute of Hermann Weyl's proximity. What I liked about it especially is that one had to look to the skew-symmetric expressions under W before one found a simple algorithm for the manifestly W-invariant expression χ_V. In contemporary geometric terms, the square roots which occur here really have to do with a *spin structure* on K/T, and the difficulty of taking the square root of $\prod_{\alpha > 0} \alpha$ in U_n, say, is a hint at the difficulty of choosing equivariant spin structures in the loop group situation.

But I am again looking ahead. Beautiful as this expression for χ_V is, it has the drawback of all "pure character formulae". We would also like to "see" the representation, preferably created out of the group itself. Actually this addendum to (17) was "just in the air" in the early fifties due to the whole development of sheaf theory, line bundles, Riemann-Roch, Kodaira vanishing theorems, and so on, and is known by the name of the Borel-Weil theorem. (In fact, it is sometimes in my presence even referred to by very kind people as the Borel-Weil-Bott theorem.)

But it took the genius of A. Weil and Borel to see all the pieces and formulate the result properly—and so to give us the "Borel-Weil" theorem. It appeared in the May 1954 Bourbaki Séminaire report by J.-P. Serre. His account starts from the imbedding which can be associated to an irreducible representation of K on V with maximal weight λ. Indeed, choose a maximal weight vector v_λ and consider the orbit it sweeps out in $P(V)$, the projective space of 1-dimensional subspaces of V. This orbit which is of course K/P_λ where P_λ is the stabilizer of v_λ, turns out to be a nonsingular complex submanifold—and hence algebraic!— and one can now reinterpret the representation in terms of the pullback of the canonical line bundle from $P(V)$.

But let me here rather start from the point of view that the Borel-Weil construction amounts to combining the very classical notion of "induction" with that of complex structures. Precisely, recall that if $H \subset K$ is a closed subgroup and λ a representation of H on $\operatorname{End} W$, then λ defines a vector bundle

$$(18) \qquad W_\lambda = K \times_H W$$

over K/H, with fiber W, whose space of sections

$$(19) \qquad \Gamma(W_\lambda) \sim \{f : K \to W | f(kh) = \lambda^{-1}(h) \cdot f(k)\}$$

coincides with the "induced representation space of λ" in the realm of finite groups. In our context we take $\Gamma(W_\lambda)$ to be C^∞ so that this "induced representation space" is infinite-dimensional, but it is in any case naturally a K-module under the action

$$(20) \qquad f(x) \rightsquigarrow f(k^{-1}x)$$

of left multiplication of $k \in K$.

We turn next to the complex analytic structure of the space Γ associated to the particular situation: $T \subset K$, and $\lambda : T - \operatorname{End} \mathbf{C}$ simply a character on T. The induced bundle $\mathbf{C}_\lambda(K/T)$ is then just a *line-bundle*, L_λ, over K/T and it is then easy to see that a choice of positive roots for K on T not only describes a complex structure on K/T but also on L_λ, so that one can speak meaningfully of the subspace of *holomorphic sections* $\Gamma_{\mathrm{hol}}(L_\lambda) \subset \Gamma(L_\lambda)$.

This space is *finite-dimensional* by compactness and ellipticity arguments, is naturally a K-module under the action described earlier, and is in fact the Borel-Weil module which realizes the character formula (17):

Precisely one has the following:

THEOREM 2 (BOREL-WEIL). *For $\lambda\rho$ a positive character of T, the K-module $\Gamma_{\mathrm{hol}}(L_\lambda)$ constitutes an irreducible representation of K, whose character is given by (17), and as $\lambda\rho$ ranges over the positive characters of $\hat{T}$ these modules range over all the irreducibles of K in a 1-1 manner.*

Let me now turn to my favorite proof of why the representation of K on $\Gamma_{\mathrm{hol}}(L_\lambda)$ has the expression (17) for its character. This proof goes via a fixed point theorem which Atiyah and I noticed in 1964 [**AB**]. Our formula is very conceptual and broadly applicable to elliptic complexes in general, but was also very much "in the air" in the holomorphic context at that time, in particular in the work of Shimura and Eichler.

Let us start from the following "obvious" (and therefore correspondingly fundamental) remark concerning characters of finite groups. Namely, suppose G is represented as permutations of a finite set X and that $V = \mathscr{F}(X)$ is the corresponding G-module of real-valued functions on X. Then the value of χ_V on g is given by:

$$(21) \qquad \chi_V(g) = \# \text{ of fixed points of } g \text{ in } X.$$

Indeed in the obvious basis for $\mathscr{F}(X)$ furnished by the δ-functions on X the matrix representing g will have diagonal entries equal to $+1$ precisely at the fixed points of g.

Now Atiyah and I observed first of all that there is a very natural extension of this principle in the C^∞-category. Namely, suppose that $\varphi\colon M \to M$ is a diffeomorphism of a manifold M, and that we are interested in the trace of the induced map φ^* on $\mathscr{F}(M)$, the C^∞-functions on M.

Then (21) has a very plausible extension *provided only* that the graph of φ is "transversal to the identity" and that the set of fixed points is finite. Under these conditions, we argued, one has:

$$(22) \qquad\qquad \text{trace}\,\varphi^* = \sum_{\varphi(p)=p} \frac{1}{|\det(1 - d\varphi_p)|}.$$

The rationale for (22) is that on the one hand smooth endomorphisms of $\mathscr{F}(M)$ of the type $f(x) \rightsquigarrow \int K(x,y)f(y)\,dy$ clearly have $\int K(x,x)\,dx$ for their trace (if this integral makes sense), while on the other, our φ^* has for its "distributional kernel" $K(x,y) = \delta(x,\varphi(x))$, the "delta function centered on the graph of φ", so that the trace of φ^* should be given by

$$(23) \qquad\qquad \text{trace}\,\varphi^* = \int \delta(x,\varphi(x))\,dx,$$

and this integral immediately leads to (22)—provided only that the graph of φ is transversal to the diagonal—as any physicist worth his salt will tell you. In more modern terms, this is an example of the fact that one has a well-defined "*product*" for distributions whose wave-front sets are disjoint.

Granting (22), let us consider the action of a generic element $t \in T$ on the induced representation space $\Gamma(L_\lambda)$. This action covers the geometric action of t on K/T and hence is a slightly twisted form of the φ^* just described, to which the expression (22) easily extends. Hence we need first to understand the fixed *points* of $t \in T$. However these, as I believe A. Weil was the first to point out, are easy to find and correspond 1 to 1 with the Weyl group W of K! Indeed for a coset kT to be fixed by $t \in T$ we must have

$$t \cdot k \cdot T = kT$$

which implies that $k^{-1}tk \subset T$ whence, because t is generic in T, $k^{-1}Tk \subset T$— i.e., $kT \in W(K)$.   QED

Furthermore, in this situation these fixed points are all transversal! In short, for generic t, one obtains the formula:

$$(24) \qquad\qquad \text{trace}\,t \text{ on } \Gamma(L_\lambda) = \sum_{w \in W} \frac{\lambda^w(t)}{|\det(1 - dt_{K/T})|^w}.$$

Here the numerator $\lambda^w(t)$ comes from the twist that distinguishes $\Gamma(L_\lambda)$ from $\mathscr{F}(K/T)$ and by $dt_{K/T}$ we mean the differential of the action of t on the tangent space to K/T at T. But the differential of this action just rotates the root spaces E_α by $\alpha(t)$ so that:

$$|\det(1 - dt_{K/T})| = \left| \prod_{\alpha>0}(1 - \alpha(t)) \right|^2.$$

In short this factor is simply the value of Hermann Weyl's $|A|^2$ at t, so that (22) also takes the form:

$$\text{(25)} \qquad \text{character of } \Gamma(L_\lambda) \mid \text{generic } T = |A|^{-2}\Big(\sum_{w\in W} \lambda^w \Big).$$

We seem to be off by a square root of $|A|^2$ from the Hermann Weyl formula (17) but that is all for the good, because the expression we have before us is a heuristic character for *the whole representation induced by λ*, whereas for the Borel-Weil Theorem we are seeking only the character of t acting on the *holomorphic part of $\Gamma(L_\lambda)$*. As we will see in a moment, by taking this into account we will indeed find a square root of $|A|^2$—however, a different one from the one Weyl chose.

For this last step we have to come to grips with the manner in which the holomorphic part of $\Gamma(L_\lambda)$ is "cut out" of $\Gamma(L_\lambda)$, and this consideration brings us inevitably into the realm of "elliptic complexes".

Recall then that the $\bar\partial$ operator is "elliptic" only in its *entirety*—that is, acting on *all* the $(0,q)$ forms $\Omega^{0,q}(M;E)$ with values in a holomorphic vector bundle E. In our case then the pertinent elliptic complex is given by

$$\text{(26)} \qquad 0 \to \Omega^{0,0}(L_\lambda) \xrightarrow{\bar\partial} \Omega^{0,1}(L_\lambda) \to \cdots \to \Omega^{0,m}(L_\lambda) \to 0.$$

It has $\bar\partial^2 = 0$ and hence gives rise to cohomology groups $H^{0,q}(K/\Gamma; L_\lambda)$, which we denote by $H^q(L_\lambda)$ for simplicity's sake. Our group K acts naturally on everything in sight, and by ellipticity the $H^q(L_\lambda)$ are all finite-dimensional K-modules, with $H^0(L_\lambda)$ identical to the group $\Gamma_{\text{hol}}(L_\lambda)$ of the Borel-Weil theory.

From this viewpoint it is therefore more natural to seek a character formula for all of $H^*(L_\lambda)$ and in particular the Lefschetz Principle should apply. Namely, *the character of the virtual module $\sum(-1)^q H^q(L_\lambda)$ should equal that of the virtual module $\sum(-1)^q \Omega^{0,q}(L_\lambda)$.*

Now the fixed point formula again applies to each of the K-modules $\Omega^{0,q}(L_\lambda)$ for *they also are induced by representations of T*. Namely $\Omega^{0,q}$ is induced by the natural representation of T on $\mathbf{C} \otimes \Lambda^{0,q}(\mathfrak{k}/\mathfrak{t})$ given by $\lambda \otimes \Lambda^{0,q}$, and these exterior power representations now serve to cancel one half (the antiholomorphic half) of the $|\det(1 - dt)|^2$ factor in $|A|^2$. This comes about through the well-known identity:

$$\text{(27)} \qquad \det(1 - A) = \sum(-1)^q \text{ Trace } \Lambda^q A.$$

In short, one now obtains the relation:

$$\text{(28)} \qquad \text{char} \sum(-1)^q H^{0,q}(L_\lambda) \mid \text{gen } T = \sum_{w\in W} \Big\{ \frac{\lambda}{\prod_{\alpha<0}(1 - \alpha)} \Big\}^w.$$

At this point the r.h.s. is seen to be the r.h.s. of (17). Indeed from the identity $1 - \alpha^{-1} = \alpha^{-1/2}(\alpha^{1/2} - \alpha^{-1/2})$ it follows that

$$\text{(29)} \qquad \prod_{\alpha>0}(1 - \alpha^{-1}) = \rho^{-1} \cdot A,$$

and hence that

$$(30) \qquad \sum \left\{ \frac{\lambda}{\prod(1 - \alpha^{-1})} \right\}^w = A^{-1} \left(\sum (-1)^w (\lambda \rho)^w \right),$$

as was to be shown.

Finally the Kodaira vanishing theorem comes into play for the l.h.s., to complete the story. For the character $\lambda \rho$ "positive" it is seen to imply that all the higher terms in (28) *vanish*, so that in that case the virtual module on the left reduces to the honest module $H^0(L_\lambda)) = \Gamma_{\text{hol}}(L_\lambda)$.   QED.

Two very natural questions now come to mind. The first is: what happens when $\lambda \rho$ is not positive? The relation (26) is of course still valid, but the Kodaira vanishing theorem is not directly applicable. It was this question which motivated my paper on homogeneous vector bundles in 1958, and the answer turned out to be the best one possible compatible with the relation (26).

The result is as follows: *Let* $\lambda \in T^*$ *be given. Then two possibilities arise*:

(a) $(\lambda \rho)$ *is on the boundary of a fundamental chamber of* W *in* T^*. *In that case all the* $H^q(L_\lambda) = 0$.

(b) *If* $\lambda \rho$ *is in the interior of some chamber then there is a unique element* $w \in W$ *which moves it into the positive chamber. Thus* $(\lambda \rho)^w = \pi$, *with* $\pi > 0$. *In this situation let* $q = |w|$ *denote the number of positive roots changed by* w *into negative ones. With that understood, the* K-*module* $H^q(L_\lambda)$ *is given by*:

$$H^q(L_\lambda) \simeq H^0(L_\pi)$$

while all others $H^p(L_\lambda) \equiv 0$.

My paper is a bit heavy-handed and by now much better versions of the matters initiated there are available: notably papers of Demazure [**D**] and Kostant [**K**].

The second question arising naturally from my account is whether the Hermann Weyl A, that is, "his" square root of the measure $d\mu$, has a natural geometric interpretation as "our" square root $\prod_{\alpha > 0}(1 - \alpha^{-1})$ had in terms of the $\bar{\partial}$ operator. The answer is very much 'yes' and very *appropriately* so. Namely, K/T always admits a homogeneous *spin structure* and therefore has a corresponding Dirac-Weyl operator

$$\not{\partial}: S^+ \to S^-.$$

This operator is elliptic, its kernel and cokernel $\mathscr{H}^+$ and $\mathscr{H}^-$ are therefore finite-dimensional K-modules, the fixed point formula applies and this time produces precisely Hermann Weyl's denominator! How nice it would have been to explain these matters to him! I expect he would have been both surprised and pleased at this new guise of his old measure. But then who knows—maybe he knew it all along!

So much then for a "sixties" view of the character formula. If I look at it now—another twenty years later—I am aware of two distinct and joyous surprises which were in store for me. The first was the "geometric quantization" point of view towards these representations which has emerged in the work of Kirillov,

Kostant, Suriau, Sternberg, Blattner, Guillemin, Weinstein, Marsden, Vergne, and many others. The second is the essentially word-for-word transcription of these results in the infinite-dimensional "loop group" situation.

Let me close with just a very brief comment on each of these developments.

In my account so far the main virtue of the maximal torus $T \subset K$ was that it sliced through the conjugacy classes of K in a generically transversal manner. But K/T is also the generic orbit of the adjoint representation of K on its Lie algebra. (The K/P_λ are the others.) This was very well known to me, and in fact the basis of my Morse-theoretic proof (in 1951) that K/T had no torsion. The gist of the argument was simply that a generic linear function f on k has nondegenerate points of only even index on all the orbits of the adjoint action. In fact, for a generic orbit these critical points are again in 1-1 correspondence with the Weyl groups W and the index of the critical point labelled by $w \in W$ is always 2 ×(the number of positive roots changed to negative ones by w).

For the energy function on the loop-space, ΩK, the Morse theory yielded a mysteriously analogous result. This time the critical points were labelled by the elements of the *affine* Weyl group $\hat{W}$, with a similar formula for the indexes in terms of the "affine roots".

Now the first maxim of the geometric quantization principle is: Never look at the orbits of the adjoint action—rather always look at the orbits of the *coadjoint action*.

The reason is first of all a functorial one. Namely the orbit of K acting on $\theta \in k^*$ naturally inherits a *pre-symplectic* form ω from θ, via the simple formula:

$$\omega_\theta(\mathscr{L}(x)\theta, \mathscr{L}(y)\theta) = -\theta([x, y]), \qquad x, y \in k,$$

which, however—the novice be warned—takes some conceptual digesting. In particular, ω is a closed form and so those θ in k^* for which ω is an *integral* cohomology class, are *naturally singled out*. In short, the coadjoint orbits are naturally "*quantized*".

Further, once we are at an *integral* θ, then there is a line bundle L_ω over the orbit whose first Chern class is ω. In this way then our L_λ of the previous pages is recreated in a new context.

For a compact K the geometry of the orbits of the adjoint action on k and k^* are of course C^∞ isomorphic and any invariant metric on k induces such an isomorphism. Nevertheless this point of view sheds insight even in this case. For instance my generic coordinate function f on k, when restricted to the orbit K/T, goes over, in the new picture, to the "hamiltonian function" associated to a generic infinitesimal rotation X of T, via the symplectic form ω: that is,

$$df = \iota_X \omega.$$

Hence the *critical* points of f indeed correspond to the *fixed* points of the action of T on K/T, and hence to W. This point of view brings us into the realm of the moment map, effective Lagrangians for dynamical systems with symmetry, and even the stability theory of Mumford.

These connections then constitute, so to speak, my "first" surprise. Let me turn finally to the second one. Here it turned out, as I learned from Graeme Segal over the years, that, properly understood, my old friend the loop-space ΩK of a compact group, is in fact again a *coadjoint orbit* but only for the *true* loop group $\mathscr{L}(K)$, rather than the naive one I had in mind in the early 50s.

First let me clarify this matter. The topologists usually think of the loop-space ΩK of a space X as the set of continuous (or smooth) maps of a pointed circle S^1 into a pointed space X. If X happens to be a group, ΩX inherits this group structure by pointwise multiplication. If X is a Lie group K the "Lie algebra" of ΩX would then correspond to maps of S^1 into the Lie algebra $\mathfrak{k}$ of K, with the corresponding pointwise induced Lie structure. This situation is the naive one and the coadjoint orbits of this ΩX are not the interesting ones.

A first step in the "right" direction is that ΩK can also be thought of as a homogeneous space of the larger group $\Lambda K = \mathrm{Map}(S^1, K)$ of *all* maps of S^1 to K. Indeed, if $\mu \colon S^1 \to K$ is a point of ΛK then the assignment

$$\mu \to \mu(1)^{-1} \cdot \mu$$

exhibits ΩK as $\Lambda K / K$ where $K \subset \Lambda K$ should be thought of as the *constant* maps in ΛK. The next step literally requires a quantum leap, for it was first indicated to us by the physicists on the infinitesimal level. The point is that the representations which they constructed of the "Lie algebra of ΛK" were only projective and so pointed to a central extension of ΛK. And indeed globally the appropriate group turns out to be a central extension $\hat\Lambda K$ of ΛK by S^1.

In fact, this extension is nontrivial, even topologically. That is, $\hat\Lambda K$ is a nontrivial circle bundle over ΛK, so that it is described by a "discontinuous cocycle"—which is correspondingly difficult to write down explicitly. However, on the Lie algebra level it can be pinned down. Explicitly: the corresponding cocycle in $H^2\{\mathrm{Map}(S^1, k)\}$ is given by

$$\eta(f, g) = \int_{S^1} f' \cdot g \, d\theta.$$

Here the prime denotes differentiation and the dot denotes the positive definite inner product on k so that $\eta(f, g) = 0$ for all g implies $f' = 0$; or, $f = \mathrm{constant}$.

The full implication of this fact is precisely what we have been searching for. Indeed the generic coadjoint orbit of $\hat\Lambda K$ turns out to be isomorphic to $\Lambda K / T$, and the orbits of the canonical "new" direction, dual to the central extension, are precisely $\Lambda K / K = \Omega K$!

Let me spell this out in a little more detail. In view of our cocycle formula the Lie algebra of $\hat\Lambda K$ is represented by pairs (f, λ), $f \in \Lambda k$, $\lambda \in \mathbf{R}$, with the bracket operation

$$[(f, \lambda), (g, \lambda')] = ([f, g], \eta(f, g)).$$

Now let θ be the linear function on Λk defined by $\theta(f, \lambda) = \lambda$. This is then a *new* direction in the dual of Λk I have been talking about.

The tangent space to the orbit of θ is of course given by $\{\mathrm{ad}(f, \lambda) \cdot \theta\}$ as f, λ range over $\hat\Lambda k$. Hence (f, λ) centralizes θ if and only if $(\mathrm{ad}(f, \lambda)\theta)(g, \lambda') \equiv 0$

for all (g, λ'). But that reduces to $\eta(f, g) = 0$ for all g, so that $f' = 0$. In short, only the point loops centralize θ, whence—formally at least—the orbit is $\Lambda K / K$. Q.E.D..

In this framework then, my old mysterious analogies are now completely explained—but that is only the beginning. The entire representation theory, character formula and fixed point theory now carries over to the representations of "*positive energy*" of $\hat{\Lambda} K$. To explain this notion we need to add one more circle to $\hat{\Lambda} K$, that is, we form the "true" loop-group

$$\mathscr{L}(K) = \hat{\Lambda} K \times S^1,$$

defined as the semidirect product of $\hat{\Lambda} K$ by S^1 with S^1 acting on $\hat{\Lambda} K$ by "rotating the loops". The positive energy representations of $\mathscr{L}(K)$ are then simply those on which the infinitesimal rotations have positive spectrum.

This whole beautiful development we owe to a large number of people. First of course are the physicists who pointed the way. On the infinitesimal side there is then the fundamental work of Kac, Lepowsky, Frankel, Garland, and others, while the global picture with its many analytic difficulties has been developed by Graeme Segal, G. Pressley, D. Quillen, and others.

For a beautiful account of all these matters I refer you to the just-published book by Pressley and Segal [**PS**] on the subject.

My time is up, without having said a single word about the far-reaching effects of the Hermann Weyl character formula in the development of the representation theory of the real forms! But that—I am certain—will be amply remedied by the speakers that follow. So let me close with a quote from Weyl which formulates— as only he could—the faith in our subject which I think we all share and which, I hope, the developments I have been describing evince. Here are his words:

> The problems of Mathematics are not problems in a vacuum. There pulses in them the life of ideas which realize themselves in concreto through our human endeavours in our historical existence, but forming an indissoluble whole transcending any particular science.

BIBLIOGRAPHY

[**B**] A. Borel, *Hermann Weyl and Lie Groups*, Hermann Weyl 1885–1985, ETH Zürich and Springer-Verlag, Berlin, 1986.

[**AB**] M. Atiyah and R. Bott, *A Lefschetz fixed point formula for elliptic complexes*, Ann. of Math. (2) **88** (1968), 451–491.

[**D**] M. Demazure, *Une démonstration d'un théorème de Bott*, Invent. Math. **5** (1968), 349–356.

[**K**] B. Kostant, *Lie algebraic cohomology and the generalized Borel-Weil Theorem*, Ann. of Math. (2) **74** (1961), 329–387.

[**PS**] A. Pressley and G. Segal, *Loop Groups*, Oxford Math. Monographs, 1986.

HARVARD UNIVERSITY

Proceedings of Symposia in Pure Mathematics
Volume 48 (1988)

Differentiable Structures on Fractal-like Sets, Determined by Intrinsic Scaling Functions on Dual Cantor Sets

DENNIS SULLIVAN

There is an easy notion of differentiable structure on a topological space. In the case of an embedded Cantor set in the line the differentiable structure is equivalent to the fine-scale geometrical structure, §1. We will discuss two examples from the theory of one-dimensional smooth dynamical systems, namely Cantor sets dynamically defined by (i) folding maps on the boundary of chaos, and by (ii) smooth expanding maps.

In example (i) there is a remarkable discovery due to M. Feigenbaum [1] and independently P. Coullet and C. Tresser [2] that there is a universality or rigidity in the fine geometric structure of the Cantor set attractor for folding maps on the boundary of chaos. Feigenbaum expressed this discovery in terms of a universal scaling function for the Cantor set. Both papers offer an explanation motivated by the renormalization group idea of physics. These discoveries were empirical, and even today after much theoretical work they are not well understood. For example, the fine structure is codified by a universal scaling function defined on a logically distinct perfect set—the dual Cantor set. The main unsolved mystery is why the renormalizations converge. We prove here the rigidity conjecture, $C(1,2)$ equivalence of the Cantor sets, assuming renormalization converges[1] (§§5, 6). We also prove a converse. The proofs use the theory of the second example and a study of nonlinearity based on the bounded geometry of the Cantor set. In §4 there is an exposition of the period doubling cascade on the boundary of chaos.

In example (ii) the Cantor set is the opposite of an attractor. It is the maximal invariant set of a $C(1,\alpha)$ expanding mapping of a 1-dimensional manifold. Now the fine structure of the Cantor set is not rigid but depends on many parameters. A complete set of invariants for the differentiable structure is again a scaling

---

1980 *Mathematics Subject Classification* (1985 *Revision*). Primary 58F14, 58F08; Secondary 54H20.

[1]In earlier unpublished work with Feigenbaum we proved the rigidity differently, assuming a definite rate of convergence. Recently, David Rand has also derived a rigidity result.

function but now the scaling function is an arbitrary Hölder continuous function on a perfect set. Here the theoretical discussion is complete, straightforward, and easy (§§1, 2, 3).

1. Differentiable structures on fractal sets. Let X be a topological space which is locally compact and can be locally embedded in $\mathbf{R}^n$. If Q denotes some adjective like smooth, real analytic, complex analytic, etc. defining a pseudogroup of local isomorphisms of $\mathbf{R}^n$, we can define *a Q-structure on X of dimension n*. Say that a collection of local embeddings in $\mathbf{R}^n$ is Q-coherent if whenever i and j are two such embeddings defined near $x \in X$ there is a local Q-isomorphism φ of $\mathbf{R}^n$ so that $\varphi \circ i = j$ near x. Then a Q-structure (of dimension n) on X is a maximal collection of Q-coherent local embeddings whose domains cover all of X.

2. Linear differentiable structures on Cantor sets. For concreteness let C denote the set of one-sided right infinite sequences of 0's and 1's with the product topology. Let $C(1,\alpha)$ denote the pseudogroup of smooth local diffeomorphisms of $\mathbf{R}$ with α-Hölder continuous derivatives, *for all α, $0 < \alpha \leq 1$*. We denote this pseudogroup $C(1,\alpha)$ (instead of the usual symbol) because α is not fixed.

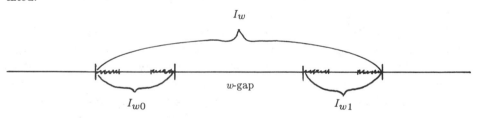

I_w

w-gap

I_{w0} I_{w1}

FIGURE 1

We will consider those $C(1,\alpha)$ structures on the Cantor set C where if $C_w = \{$sequences with initial n-segment$= w\}$ then there is a finite coordinate cover so that in a chart containing C_w we have the picture shown in Figure 1, where $I_{w'}$ denotes the smallest interval containing $C_{w'}$. In other words we want I_{w0} and I_{w1} to be disjoint.

We define in terms of the coordinate cover the *ratio geometry of w* to be the three ratios length I_{w0}/length I_w, length I_{w1}/length I_w, (length w-gap)/length I_w. Note these three positive numbers sum to unity.

DEFINITION. We say the differential structure has *bounded geometry* if in addition to the above disjointness property these ratios are bounded away from zero (uniformly in w).

LEMMA 1. *If length I_w tends exponentially fast to zero in length w, the coordinate ratio function $w \rightarrow$ ratio geometry is determined exponentially fast in length w by the differentiable structure.*

PROOF. Changes of coordinates being $C(1,\alpha)$ have exponentially small nonlinearity on intervals of exponentially small size. Q.E.D.

Now for some cover of C given by a finite system of charts deform the embeddings into $\mathbf{R}$. Namely, imagine changing the lengths of the I_w and the gaps without changing the local ordering of points of C.

THEOREM 2. *If C has bounded geometry and if the ratio functions of the deformed charts are only changed by an exponentially small error in length w, then the new charts belong to the original differentiable structure.*

PROOF. We fill in the diagram locally to construct φ

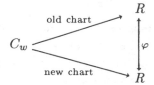

defined between the images of C. The difference quotient

$$\frac{\varphi(x) - \varphi(y)}{x - y}$$

for $x, y \in C_w$ has the form

$$\frac{a_1 + a_2 + \cdots}{b_1 + b_2 + \cdots}$$

where a_i and b_i are respective gap lengths in the two different charts and the sums are infinite. These are determined exponentially fast by the respective lengths of I_w and the ratio functions.

One sees that the difference quotient for x, y in C_w is Hölder continuous with approximate value (new chart length I_w)/(old chart length I_w).

An elementary extension lemma shows φ has local $C(1, \alpha)$ extensions for some $0 < \alpha \le 1$.   Q.E.D.

We say two ratio geometry functions are exponentially equivalent if they differ by exponentially small quantities in length w.

THEOREM 3. *There is a one-to-one correspondence between $C(1, \alpha)$ differentiable structures on C of bounded geometry with given local order on the one hand and exponential equivalence classes of bounded from zero ratio functions, $\{w\} \to$ ratio geometry, on the other.*

PROOF. One way is Theorem 2. Conversely, if an abstract ratio function is bounded away from zero one builds the Cantor set C in $\mathbf{R}$ directly satisfying the property indicated by Figure 1 and realizing the ratio function.

3. Differentiable structures with smooth magnification and scaling functions. Now we ask the question: when is the shift map $(\varepsilon_0 \varepsilon_1 \varepsilon_2 \ldots) \overset{J}{\mapsto} (\varepsilon_1 \varepsilon_2 \ldots)$ of C locally a smooth diffeomorphism of class $C(1, \alpha)$ for some given differentiable structure on C.

There is a subtlety we will not deal with here. We will only characterize the situation when one of the two equivalent properties holds:

(i) J is smooth and for some smooth metric $J' \ge \lambda > 1$, or

(ii) J is smooth and the structure has bounded geometry.

The basic fact for everything is that the nonlinearity of the composition $I_{w_1} \xrightarrow{J}$ $I_{w_2} \xrightarrow{J} I_{w_3} \xrightarrow{J} \cdots \rightarrow J_{w_n}$, where $w_{k+1} = \varepsilon w_k$, $\varepsilon = 0$ or 1, will be controlled by $\sum (\text{length } I_w)^\alpha$, which is part of a geometric series. (See Appendix 1.) This implies the ratio geometry of w stops changing exponentially fast in length w if we add arbitrary symbols to w on the left.

Thus there is a limiting ratio geometry $\sigma(\cdots \varepsilon_2 \varepsilon_1 \varepsilon_0)$ attached to each *left-infinite word*. These limit ratios are called the *scaling function of the differentiable structure*. This proves

THEOREM 4. *If the shift map on the Cantor set of right-infinite words is smooth* $(C(1, \alpha))$ *in a differentiable structure of bounded geometry, the coordinate dependent ratio function* $w \rightarrow$ *ratio geometry defines a limiting scaling function which is coordinate cover independent and attached intrinsically to the differentiable structure. The scaling function assigns to each left-infinite word a triple of positive ratios adding up to one.*

REMARK. The proof shows this scaling function is exp-continuous on $(\cdots \varepsilon_2 \varepsilon_1 \varepsilon_0)$, namely there is exponentially fast determination of the value of σ by knowledge of initial n-segment of $(\cdots \varepsilon_2 \varepsilon_1 \varepsilon_0)$. We call this property Hölder continuity of the scaling function σ.

THEOREM 5. *Conversely, if there is a Hölder continuous limiting scaling function for the differentiable structure (as in the remark) the shift is a smooth* $C(1, \alpha)$ *expanding map (in some smooth metric).*

REMARK. All Hölder continuous scaling functions on $\{\cdots \varepsilon_2 \varepsilon_1 \varepsilon_0\}$ occur in this discussion.

The proof of Theorem 5 involves exactly the same consideration as that of Theorem 2. One sees that the relevant difference quotient is Hölder using the scaling function. A standard argument shows the shift is expanding in some smooth metric because the bounded geometry implies all the derivatives at period points are greater than unity.

SUMMARY. Differentiable structures on C where the shift is a $C(1, \alpha)$ expanding map are precisely those structures which have bounded geometry and whose asymptotic ratio geometry is described by a scaling function

$$\{\cdots \varepsilon_2 \varepsilon_1 \varepsilon_0\} \xrightarrow{\sigma} \text{ratio geometry} = \{a > 0, b > 0, c > 0 \mid a + b + c = 1\}.$$

All Hölder continuous σ occur in this discussion. *There is a one-to-one correspondence between these* $C(1, \alpha)$ *structures and exponentially continuous scaling functions* (*Theorems* 3, 4, 5).

Furthermore if the structure admits a $C(k, \alpha)$ refinement so that the shift is $C(k, \alpha)$, this structure is also determined uniquely by the same scaling function $k = 1, 2, \ldots; k = \infty$; or $k = \omega$. In fact, a shift-commuting homeomorphism between structures which has a nonzero derivative at one point, already is the restriction of a $C(k, \alpha)$ equivalence (Appendix, part (ii) of Corollary).

An unsolved problem here is to determine what further properties of the scaling function σ allows higher smoothness. From earlier work we also know that if the structure is at least C^2 and for any smooth metric the second derivative of the shift is nonzero at some point of C, the scaling function itself is determined by the thermodynamics of C which we know to be determined by the underlying Lipschitz structure. By thermodynamics we mean a certain mathematical discussion whose input is the sizes of the I_w, the set of numbers obtained by taking k-fold products of σ over k-fold shifts of k-periodic sequences.

4. The period doubling attractor (informal discussion). Let us consider the simplest class of maps which allows a transition from very simple dynamics to complicated dynamics with exponential effects. These are the folding maps of an interval $I \xrightarrow{f} I$ which have a turning point c in I so that f is increasing before c and decreasing after c (Figure 2(a)). If there is a parameter t in the formula for f which raises the graph enough and the family has appropriate smoothness there will be a parameter value a where f_a has an attractive Cantor set (all but a countable sequence of points are asymptotic to C).

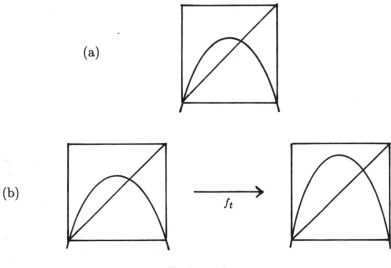

FIGURE 2

This Cantor set is the closure of the forward orbit of the turning point and is created by an infinite sequence of period doublings bifurcations of known form (Figure 3).

The forward orbit of the critical point denoted $\{1, 2, 3, \ldots\}$ increases its complexity as t increases to a. At a sequence of values $a_1, a_2, a_3, \ldots$ tending to a, the critical point has 2^n-periodicity (Figure 4).

The numbers in L at stage n are just those at stage $(n-1)$ doubled and reversed in order. The numbers in R at stage n are obtained from those in L at stage n by subtracting 1 and reversing the order (Figure 4).

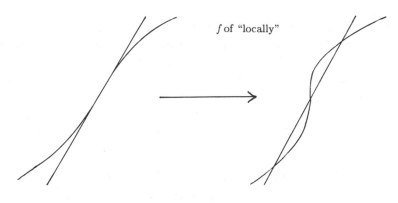

f of "locally"

FIGURE 3

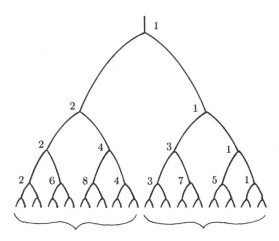

FIGURE 4

In the limiting map f_a (at the "boundary of chaos") there is a 2-adic Cantor type structure on which f acts by "adding 1". The more precise statement is that the closure of the orbit of the critical point is a Cantor set on which f is equivalent to $(\varepsilon_0\varepsilon_1\cdots) \mapsto g(\varepsilon_0\varepsilon_1\cdots)$ where $g(\varepsilon_0\varepsilon_1\cdots)$ is, "change the first zero to a one and all previous ones to zeroes" (adding "1" in the 2-adic integers).

The identification can be chosen so that the critical point is $111\ldots$ and the critical value is $000\ldots$.

Feigenbaum made the remarkable discovery that for many examples deep ratios in the Cantor set have asymptotic limits independent of the smooth family f_t. His calculations involved smooth functions with quadratic turning points. The only way to change the fine scaling in practice was to change the nature of the critical point or to introduce other critical points.

5. The Feigenbaum Rigidity Conjecture. Let us formulate a precise rigidity statement corresponding to the Feigenbaum discovery. We assume f is a folding map $f: I \to I$ satisfying

(i) f has a Lipschitz first derivative, i.e., $f \in C(1,1)$.

(ii) f has exactly one critical point c, namely, $f'c = 0$, and $f'x \neq 0$ for $x \neq c$.

(iii) f in some $C(1,1)$ coordinate system near c is just $(x-c)^2 + f(c)$.

(iv) $fc, f^2c, f^3c, \ldots$ is deployed in the interval in terms of order as described in §4, Figure 4.

RIGIDITY CONJECTURE. The closure of the forward orbit of the critical point $\{fc, f^2c, \ldots\}$ is the 2-adic Cantor set C of one-sided sequences of 0's and 1's with f acting on C by adding "1". The $C(1,\alpha)$ differentiable structure on C induced from its embedding in $\mathbf{R}$ is unique and described by a universal scaling function (§3).

A corollary of the mere existence of the scaling function for C is that the shift map J of the Cantor set is a $C(1,\alpha)$ expanding map (§3). In the 2-adic notation $x = \varepsilon_0 + \cdots + \varepsilon_i 2^i + \cdots$, $f(x) = x+1$ on C, and the shift map J is $x \rightarrow$ greatest integer in $\frac{1}{2}x = [\frac{1}{2}x]$. The calculation $\frac{1}{2}[x+1+1] = [\frac{1}{2}x]+1$ shows

$$\boxed{Jff = fJ}.$$

This is the topological form of the celebrated Cvitanović-Feigenbaum functional equation [1].

This equation can be iterated to obtain $J^n f^{2^n} = fJ^n$. Now J^n provides 2^n diffeomorphisms between I_C, the interval subtending C, and the 2^n I_w's, the intervals subtending the C_w's, length $w = n$. The I_w's are each invariant by f^{2^n} and the branches of J^{-n} provides smooth conjugacies between f^{2^n}/I_w and f/I_C, restricted to the Cantor set.

For example, let $I_{w(n)}$, where $w(n)$ has n 1's, converge down to the critical point $111111\ldots$ fixed by J. If $\alpha = J'(11111\ldots)$, then by calculus $\alpha^n J_{w(n)}^{-n}$ has a limit. Thus if $J_n = J_{w(n)}$ and $f_n = f^{2^n}/I_{w(n)}$, then

$$\alpha^n J_n f J_n^{-1} \alpha^{-n} = \alpha^n f_n \alpha^{-n}$$

has a limit. (These equations only hold on the Cantor set, a detail we will clarify in the next paper.) The limit g satisfies the Cvitanović-Feigenbaum functional equation

$$\boxed{\alpha g^2 \alpha^{-1} = g},$$

where $g = \lim_{n\to\infty} \alpha^n f^{2^n} \alpha^{-n}$, $\alpha = J'(111\ldots)$. This proves the first part of Theorem 6. The rest is explained by the argument of the next section.

THEOREM 6. *If the period-doubling Cantor set has a scaling function, then the nth renormalization of f, $\alpha^n f^{2^n} \alpha^{-n}$, converges to g, a solution of the CF functional equation $\alpha g^2 \alpha^{-1} = g$. The limit g only depends on the scaling function.*

6. The Rigidity Conjecture and renormalization. The folding maps f we are considering satisfy (by hypothesis) that $\{1, 2, 3, \ldots, 2k\}$ denoting the first $2k$ forward iterates of the critical point $k = 2^n$ are deployed as follows:

$$\underset{2 \qquad k+2}{\overset{I_2}{\vdash\!\!\!\dashv}} \quad \cdots \quad \underset{2^n \qquad 2^{n+1}}{\overset{I_{2^n}}{\vdash\!\!\!\dashv}} \quad \cdots \quad \underset{j \qquad k+j}{\overset{I_j}{\vdash\!\!\!\dashv}} \quad \cdots \quad \underset{k+1 \qquad 1}{\overset{I_1}{\vdash\!\!\!\dashv}}.$$

Consequently, $f^{2^n} = f^k$ preserves each of the indicated intervals and is a folding map of the same form. Each of these is called an nth renormalization of f.

As we observed in §5, one of these renormalizations after linear rescaling by powers of $J'(111\ldots) = \sigma(\ldots 111)$ converges, assuming the part of the conjecture about the existence of the scaling function σ.

There is a converse.

THEOREM 7. *If the renormalizations of f about the critical point converges (in the C^0 topology, to a folding map with a quadratic critical point), then the Cantor set of f has a scaling function only dependent on this renormalization limit.*

COROLLARY. *If two folding maps have the same renormalization limits there is a $C(1, \alpha)$ diffeomorphism between their Cantor sets conjugating the dynamics on the Cantor sets. (The author does not expect more smoothness even for analytic mappings.)*

PROOF OF COROLLARY. Theorem 7 and Summary of §3.

We make the proof of Theorem 7 assuming the Cantor sets have bounded geometry. We will expose our general result (valid for all maps described at the beginning of the section) on bounded geometry and general a priori estimates on the nonlinearity of renormalization in our next paper.

Now consider the measure $|dx/x|$ restricted to all the intervals at the nth level except I_{2^n} containing the critical point. Here $x = 0$ is the critical point. By induction on n we prove two properties:

(i) The density of the measure is quasiconstant on each interval.

(ii) The total mass is controlled independent of n.

Passing from level n to $n + 1$, we cut away a middle piece from each interval, which by (i) and the bounded geometry reduces the total mass by a definite factor (and keeps property (i)). We also add a new interval near the critical point. This only adds a new term of bounded mass and quasiconstant density because the interval is nicely situated with respect to the critical point by the bounded geometry assumption on C. This completes the induction.

Now the nonlinearity of f (the measure $(f''/f')\,dx$) is controlled by a bounded density measure away from the critical point and the measure $|dx/x|$ near the critical point. Thus $|f''/f'|\,|dx|$ is controlled by a measure satisfying (i) and (ii) since $|dx/x|$ and bounded measures satisfy (i) and (ii).

Now consider the ratio geometry associated with a long word w (of length r, say). Fix j. We can keep the j-segment on the right fixed and change the other digits to 1's by applying f no more than 2^{r-j} times. The ratio geometry of w is that of something at level j inside some interval at depth $r - j$. We transform this over to the critical point interval by applying f no more than 2^{r-j} times.

These iterates all have bounded nonlinearity by (ii) (applied to level $r-j$). We care about the distortion of an object j levels deeper. This object is exponentially smaller in j, relatively (by bounded geometry). The nonlinearity measure (by (i) at level $(r - j)$) we see is exponentially smaller in j. Thus the distortion of the appropriate iterate of f restricted to the smaller object is exponentially small in j. Thus the ratio geometry of w of length r is the same as that of the word beginning with $r - j$ ones and ending with the same last j-segment as w with an exponentially small in j error. Thus much follows from just the bounded geometry assumption on the Cantor set.

Now let r increase, still keeping the final j-segment of w fixed. The structure of the ratio geometry of $1111111\ldots 111$ (final j segment of w) only depends on the 2^j forward orbit of the renormalized map which is converging in C^0 to a folding map f_∞. Thus we can define the scaling function at arguments $\ldots 1111111$ (word of length j) in terms of the first 2^j iterates of f_∞. Since j was unrestricted in the argument, we have defined the scaling function at all left-infinite words which are all 1's eventually. Moreover, by the first part of the argument these values are determined exponentially fast by the initial segments on the right. This proves Theorem 7 assuming bounded geometry of C.

Appendix (Composition of contractions in $C(k, \alpha)$). Consider a composition g of diffeomorphisms

$$I_1 \xrightarrow{f_1} I_2 \xrightarrow{f_2} \cdots \xrightarrow{f_n} I_{n+1},$$

where $|f_i'| \le \lambda < 1$.

If x_1 and y_1 are two points in I_1 let $(x_{i+1}, y_{i+1}) = (f_i(x_i), f_i(y_i))$.

Note that $|x_i - y_i| \le \lambda^{i-1}|x_1 - y_1|$. If φ_i is a function on I_i satisfying $|\varphi_i(x) - \varphi_i y| \le C|x_i - y_i|^\alpha$, $0 < \alpha \le 1$, then $\varphi_1(x_1) + \varphi_2(x_2) + \cdots$ is also Hölder continuous with constant $C(1/1 - \lambda^\alpha)$ and same exponent α.

We apply this to $\varphi_i = D^k \log f_i'$, $k = 0, 1, 2, \ldots$, to see that if the f_i's satisfy

$$|D^k \log f_i'(x) - D^k \log f_i(y)| \le C|x - y|^\alpha$$

then so does $D^k \log g'(x)$ for the same α and the new C as above.

COROLLARY. (1) *A composition of uniform contractions which are individually bounded in $C(k, \alpha)$ (as diffeomorphisms) is also in $C(k, \alpha)$ (same α and new constant).*

(2) *If a sequence of such compositions is renormalized by postcomposition with linear maps to obtain mappings between unit intervals, the sequence is precompact in $C(k, \alpha)$.*

REFERENCES

1. M. Feigenbaum, *The universal metric properties of nonlinear transformations*, J. Statist. Phys. **19** (1978), 25–52; **21** (1979), 669–706.

2. P. Coullet and J. Tresser, *Iterations d'endomorphisms et groupe de renormalisation*, C. R. Acad. Sci. Paris **287** (1978), 577; J. Physique **C5** (1978), 25.

GRADUATE SCHOOL AND UNIVERSITY CENTER, CITY UNIVERSITY OF NEW YORK

Proceedings of Symposia in Pure Mathematics
Volume 48 (1988)

Representation Theory and Arithmetic

R. P. LANGLANDS

Although some of the books of Hermann Weyl, especially those dealing with algebraic matters, are notoriously difficult, the papers on geometry and analysis were often models of ease and transparency, as much in the incidental papers as in the major ones, like those on the spectral theory of ordinary differential equations or the representation theory of compact Lie groups.

This lecture is a brief introduction to some problems in the contemporary theory of automorphic forms, a part of the spectral theory of group actions, a topic that perhaps began with the theorem of Peter-Weyl on the representation theory of general compact groups; but the clue to the present investigations, and indirectly the major link to Hermann Weyl, is provided by the spectral theory of Harish-Chandra for non-compact semisimple groups. The influence of Weyl's techniques for studying characters and of the spectral theory of ordinary differential equations is manifest throughout the work of Harish-Chandra. Specifically, however, the clue is given by the geometrical and cohomological properties of the discrete series.

None the less our major concerns will be arithmetical and owe more to Weyl's fellow student Hecke than to Weyl himself, for two subjects that began with Hecke play the principal roles, the extension of the theory of complex multiplication to higher-dimensional varieties, a subject that has become the theory of Shimura varieties, and the theory of Hecke operators and the associated L-series. Even so, Weyl was fascinated by arithmetic from the beginning of his career, Hilbert's *Klassenkörperbericht* being one of the first papers he read as a student, and, as his monograph on ideal theory and other papers testify, it continued to attract his interest until the end.

I begin by recalling some familiar, but fundamental and ultimately very difficult concepts. We begin with a smooth projective variety V over a finite field κ. If κ_n is the extension of κ of degree n, let N_n be the number of points on V

1980 *Mathematics Subject Classification* (1985 *Revision*). Primary 11F70; Secondary 14K15.

with coefficients in κ_n, and form

$$Z(t,V) = \exp\left(\sum_{n=1}^{\infty} \frac{N_n}{n} t^n\right).$$

It is the zeta-function of V introduced by Weil.

If, for example, κ has q elements and V is the projective line, then $N_n = q^n + 1$ and

$$Z(t,V) = \exp\left(\sum_{n=1}^{\infty} \frac{(qt)^n}{n} + \sum_{n=1}^{\infty} \frac{t^n}{n}\right) = \frac{1}{(1-qt)(1-t)}.$$

It is by now very well known that then for any variety V the function $Z(t,V)$ is a rational function of t of the form

$$Z(t,V) = \prod_{0 \leq i \leq 2\dim V} L_i(t,V)^{(-1)^{i-1}},$$

where $L_i(t,V)$ is a polynomial

$$L_i(t,V) = \prod_{j=1}^{d_i}(1 - \alpha_{ij}t).$$

In addition, $|\alpha_{ij}| = q^{i/2}$ and d_i has cohomological significance.

If we take a variety V over a global field F, in particular over $\mathbf{Q}$, then V will be defined by a finite number of equations with coefficients that are integral outside a finite set S of primes, and thus can be reduced modulo any prime not in S, and if S is taken to be sufficiently large will even give upon reduction a smooth variety over the residue field and thus a zeta-function

$$Z(t,V;\mathfrak{p}) = \prod_i L_i(t,V;\mathfrak{p})^{(-1)^{i-1}}.$$

It has been suggested, somewhat casually and in specific cases by Hasse and then systematically, and independently, by Weil, that the Euler products

$$L_i(t,V;S) = \prod_{\mathfrak{p} \notin S} \frac{1}{L_i(N\mathfrak{p}^{-s},V)}$$

would be of interest. For example, if V is just a point, the global field is $\mathbf{Q}$, and if S is empty then $L_0(t,V;S)$ is simply the Riemann zeta-function.

In general these functions are of interest for at least two reasons.

(i) They pose an obvious problem of analytic continuation.

(ii) Although the functions are defined in terms of local data, they yield information about the global arithmetic of the variety. For example, for varieties of dimension zero this is expressed by the classical class-number formulas and for elliptic curves by the conjectures of Birch and Swinnerton-Dyer.

The problem (i) is of course patent and in comparison with those posed by the ideas implicit in (ii) puerile. None the less it leads not only to serious analytic questions but also to serious arithmetic questions. Even for varieties of dimension 0 it requires class-field theory to solve it even in part.

Depth aside, it is certain that the problem is solved in very few cases:

(i) varieties of dimension zero associated to abelian extensions;

(ii) abelian varieties with complex multiplication, in particular, for elliptic curves with complex multiplication but not, except for a few isolated examples, for other elliptic curves.

Thus even for curves there is a great deal left to do. There is one class of curves for which much is known, the modular curves, and more generally Shimura curves. A fairly general family of modular curves is obtained by dividing the upper half-plane by the discrete groups

$$\Gamma_N = \left\{ \gamma \in \mathrm{SL}(2,\mathbf{Z}) | \gamma \equiv \begin{pmatrix} 1 & * \\ 0 & 1 \end{pmatrix} \,(\mathrm{mod}\,N) \right\}.$$

The associated complex algebraic curve Sh_N can be made projective by adding a finite number of points and then given a structure over $\mathbf{Q}$.

It is possible to show that $L_1(s, \mathrm{Sh}_N; S)$ can be analytically continued by showing that it is a product of the L-functions attached by Hecke to automorphic forms on the upper half-plane, which are of the form

$$L_S(s, \pi) = \prod_{p \notin S} \frac{1}{(1 - \alpha_p/p^s)(1 - \beta_p/p^s)},$$

and thus of degree two. Here π denotes the form or, what amounts to the same thing, the associated representation. Thus

(1) $$L_1(s, \mathrm{Sh}_N; S) = \prod_{\pi} L_S(s - 1/2, \pi),$$

only a finite number of π, and these not necessarily distinct, intervening in the product.

Such a result poses further problems, for if one of the curves Sh_N appears as a ramified covering of some curve C, not itself an Sh_N, then one may hope and expect to deduce from (1) a similar representation for $L_1(s, C; S)$, and thus verify that it too can be analytically continued. This is the method that has been proposed—for very good reasons—for dealing with elliptic curves. In order to deal with other base fields, one needs a theory of base change for automorphic forms, but that is only partially developed [L2] and not pertinent to this lecture. It is more important to stress that the methods that lead to (1) and its refinements are also important for apparently quite different arithmetic problems, like the structure of the ideal-class group of cyclotomic fields [M-W].

There is another class of varieties for which an analogue of (1) is valid, those attached to the names of Hilbert and Blumenthal. They can be of any dimension, but the surfaces of this type—associated to real quadratic fields—are perhaps of most interest at the moment because for them a number of important conjectures can be tested with the help of (1), the conjectures of Tate relating algebraic cycles to the Galois action on étale cohomology and to the order of the poles of the Hasse-Weil zeta-function [HLR] and the conjectures of Beilinson [Ra].

All this is by way of preface to stress the importance of the problem of analytic continuation and to observe that its solution even for what appear to be very special varieties can lead to unpredictable and valuable arithmetic consequences.

The one class of varieties that offers hope for substantial advances is that of Shimura varieties. There are several problems involved and on all but one progress was being made, especially by R. Kottwitz, but there is one central obstacle that it was not clear would be removed in the near future, so that I feared that like Jean Débardeur we would remain "toujours à terre, jamais au large"; but the obstacle has now been removed by Kottwitz himself [**K6**], and by H. Reimann and T. Zink as well [**RZ**]. These are important developments, and the purpose of this lecture is to draw attention to them.

There are three types of Shimura varieties to be distinguished:

(a) the general type;

(b) those associated to a moduli problem for abelian varieties with endomorphism algebra and polarization;

(c) those associated to the Siegel upper half-spaces.

The problems can be posed for all of them, but it is often a major step to pass from the solution for those of type (b) to the general solution for those of type (a). At the moment one is attempting only to deal with those of type (b). The methods that work for those of type (c) usually work for those of type (b) with little change. Thus I confine myself to type (c).

The Shimura varieties associated to the Siegel upper half-spaces are, properly speaking, attached to the group of symplectic similitudes, the group G of $2n \times 2n$ matrices U for which

$$^{t}U J U = \lambda J,$$

λ a scalar, and

$$J = \begin{pmatrix} 0 & I \\ -I & 0 \end{pmatrix}.$$

To describe, even approximately, the form that (1) is expected to take we have to introduce at the same time the L-group ^{L}G of G. The group G is a group over $\mathbf{Q}$; the L-group is in contrast a group over $\mathbf{C}$. It is the Clifford group attached to the orthogonal form in $2n + 1$ complex variables. The spin representation of the corresponding orthogonal group is of dimension 2^n and ^{L}G consists of all matrices that can be written as the product of a scalar matrix and an element of the spin group, so that ^{L}G has a natural representation r of degree 2^n.

According to the general definition of L-functions associated to automorphic forms there is attached to every finite-dimensional representation ρ of ^{L}G and every automorphic representation π of G an Euler product

$$(2) \qquad\qquad L_S(s, \pi, \rho).$$

Here S is some large finite set of primes of $\mathbf{Q}$.

The Euler products attached to $\rho = r$ are of particular importance for the zeta-functions of the Shimura varieties attached to G. Questions of completeness and connectedness aside, these are as complex manifolds essentially quotients

$\Gamma\backslash H$, where Γ is a congruence subgroup of $G(\mathbf{Z})$ and H is the set of all complex symmetric matrices $Z = X + iY$ with $Y > 0$ and

$$\gamma = \begin{pmatrix} A & B \\ C & D \end{pmatrix} : Z \to (AZ + B)(CZ + D)^{-1}.$$

The structure of these varieties over $\mathbf{Q}$, or over a number field if that is appropriate, is given by the theory of Shimura, completed by Deligne [**D**].

To obtain a Shimura variety in the proper sense, one must in fact take the disjoint union of several of these varieties, obtaining for this particular group varieties over $\mathbf{Q}$. The question of completeness is more vexing, and forces us to enlarge the notion of a zeta-function with the help of intersection cohomology to deal with singular varieties. The conjecture of Zucker, proved by Looijenga and Saper-Stern, allows one for many purposes to argue as though the quotients $\Gamma\backslash H$ were compact, and in order to arrive without too much delay at the problems that have actually been settled we do so here.

The bulk of the cohomology of the Shimura variety Sh is contained in the middle dimension $q = n(n + 1)/2$ and if calculated by means of the theory of continuous cohomology [**BW**] is given by the discrete-series representations of $G(\mathbf{R})$ that annihilate the Casimir operator. The set Π_∞ of such representations V has 2^{n-1} elements. If π_∞ is one of them, and if K is the open compact subgroup of the adelic group $G(\mathbf{A}_f)$ that must be introduced when Sh is defined completely, then each time that an automorphic representation $\pi = \pi_\infty \otimes \pi_f$, π_f being an irreducible representation of $G(\mathbf{A}_f)$, occurs in $L^2(G(\mathbf{Q})Z(\mathbf{R})\backslash G(\mathbf{A}))$ there is a contribution to the cohomology in degree q of dimension $2d(\pi_f^K)$. We denote by $d(\pi_f^K)$ the dimension of the space of vectors fixed by K under π_f. The critical observation is that $2 \cdot 2^{n-1} = 2^n$, the dimension of r and thus the degree of the Euler product $L_S(s, \pi, r)$.

If $\pi' = \pi'_\infty \otimes \pi_f$, where $\pi'_\infty \in \Pi_\infty$ then, by definition,

$$L_S(s, \pi', r) = L_S(s, \pi, r).$$

(This would be valid even if Γ-factors had been incorporated into the L-functions.) Thus if, as is often but not always the case, whenever $\pi_\infty \otimes \pi_f$ occurs in $L^2(G(\mathbf{Q})Z(\mathbf{R})\backslash G(\mathbf{A}))$ then $\pi'_\infty \otimes \pi_f$ also occurs for any $\pi'_\infty \in \Pi_\infty$ then the representations $\{\pi_\infty \otimes \pi_f | \pi_\infty \in \Pi_\infty\}$ contribute a space of dimension $2^n d(\pi_f^K)$ to the cohomology each time that they occur, and thus should contribute a factor of degree $2^n d(\pi_f^K)$ to the L-function $L_q(s, \mathrm{Sh}; S)$. If the Eichler-Shimura theory for the upper half-plane which leads to (1) is kept in mind, then a natural guess is that this factor is

(3) $$L_S(s - q/2, \pi_f, r)^{d(\pi_f^K)}.$$

The shift by $q/2$ is to account for the absolute value of the roots of the local L-functions.

There are two distinct questions implicit here: (a) can the Euler products (2) be analytically continued; (b) can the zeta-function of the variety Sh really be expressed in terms of these functions? These are two very different aspects of the problems posed by the introduction of the general Euler products into the theory of automorphic forms. The problem of analytic continuation can be approached in various ways [GS] and is in particular tied to functoriality, so that although a great deal remains to be done, it is clear that we are dealing with promising material and promising methods [AC].

The question (b) emphasizes a distinct consideration. Even if the functions (2) have interesting analytic properties and lead to an internally rich theory of automorphic forms, is it a theory that bears on other domains of mathematics, in particular, on arithmetic? A first, after the Eichler-Shimura theory, and almost but not quite decisive response to this question is to show that the zeta-functions of Shimura varieties can be expressed in terms of these functions, for then we may hope that even those varieties not defined by groups have zeta-functions that can be so expressed.

We are here concerned with question (b), which requires that we give a precise expression for the zeta-function as a product of the functions (2) (and their inverses) and that we prove it. Since the precise expression is not so important, simply whatever the proof yields, it is the strategy of the proof that counts, and that is elaborate. It has to be recognized immediately that the occurrence of $\pi_\infty \otimes \pi_f, \pi_\infty \in \Pi_\infty$, in $L^2(G(\mathbf{Q})Z(\mathbf{R})\backslash G(\mathbf{A}))$ does not always entail the occurrence of $\pi'_\infty \otimes \pi_f$ with the same multiplicity. This is the subject of endoscopy and the stable trace formula, which have only begun to be developed [K1, K2, L3, LS, Ro]. Our experience so far [L1] suggests that there are subgroups $^LH \hookrightarrow {}^LG$, attached to groups H over $\mathbf{Q}$, and that $r' = r|{}^LH$ decomposes into a direct sum $\bigoplus r_i$ of irreducible representations, so that for a representation π obtained by functoriality from a representation π' of $H(\mathbf{A})$ there is a factorization

$$L(s - q/2, \pi, r) = L(s - q/2, \pi', r') = \prod_i L(s - q/2, \pi', r_i)$$

and that it is not $L(s - q/2, \pi, r)$ that occurs in the zeta-function but only some of the factors $L(s - q/2, \pi', r_i)$.

To compare two L-functions, and that is what one is attempting, it is simpler to compare their logarithms, or rather for each p and n the coefficients of $1/p^{ns}$ in the expansion of their logarithms.

On one side, for the product of automorphic L-functions, this will turn out to be a sum

$$(4) \qquad\qquad \sum_H c_H \mathrm{ST}(f_H),$$

where f_H is a function in $H(\mathbf{A})$ that depends on p and n and ST denotes the stable trace.

For the zeta-function this is, apart from difficulties with the cusps,

$$(5) \qquad\qquad N_{p,n}$$

the number of points on the variety with coefficients from $\mathbf{F}_{p^n}$.

To compute (4) we use the stable trace formula, which in principle expresses (4) as a sum over stable conjugacy classes in the various H and thus as a sum over conjugacy classes in G. Thus to make the comparison we need a method of calculating $N_{p,n}$ as a similar sum.

Now, to reach this stage, we have had to proceed as though some developments that were only beginning had been carried successfully to completion, but at least they have been inching forward. Until the recent work of Kottwitz and Zink, however, $N_{p,n}$ offered quite different difficulties, and there were some who felt that we were dealing with a problem that would remain for the foreseeable future intractable.

There are two things to be done: (i) to find a group-theoretical description of the points on the variety with coefficients in $\mathbf{F}_{p^n}$ that allows one to calculate $N_{p,n}$ in terms of G; (ii) to put the resulting expression in a form that can be compared term-by-term with the expansion of (4). Kottwitz had already shown that step (ii) could be effected by the fundamental lemma for the endoscopic groups for base change [K4], and thus reduced to a problem in harmonic analysis for which at least some serious progress could be made [K5, AC]. In addition he had isolated the algebro-geometrical problem that has to be regarded as the irreducible form of (i), namely to show that an invariant introduced by him, and referred to in [LR] as the Kottwitz invariant, was 1 for abelian varieties over finite fields. Only recently have Kottwitz himself [K6] and Reimann-Zink [RZ] succeeded in showing that this is so, thus overcoming what seemed to me the major obstacle to a successful treatment of the zeta-function of Shimura varieties, so that, in spite of the many difficulties that remain and that I hope have not been slighted here, we can at last be sanguine about the prospect of obtaining utilizable results in the not-too-distant future.

The Kottwitz invariant for the group of symplectic similitudes G is attached to a triple $(\gamma, \delta, \varepsilon)$. Here ε lies in $G(\mathbf{Q})$, is elliptic in $G(\mathbf{R})$, and

$$\langle \varepsilon x, \varepsilon y \rangle = c(\varepsilon)\langle x, y \rangle, \qquad |c(\varepsilon)|_p = |q|_p,$$
$$q = p^r, \quad r > 0, \quad \langle x, y \rangle = {}^t x J_y.$$

Moreover $\gamma = \{\gamma_l | l \neq p\}$, $\gamma_l \in G(\mathbf{Q}_\lambda)$, and γ_l is conjugate to ε in $G(\overline{\mathbf{Q}}_l)$ for all l and in $G(\mathbf{Q}_l)$ for almost all l. If F is the unramified extension of $\mathbf{Q}_p$ of degree r and σ the Frobenius element in $\mathrm{Gal}(F/\mathbf{Q}_p)$ then $\delta \in G(F)$ and

$$\delta \sigma(\delta) \cdots \sigma^{r-1}(\delta)$$

is conjugate to ε in $G(\overline{\mathbf{Q}}_p)$.

The associated invariant $\kappa(\gamma, \delta; \varepsilon)$ is of cohomological nature, and is most easily defined when the centralizer of ε in G is a torus I. Suppose $\Gamma = \mathrm{Gal}(\overline{\mathbf{Q}}/\mathbf{Q})$. The invariant takes values in the dual of $\pi_0(\hat{I}^\Gamma)$, the connected component of the group of Γ-invariant elements in $\hat{I}$. The group $\hat{I}$ is that complex torus on which Γ acts in such a way that

$$\mathrm{Hom}(\hat{I}, G_m) \simeq \mathrm{Hom}(G_m, I)$$

is a homomorphism of Γ-modules.

If v is a place of $\mathbf{Q}$ let $\Gamma_v \subseteq \Gamma$ be $\mathrm{Gal}(\overline{\mathbf{Q}}_v/\mathbf{Q}_v)$. The invariant is a product $\prod_v \beta(v)$, where $\beta(v)$ is a homomorphism from $\hat{I}^{\Gamma_v}$ to $\mathbf{C}^\times$, or properly speaking the restriction of such a homomorphism to $\hat{I}^\Gamma$. In the definition of $\beta(v)$ three types of places are distinguished.

(i) If $v = l \neq p$, then

$$\gamma_l = c\varepsilon c^{-1}, \qquad c \in G(\overline{\mathbf{Q}}_l).$$

Since both γ_l and ε lie in $G(\mathbf{Q}_l)$, the cochain

$$\{c^{-1}\sigma(c)\}$$

defines an element of $H^1(\mathbf{Q}_l, I)$ and thus by Tate-Nakayama theory a homomorphism from $\hat{I}^{\Gamma_v}$ to $\mathbf{C}^\times$.

(ii) For $v = p$ we write

$$\delta\sigma(\delta)\cdots\sigma^{r-1}(\delta) = c\varepsilon c^{-1}, \qquad c \in G(\mathbf{Q}_p^{un}),$$

and then

$$b = c^{-1}\delta\sigma(c) \in I(\mathbf{Q}_p^{un}).$$

In [**K3**] Kottwitz associates to this b a coweight of $\hat{I}^{\Gamma_v}$. It is taken as $\beta(v)$.

(iii) If $v = \infty$ then $I(\mathbf{R}) \cap G_{sc}(\mathbf{R})$ is a compact Cartan subgroup of $G_{sc}(\mathbf{R})$, the symplectic group. All of these are conjugate and possess a standard coweight that is used to define $\beta(\infty)$.

Precise general definitions can be found in [**K6**] and [**LR**]. To pass from the Kottwitz invariant for triples to the Kottwitz invariant for abelian varieties with polarization, observe that if the variety and the polarization are defined over a field with q elements then the l-adic cohomology together with the Frobenius endomorphism yields γ_l, $l \neq p$, so that γ is defined. The element δ is provided by the Dieudonné module attached to the variety. All the γ_l have the same eigenvalues. They are algebraic numbers and there is at least one element of $G(\mathbf{Q})$ with these eigenvalues. Any such element serves as ε, and the geometric theorem essential to the calculation of the $N_{p,n}$ is that for triples arising in this way the invariant is 1.

The argument of Kottwitz has a strong functorial flavor and uses Fontaine's theory for Galois modules attached to p-divisible groups, while Reimann and Zink use more explicit methods based on classifications of group schemes over finite fields due to Raynaud.

References

[**AC**] J. Arthur and L. Clozel, *Simple algebras, base change, and the advanced theory of the trace formula*, Ann. of Math. Studies (to appear).

[**BW**] A. Borel and N. Wallach, *Continuous cohomology, discrete subgroups, and representations of reductive groups*, Ann. of Math. Studies (1980).

[**D**] P. Deligne, *Variétés de Shimura: Interprétation modulaire, et techniques de construction de modèles canoniques*, Proc. Sympos. Pure Math., vol. 33, part 2, Amer. Math. Soc., Providence, R.I., 1979, pp. 247–289.

[**GS**] S. Gelbart and F. Shahidi, *Analytic properties of automorphic L-functions* (to appear).

[HLR] G. Harder, R. P. Langlands, and M. Rapoport, *Algebraische Zyklen auf Hilbert-Blumenthal-Flächen*, J. Reine Angew. Math. **366** (1986).

[K1] R. Kottwitz, *Stable trace formula: cuspidal tempered terms*, Duke Math. J. **57** (1984).

[K2] ____, *Stable trace formula: elliptic singular terms*, Math. Ann. **275** (1986).

[K3] ____, *Isocrystals with additional structure*, Compositio Math. **56** (1985).

[K4] ____, *Shimura varieties and twisted orbital integrals*, Math. Ann. **269** (1984).

[K5] ____, *Base change for unit elements of Hecke algebras*, Compositio Math. **60** (1986).

[K6] ____, in preparation.

[KS] R. Kottwitz and D. Shelstad, in preparation.

[L1] R. P. Langlands, *On the zeta-functions of some simple Shimura varieties*, Canad. J. Math. **31** (1979).

[L2] ____, *Base change for* GL(2), Ann. of Math. Studies (1980).

[L3] ____, *Les débuts d'une formule des traces stable*, Publ. de l'Univ. Paris VII, No. 13 (1983).

[LR] R. P. Langlands and M. Rapoport, *Shimuravarietäten und Gerben*, J. Reine Angew. Math. **378** (1987).

[LS] R. P. Langlands and D. Shelstad, *On the definition of transfer functors*, Math. Ann. **278** (1987).

[MW] B. Mazur and A. Wiles, *Class fields and abelian extensions of* **Q**, Invent. Math. **76** (1984).

[Ra] D. Ramakrishnan, *Valeurs de fonctions L des surfaces d'Hilbert-Blumenthal en s = 1*, C. R. Acad. Sci. Paris Sér. I Math. **301** (1985).

[RZ] H. Reimann and T. Zink, *Der Dieudonnémodul einer polarisierten abelschen Varietät vom CM-Typ*, Ann. of Math. (2) (to appear).

[Ro] J. Rogawski, *Automorphic representations of unitary groups in three variables*, in preparation.

INSTITUTE FOR ADVANCED STUDY

Proceedings of Symposia in Pure Mathematics
Volume 48 (1988)

Noncommutative Algebras
and Unitary Representations

DAVID A. VOGAN, JR.

1. Introduction. The problem of quantization—roughly, of finding quantum-mechanical models of classical physical theories—has been a central theme of mathematical physics in this century. The problem of understanding unitary group representations holds a less exalted position, but it is certainly of great importance both in physics and in mathematics. The problem of deformation of algebras is not as glamorous as either of the others, but it is related to interesting mathematics (such as the theory of moduli spaces). That each of the last two problems has something to do with the first has been understood at least since the 1930s. An obvious consequence is that group representations and deformation of algebras might have something to do with each other. The purpose of this article is to explore that possibility, particularly for the light that it might shed on representation theory. What we achieve is an improved (but still incomplete) definition of unipotent representations.

The first three sections present somewhat more precise formulations of the three problems, together with some simple examples. The formulation of the first particularly is meant only to further my very limited purposes, and should not be taken seriously as mathematical physics. (There was perhaps no real danger of that, but it is necessary in mathematics to consider every possibility.) The second problem (group representations) I have taken to be synonymous with the "orbit method" of Kirillov and Kostant. This point of view is of course incomplete, but it can be made to include much of the subject. For the third (deformation theory) I have simply reproduced the framework established by Gerstenhaber, along with a few of the elaborations introduced by mathematical physicists (beginning with Moyal).

The fourth section outlines some of the ideas of geometric quantization. Perhaps the main point of this is to isolate the least tractable cases of the orbit

1980 *Mathematics Subject Classification* (1985 *Revision*). Primary 22E45; Secondary 17B35, 53C15.

The author is a Sloan Fellow. Research supported in part by NSF grant DMS-8504029.

method—those in which current technology is not able to produce unitary representations. The last two sections examine those cases and present some conjectures suggested by the various philosophies expounded earlier.

A less explicit theme is the interplay among symplectic geometry, Lagrangians, and "quantization" in various senses. A submanifold L of a symplectic manifold X is called *Lagrangian* if the symplectic form vanishes on L, and the dimension of L is half the dimension of X. (In terms of the Poisson structures emphasized here, a Lagrangian is a Poisson submanifold of minimal dimension.) It is fairly well known that Lagrangian submanifolds and related objects are crucial to various asymptotic approximations: geometric optics as an approximation to physical optics; short-wave asymptotics for the Schrödinger equation; and a significant fraction of the C^∞ theory of linear partial differential equations. (Many of these topics are discussed in [Guillemin-Sternberg, 1978].) What is not quite so well known is that there are interesting cases in which these approximations may be replaced by perfect identifications between a classical geometric picture and some "quantized" version. Perhaps the best example is the analytic theory of holonomic systems of differential equations (see for example [Kashiwara-Kawai, 1983]). Group representations should have a similar behavior, but the theory is not yet as complete.

Partly because so much of the material mentioned in this paper is outside of my expertise, I have relied heavily on discussions with many people. I am particularly grateful to Mike Artin, Dan Barbasch, Bert Kostant, Monty McGovern, Richard Montgomery, Iwan Pranata, Jim Schwartz, and Gregg Zuckerman. Section 5 is based on joint work with Dan Barbasch. It is not possible to say anything sensible about unipotent representations without taking his work into account (as I have learned in the obvious way).

1. Quantization. We will begin this section with a description of the Hamiltonian formulation of classical mechanics (on the cotangent bundle of configuration space). This formulation has a simple quantum-mechanical version. The process of passing from the classical system to the quantum one is called *quantization*. We then generalize the classical Hamiltonian formalism to the setting of a general symplectic manifold, and formulate the problem of quantizing such a system.

A classical mechanical system is described in part by a *configuration space* M whose points represent possible arrangements of the bodies constituting the system. The configuration space of a system of three free particles is a product of three copies of Euclidean space—$\mathbf{R}^9$, if coordinates are introduced. The configuration space of a pendulum is a circle. We will always take a configuration space to be a smooth manifold. A *motion* of the mechanical system is represented by a map q from $\mathbf{R}$ to M. Here the variable in $\mathbf{R}$ (denoted t) represents time, and $q(t)$ is the configuration of the system at time t. We will assume that motions are smooth maps. What physics seeks to do in this setting is understand what mappings q—what motions of the system—are allowed. Newton discovered that the constraint is very often a second-order differential equation

$$(1.1) \qquad (d^2q/dt^2)(t) = F(q(t), (dq/dt)(t)).$$

This says that the acceleration of a body in the system depends only on its position and velocity.

To say more, we need to make more assumptions on (1.1). The derivative dq/dt takes values in the tangent space $T_{q(t)}M$ to M at $q(t)$. Consequently (1.1) may be interpreted as a first-order differential equation for the motion of the point $(q(t), (dq/dt)(t))$ in TM. Often there is a notion of "kinetic energy," which is some kind of sum of squares of velocities multiplied by masses. Mathematically this is idealized as a Riemannian structure on M (a smoothly varying positive definite inner product $\langle\,,\,\rangle_q$ on each tangent space T_qM). Such an inner product provides an identification of the tangent space T_qM with the cotangent space T^*M. The velocity vector dq/dt is therefore identified with a vector $p(t)$ in T^*M called the *momentum*:

$$(1.2) \qquad p(t)(v) = \langle (dq/dt)(t), v\rangle_{q(t)} \qquad (v \in T_{q(t)}M).$$

The cotangent bundle is called the *phase space* of the mechanical system, and (1.1) becomes a first-order differential equation for the motion of $(q(t), p(t))$ in phase space.

Recall that a *symplectic manifold* X is a manifold endowed with a smooth closed 2-form ω, which is required to be nondegenerate (as a skew-symmetric bilinear form) on each tangent space. The cotangent bundle T^*M has a natural symplectic structure, defined as follows. We begin by defining a 1-form ω^1 on T^*M. To do this, suppose ξ is a tangent vector to T^*M at (q, p):

$$(1.3a) \qquad \xi \in T_{(q,p)}(T^*M).$$

(Here q is a point in M and p is a cotangent vector to M at q.) Write π for the natural projection from T^*M to M. The differential $d\pi$ maps $T_{(q,p)}(T^*M)$ to T_qM, and so takes ξ to a tangent vector at q. Since p is a linear functional on tangent vectors at q, we can define

$$(1.3b) \qquad \omega^1(\xi) = p(d\pi(\xi)).$$

Set

$$(1.3c) \qquad \omega = d\omega^1.$$

This is obviously a smooth closed 2-form, and a calculation in local coordinates shows that it is nondegenerate.

Any bilinear form ω on a vector space V (for us a tangent space of X) gives a linear map J from V to V^*, sending a vector ξ in V to the linear functional $J\xi$ defined by

$$(1.4a) \qquad (J\xi)(\eta) = \omega(\eta, \xi).$$

When the form is nondegenerate, there is a well-defined inverse I, satisfying

$$(1.4b) \qquad \lambda(\eta) = \omega(\eta, I\lambda).$$

A *Hamiltonian vector field* on a symplectic manifold X is the image under I of an exact 1-form. That is, if H is a smooth function on X, the corresponding vector field is

$$(1.5a) \qquad \xi_H = I(dH).$$

Equivalently, we require that for any tangent vector field η,

$$(1.5b) \qquad \omega(\eta, \xi_H) = \eta \cdot H;$$

on the left η is regarded as a first-order differential operator.

Recall that we have identified (1.1) with a system of first-order differential equations for the motion of a point $x = (q, p)$ in the phase space T^*M. The mechanical system is called *Hamiltonian* if there is a function H on T^*M (called the *Hamiltonian function*) so that (1.1) is the equation for an integral curve of the Hamiltonian vector field:

$$(1.6) \qquad dx/dt = \xi_H.$$

Any mechanical system consisting of free bodies interacting through conservative forces and Newton's laws turns out to be Hamiltonian. In this case the Hamiltonian function is the sum of the kinetic energy and the potential energy V. The kinetic energy is just the bilinear form $\langle \ , \ \rangle_q$ coming from the Riemannian structure (regarded now as defined on T^*M instead of TM). The potential energy depends only on the configuration (and not on the momentum). That is,

$$(1.7) \qquad H(q, p) = \langle p, p \rangle_q + V(q).$$

A quantum state of the system with (Riemannian) configuration manifold M is something fuzzy; if we observe the same state several times, we may see different configurations. Roughly speaking, a state is a probability distribution on M, which describes how likely we are to observe any particular set of configurations. More precisely, we introduce the complex Hilbert space

$$(1.8) \qquad \mathscr{H} = L^2(M).$$

A *state* is defined to be a line in $\mathscr{H}$, or a unit vector defined up to multiplication by a complex number of modulus one. If ψ is such a unit vector, then $|\psi(q)|^2 dq$ is a measure on M of total mass one. It is the probability distribution mentioned above.

A *motion* of the quantum-mechanical system is given by a map ψ from $\mathbf{R}$ to $\mathscr{H}$; that is, by a function on $\mathbf{R} \times M$. The analogue of (1.6) and (1.7) is the *Schrödinger equation*

$$(1.9) \qquad i\hbar(d\psi/dt) = -\hbar^2 \Delta \psi + V(q)\psi.$$

Here $\hbar$ is a real constant (Planck's constant), Δ is the Laplace operator on M, and V is the potential energy function from (1.7). To see why (1.9) is at least formally analogous to (1.6), recall that a kth-order differential operator D on M has a *symbol* $\sigma_k(D)$, which is a function on T^*M that is a homogeneous polynomial of degree k on each fiber. The symbol of Δ is the Riemannian length function:

$$(1.10) \qquad \sigma_2(\Delta)(q,p) = \langle p,p \rangle_q.$$

Now the two terms on the right in (1.7) correspond formally to the two terms on the right in (1.9).

In general, we take the states of a quantum-mechanical system to be lines in a complex Hilbert space $\mathscr{H}$ (or as unit vectors in $\mathscr{H}$, defined up to multiplication by a complex number of absolute value one). A motion of the system is represented by a map ψ from $\mathbf{R}$ to $\mathscr{H}$. The dynamics of the system are encoded by a first-order differential equation (the Schrödinger equation)

$$(1.11) \qquad i\hbar(d\psi/dt) = S\psi.$$

Here S is a self-adjoint operator on $\mathscr{H}$. Quantizing a classical mechanical system will mean (at least) finding an appropriate Hilbert space $\mathscr{H}$ and Schrödinger operator S. ("Appropriate" should be judged by predictive power; but since this is secretly mathematics, we have no experiments to consider. We will ask instead that our constructions be compatible with the example (1.8)–(1.9), and that symmetries of the classical system induce symmetries of the quantum one.)

We return for a while to classical mechanics. Suppose that X is an arbitrary symplectic manifold (not necessarily a cotangent bundle). We fix a *Hamiltonian function H* on X, and consider motions x (functions on $\mathbf{R}$ with values in X) satisfying Hamilton's equation

$$(1.12) \qquad dx/dt = \xi_H.$$

We think of X as the phase space of a mechanical system, but it no longer makes sense to speak of configurations or momenta separately. An *observable* is a smooth function F on X. (In the example T^*M considered earlier, this means that an observable is something depending only on the configuration and momenta of the bodies in the system. Examples are positions, velocities, angular momenta, kinetic energy, and so on.) A fundamental question about an observable F is whether it is *conserved*; that is, whether it is constant along each motion of the system. If x is such a motion, then

$$(1.13) \qquad \begin{aligned} d(F \circ x)/dt &= (dx/dt)F \\ &= \xi_H F \\ &= \omega(\xi_H, \xi_F). \end{aligned}$$

(On the right side in the first two lines we use the action of tangent vectors on functions. The second follows from (1.12), and the third from (1.5b).) It follows that the observable F is conserved in (all motions of) the system (1.12) if and only if $\omega(\xi_H, \xi_F)$ is identically zero. In particular the Hamiltonian function H itself is conserved; because of the example (1.7), we call it the *energy*.

These considerations suggest the importance of the following definition. Suppose X is a symplectic manifold, and F and H are smooth functions on X. The

Poisson bracket of F and H is the function

$$(1.14) \qquad (F, H) = \omega(\xi_H, \xi_F) = \xi_H F = -\xi_F H$$

(notation (1.5)). Perhaps the most important properties of the Poisson bracket are the Leibniz rule and the Jacobi identity:

$$(1.15a) \qquad \{F, GH\} = \{F, G\}H + \{F, H\}G,$$

$$(1.15b) \qquad \{F, \{G, H\}\} = \{\{F, G\}, H\} + \{G, \{F, H\}\}.$$

(The first of these is an easy consequence of (1.14), but the second requires some effort.) The first equation in (1.14) gives at once

$$(1.15c) \qquad \{F, H\} = -\{H, F\},$$

so the Poisson bracket defines a Lie algebra structure on $C^\infty(X)$.

It is worth observing that the formulation (1.6) of classical mechanics can be made to depend only on the Poisson bracket (and not directly on the symplectic structure). To do that, we first define a *Poisson algebra* to be a commutative algebra $\mathscr{A}$ endowed with a bilinear bracket $\{\ ,\ \}$ subject to the conditions in (1.15). For H in $\mathscr{A}$, we can define a derivation ξ_H by

$$(1.16) \qquad \xi_H F = \{F, H\}.$$

This defines a homomorphism from the Lie algebra $(\mathscr{A}, \{\ ,\ \})$ to the Lie algebra of derivations of $\mathscr{A}$. A *Poisson manifold* is a manifold X together with a Poisson structure on $C^\infty(X, \mathbf{R})$. Now (1.16) allows us to write Hamilton's equation (1.6) on a Poisson manifold.

Poisson manifolds are in some sense degenerate symplectic manifolds, but they are more interesting than those obtained by the more obvious method of allowing the 2-form to be degenerate. (Very roughly speaking, a Poisson structure on X amounts to a foliation of X by symplectic manifolds. An interesting family of examples is given in (2.5).) Similarly one can define Poisson algebraic varieties; there is no need for these to be smooth.

Now we can formulate the first of our three problems. (It should be emphasized that this is a philosophical goal and not in any sense a conjecture; it cannot be true in anything like this generality.)

1.17 *The problem of quantization.* Suppose X is a Poisson manifold (the phase space), H is a real-valued smooth function on X (the Hamiltonian), and $\mathscr{A}$ is an algebra of complex-valued smooth functions on X containing H (the algebra of observables). We assume that $\mathscr{A}$ is closed under Poisson bracket. The problem is to associate to $(X, H, \mathscr{A})$ a Hilbert space $\mathscr{H}$ (the state space), a self-adjoint operator S (the Schrödinger operator), and an algebra $\mathscr{B}$ of operators on $\mathscr{H}$ containing S (the algebra of observables). The commutative algebra $\mathscr{A}$ should be approximately isomorphic to the noncommutative algebra $\mathscr{B}$. This approximate isomorphism should send $\hbar$ times the Poisson bracket in $\mathscr{A}$ to the commutator of operators in $\mathscr{B}$. (Here $\hbar$ is Planck's constant.)

The last requirement is not intended to be precise. As an example of what it might mean, let us return to the case of T^*M. Take for $\mathscr{A}$ the algebra of all smooth functions that are polynomial of bounded degree on each cotangent space. We know that $\mathscr{H}$ should be $L^2(M)$. Let $\mathscr{B}$ be the algebra of differential operators on M. Suppose D_1 and D_2 are two such operators, of degrees j and k. Then the symbols $\sigma_k(D_1)$ and $\sigma_j(D_2)$ (discussed before (1.10)) belong to $\mathscr{A}$. We have

$$(1.18) \qquad \sigma_{k+j}(D_1 D_2) = \sigma_k(D_1)\sigma_j(D_2),$$
$$\sigma_{k+j-1}(D_1 D_2 - D_2 D_1) = \{\sigma_k(D_1), \sigma_j(D_2)\}.$$

This provides a version of the requirement in (1.16); a differential operator D of degree k should correspond to something like $h^{-k}\sigma_k(D)$ in the approximate isomorphism.

2. Group representations. In this section we recall the outlines of the Kirillov-Kostant philosophy of coadjoint orbits. (More details may be found for example in [Kostant, 1970], [Kirillov, 1976], or [Guillemin-Sternberg, 1978].) For our purposes the goal of this philosophy is to attach unitary representations of G to certain geometric objects (Poisson G-spaces). We will make no attempt to explain the evidence for this philosophy (even by enumerating its successes); but we will try to emphasize the analogy with the problem of quantization. To that end, we will violate the historical development of the subject by defining Poisson G-spaces before unitary representations.

Suppose G is a Lie group. Write

$$(2.1) \qquad \mathfrak{g} = \mathrm{Lie}(G), \qquad \mathfrak{g}_{\mathbf{C}} = \mathfrak{g} \otimes_{\mathbf{R}} \mathbf{C}, \qquad \mathfrak{g}^* = \mathrm{Hom}_{\mathbf{R}}(\mathfrak{g}, \mathbf{R}).$$

If G acts on a manifold M, then each element Y of $\mathfrak{g}$ defines a vector field ∂_Y on M. This map is a G-equivariant Lie algebra homomorphism of $\mathfrak{g}$ into the Lie algebra of vector fields on M. Just as classical mechanics is concerned with certain special manifolds, the classical analogue of representation theory is the study of certain special manifolds with an action of G.

Recall from after (1.16) the definition of a Poisson manifold. A *Poisson G-space* is a Poisson manifold X endowed with an action of G (preserving the Poisson structure) and a G-equivariant Lie algebra homomorphism

$$(2.2a) \qquad \mu^*: \mathfrak{g} \to C^\infty(X, \mathbf{R}).$$

Here the term on the right (the real-valued smooth functions on X) is given the Lie algebra structure coming from the Poisson bracket (cf. (1.14)), and the G action by translation of functions. We require also that for every Y in $\mathfrak{g}$,

$$(2.2b) \qquad \partial_Y = \xi_{\mu^*(Y)}$$

(notation (1.5)). The *moment map* of the Poisson G-space is a smooth map

$$(2.3a) \qquad \mu: X \to \mathfrak{g}^*,$$

defined as follows. For each x in X, we must define a linear functional $\mu(x)$ on $\mathfrak{g}$. We must therefore specify the value of that linear functional at each element Y of $\mathfrak{g}$. The formula is

$$(2.3b) \qquad \mu(x)(Y) = \mu^*(Y)(x).$$

Finally, a *Hamiltonian G-space* is a Poisson G-space X for which the Poisson structure comes from a G-invariant symplectic structure.

The most obvious examples of Poisson G-spaces arise in analogy with Hamiltonian mechanics. Suppose M is any manifold on which G acts. We have explained in (1.3) how T^*M acquires a symplectic structure. For Y in $\mathfrak{g}$, the vector field ∂_Y on M may be regarded as a function $\mu^*(Y)$ on T^*M; if ∂_Y is regarded as a first-order differential operator on M, then

$$(2.4) \qquad \mu^*(Y) = \sigma_1(\partial_Y),$$

the symbol considered before (1.10). One can check that this definition of μ^* makes T^*M a Hamiltonian G-space. This is an important class of examples, and we will return to it later.

Here is an example with a completely different flavor. We want to endow the dual $\mathfrak{g}^*$ of $\mathfrak{g}$ with the structure of a Poisson G-space. The group G acts on $\mathfrak{g}^*$ by the coadjoint action Ad^*:

$$(2.5a) \qquad [\mathrm{Ad}^*(g)\lambda](Y) = \lambda(\mathrm{Ad}(g^{-1})Y).$$

Since $\mathfrak{g}^*$ is a vector space, each cotangent space is naturally identified with the dual of $\mathfrak{g}^*$, which is just $\mathfrak{g}$. If F is any smooth real-valued function on $\mathfrak{g}^*$, we can therefore regard dF as a smooth map from $\mathfrak{g}^*$ to $\mathfrak{g}$. We can therefore define a vector field on $\mathfrak{g}^*$ by

$$(2.5b) \qquad \xi_F(\lambda) = [\partial_{dF(\lambda)}](\lambda).$$

Here the vector field on the right is the one induced by the coadjoint action. In particular, suppose Y is an element of the Lie algebra $\mathfrak{g}$; regard Y as a (linear) function on $\mathfrak{g}^*$. Then its differential is the constant map taking every point of $\mathfrak{g}^*$ to Y. Consequently

$$(2.5c) \qquad \xi_Y = \partial_Y.$$

The Poisson bracket on smooth functions is defined by

$$(2.5d) \qquad \{H, F\} = \xi_F H.$$

Its restriction to the space $\mathfrak{g}$ of linear functions on $\mathfrak{g}^*$ coincides with the Lie bracket on $\mathfrak{g}$.

The Poisson G-space $\mathfrak{g}^*$ has a universality property that is worth stating separately.

PROPOSITION 2.6. *A Poisson G-space is a Poisson manifold X equipped with an action of G and a G-equivariant map of Poisson manifolds $\mu\colon X \to \mathfrak{g}^*$.*

COROLLARY 2.7. *A Poisson G-space on which G acts transitively is necessarily Hamiltonian; it is a covering of an orbit of the coadjoint action of G on $\mathfrak{g}^*$.*

We turn now to the quantized side of the picture; that is, from manifolds and functions to Hilbert spaces and operators. Recall that a *unitary representation* of G is a pair $(\pi, \mathcal{H})$, with $\mathcal{H}$ a complex Hilbert space and π a (weakly continuous) homomorphism of G into the group of unitary operators on $\mathcal{H}$. We say that $(\pi, \mathcal{H})$ is *irreducible* if $\mathcal{H}$ is not zero and there is no proper closed subspace of $\mathcal{H}$ invariant under all the operators $\pi(g)$ (for g in G). The real "problem of unitary representations" is to find all the irreducible unitary representations of type I Lie groups. (Type I Lie groups include all algebraic Lie groups. The problem makes sense for general Lie groups, but for technical reasons it is much less interesting in that setting.) This (unsolved) problem has two parts: finding all the representations (construction) and proving that you have done so (exhaustion). The construction part is usually very interesting and the exhaustion very dull; so we concentrate on the construction. It may in turn be divided into constructions by deformation ("complementary series") and everything else ("everything else"). In this division it is the first part that is dull. Experience indicates that the second part may be subsumed by the following problem.

2.8. *The problem of unitary representations.* Suppose G is a Lie group and X is a Poisson G-space. The problem is to associate to X a unitary representation of G. If the action of G on X is transitive, then the representation should be almost irreducible.

Notice that this problem is more or less contained in the quantization problem for Poisson manifolds. Suppose we had a functorial solution to Problem 1.17. (This means that maps of Poisson manifolds should give rise to maps of Hilbert spaces; recall that this is considerably too much to ask.) In the setting of Problem 2.8, we would attach to X the Hilbert space $\mathcal{H}$ of Problem 1.17. Each element g of G defines an automorphism of X, and would therefore (by functoriality) define a unitary automorphism $\pi(g)$ of $\mathcal{H}$. The various $\pi(g)$ would constitute the desired representation.

This analogy with quantization suggests that we should seek two more things: Hamiltonians and noncommutative algebras of observables. Gregg Zuckerman has conveyed to me his understanding that the role of the Hamiltonian is played by the entire center of the universal enveloping algebra; we will not pursue this point. As for an algebra of observables, the universal enveloping algebra $U(\mathfrak{g})$ seems to be such a natural candidate that nothing else has usually been considered. (Such a statement seems to slight the group algebra $L^1(G)$, endowed with the convolution product, and its various analogues in distribution theory. The difference between $U(\mathfrak{g})$ and $L^1(G)$ is merely the difference between algebra and analysis. At the philosophical level of the present discourse, such distinctions are of no importance.) However, the work of Joseph and others on ideal theory in $U(\mathfrak{g})$ shows the usefulness of a variety of slightly larger algebras. Here is a reformulation of Problem 2.8 taking some of this into account.

2.9. *The problem of unitary representations and Dixmier algebras.* Suppose X is a Poisson G-space; write

$$\mu \colon X \to \mathfrak{g}^*$$

for the moment map (cf. (2.3a)), and

$$\mu^* \colon S(\mathfrak{g}) \to C^\infty(X, \mathbf{R})$$

for the induced map of Poisson algebras. Fix a G-invariant algebra $\mathscr{A}$ of smooth functions on X containing the image of $S(\mathfrak{g})$:

$$\mu^* \colon S(\mathfrak{g}) \to \mathscr{A}.$$

The problem is to associate to $(X, \mathscr{A})$ a unitary representation $(\pi, \mathscr{H})$ of G on a Hilbert space $\mathscr{H}$, and an algebra $\mathscr{B}$ of operators on $\mathscr{H}$ containing the operators $\pi(Y)$ for Y in $\mathfrak{g}$. We would therefore have a map

$$\pi \colon U(\mathfrak{g}) \to \mathscr{B}.$$

The commutative algebra $\mathscr{A}$ should be approximately isomorphic to the non-commutative algebra $\mathscr{B}$, in a way respecting the approximate isomorphism of $S(\mathfrak{g})$ with $U(\mathfrak{g})$. The term "Dixmier algebra," which was coined in [McGovern, 1987], refers to the algebra $\mathscr{B}$. We will not define Dixmier algebras precisely in general (but see Definition 5.4).

We will postpone discussion of any examples of this Problem 2.9 until §4.

3. Deformation of algebras. In this section we recall from [Gerstenhaber] the framework of the theory of deformations of algebras. Of course an algebra A (say over $\mathbf{C}$) is a vector space over $\mathbf{C}$ endowed with a bilinear map

(3.1) $$P \colon A \times A \to A$$

called the *product.* If the algebra is associative, we usually write

$$P(a, b) = a \cdot b \quad \text{(or even } ab\text{)}.$$

If A is a Lie algebra, we may write

$$P(a, b) = [a, b].$$

Roughly speaking, a deformation of (A, P) should be a one-parameter family of algebras (A, P_t) depending analytically on a parameter t in $\mathbf{C}$, with P_0 equal to P. The difficulty with making this a definition is that most interesting algebras are infinite dimensional as vector spaces. It is therefore not obvious what it should mean for P_t to depend analytically on t.

In the end this will be a serious question, because we really want the deformations to exist as algebras. To start the theory, however, we consider simply formal power series instead of analytic functions. That is, a *deformation* of (A, P) is a collection

(3.2a) $$\{M_n \mid n \in \mathbf{Z}\}$$

of bilinear maps from $A \times A$ to A that satisfies

(3.2b) $$M_0 = P.$$

We can regard $\{M_n\}$ as an algebra structure P_* on the space $A[[t]]$ of formal power series in t with coefficients in A, by

(3.2c) $$P_*\left(\sum a_q t^q, \sum b_r t^r\right) = \sum M_p(a_q, b_r) t^{p+q+r}.$$

In particular, for a and b in A,

(3.2d) $$P_*(a, b) = \sum M_n(a, b) t^n.$$

Especially when the deformed algebra is associative, we may write $a *_t b$ in place of $P_*(a, b)$. Then

(3.3) $$a *_t b = a \cdot b + t M_1(a, b) + t^2 M_2(a, b) + \cdots.$$

There is an uninteresting deformation taking any algebra structure on A to any other, by linear interpolation. What makes it uninteresting is that the intermediate algebras have no interesting properties; they will not usually be associative even if the endpoints are, for example. We are therefore interested in detecting things like associativity in terms of $\{M_n\}$. Obviously the algebra $(A[[t]], P_*)$ is commutative if and only if the bilinear maps N_m are symmetric. Other properties are encoded by the sequence $\{M_n\}$ in a moderately complicated way. For example, P_* is associative if and only if

(3.4a) $$\sum_{p+q=n} M_p(M_q(a, b), c) = \sum_{p+q=n} M_p(a, M_q(b, c))$$

for every n. The first of these identities is just

(3.4b) $$M_0(M_0(a, b), c) = M_0(a, M_0(b, c)),$$

which just says that (A, P) is associative. The next two are (writing the product in A as ab)

(3.4c) $$M_1(a, b)c + M_1(ab, c) = aM_1(b, c) + M_1(a, bc),$$

(3.4d) $$M_2(a, b)c + M_1(M_1(a, b), c) + M_2(ab, c)$$
$$= aM_2(b, c) + M_1(a, M_1(b, c)) + M_2(a, bc).$$

Although such identities are considered in the cohomology theory of algebras, they are difficult to interpret in familiar terms. We will return to this point in Proposition 3.10.

Here is a formulation of deformation theory.

3.5. *The problem of deformation of algebras.* Suppose A is an associative algebra. What are the possible associative deformations of A? Suppose M_1 is a fixed bilinear map from $A \times A$ to A. Under what conditions is M_1 the first-order term of an associative deformation of A? If such a deformation exists, is it in any sense unique?

This problem must be made substantially more specific before it is tractable or interesting. The following class of examples will suggest some reasonable conditions to add.

EXAMPLE 3.6. Suppose $(B, *)$ is an associative algebra with an increasing filtration $\{B_n\}$ parametrized by $\mathbf{N}$. Write

$$(A, \cdot) = \mathrm{gr}(B, *)$$

for the associated graded algebra. The nth level of the gradation is

$$A_{(n)} = B_n / B_{n-1}.$$

We claim that B is the value at 1 of a convergent deformation of A through associative algebras. To see this, choose linear splittings $\psi_n \colon A_{(n)} \to B_n$ for the natural quotient maps. These define a linear isomorphism $\psi \colon A \to B$. Fix a in $A_{(p)}$ and b in $A_{(q)}$. Then $\psi(a) * \psi(b)$ belongs to B_{p+q}. We may therefore define maps

$$M_n \colon A_p \times A_q \to A_{p+q-n}$$

by the requirement that

$$\psi^{-1}(\psi(a) * \psi(b)) = \sum M_n(a, b).$$

We extend M_n by bilinearity to all of A. Suppose c belongs to $A_{(r)}$. Then the associativity of $*$ guarantees that

$$\sum M_n(M_m(a, b), c) = \sum M_n(a, M_m(b, c)).$$

The summand labelled by n and m belongs to $A_{(p+q+r-n-m)}$. The identity is therefore still true if we restrict to n and m with a fixed sum. This is the condition (3.4) for an associative deformation.

As a special case, this example shows that the algebra $C^\infty(T^*M)_{\mathrm{pol}}$ (of smooth functions on a cotangent bundle that are polynomial of bounded degree in the fiber directions) may be deformed to the algebra of differential operators on M. This suggests that deformation theory may sometimes be able to provide the notion of "approximate isomorphism" needed to formulate the problem of quantization (Problem 1.17). The example also suggests a way to deal with the convergence problem in the definition of deformation.

Suppose A is a filtered algebra. A deformation $\{M_n\}$ of A is called *filtered* if

$$(3.7) \qquad\qquad M_n(A_p, A_q) \subset A_{p+q-n}.$$

(Similarly we can define a *graded deformation*.) An immediate consequence of the definition is that (for a and b fixed) $M_n(a, b)$ is zero for large n. Consequently the formal power series (3.3) defining $a *_t b$ converges for any complex number t, and we get a family of products $*_t$ on A; each of them respects the filtration of A, and all have the same associated graded algebra.

Next we consider commutativity properties. The problem of quantization suggests that we are most interested in the deformation of commutative algebras to strictly noncommutative ones. We say that a deformation is *commutative to order k* if

$$(3.8) \qquad M_n(a,b) = M_n(b,a) \qquad (n \le k).$$

Considering a commutative algebra as base point means considering deformations commutative to order 0. To get noncommutative algebras it is natural to take the first-order part of the deformation to be as nonsymmetric as possible. We say that a deformation is *skew to first order* if

$$(3.9) \qquad M_1(a,b) = -M_1(b,a).$$

With an assumption like this, the identities (3.4) become a little more comprehensible.

PROPOSITION 3.10. *Suppose $\{M_n\}$ is an associative deformation of the commutative algebra A that is skew to first order (cf. (3.9)). Then M_1 defines a Poisson algebra structure on A.*

PROOF. We must check that M_1 satisfies (a), (b), and (c) of (1.15). Of course (1.15c) is just the assumption (3.9). For (1.15a), sum the identities (3.4c) for (a,b,c) and (c,a,b), and subtract the identity for (b,c,a). For (1.15b), form the alternating sum of the identities (3.4d) over permutations of (a,b,c). Q.E.D.

In light of the definition in (3.7), Proposition 3.10 suggests the following definition. Suppose A is a filtered commutative algebra. A Poisson algebra structure $\{\,,\,\}$ on A is called *filtered* if

$$(3.11) \qquad \{A_p, A_q\} \subset A_{p+q-1}.$$

Similarly we can define a *graded* Poisson structure on a graded commutative algebra. The Poisson algebra structure on $S(\mathfrak{g})$ introduced in (2.5) is graded.

Here is a particular version of the problem of deformation of algebras.

PROBLEM 3.12. Suppose A is a finitely generated filtered commutative algebra with 1 over $\mathbf{C}$. Let $\{\,,\,\}$ be a filtered Poisson structure on A (cf. (3.11)). Under what conditions does there exist a filtered associative deformation $\{M_n\}$ of A (cf. (3.7)) such that

$$(a) \qquad M_1(a,b) = \tfrac{1}{2}\{a,b\}?$$

Of course one could easily replace "filtered" by "graded".

Analogous problems for C^∞ structures (instead of algebraic ones) have been intensively studied for the past ten or fifteen years (see for example [Bayen et al.] and [Sternheimer] and the references therein). The most striking result is that for symplectic manifolds there are no obstructions to the existence of formal associative deformations beginning with the Poisson bracket (see [De Wilde–Lecomte]; their work relies heavily on partial results of Vey, Lichnerowicz, Neroslavsky, and Vlassov).

4. Geometric quantization. In this section we will outline a few of the ideas developed by Auslander, Blattner, Kirillov, Kostant, Souriau, and others to solve Problem 2.8. (In addition to the references mentioned at the beginning of §2, one can consult [Kostant, 1983].) Since the point is largely to explain why these ideas do not always apply (more precisely, to motivate Problem 4.17) we can afford to be rather sketchy. The general technique is called *geometric quantization.* Inelegantly stated, the idea to this. There are various known ways to construct a unitary representation $\pi(D)$ from some data D. One seeks then to construct a Poisson G-space $X(D)$ from the same data. Given an abstract Poisson G-space X, one seeks an isomorphism of X with some $X(D)$. If such an isomorphism can be found, we declare that $\pi(D)$ solves Problem 2.8.

The guiding principle here is the example of cotangent bundles (cf. (2.4)). If X is the cotangent bundle of a manifold M on which G acts, then the analogy with quantization suggests that we should attach to X the unitary representation of G on something like $L^2(M)$. To make sense of that, we need a G-invariant measure on M; and such a measure need not exist. We therefore use (instead of $L^2(M)$) the Hilbert space $L^2(M, |\Lambda^{1/2}|)$ of square-integrable half-densities on M. (The line bundle $|\Lambda^{1/2}|$ of half-densities is characterized by the fact that sections transform according to the square root of the absolute value of the Jacobian under coordinate changes.)

Very few Poisson G-spaces look like cotangent bundles. To generalize this example, suppose that G acts on a Hermitian line bundle $\mathscr{L}$ over M. This setting does give rise to a unitary representation of G (on square-integrable sections of $\mathscr{L} \otimes |\Lambda^{1/2}|$). It is less obvious how to attach a Poisson G-space; but this can be done in the following way. Recall that a *connection* on a line bundle $\mathscr{L}$ is a map ∇ that assigns to each vector field Y on M a first-order differential operator ∇_Y on sections of $\mathscr{L}$. The map ∇ is required to be $C^\infty(M)$-linear in Y, and to satisfy

$$(4.1) \qquad \nabla_Y(fs) = (Yf)s + f\nabla_Y(s)$$

for f a smooth function and Y a smooth section of L. (This amounts to requiring that the operator ∇_Y have principal symbol equal to Y (regarded as a function on T^*M).) It follows that the difference $\nabla - \nabla'$ of two connections is a $C^\infty(M)$-linear map g from vector fields to zeroth-order differential operators. A zeroth-order differential operator on sections of $\mathscr{L}$ is multiplication by a function. A $C^\infty(M)$-linear map from vector fields to functions is just a section of the complexified cotangent bundle. This argument (judiciously sprinkled with local coordinates) proves

LEMMA 4.2. *Suppose $\mathscr{L}$ is a complex line bundle on the real manifold M. Then the space of connections on $\mathscr{L}$ may be identified with the space of smooth sections of a fiber bundle $T^*(M, \mathscr{L})_{\mathbf{C}}$. This bundle is an affine bundle for the complexified cotangent bundle.*

A point of $T^*(M, \mathscr{L})_{\mathbf{C}}$ is a pair $(q, \nabla(q))$. Here q is a point of M, and $\nabla(q)$ is a map from $T_q(M)$ to first-order differential operators on sections of $\mathscr{L}$ at the point q. (Such operators are distributions supported at q on the space of sections of $\mathscr{L}$, taking values in the fiber of $\mathscr{L}$ at q.) The map $\nabla(q)$ is required to have the following property: suppose ξ is a tangent vector at q, and f is a smooth function such that $f(q)$ and ξf are both zero. Then

$$(4.3) \qquad \nabla_\xi(q)(fs) = 0$$

for any section s of $\mathscr{L}$.

Suppose now that the line bundle $\mathscr{L}$ is Hermitian. We can define the *skew-adjoint* ∇^h of a connection ∇ by

$$(4.4) \qquad \langle \nabla_X s, t \rangle + \langle s, \nabla_X^h t \rangle = X\langle s, t \rangle$$

for a real vector field X. (Here $\langle \, , \, \rangle$ is the Hermitian pairing from sections of $\mathscr{L}$ to functions.) A connection is called *purely imaginary* if it is equal to its skew-adjoint. If $\mathscr{L}$ is trivial, a skew-adjoint connection is one of the form

$$(4.5) \qquad (\nabla_X)s = Xs + i\lambda(X)s,$$

with λ a section of the (real) cotangent bundle.

PROPOSITION 4.6. *Suppose $\mathscr{L}$ is a Hermitian line bundle on the real manifold M. Then the space of purely imaginary connections on $\mathscr{L}$ may be identified with the space of smooth sections of a fiber bundle $T^*(M, \mathscr{L})$. This bundle is an affine bundle for the cotangent bundle. It carries a natural symplectic structure $\omega(\mathscr{L})$. If G acts on $(M, \mathscr{L})$, then $T^*(M, \mathscr{L})$ is a Hamiltonian G-space.*

The first part requiring serious thought is the definition of $\omega(\mathscr{L})$. For this, we confine attention to a neighborhood of a point q in M. Choose a section s_0 of $\mathscr{L}$ that does not vanish near q. Then s_0 defines a local trivialization of $\mathscr{L}$ near q, and so an identification of $T^*(M, \mathscr{L})$ with T^*M near q. Now define ω^1 and $\omega(\mathscr{L}) = d\omega^1$ near q exactly as in (1.3). If s_0 is replaced by another section fs_0 in this definition, then one calculates that ω^1 is changed by $d(f \circ \pi)/f \circ \pi$ (the differential of $\log(f)$); so $\omega(\mathscr{L})$ is unchanged.

To specify the Hamiltonian G-space structure in Proposition 4.6, we must send each element Y of $\mathfrak{g}$ to a function $\mu^*(Y)$ on $T^*(M, \mathscr{L})$. Suppose $(q, \nabla(q))$ is a point of $T^*(M, \mathscr{L})$. The action of G on $\mathscr{L}$ assigns to Y a first-order differential operator $\partial_Y^{\mathscr{L}}$ on sections of $\mathscr{L}$. On the other hand, $\nabla(q)$ sends the tangent vector $\partial_Y(q)$ to a first-order differential operator on sections at q. These two operators differ by a zeroth-order operator; that is, by multiplication by a constant. This constant is i times the value of $\mu^*(Y)$ at $(q, \nabla(q))$.

We can now describe the first case of geometric quantization. Given a Hamiltonian G-space X, one tries to find an action of G on a Hermitian line bundle $\mathscr{L}$ over a manifold M, with the property that

$$(4.7) \qquad X \cong T^*(M, \mathscr{L})$$

as a Hamiltonian G-space. One then attaches to X the unitary representation of G on $L^2(M, \mathscr{L} \otimes |\Lambda^{1/2}|)$.

Almost any extension of geometric quantization beyond this setting leads to thorny technical problems. Nevertheless, one such extension deserves a brief mention.

Suppose X is a (possibly indefinite) Kähler manifold, and $\mathscr{L}$ is a Hermitian holomorphic line bundle over X. Assume that G acts on X and $\mathscr{L}$, preserving the holomorphic structures, the Kähler form, and the metric on $\mathscr{L}$. The most obvious candidate for a unitary representation attached to these data is the one on the space $H_2^0(X, \mathscr{L})$ of L^2 holomorphic sections of $\mathscr{L}$. Unfortunately this space is zero in a great many interesting cases. An appropriate generalization is the L^2 Dolbeault cohomology space $H_2^p(X, \mathscr{L})$, which consists of square-integrable harmonic $\mathscr{L}$-valued $(0, p)$-forms on X. This space is defined whenever the Kähler form is definite. When the Kähler form is indefinite (and this still includes important examples) the "inner product" on forms is indefinite, and it becomes very difficult even to define $H_2(X, \mathscr{L})$ as a unitary representation (see [Rawnsley-Schmid-Wolf, 1983]).

Despite these problems with the Hilbert space side of the picture, we can explain in reasonable generality how X may be regarded as a Hamiltonian G-space. Throughout this construction, keep in mind the example of $T^*(M, \mathscr{L})$, to which X is analogous. The (complex) antiholomorphic tangent directions on X play the role of the directions along the fibers in $T^*(M, \mathscr{L})$. The analogue of the half-density bundle on M would be a Hermitian line bundle with curvature form (defined below) equal to the negative of the symplectic form on X. Such a bundle may not exist; so we essentially pretend that it is already built into the line bundle we are given. To maintain the analogy with the real case, we therefore put a shift into the definition of $\omega(\mathscr{L})$ below.

One can associate to $\mathscr{L}$ a two-form $\omega(\mathscr{L})$ (real and of type $(1,1)$) in the following way. Fix an open set U in X over which $\mathscr{L}$ has a nowhere-vanishing holomorphic section s. Then $\langle s, s \rangle$ is a real-valued positive smooth function on U. Set

$$(4.8) \qquad \omega_c(\mathscr{L}) = i\partial\overline{\partial}(\log\langle s, s \rangle).$$

Here ∂ and $\overline{\partial}$ are the components of the de Rham differential d having bidegree $(1,0)$ and $(0,1)$ respectively. (One can think of the subscript c as standing for Chern or for curvature or even for canonical; our ω_c differs from the curvature form by a constant multiple, however.) In local holomorphic coordinates, this is

$$(4.8)' \qquad \omega_c(\mathscr{L}) = i\sum \frac{\partial^2 \log\langle s, s \rangle}{\partial z_i \partial \bar{z}_j} dz_i \wedge d\bar{z}_j.$$

Replacing s by another holomorphic section adds to $\log\langle s, s \rangle$ a term $\log|f|^2$, with f a holomorphic function. Such a term is annihilated by $\partial\overline{\partial}$ (by an easy calculation; a special case is the fact that the log of the absolute value of a holomorphic function of one variable is harmonic). The 2-form $\omega_c(\mathscr{L})$ is therefore well-defined on all of X. That it is closed is immediate from (4.8). Finally, define

$$(4.9) \qquad \omega(\mathscr{L}) = \omega_c(\mathscr{L}) + \omega_0,$$

with ω_0 the (nondegenerate) 2-form defined by the Kähler structure. The form $\omega(\mathscr{L})$ may be degenerate in general; we will consider only the case when it is not.

We make X into a Hamiltonian G-space in the following way. First, we need the structure

(4.10a) $$\mu_0^*: \mathfrak{g} \to C^\infty(X, \mathbf{R})$$

of a Hamiltonian G-space for the symplectic structure ω_0 on X (cf. (2.2)). (To see that this is really needed, and cannot somehow be constructed naturally, consider the group $G = \mathbf{R}$ acting on $X = \mathbf{C}$ by translation in the real direction. The image of μ^* must consist of multiples of a function of the form

$$f(x + iy) = y + c,$$

with c a real constant. There is no preferred value of c.) Now we use the line bundle $\mathscr{L}$ to define a map

(4.10b) $$\mu_c^*: \mathfrak{g} \to C^\infty(X, \mathbf{R}).$$

This we do locally, using a nonvanishing holomorphic section s of $\mathscr{L}$. Fix Y in $\mathfrak{g}$. Write ∂_Y for the corresponding vector field on X, and $\partial_Y^{\mathscr{L}}$ for the differential operator on sections of $\mathscr{L}$. Write ∂_Y^h for the holomorphic component of ∂_Y (with respect to the decomposition of each complex tangent space of X into holomorphic and antiholomorphic parts). Put

(4.11a) $$\mu_c^*(Y) = -i\left\{\frac{\partial_Y^h\langle s,s\rangle}{\langle s,s\rangle} - \frac{\langle\partial_Y^{\mathscr{L}}s,s\rangle}{\langle s,s\rangle}\right\}.$$

The key properties of this map are that it is independent of the choice of holomorphic section s, and that for any vector field ξ on X,

(4.11b) $$\omega_c(\mathscr{L})(\xi, \partial_Y) = \xi\mu_c(Y).$$

If we define

(4.12a) $$\mu^* = \mu_0^* + \mu_c^*,$$

then

(4.12b) $$\omega(\mathscr{L})(\xi, \partial_Y) = \xi\mu^*(Y).$$

When $\omega(\mathscr{L})$ is nondegenerate, (4.12b) shows that μ^* defines the structure of a Hamiltonian G-space on X. (When $\omega(\mathscr{L})$ is degenerate, it suggests that X should be replaced by (the maximal spectrum of) the Poisson algebra of functions on X that are killed by all tangent vectors in the radical of $\omega(\mathscr{L})$. We will not pursue this, however.)

It is possible to generalize geometric quantization still further by considering spaces intermediate between cotangent bundles and Kähler manifolds. We will not give a detailed presentation of what is known about constructing unitary representations from such spaces. What matters for us is that they do not exhaust the interesting examples even of Hamiltonian G-spaces. For reductive groups, there is a nice structure theory for Hamiltonian homogeneous spaces that isolates the remaining examples rather precisely. We will conclude this section with an account of that theory.

Suppose G is a reductive Lie group. An element λ in $\mathfrak{g}^*$ is called *nilpotent* if either of the following equivalent conditions is satisfied:

(4.13a)   the restriction of λ to the isotropy algebra $\mathfrak{g}^\lambda$ is zero; or

(4.13b)   there is a Borel subalgebra $\mathfrak{b}$ of $\mathfrak{g}_{\mathbb{C}}$ such that $\lambda|_{\mathfrak{b}}$ is zero.

We say that λ is *semisimple* if the isotropy algebra $\mathfrak{g}^\lambda$ is reductive in $\mathfrak{g}$. The analogue of the Jordan decomposition in $\mathfrak{g}^*$ is

PROPOSITION 4.14. *Suppose G is a reductive Lie group. Then the number of nilpotent G-orbits in $\mathfrak{g}^*$ is finite, and the set of semisimple elements is open and dense. Any element λ in $\mathfrak{g}^*$ has a unique Jordan decomposition $\lambda = \lambda_s + \lambda_n$ as a sum of a semisimple and a nilpotent element, characterized by the property that the isotropy group for λ is contained in those for λ_s and λ_n.*

A Poisson G-space (for a reductive group) is called *unipotent* if the image of its moment map consists of nilpotent elements in $\mathfrak{g}^*$ (cf. (2.3) and Corollary 2.7). It is called *semisimple* if the image of the moment map consists of semisimple elements.

Here is the first part of the structure theorem we want. (It is just a fancy restatement of the Jordan decomposition.)

THEOREM 4.15. *Suppose G is a reductive group and X is a Hamiltonian homogeneous space for G. Then there is a canonical fibration*

$$X \to X_s$$

of X over a semisimple Hamiltonian homogeneous space X_s. The fiber over a point z in X_s is a unipotent Hamiltonian homogeneous space for the (reductive) isotropy group of z.

The second part of the structure theorem relates semisimple Hamiltonian homogeneous spaces to cotangent bundles and Kähler manifolds.

THEOREM 4.16. *Suppose G is a reductive group and X is a semisimple Hamiltonian homogeneous space for G. Then there is a canonical fibration*

$$X \to X_h$$

of X over another semisimple Hamiltonian G-space X_h. The base space X_h is isomorphic to $T^*(M, \mathscr{L})$ (cf. Proposition 4.6) for some homogeneous Hermitian line bundle $\mathscr{L}$ over a homogeneous space X. The fiber over a point z in X_h is a homogeneous (possibly indefinite) Kähler manifold for the (reductive) isotropy group of z.

The subscript h stands for "hyperbolic." The analogy with Theorem 4.15 may be sharpened by defining first hyperbolic and elliptic elements in $\mathfrak{g}^*$, then hyperbolic and elliptic Poisson G-spaces. The two theorems may then be reorganized as follows. First, any Hamiltonian homogeneous space (still for a reductive group) is built from nilpotent, elliptic, and hyperbolic ones by successive fibrations. For many purposes it is then enough to treat the three classes separately. Next, any elliptic Hamiltonian homogeneous space is Kähler, and any hyperbolic Hamiltonian homogeneous space is of the form $T^*(M, \mathscr{L})$.

In light of this analysis, one can believe that the following special case of Problem 2.8 is of particular importance.

4.17. *The problem of unipotent representations.* Suppose G is a reductive group, and X is a unipotent Poisson G-space (defined after Proposition 4.14). Attach to X a unitary representation of G.

Problem 2.9 suggests that we should first try to construct appropriate algebras of observables. Precise requirements for such algebras will be specified in §5. In §6 we will consider the problem of finding the representations.

5. Unipotent Dixmier algebras. Problem 4.17 isolates a particularly interesting class of Hamiltonian homogeneous spaces. In this section we will discuss a nonconstructive method for attaching algebras of observables to them (Definition 5.4; this definition is closely related to the Dixmier conjecture formulated in [Vogan, 1986]). Problem 3.12 suggests that such an algebra might be attached to a Poisson algebraic variety. Fix now a connected complex semisimple group $G_{\mathbf{C}}$. We defined a Poisson G-space in (2.2) as a manifold endowed with certain additional structure; we can define a *Poisson $G_{\mathbf{C}}$-variety* as an algebraic variety endowed with the analogous algebraic structures. (This makes sense because $G_{\mathbf{C}}$ is an algebraic group.)

DEFINITION 5.1. A *unipotent Poisson variety* is an irreducible affine algebraic variety X that is at the same time a Poisson $G_{\mathbf{C}}$-variety. We require in addition that

(a) the moment map μ is a finite map;

(b) the image of μ is the closure X_0 of a nilpotent coadjoint orbit $\mathscr{O}_{\mathbf{C}}$ in $\mathfrak{g}_{\mathbf{C}}^*$ (cf. (4.13)); and

(c) X is a normal algebraic variety.

We say that X is *attached* to $\mathscr{O}_{\mathbf{C}}$ or X_0.

Because the nilpotent coadjoint orbits are even-dimensional symplectic cones, it is rather easy to classify the unipotent Poisson varieties. If G/H is any homogeneous space for a Lie group G we (rather pompously) define the *equivariant fundamental group* of G/H to be

$$(5.2) \qquad \pi_1^G(G/H) = H/H_0;$$

here H_0 is the identity component of H. If G is connected and simply connected, this is the ordinary fundamental group. In general its subgroups correspond to the homogeneous spaces for G that cover G/H.

THEOREM 5.3. *Suppose $G_{\mathbf{C}}$ is a complex connected semisimple Lie group, and $\mathscr{O}_{\mathbf{C}}$ is a nilpotent coadjoint orbit in $\mathfrak{g}_{\mathbf{C}}^*$. Then the isomorphism classes of* *unipotent Poisson varieties attached to $\mathscr{O}_{\mathbf{C}}$ (Definition 5.2) are in one-to-one correspondence with conjugacy classes of subgroups of $\pi_1^{G_{\mathbf{C}}}(\mathscr{O}_{\mathbf{C}})$. Fix such a subgroup S, and let X be the corresponding Poisson variety.*

(a) *Write $N(S)$ for the normalizer of S in $\pi_1^{G_{\mathbf{C}}}(\mathscr{O}_{\mathbf{C}})$. Then the quotient $N(S)/S$ acts by automorphisms on X, respecting both the action of $G_{\mathbf{C}}$ and the moment map of X to X_0.*

(b) *The ring of functions $R(X)$ carries a natural $G_{\mathbf{C}}$-invariant grading $\{R_p(X)\}$ indexed by $k^{-1}\mathbf{N}$, for some positive integer k. This grading is compatible with the map μ^* from $S(\mathfrak{g}_{\mathbf{C}})$ to $R(X)$ corresponding to the moment map; that is, μ^* takes polynomials homogeneous of degree n into the nth level of the gradation.*

(c) *If f and h belong to $R_p(X)$ and $R_q(X)$ respectively, then the Poisson bracket $\{f, h\}$ belongs to R_{p+q-1}.*

We will not give a detailed proof. The grading on $R(X)$ amounts to an action of the k-fold cover of $\mathbf{C}^\times$ (which is again isomorphic to $\mathbf{C}^\times$) on X. The Poisson structure comes from a symplectic structure on the dense open subset $\mu^{-1}(\mathscr{O}_{\mathbf{C}})$. Each of these structures may be lifted in a routine way from the corresponding one on $\mathscr{O}_{\mathbf{C}}$ itself.

We can now specify the algebras of observables we want, at least conjecturally.

DEFINITION 5.4. Suppose X is a unipotent Poisson variety attached to a nilpotent coadjoint orbit $\mathscr{O}_{\mathbf{C}}$. Use the notation of Theorem 5.3. A *unipotent Dixmier algebra* attached to X is an associative algebra A filtered by $k^{-1}\mathbf{N}$, on which $G_{\mathbf{C}}$ acts by automorphisms in a locally algebraic, degree-preserving way. This action is denoted Ad. We require in addition that

(a) The associated graded algebra $\operatorname{gr} A$ is isomorphic to $R(X)$ as a graded algebra with an action of $G_{\mathbf{C}}$.

(b) Suppose a belongs to A_p and b to A_q. Write $\operatorname{gr}(a)$ and $\operatorname{gr}(b)$ for the elements of $R(X)_p$ and $R(X)_q$ corresponding to the classes of a and b in $\operatorname{gr} A$. Then

$$\operatorname{gr}(ab - ba) = \{\operatorname{gr}(a), \operatorname{gr}(b)\}.$$

Here the structure on the right is the Poisson bracket on $R(X)$.

(c) There is a map

$$\phi\colon U(\mathfrak{g}_{\mathbf{C}}) \to A$$

of filtered algebras with $G_{\mathbf{C}}$ action. The differential ad of the action Ad on A satisfies

$$\operatorname{ad}(Y)a = \phi(Y)a - a\phi(Y) \qquad (a \in A, Y \in \mathfrak{g}_{\mathbf{C}}).$$

(d) The following diagram of maps of graded algebras is commutative.

$$
\begin{array}{ccc}
\operatorname{gr} U(\mathfrak{g}_{\mathbf{C}}) & \xrightarrow{\ \operatorname{gr}\phi\ } & \operatorname{gr} A \\
\downarrow & & \downarrow \\
S(\mathfrak{g}_{\mathbf{C}}) & \xrightarrow{\ \mu^*\ } & R(X)
\end{array}
$$

The vertical isomorphisms are Poincaré-Birkhoff-Witt and the isomorphism of (a) above.

(e) It follows from the preceding assumptions that the kernel of ϕ contains a maximal ideal $\mathscr{I}$ in the center of $U(\mathfrak{g}_{\mathbf{C}})$. We assume that the Harish-Chandra homomormophism (see [Humphreys]) identifies $\mathscr{I}$ with a weight λ in the real span of the roots.

(f) The length of the weight λ defined in (e) (with respect to the Killing form) is minimal among all weights arising for algebras A' satisfying (a)–(e).

Using Example 3.6, it is easy to see that unipotent Dixmier algebras are closely related to deformations of $R(X)$. Here is a precise statement.

LEMMA 5.5. *Suppose X is a unipotent Poisson variety. Let $\{M_n\}$ be a graded $G_{\mathbf{C}}$-equivariant associative deformation of $R(X)$ such that*

$$M_1 = \tfrac{1}{2}\{\ ,\ \}.$$

Then the deformed algebra at $t = 1$ satisfies conditions (a)–(d) *of Definition* 5.4.

Conversely, suppose A is a filtered associative algebra satisfying (a)–(d) *of Definition* 5.4. *Construct a deformation $\{M_n\}$ as in Example* 3.6, *using $G_{\mathbf{C}}$-equivariant splittings ψ_n. Then $\{M_n\}$ is a graded $G_{\mathbf{C}}$-equivariant deformation of $R(X)$.*

The existence of unipotent Dixmier algebras is therefore more or less a special case of Problem 3.12. That problem should have an excellent solution in this case.

CONJECTURE 5.6. Suppose X is a unipotent Poisson variety (Definition 5.1). Then there is a unique (up to isomorphism) Dixmier algebra $A(X)$ associated to X (Definition 5.4). The group $N(S)/S$ of Theorem 5.3 acts by automorphisms on $A(X)$; this action commutes with that of $G_{\mathbf{C}}$. The kernel of the map ϕ of Definition 5.4(c) is a weakly unipotent maximal ideal in $U(\mathfrak{g}_{\mathbf{C}})$. (The term "weakly unipotent" is defined in [Vogan, 1984]; this condition is very important technically, but can be ignored for our present purposes.)

One of the simplest examples of a unipotent Poisson variety is the full cone $\mathscr{N}^*$ of nilpotent elements in $\mathfrak{g}_{\mathbf{C}}^*$. (This is normal by a theorem of Kostant.) Related results of Kostant make it easy to find all the algebras A satisfying (a)–(d) of Definition 5.3. They are exactly the quotients $U(\mathfrak{g}_{\mathbf{C}})/U(\mathfrak{g}_{\mathbf{C}})\mathscr{I}$, with $\mathscr{I}$ any maximal ideal in the center of $U(\mathfrak{g}_{\mathbf{C}})$. The weight λ appearing in Definition 5.3(e) can therefore be arbitrary. Conditions (e) and (f) require that we choose λ of minimal length in the real span of the roots; that is, that we take λ equal to zero. This is *not* the augmentation ideal, incidentally. Rather it is the annihilator in the center of $U(\mathfrak{g}_{\mathbf{C}})$ of the Verma module M of highest weight $-\rho$, with ρ half the sum of the positive roots. The algebra A may therefore be realized as an algebra of operators on M.

If Y is a unipotent Poisson variety that covers another one, X, then $R(X)$ may be realized as a subalgebra of $R(Y)$. It is reasonable to suspect that unipotent Dixmier algebras behave analogously, but they do not. As an example, suppose

$G_{\mathbf{C}}$ is $\mathrm{SL}(n,\mathbf{C})$. The equivariant fundamental group (cf. (5.2)) of the principal nilpotent orbit $\mathscr{O}_{\mathbf{C}}$ turns out to be $\mathbf{Z}/n\mathbf{Z}$. Let X be the unipotent Poisson variety attached to $\mathscr{O}_{\mathbf{C}}$ and the trivial subgroup S (Theorem 5.3); it is an n-fold ramified cover of $\mathscr{N}^*$. It turns out that there is at most one algebra A satisfying parts (a)–(d) of Definition 5.4; if it exists, it is a unipotent Dixmier algebra, and the corresponding λ is ρ/n. The algebra may be realized as operators on a Whittaker module for $\mathfrak{g}_{\mathbf{C}}$ (see [Kostant, 1978]).

For another fundamental example, suppose $G_{\mathbf{C}}$ is $\mathrm{Sp}(2n,\mathbf{C})$, and X is $\mathbf{R}^{2n}$ (often written as $\mathbf{R}^n \times \mathbf{R}^n$). Endow X with the standard symplectic structure

$$\omega((x,y),(v,w)) = \langle x,w \rangle - \langle y,v \rangle.$$

Here we identify each tangent space with $\mathbf{R}^{2n}$, and write $\langle \ , \ \rangle$ for the inner product on $\mathbf{R}^n$. The group $G_{\mathbf{C}}$ acts by linear transformations on X. This action is Hamiltonian; we will not write the map μ^* as in (2.2) explicitly, but its image consists of homogeneous polynomial functions of degree 2 on X. The corresponding moment map μ from X to $\mathfrak{g}^*$ is therefore homogeneous quadratic. Its image is a cone consisting of nilpotent elements, and μ is a two-to-one covering except at the origin. It follows that X is a unipotent Poisson variety. Conjecture 5.6 is true in this case; $A(X)$ is the Weyl algebra of polynomial coefficient differential operators in n variables, filtered by one half the total order. (The total order refers to the sum of the orders of the differential operator and the polynomial coefficient.) The map ϕ sends $\mathfrak{g}_{\mathbf{C}}$ to certain even differential operators of total order two.

For further examples of unipotent Dixmier algebras and related matters, we refer to [Vogan, 1986].

6. Unipotent representations. Throughout this section, we fix a semisimple Lie group G, assumed to have finitely many connected components and finite center. Fix also a connected complex semisimple group $G_{\mathbf{C}}$ with Lie algebra $\mathfrak{g}_{\mathbf{C}}$. We do not require $G_{\mathbf{C}}$ to contain G (because we want to allow nonlinear groups), but we assume that the adjoint representation of G factors through a homomorphism

$$(6.1a) \qquad\qquad\qquad G \to G_{\mathbf{C}}.$$

The problem of finding unitary representations of G has been reduced completely to algebra by the work of Harish-Chandra, and we want to take full advantage of this reduction. We will not describe Harish-Chandra's results in detail, but we will recall the relevant algebraic objects. Fix a maximal compact subgroup K of G, and write $K_{\mathbf{C}}$ for its complexification. The restriction to K of the map (6.1a) (like any map of K into a complex group) extends automatically to

$$(6.1b) \qquad\qquad\qquad K_{\mathbf{C}} \to G_{\mathbf{C}}.$$

The *Cartan involution* θ of G is the unique automorphism of G of order 2 with fixed point set equal to K. We assume that θ extends to an automorphism of

$G_{\mathbf{C}}$. We will write θ also for the differential of θ, an automorphism of $\mathfrak{g}$ or $\mathfrak{g}_{\mathbf{C}}$. Write

(6.2a) $$\mathfrak{p} = -1 \text{ eigenspace of } \theta \text{ on } \mathfrak{g}.$$

The group K acts on $\mathfrak{p}$, so $K_{\mathbf{C}}$ acts on $\mathfrak{p}_{\mathbf{C}}$. We can identify $\mathfrak{p}_{\mathbf{C}}^*$ with the linear functionals on $\mathfrak{g}_{\mathbf{C}}$ vanishing on $\mathfrak{k}_{\mathbf{C}}$, and write

(6.2b) $$\mathfrak{p}_{\mathbf{C}}^* \subset \mathfrak{g}_{\mathbf{C}}^*.$$

EXAMPLE 6.3. Suppose G is $\mathrm{SL}(n, \mathbf{R})$, the group of n-by-n real matrices of determinant one. Then we may take K to be the group $\mathrm{SO}(n)$ of n-by-n real orthogonal matrices of determinant one; $K_{\mathbf{C}}$ is $\mathrm{SO}(n, \mathbf{C})$, and

$$\theta_g = {}^tg^{-1}, \qquad \theta Z = -{}^tZ$$

for g in G and Z in $\mathfrak{g}$. The subspace $\mathfrak{p}$ consists of symmetric matrices.

Here are the algebraic objects in Harish-Chandra's theory. A $(\mathfrak{g}, K)$-*module* is a complex vector space V carrying representations of both $\mathfrak{g}$ and K. These are required to satisfy the following conditions (for any v in V, Y in $\mathfrak{g}$, and k in K):

(6.4a) The subspace $\langle K \cdot v \rangle$ spanned by the K-translates of v is finite-dimensional.

(6.4b) K acts smoothly on $\langle K \cdot v \rangle$, and the differential of this action is the restriction to $\mathfrak{k}$ of the specified $\mathfrak{g}$ action.

(6.4c) $$k \cdot (Y \cdot v) = [\mathrm{Ad}(k)Y] \cdot (k \cdot v).$$

For k in the identity component of K, condition (c) is a consequence of (a) and (b). Any representation of $\mathfrak{g}$ on a complex vector space extends to $\mathfrak{g}_{\mathbf{C}}$, and the representation of K extends to $K_{\mathbf{C}}$ because of (a) and (b). We may therefore speak of V as a $(\mathfrak{g}_{\mathbf{C}}, K_{\mathbf{C}})$-module. In this guise, it satisfies conditions analogous to (6.4). Finally, we may extend the representation of the Lie algebra $\mathfrak{g}_{\mathbf{C}}$ to a representation of the associative algebra $U(\mathfrak{g}_{\mathbf{C}})$, and speak of a $(U(\mathfrak{g}_{\mathbf{C}}), K_{\mathbf{C}})$-module.

The real form $\mathfrak{g}$ determines an antiautomorphism of $U(\mathfrak{g}_{\mathbf{C}})$, defined for Y and Z in $\mathfrak{g}$ by

(6.5a) $$(Y + iZ)^h = -Y + iZ.$$

Its restriction to $\mathfrak{k}_{\mathbf{C}}$ is the differential of an antiautomorphism of $K_{\mathbf{C}}$, satisfying

(6.5b) $$K = \{k \in K_{\mathbf{C}} \mid k^h = k^{-1}\}.$$

A Hermitian form $\langle \ , \ \rangle$ on the $(\mathfrak{g}, K)$-module V is called *invariant* if whenever v and w are in V, Y is in $\mathfrak{g}$, and k is in K,

(6.6) $$\langle Y \cdot v, w \rangle = -\langle v, Y \cdot w \rangle, \qquad \langle k \cdot v, w \rangle = \langle v, k^{-1} \cdot w \rangle.$$

It is equivalent to require

(6.6)' $$\langle Y \cdot v, w \rangle = \langle v, Y^h \cdot w \rangle, \qquad \langle k \cdot v, w \rangle = \langle v, k^h \cdot w \rangle$$

for Y in $\mathfrak{g}_{\mathbf{C}}$ and k in $K_{\mathbf{C}}$.

Here is the algebraic translation of unitary representation theory.

THEOREM 6.7 (HARISH-CHANDRA). *Suppose G is a real semisimple Lie group. Then there is a one-to-one correspondence between the set of (equivalence classes of) irreducible unitary representations of G and the set of (equivalence classes of) irreducible $(\mathfrak{g}, K)$-modules admitting a positive-definite invariant Hermitian form (cf. (6.4) and (6.6)).*

Next we want to bring our proposed algebra of observables into the picture. Suppose therefore that A is a unipotent Dixmier algebra attached to the unipotent Poisson variety X (Definition 5.4). The map (6.1b) (together with Definition 5.4) provides an action that we can call Ad of $K_{\mathbf{C}}$ of A. Using this action, we can define an $(A, K_{\mathbf{C}})$-*module* in analogy with (6.4).

PROBLEM 6.8. Suppose A is a unipotent Dixmier algebra and $K_{\mathbf{C}}$ is the complexified maximal compact subgroup of a real form of $G_{\mathbf{C}}$. Give a geometric description of the category of $(A, K_{\mathbf{C}})$-modules.

The remarks after Proposition 6.11 below describe a little more precisely what kind of geometric description should be possible. It seems likely that this problem admits a rather complete solution, although the geometry involved may turn out to be subtle. One reason to believe this is the analogy with the results of [Beilinson-Bernstein, 1981]. Another is that there is a little experimental evidence for

CONJECTURE 6.9. In the setting of Problem 6.8, any $(A, K_{\mathbf{C}})$-module is completely reducible.

We want now to describe the geometry that should figure in a solution of Problem 6.8. By Definition 5.1 we have a map μ from X (the unipotent Poisson variety) to $\mathfrak{g}_{\mathbf{C}}^*$. We define the $K_{\mathbf{C}}$-*Lagrangian in* X to be the inverse image of $\mathfrak{p}_{\mathbf{C}}^*$:

$$(6.10) \qquad\qquad L = \{x \in X \mid \mu(x) \in \mathfrak{p}_{\mathbf{C}}^*\}.$$

L is clearly a Zariski-closed subvariety of X.

PROPOSITION 6.11. *In the setting of Problem 6.8 and (6.10), the ideal $I(L)$ of functions in $R(X)$ vanishing on L is an ideal for Poisson bracket:*

$$\{R(X), I(L)\} \subset I(L).$$

The dimension of L is at most half the dimension of X. If the nilpotent coadjoint orbit $\mathcal{O}$ to which X is attached (Definition 5.1) meets $\mathfrak{p}_{\mathbf{C}}^*$, then

$$\dim L = \tfrac{1}{2} \dim X.$$

Suppose V is a finitely generated $(A, K_{\mathbf{C}})$-module. Fix a good $K_{\mathbf{C}}$-invariant filtration of V with respect to the filtration of A (Definition 5.4). Then $\operatorname{gr} V$ is a finitely generated graded $(R(X), K_{\mathbf{C}})$-module supported on L.

Here of course $(R(X), K_{\mathbf{C}})$-modules are defined in analogy with (6.4). The nontrivial assertions in the Proposition (about dimension) follow from [Kostant-Rallis, 1971].

What a solution to Problem 6.8 should provide is a description of $(A, K_{\mathbf{C}})$-modules in terms of graded $(R(X), K_{\mathbf{C}})$-modules supported on L. This suggests at once that we seek graded deformations on $(R(X), K_{\mathbf{C}})$-modules to $(A, K_{\mathbf{C}})$-modules. Examples show that such deformations need not exist. Since I can say nothing serious about the obstructions to their existence, I will not formulate the obvious definition of these deformations.

The unipotent Dixmier algebra A is said to be *defined over* G if it is endowed with a sesquilinear antiautomorphism having the properties

(6.12a) $$\phi(u^h) = \phi(u)^h \qquad (u \in U(g_{\mathbf{C}})),$$

(6.12b) $$[\mathrm{Ad}(g)a]^h = \mathrm{Ad}(g)(a^h) \qquad (g \in G, \ a \in A).$$

In conjunction with Definition 5.4(c), the first of these properties implies the second for g in the identity component of G. Assuming (6.12a), property (6.12b) is equivalent to

(6.12b)′ $$[\mathrm{Ad}(k)a]^h = \mathrm{Ad}((k^h)^{-1})(a^h) \qquad (k \in K_{\mathbf{C}}, \ a \in A).$$

A Hermitian form $\langle \ , \ \rangle$ on an $(A, K_{\mathbf{C}})$-module V is called *invariant* if

(6.13) $$\langle a \cdot v, w \rangle = \langle v, a^h \cdot w \rangle, \qquad \langle k \cdot v, w \rangle = \langle v, k^h \cdot w \rangle$$

for a in A and k in $K_{\mathbf{C}}$.

DEFINITION 6.14. Suppose A is a unipotent Dixmier algebra defined over G (cf. (6.12)). An *almost unipotent representation* attached to A is an $(A, K_{\mathbf{C}})$-module carrying a positive definite invariant Hermitian form.

Theorem 6.7 attaches unitary representations of G to these objects. The "almost" reflects a weakness in the definition. One should not have to impose the existence of a positive definite invariant Hermitian form; it should be a consequence of some geometric condition. This geometric condition may in turn exclude some representations that have positive definite invariant Hermitian forms, and we would not wish to call the excluded representations unipotent.

REFERENCES

F. Bayen, M. Flato, C. Fronsdal, A. Lichnerowicz, and D. Sternheimer, *Deformation theory and quantization. I. Deformation of symplectic structures*, Ann. Physics 111 (1978), 61–110.

A. Beilinson and J. Bernstein, *Localisation de g-modules*, C. R. Acad. Sci. Paris 292 (1981), 15–18.

M. De Wilde and P. Lecomte, *Existence of star-products and formal deformations of the Poisson Lie algebra of arbitrary symplectic manifolds*, Lett. Math. Phys. 7 (1983), 487–496.

M. Gerstenhaber, *On the deformation of rings and algebras*, Ann. of Math. (2) 79 (1964), 59–103.

V. Guillemin and S. Sternberg, *Geometric asymptotics*, Math. Surveys, vol. 14, Amer. Math. Soc., Providence, R. I., 1978.

J. E. Humphreys, *Introduction to Lie algebras and representation theory*, Springer-Verlag, Berlin/Heidelberg/New York, 1972.

M. Kashiwara and T. Kawai, *Microlocal analysis*, preprint, Res. Inst. Math. Soc., 1983.

A. Kirillov, *Elements of the theory of representations*, translated by E. Hewitt, Springer-Verlag, Berlin/Heidelberg/New York, 1976.

B. Kostant, *Quantization and unitary representations*, Lectures in Modern Analysis and Applications (C. Taam, ed.), Lecture Notes in Math., vol. 170, Springer-Verlag, Berlin/Heidelberg/New York, 1970.

____, *On Whittaker vectors and representation theory*, Invent. Math. **48** (1979), 101–184.

____, *Coadjoint orbits and a new symbol calculus for line bundles*, Conference on Differential Geometric Methods in Theoretical Physics (G. Denardo and H. D. Doebner, eds.), World Scientific, Singapore, 1983.

B. Kostant and S. Rallis, *Orbits and representations associated with symmetric spaces*, Amer. J. Math. **93** (1971), 753–809.

W. McGovern, *Primitive ideals and nilpotent orbits in complex semisimple Lie algebras*, Ph.D. dissertation, M. I. T., 1987.

J. Rawnsley, W. Schmid, and J. Wolf, *Singular unitary representations and indefinite harmonic theory*, J. Funct. Anal. **51** (1983), 1–114.

D. Sternheimer, *Phase-space representations*, Lectures in Applied Math., vol. 21, Amer. Math. Soc., Providence, R. I., 1985.

D. Vogan, *Unitarizability of certain series of representations*, Ann. of Math. (2) **120** (1984), 141–187.

____, *The orbit method and primitive ideals for semisimple Lie algebras*, Lie Algebras and Related Topics, Canad. Math. Soc. Conf. Proc., vol. 5 (D. Britten, F. Lemire, and R. Moody, eds.), Amer. Math. Soc. for CMS, Providence, R. I., 1986.

____, *Unitary representations of reductive Lie groups*, Ann. of Math. Studies, Princeton Univ. Press, Princeton, N. J., 1987.

MASSACHUSETTS INSTITUTE OF TECHNOLOGY

Proceedings of Symposia in Pure Mathematics
Volume 48 (1988)

The Oscillator Semigroup

ROGER HOWE

CONTENTS

0. Introduction
1. Gaussian Functions
2. Integral Operators
3. Gaussian Kernel Operators : The Oscillator Semigroup
4. Relations with Familiar Operators
5. An Example : The Hermite Semigroup and the Plancherel Theorem
6. Description of Infinitesimal Generators
7. Twisted Convolution
8. The Oscillator Semigroup and Twisted Convolution
9. The Symplectic Group
10. The Symplectic Lie Algebra and Cayley Transform
11. Normalization in Ω
12. Homomorphism to the Complex Symplectic Group
13. Comparison of Twisted Convolution and Integral Operator Realizations
14. Positive Definite Conjugacy Classes in $\mathfrak{sp}_{2n}(\mathbf{R})$
15. The Contraction Property
16. Closure : The Metaplectic Group
17. Group Laws are Determined by Three-Quarters Majority
18. Extending by Twisted Translations
19. The Conjugation Property
20. Action on Gaussian Densities/Coherent States/Wave Packets
21. The Grassmannian of Polarizations
22. Quantization of Semipositive Polarizations
23. Characterization of $\operatorname{im}\omega^{-1}$
24. Spectrum and Positivity
25. Polar Decomposition
26. Estimates I : The Calderon-Vaillancourt $(0,0)$-estimate
27. Symbolic Composition
28. Estimates II : Symbols of Beals-Hörmander-Unterberger (ρ,ρ) type
29. Fourier Integral Operators

0. Introduction. This paper describes a semigroup of integral operators on $L^2(\mathbf{R}^n)$; we call it the *oscillator semigroup*. This semigroup has numerous

1980 *Mathematics Subject Classification* (1985 *Revision*). Primary 22A20; Secondary 22E15, 22E30, 22E45, 32M10, 35J10, 35S05, 42B20, 42C99, 43A80, 45A05, 45E99, 46F12, 46H30, 47G05.

Research partially supported by NSF Grant No. DMS8506130.

points of contact with very well-known mathematics. On one hand, the closure of the oscillator semigroup contains the "metaplectic group", the group of unitary operators isomorphic to $\widetilde{\mathrm{Sp}}_{2n}$, the 2-fold cover of the real symplectic group, constructed by Shale [**Sh**] and Weil [**Wi**]. In particular, study of the oscillator semigroup yields a construction of $\widetilde{\mathrm{Sp}}_{2n}$.

On another hand, the closure of the oscillator semigroup contains several operators of classical importance in harmonic analysis including:

(i) linear transformations,

(ii) Fourier transform,

(iii) Gaussian convolution operators (hence heat semigroups), and

(iv) The Hermite semigroup (generated by the Hamiltonian for the quantum harmonic oscillator).

The first two examples are already in $\widetilde{\mathrm{Sp}}_{2n}$. Example (iv) is intimately related to the structure of our semigroup, whence the name oscillator semigroup. Further, we may extend our semigroup by translations together with multiplications by unitary characters of $\mathbf{R}^n$. The structure of the resulting extended semigroup captures the essential aspects of the estimates on pseudodifferential operators proven in the 1970s by Beals and others [**Be, Hr, U2**]. We will show how to use the oscillator semigroup to derive estimates of that type. Our approach has similarities to that of Unterberger [**U1, U2**].

Remarkably, the oscillator semigroup seems to have almost escaped notice in the literature. An exception is the paper [**DB2**] of DeBruijn, which treats the one-dimensional case. Most of the results in this paper were obtained before I learned of [**DB2**]. In fact, the ideas involved in this paper are of interest to several groups of investigators with varying motivations and perspectives. These groups have tended to develop results in parallel rather than under mutual influence. Perhaps this is a reasonable place to note some of the noninteractions. This paper is a descendent of Weil [**Wi**], which is not quoted in [**DB2**]. And [**U1**], which seems tantalizingly close to finding the oscillator semigroup, quotes DeBruijn [**DB1**] but not [**DB2**]. The paper [**CF**], which also comes close to the oscillator semigroup (see §20 for the exact connection), and which also bears on §§26–29, was likewise not known to me until the basic constructions of this paper were made. There may be other literature of which I am still unaware. For connections of many topics here, especially §5, with earlier literature we refer to [**DB2**].

1. Gaussian functions.

1.1. *Siegel upper half-space.* Let $\varsigma^2(\mathbf{R}^n)$ denote the vector space of symmetric matrices with real entries and $\varsigma^2(\mathbf{C}^n)$ the space of symmetric matrices with complex entries. Given $A \in \varsigma^2(\mathbf{C}^n)$ we can break it into its real and imaginary parts:

$$(1.1) \qquad A = \mathrm{Re}\,A + i\,\mathrm{Im}\,A, \qquad \mathrm{Re}\,A,\ \mathrm{Im}\,A \in \varsigma^2(\mathbf{R}^n).$$

Denote by $\varsigma^{2+}(\mathbf{R}^n)$ the (open) cone of positive-definite matrices. Denote by $\mathfrak{S}_n$ the (open) cone of $A \in \varsigma^2(\mathbf{C}^n)$ such that $\operatorname{Re} A \in \varsigma^{2+}(\mathbf{R}^n)$; the cone $\mathfrak{S}_n$ is known as the *Siegel upper half-space* of rank n [Si].

1.2. *Gaussians.* For a matrix $A \in \mathfrak{S}_n$, we associate the function γ_A on $\mathbf{R}^n$, by the formula

(1.2.1) $$\gamma_A(x) = e^{-\pi x^T A x}, \qquad x \in \mathbf{R}^n.$$

Here x^T is the transpose of x: we follow the common convention that x is a column vector, x^T a row vector, so $y^T x$ is the usual inner product of $x, y \in \mathbf{R}^n$. We call γ_A the *Gaussian function with exponent A*. Note that γ_A along with all its derivatives vanishes very rapidly at ∞. For the set of γ_A, we use the notation

(1.2.2) $$\gamma_n^{\mathfrak{S}} = \{t\gamma_A : t \in \mathbf{C}^x, \ A \in \mathfrak{S}_n\}.$$

1.3. *Integration formulas.* The basic formula of integration of Gaussians is the famous Liouville formula

(1.3.1) $$\int_{-\infty}^{\infty} e^{-\pi t^2} \, dt = 1.$$

Taking an n-fold product yields the formula

(1.3.2) $$\int_{\mathbf{R}^n} e^{-\pi x^T x} \, dx = 1.$$

Given $A \in \varsigma^{2+}(\mathbf{R}^n)$, we can take its square root $A^{1/2}$, which is also in $\varsigma^{2+}(\mathbf{R}^n)$. Making the change of variables $y = A^{1/2}x$ gives the formula

(1.3.3) $$\int_{\mathbf{R}^n} \gamma_A(x) \, dx = \int_{\mathbf{R}^n} e^{-\pi x^T A x} \, dx = (\det A)^{-1/2}.$$

By analytic continuation, this formula holds for $A \in \mathfrak{S}_n$. (Note that $A \in \mathfrak{S}_n$ is invertible: for we may write

$$A = \operatorname{Re} A + i \operatorname{Im} A = (\operatorname{Re} A)^{1/2}(1 + iB)(\operatorname{Re} A)^{1/2},$$

where $B \in \varsigma^2(\mathbf{R}^n)$. Clearly A is invertible if and only if $1 + iB$ is invertible; and $1 + iB$ has eigenvalues of the form $1 + ib$, $b \in \mathbf{R}$, and thus is invertible.)

The same procedure of factoring $A = (A^{1/2})^2$, combined with completing the square, and followed by analytic continuation yields the formula

(1.3.4) $$\int_{\mathbf{R}^n} \gamma_A(x) e^{2\pi u^T x} \, dx = (\det A)^{-1/2} e^{\pi u^T A^{-1} u}.$$

Although x is always real, we may let u, as well as A, be complex in formula (1.3.4).

2. Integral operators.

2.1. *The operator attached to a kernel.* We use the customary notation $L^2(\mathbf{R}^n)$ to denote the Hilbert space of complex-valued, square integrable functions on $\mathbf{R}^n$. We let $\mathscr{S}(\mathbf{R}^n)$ denote the Schwartz space of smooth rapidly decreasing complex-valued functions on $\mathbf{R}^n$ [L2, p. 361], with its usual locally convex topology. The Schwartz space is a dense subspace of $L^2(\mathbf{R}^n)$.

Let $K(\cdot, \cdot)$ be a reasonable function on $\mathbf{R}^n \times \mathbf{R}^n$. For example, for the definition we are about to give, it would suffice that K be continuous and of polynomial growth. We define T_K, the *integral operator with kernel K*, by

$$(2.1) \qquad T_K(f)(x) = \int_{\mathbf{R}^n} K(x,y) f(y)\, dy, \qquad f \in \mathscr{S}(\mathbf{R}^n).$$

2.2. *Composition.* If K is continuous and of polynomial growth, it is elementary to check that $T_K(f)$ is likewise. If K belongs to $\mathscr{S}(\mathbf{R}^n \times \mathbf{R}^n)$, it is equally simple to check that $T_K(f)$ is again in $\mathscr{S}(\mathbf{R}^n)$. Thus for a Schwartz space kernel K, the operator T_K is an endomorphism of $\mathscr{S}(\mathbf{R}^n)$. Hence, if we have K_1, K_2 in $\mathscr{S}(\mathbf{R}^n \times \mathbf{R}^n)$, we may form the composite operator $T_{K_2} T_{K_1}$. A straightforward calculation using Fubini's Theorem shows that

$$(2.2a) \qquad T_{K_2} T_{K_1} = T_{K_3},$$

where

$$(2.2b) \qquad K_3(x,y) = \int_{\mathbf{R}^n} K_2(x,z) K_1(z,y)\, dz$$

is again in $\mathscr{S}(\mathbf{R}^n \times \mathbf{R}^n)$.

2.3. *Adjoint.* On $L^2(\mathbf{R}^n)$, we have the usual inner product

$$(2.3.1) \qquad (f_1, f_2) = \int_{\mathbf{R}^n} f_1(x)\overline{f_2(x)}\, dx, \qquad f_1, f_2 \in L^2(\mathbf{R}^n).$$

Given an integral operator T_K as in (2.1), we can define its formal adjoint T_K^* by

$$(2.3.2) \qquad (f_1, T_K^*(f_2)) = (T_K f_1, f_2), \qquad f_i \in \mathscr{S}(\mathbf{R}^n).$$

Obviously, if T_K defines a bounded operator on $L^2(\mathbf{R}^n)$, this coincides with the usual notion of adjoint for Hilbert space operators. A formal computation shows that

$$(2.3.3a) \qquad (T_K)^* = T_{K^*}.$$

where

$$(2.3.3b) \qquad K^*(x,y) = \overline{K(y,x)}.$$

Here $\overline{}$ indicates complex conjugation.

3. Gaussian kernel operators. The oscillator semigroup.

3.1. We want to consider integral operators corresponding to kernels which are Gaussians $\gamma_{\mathscr{A}}$, $\mathscr{A} \in \mathfrak{S}_{2n}$. To be explicit, we partition $\mathscr{A}$ into $n \times n$ blocks

$$(3.1.1) \qquad \mathscr{A} = \begin{bmatrix} A & B \\ B^T & D \end{bmatrix}.$$

Then

$(3.1.2)$

$$\gamma_{\mathscr{A}}(x,y) = \exp\left(-\pi \begin{bmatrix} x \\ y \end{bmatrix}^T \mathscr{A} \begin{bmatrix} x \\ y \end{bmatrix}\right) = \exp(-\pi(x^T A x + 2x^T B y + y^T D y)), \qquad x, y \in \mathbf{R}^n.$$

We will abbreviate (cf. (2.1))

$$T_{\gamma_{\mathscr{A}}} = T_{\mathscr{A}}, \qquad \mathscr{A} \in \mathfrak{S}_{2n}.$$

3.2. Here is the main observation of this paper.

THEOREM. *The set of operators*

$$(3.2.1) \qquad \Omega = \{sT_{\mathscr{A}} : s \in \mathbf{C}^x, \ \mathscr{A} \in \mathfrak{S}_{2n}\}$$

is a semigroup (under composition of operators). Moreover, Ω is self-adjoint, i.e., closed under conjugation. Explicitly we have

$$(3.2.2a) \qquad T_{\mathscr{A}_1} T_{\mathscr{A}_2} = \det(D_1 + A_2)^{-1} T_{\mathscr{A}_3},$$

where
(3.2.2b)

$$\mathscr{A}_3 = \begin{bmatrix} A_3 & B_3 \\ B_3^T & D_3 \end{bmatrix} = \begin{bmatrix} A_1 - B_1(D_1 + A_2)^{-1} B_1^T & -B_1(D_1 + A_2)^{-1} B_2 \\ -B_2^T(D_1 + A_2)^{-1} B_1^T & D_2 - B_2^T(D_1 + A_2)^{-1} B_2 \end{bmatrix}.$$

Also

$$(3.2.3a) \qquad (T_{\mathscr{A}})^* = T_{\check{\mathscr{A}}},$$

where

$$(3.2.3b) \qquad \check{\mathscr{A}} = \begin{bmatrix} \check{A} & \check{B} \\ \check{B}^T & \check{D} \end{bmatrix} = \begin{bmatrix} \overline{D} & \overline{B}^T \\ \overline{B} & \overline{A} \end{bmatrix}.$$

PROOF. The qualitative statements of the theorem are direct consequences of the formulas (3.2.2) and (3.2.3). These in turn are straightforward computations using formulas (1.3.3), (1.3.4), (2.2), (2.3.3) and (3.1.2).

REMARK. The structure of the semigroup law (3.2.2) is probably not completely transparent to the reader. In fact, the author hopes it is not, because a major part of the rest of this paper is devoted to delineating the structure of Ω. However, it is clear that the functions describing $\mathscr{A}_3$ in (3.2.2b) are rational functions of $\mathscr{A}_1$ and $\mathscr{A}_2$. In particular, they are holomorphic. Thus Ω is a holomorphic, self-adjoint semigroup.

We call Ω the *oscillator semigroup*.

4. Relations with familiar operators. The oscillator semigroup has very close relations with some very familiar families of operators. We list some examples.

4.1. *Linear transformations.* Given $g \in \mathrm{GL}_n(\mathbf{R})$, define an operator $\omega(g)$ on $L^2(\mathbf{R}^n)$ by the familiar formula

$$(4.1.1) \qquad \omega(g)(f)(x) = |\det g|^{-1/2} f(g^{-1}(x)), \qquad f \in L^2(\mathbf{R}^n), \ x \in \mathbf{R}^n.$$

It is trivial to check that $\omega(g)$ is unitary. Another straightforward formal computation shows that for $\mathscr{A} \in \mathfrak{S}_{2n}$

$$(4.1.2a) \qquad \omega(g_1)^{-1} T_{\mathscr{A}} \omega(g_2) = |\det(g_1 g_2)|^{1/2} T_{\mathscr{A}'},$$

where

$$
(4.1.2b) \quad
\begin{aligned}
\mathscr{A}' &= \begin{bmatrix} A' & B' \\ B'^T & D' \end{bmatrix} = \begin{bmatrix} g_1^T A g_1 & g_1^T B g_2 \\ g_2^T B^T g_1 & g_2^T D g_2 \end{bmatrix} \\
&= \begin{bmatrix} g_1^T & 0 \\ 0 & g_2^T \end{bmatrix} \begin{bmatrix} A & B \\ B^T & D \end{bmatrix} \begin{bmatrix} g_1 & 0 \\ 0 & g_2 \end{bmatrix}.
\end{aligned}
$$

Thus Ω is invariant under multiplication on the left or the right by $\omega(\mathrm{GL}_n(\mathbf{R}))$.

4.2. *Multiplication operators.* For $\phi \in L^\infty(\mathbf{R}^n)$, we can define the multiplication operator M_ϕ on $L^2(\mathbf{R}^n)$ by

$$(4.2.1) \qquad M_\phi(f)(x) = \phi(x)f(x), \qquad f \in L^2(\mathbf{R}^n),\ x \in \mathbf{R}^n.$$

We have the obvious estimate

$$(4.2.2) \qquad \|M_\phi\|_{\mathrm{op}} = \|\phi\|_\infty,$$

where $\| \ \|_{\mathrm{op}}$ denotes the norm of an operator on $L^2(\mathbf{R}^n)$, and $\| \ \|_\infty$ denotes the standard supremum norm on $L^\infty(\mathbf{R}^n)$. In particular, if $\phi = \gamma_A$ for $A \in \mathfrak{S}_n$, then $\|M_{\gamma_A}\|_{\mathrm{op}} \leq \|\gamma_A\|_\infty = \gamma_A(0) = 1$.

Multiplication on the left or right by M_{γ_A} preserves Ω. Explicitly we have

$$(4.2.3a) \qquad M_{\gamma_{A_1}} T_{\mathscr{A}} M_{\gamma_{A_2}} = T_{\mathscr{A}'},$$

where

$$(4.2.3b) \qquad \mathscr{A}' = \begin{bmatrix} A' & B' \\ B'^T & D' \end{bmatrix} = \begin{bmatrix} A_1 + A & B \\ B^T & D + A_2 \end{bmatrix}.$$

4.3. *Convolution operators.*

4.3.1. If $\psi \in L^1(\mathbf{R}^n)$, we can let ψ act on $L^2(\mathbf{R}^n)$ by convolution. Define
$$(4.3.1.1)$$
$$C_\psi(f)(x) = (\psi * f)(x) = \int_{\mathbf{R}^n} \psi(x - y)f(y)\,dy, \qquad f \in L^2(\mathbf{R}^n),\ x \in \mathbf{R}^n.$$

We have the standard elementary estimate (cf. [**L2**, p. 355])

$$(4.3.1.2) \qquad \|C_\psi\|_{\mathrm{op}} \leq \|\psi\|_1,$$

where $\| \ \|_1$ is the standard norm on $L^1(\mathbf{R}^n)$. In particular, if $\psi = \gamma_A$, $A \in \mathfrak{S}_n$, then according to formula (1.3.3)

$$(4.3.1.3) \qquad \|C_{\gamma_A}\| \leq (\det \operatorname{Re} A)^{-1/2}, \qquad A \in \mathfrak{S}_n.$$

We call the C_{γ_A} *Gaussian convolution operators.*

4.3.2. As is well known, convolution operators are integral operators:

$$(4.3.2.1a) \qquad C_\psi = T_{K_\psi},$$

where

$$(4.3.2.1b) \qquad K_\psi(x, y) = \psi(x - y).$$

This can be read off from (4.3.1.1). In particular, a Gaussian convolution operator C_{γ_A} is an integral operator corresponding to the kernel

$$(4.3.2.2a) \qquad K_{\gamma_A} = \gamma_{\tilde{A}},$$

where

(4.3.2b) $$\gamma_{\tilde{A}}(x,y) = e^{-\pi(x-y)^T A(x-y)}$$

or

(4.2.3c) $$\tilde{A} = \begin{bmatrix} A & -A \\ -A & A \end{bmatrix}.$$

The $2n \times 2n$ symmetric matrix $\text{Re}\,\tilde{A}$ is not positive definite: it yields zero when applied to vectors $\begin{bmatrix} x \\ x \end{bmatrix}$, $x \in \mathbf{R}^n$. However, it is positive semidefinite. Thus if we set

(4.3.2.3) $$\tilde{A}_\varepsilon = \begin{bmatrix} A + \varepsilon I & -A \\ -A & A + \varepsilon I \end{bmatrix}, \qquad \varepsilon \geq 0,$$

where I denotes the $n \times n$ identity matrix, then $\tilde{A}_\varepsilon \in \mathfrak{S}_{2n}$ for $\varepsilon > 0$, and clearly $\tilde{A}_\varepsilon \to \tilde{A}$ as $\varepsilon \to 0$. According to formula (4.2.3) we have a factorization

(4.3.2.4) $$T_{\tilde{A}_\varepsilon} = M_{\gamma_{\varepsilon I}} C_{\gamma_A} M_{\gamma_{\varepsilon I}}.$$

From this formula, it is very easy to see that as $\varepsilon \to 0$, the operator $T_{\tilde{A}_\varepsilon}$ approaches C_{γ_A}, in the sense that for $\varphi \in \mathscr{S}(\mathbf{R}^n)$, the images $T_{\tilde{A}_\varepsilon}(\varphi)$ approach $C_{\gamma_A}(\varphi)$ in $\mathscr{S}(\mathbf{R}^n)$; or likewise for φ in $L^2(\mathbf{R}^n)$. (For a more detailed discussion of this convergence see §5.4 below.)

4.3.3. Multiplication on the left or right by Gaussian convolutions preserves Ω. The formula for the products can be computed directly, but in view of the interpretation in §4.3.2 of the convolution operators as limits of elements of Ω, we may also simply specialize the general formula (3.2.2) to find what we want. The result is

(4.3.3.1a) $$C_{\gamma_{A_1}} T_{\mathscr{A}_2} = \det(A_1 + A_2)^{-1/2} T_{\mathscr{A}_3}.$$

where

(4.3.3.1b) $$\mathscr{A}_3 = \begin{bmatrix} A_1 - A_1(A_1 + A_2)^{-1}A_1 & A_1(A_1 + A_2)^{-1}B_2 \\ B_2^T(A_1 + A_2)^{-1}A_1 & D_2 - B_2^T(A_1 + A_2)^{-1}B_2 \end{bmatrix}$$

and

(4.3.3.1c) $$T_{\mathscr{A}_2} C_{\gamma_{A_1}} = \det(D_2 + A_1)^{-1/2} T_{\mathscr{A}_4},$$

where

(4.3.3.1d) $$\mathscr{A}_4 = \begin{bmatrix} A_2 - B_2(D_2 + A_1)^{-1}B_2^T & B_2(D_2 + A_1)^{-1}A_1 \\ A_1(D_2 + A_1)^{-1}B_2^T & A_1 - A_1(D_2 + A_1)^{-1}A_1 \end{bmatrix}.$$

4.3.4. If we also let $\mathscr{A}_2$ approach $\tilde{A}_2$, as in (4.3.2.3), we will obtain in the limit the classical semigroup law for Gaussian convolution operators. (I tried to find a reference for this formula, but it seems to be so well known that no one writes it down.)

(4.3.4.1a) $$C_{\gamma_{A_1}} C_{\gamma_{A_2}} = \det(A_1 + A_2)^{-1/2} C_{A_3}$$

where

$$(4.3.4.1b) \quad \begin{aligned} A_3 &= A_1 - A_1(A_1 + A_2)^{-1}A_1 = (A_1 + A_2 - A_1)(A_1 + A_2)^{-1}A_1 \\ &= A_2(A_1 + A_2)^{-1}A_1 = (A_1^{-1} + A_2^{-1})^{-1}. \end{aligned}$$

This may be put in a nicer form. Set

$$(4.3.4.2a) \qquad\qquad \Gamma_A = (\det A)^{-1/2} C_{\gamma_{A^{-1}}}.$$

Then the multiplication law (4.3.3.2) just says

$$(4.3.4.2b) \qquad\qquad \Gamma_{A_1}\Gamma_{A_2} = \Gamma_{A_1 + A_2}.$$

Thus the Gaussian convolution operators are exhibited as an abelian semigroup isomorphic to the cone $\mathfrak{S}_n$. Further, estimate (4.3.1.3) says if A is real, then Γ_A is a contraction, that is, its norm, as an operator on $L^2(\mathbf{R}^n)$, is not greater than 1. From formula (3.2.3) we know that for A complex

$$(4.3.4.3) \qquad\qquad (\Gamma_A)^* = \Gamma_{\overline{A}}.$$

Hence $(\Gamma_A)^*(\Gamma_A) = \Gamma_{\overline{A}+A}$ is a contraction; so Γ_A itself is a contraction. Thus the Γ_A form a commutative, holomorphic, self-adjoint semigroup of contraction operators, contained in the closure of Ω.

REMARK. For any fixed $A \in \mathfrak{S}_n$, the operators Γ_{tA}, $t \geq 0$, form a one-parameter semigroup. One can verify by direct computation that for $\varphi \in \mathscr{S}(\mathbf{R}^n)$, the function

$$(4.3.4.4a) \qquad\qquad \Phi(x, t) = (\Gamma_{tA} \cdot \varphi)(x)$$

satisfies the differential equation

$$(4.3.4.4b) \qquad\qquad \frac{\partial \Phi}{\partial t} - \frac{\nabla^T A \nabla}{4\pi}(\Phi),$$

where

$$(4.3.4.4c) \qquad\qquad \nabla = \begin{bmatrix} \dfrac{\partial}{\partial x_1} \\ \vdots \\ \dfrac{\partial}{\partial x_n} \end{bmatrix}$$

is the gradient operator, so that

$$(4.3.4.4d) \qquad\qquad \nabla^T A \nabla = \sum_{ij} a_{ij} \frac{\partial^2}{\partial x_i \partial x_j}$$

if a_{ij} are the entries of A. The equation (4.3.4.4) is a general version of the heat equation: the standard heat equation is for $A = 4\pi I$, where I is the identity matrix. Thus the Γ_{tA} form a general kind of "heat semigroup". Special cases of (4.3.4.4) essentially go back to Fourier.

4.4. *Polynomial coefficient differential operators.* The algebra of polynomial coefficient differential operators, or Weyl algebra, is the algebra generated by the operators M_{x_i}, multiplications by the coordinate functions, and $\partial/\partial x_j$, partial differentiations. The linear span of these operators are of the form

(4.4.1) $$M_{y^T x} + \partial_u, \qquad y, u \in \mathbf{R}^n,$$

where $y^T x$ signifies the linear function $x \to y^T x$, $x \in \mathbf{R}^n$, and ∂_u indicates the directional derivative in the direction u:

(4.4.2) $$\partial_u = \sum_{i=1}^n u_i \frac{\partial}{\partial x_i}$$

if u_i are the coordinates of u.

Left or right multiplication by operators (4.4.1) does not preserve Ω, but Ω and operators (4.4.1) do satisfy a conjugation relation. Straightforward computations show that

(4.4.3) $$\begin{aligned} (\partial_u + 2\pi M_{(Au)^T x})T_{\mathscr{A}} &= T_{\mathscr{A}}(-2\pi M_{(B^T u)^T x}), & \mathscr{A} \in \mathfrak{S}_{2n}, \\ 2\pi M_{(Bu)^T x}T_{\mathscr{A}} &= T_{\mathscr{A}}(\partial_u - 2\pi M_{(Du)^T x}), & u \in \mathbf{R}^n. \end{aligned}$$

Some linear algebra applied to these relations yields

(4.4.4a) $$(2\pi M_{u'^T x} + \partial_{v'})T_{\mathscr{A}} = T_{\mathscr{A}}(2\pi M_{u^T x} + \partial_v),$$

where

(4.4.4b) $$\begin{bmatrix} u' \\ v' \end{bmatrix} = \begin{bmatrix} -A(B^T)^{-1} & B - A(B^T)^{-1}D \\ -(B^T)^{-1} & -(B^T)^{-1}D \end{bmatrix} \begin{bmatrix} u \\ v \end{bmatrix}$$

or

(4.4.4c) $$\begin{bmatrix} u \\ v \end{bmatrix} = \begin{bmatrix} -DB^{-1} & DB^{-1}A - B^T \\ B^{-1} & -B^{-1}A \end{bmatrix} \begin{bmatrix} u' \\ v' \end{bmatrix}.$$

We see from these formulas that, providing the matrices in equations (4.4.4b,c) are defined, i.e., providing B is invertible, the operator $T_{\mathscr{A}}$ normalizes the algebra of polynomial coefficient differential operators. More precisely, $T_{\mathscr{A}}$ normalizes the complexification of the real linear space of operators (4.4.1).

5. An example: The Hermite semigroup and the Plancherel theorem.
To illustrate the formulas of §§3, 4, we consider an example. For simplicity of notation we work in one dimension, but the example is valid in n dimensions with virtually no change.

5.1. Consider the matrices

(5.1.1) $$\mathscr{A}_z = \begin{bmatrix} \coth z & -\operatorname{csch} z \\ -\operatorname{csch} z & \coth z \end{bmatrix},$$

where coth and csch are the hyperbolic cotangent and cosecant respectively. Explicitly

(5.1.2) $$\cosh z = \frac{e^z + e^{-z}}{2}, \qquad \sinh z = \frac{e^z - e^{-z}}{2},$$
$$\coth z = \frac{\cosh z}{\sinh z}, \qquad \operatorname{csch} z = \frac{1}{\sinh z}.$$

The identity

(5.1.3) $$\coth^2 z - \operatorname{csch}^2 z = 1$$

shows that $\mathscr{A}_z$ is in $\mathfrak{S}_2$ for z real and positive. This identity together with

(5.1.4) $$\coth z - \operatorname{csch} z = \frac{e^z - 1}{e^z + 1} = \tanh(z/2)$$

and the fact that $|e^z| > 1$ for $\operatorname{Re} z > 0$ shows that

$$\operatorname{Re}(\coth z) > |\operatorname{Re}(\operatorname{csch} z)|, \qquad \operatorname{Re} z > 0,$$

and hence that $\mathscr{A}_z \in \mathfrak{S}_2$ for all $z \in \mathbf{C}$ with $\operatorname{Re} z > 0$.

5.2. Define operators

(5.2.1) $$\mathscr{F}_z = (\operatorname{csch} z)^{1/2} T_{\mathscr{A}_z}, \qquad \operatorname{Re} z > 0.$$

The formula (3.2.2) combined with identity (5.1.3) and the addition formulas

(5.2.2a) $$\coth(z + w) = \frac{\coth z \coth w + 1}{\coth z + \coth w},$$

(5.2.2b) $$\operatorname{csch}(z + w) = \frac{\operatorname{csch} z \operatorname{csch} w}{\coth z + \coth w}$$

give us the composition law

(5.2.3) $$\mathscr{F}_z \mathscr{F}_w = \mathscr{F}_{z+w}.$$

Thus the $\mathscr{F}_z$ form a holomorphic one-parameter semigroup. Furthermore, formula (3.2.3) tells us that

(5.2.4) $$(\mathscr{F}_z)^* = \mathscr{F}_{\bar{z}}$$

so this one-parameter semigroup is self-adjoint, and $\mathscr{F}_z$ is itself self-adjoint when z is real.

5.3. Using formulas (4.2.3), (4.3.4.2), and (5.1.4), we can verify we have a factorization, valid when $\operatorname{Re}(\operatorname{csch} z) > 0$,

(5.3.1) $$\mathscr{F}_z = M_{\gamma_{\tanh(z/2)}} \Gamma_{\sinh z} M_{\gamma_{\tanh(z/2)}}.$$

It follows from this factorization and estimates (4.2.2), (4.3.1.3) that for z real, $\|J_z\|_{\mathrm{op}} \le 1$, i.e., J_z is a contraction. If z is complex, then $(\mathscr{F}_z)^* \mathscr{F}_z = \mathscr{F}_{\bar{z}} \mathscr{F}_z = \mathscr{F}_{z+\bar{z}}$ is a contraction. Hence J_z itself is a contraction. We summarize what we know so far.

PROPOSITION. *The operators $\mathscr{F}_z$ of (5.2.1) form a holomorphic, one-parameter, self-adjoint semigroup of contraction operators on $L^2(\mathbf{R})$.*

We call this semigroup the *Hermite semigroup*. It is well known classically [**ML, DB1, Bk**]. (The Hermite semigroup is in somewhat the same position as formula (4.3.4.1): Beckner refers to the "Mehler kernel" and the "Hermite semigroup" without reference.)

5.4. We want now to extend this semigroup from the open right half-plane to the closed right half-plane.

5.4.1. As a prelude to the extension, we review the notion of convergence we will use. Let I be a directed set and $\{A_i,\ i \in I\}$ a net (cf. [Kl]) of operators on a Hilbert space $\mathscr{H}$ indexed by I. Then we may say A_i converges to A with respect to the *bounded strong*[*] topology (abbreviated BS[*]) if

(i) The A_i are uniformly bounded: $\|A_i\|_{\mathrm{op}} \leq M$, for some number M, independent of i.

(ii) For every $v \in \mathscr{H}$, $A_i v$ converges to Av in $\mathscr{H}$ (with respect to the Hilbert space norm on $\mathscr{H}$). In symbols, $A_i v \to Av$.

(iii) Also, $A_i^* v$ converges to $A^* v$.

5.4.2. A few elementary observations about this notion of convergence will be helpful.

(a) If condition 5.4.1(i) holds for some net $\{A_i\}$ of operators, then condition 5.4.1(ii) will hold if it holds on a dense set of $v \in \mathscr{H}$. In particular, when $\mathscr{H} = L^2(\mathbf{R}^n)$, it would suffice to check $A_i v \to Av$ for $v \in \mathscr{S}(\mathbf{R}^n)$, or even $C_c^\infty(\mathbf{R}^n)$.

(b) By the uniform boundedness principle ([L2, p. 214] or [KN]), if $I = \mathbf{Z}^+$, the natural numbers, that is, if $\{A_i\}$ is a sequence of operators, then condition 5.4.1(ii) or (5.4.2(iii)) implies 5.4.1(i).

(c) If 5.4.1(ii) holds, then $(A_i v, u) \to (Av, u)$ for any $u \in \mathscr{H}$, where $(\ ,\ )$ indicates the inner product on $\mathscr{H}$. Hence $(v, A_i^* u) \to (v, A^* u)$. Therefore, if $A_i^* u$ converges, it must converge to $A^* u$.

(d) BS[*] convergence preserves the algebra of operators. Thus if $\{A_i,\ i \in I\}$ converges BS[*] to A and $\{B_j,\ j \in J\}$ converges BS[*] to B, then $\{A_i + B_j,\ (i,j) \in I \times J\}$ converges BS[*] to $A + B$, and $\{A_i B_j,\ (i,j) \in I \times J\}$ converges BS[*] to AB. Also $\{A_i^*,\ i \in I\}$ converges BS[*] to A^*, and if s is a scalar $\{sA_i,\ i \in A\}$ converges BS[*] to sA. These assertions are routine to check. The convergence of products is not valid for the weaker notions of strong or weak convergence (e.g., condition 5.4.1(ii) above, or condition (c) above (see [KR])); convergence of adjoints is valid for weak convergence, but not for strong convergence.

5.4.3. Here are a couple of technical lemmas that are useful for checking BS[*] convergence of integral operators. The simple proofs are left as exercises for the reader.

LEMMA 5.4.3.1. *Suppose $\{v_i\}$ is a sequence of functions in $\mathscr{S}(\mathbf{R}^n)$ which form a bounded set in $\mathscr{S}(\mathbf{R}^n)$, and $v_i \to v$ uniformly on compact sets. Then $v \in \mathscr{S}(\mathbf{R}^n)$ and $v_i \to v$ in $\mathscr{S}(\mathbf{R}^N)$. In particular $v_i \to v$ in $L^2(\mathbf{R}^n)$.*

LEMMA 5.4.3.2. *Suppose $K_i(\cdot, \cdot)$ are a sequence of continuous functions on $\mathbf{R}^n \times \mathbf{R}^n$. Suppose for $v \in \mathscr{S}(\mathbf{R}^N)$, the functions $T_{K_i}(v)$ form a bounded set in $\mathscr{S}(\mathbf{R}^n)$, and also $\|T_{K_i}\|_{\mathrm{op}} \leq M$ for some number M. Suppose $K_i \to K$ uniformly on compacta. Then $\|T_K\|_{\mathrm{op}} \leq M$, and T_{K_i} converges BS[*] to T_K.*

5.4.4. For $y \in \mathbf{R}$, we define

$$(5.4.4.1) \qquad \mathscr{F}_{iy} = \text{BS}^* \text{ limit of } \mathscr{F}_{iy+\varepsilon} \text{ as } \varepsilon \to 0,\ \varepsilon > 0,$$

whenever this limit exists. Taking the limit of formula (5.2.1), if $\mathscr{F}_{iy}$ exists, we see we must have

$$(5.4.4.2) \qquad \mathscr{F}_{iy}(f)(t) = \frac{\sqrt{2}}{2}(1 + i\,\mathrm{sgn}(y))|\sin y|^{1/2}$$

$$\times \int_0^\infty \exp(\pi i((t^2 + s^2)\cot y - 2st\csc y))f(s)\,ds$$

whenever this expression makes sense, that is, outside of $y = n\pi$, $n \in \mathbf{Z}$, the poles of $\sin y$. Here sgn is the usual sign function:

$$(5.4.4.3) \qquad\qquad \mathrm{sgn}(y) = \begin{cases} -1 & \text{if } y > 0, \\ 0 & \text{if } y = 0, \\ 1 & \text{if } y > 0. \end{cases}$$

On the other hand, we know from §5.3 that $\|\mathscr{F}_{iy}\|_{\mathrm{op}} \le 1$. This fact and formulas (4.4.4) tell us Lemma 5.4.3.2 is applicable. This lemma tells us that for $y \ne n\pi$, the operator J_{iy} does exist, has norm at most 1, and is given by expression (5.4.4.2).

REMARK. It can be tedious to show a set in $\mathscr{S}(\mathbf{R}^n)$ is bounded; however, since the topology in $\mathscr{S}(\mathbf{R}^n)$ is precisely that needed to make the polynomial coefficient differential operators continuous, boundedness of the $\mathscr{F}_{iy+\varepsilon}(f)$ is virtually immediate from uniform boundedness of $\|\mathscr{F}_{iy+\varepsilon}\|_{\mathrm{op}}$ and the conjugation relation (4.4.4). So in fact, convergence of the $\mathscr{F}_{iy}$ is in the strong* topology for operators in $\mathscr{S}(\mathbf{R}^n)$, much better than simply on $L^2(\mathbf{R}^n)$.

For $y = n\pi$, $n \in \mathbf{Z}$, formula (5.4.4.2) does not make sense and the above argument is invalid. However, because BS* convergence preserves products, we will have the semigroup laws

$$(5.4.4.4\mathrm{a}) \qquad\qquad \mathscr{F}_{x+iy} = \mathscr{F}_x\mathscr{F}_{iy}, \qquad x \in \mathbf{R}^+,\ y \in \mathbf{R},$$

$$(5.4.4.4\mathrm{b}) \qquad\qquad \mathscr{F}_{iu}\mathscr{F}_{iv} = \mathscr{F}_{i(u+v)}, \qquad u, v \in \mathbf{R},$$

whenever $u, v, u + v$ are not multiples of π. Further, if u, v are not multiples of $n\pi$, but $u + v$ is, we have, for $\varepsilon > 0$,

$$\mathscr{F}_{i(u+v)+2\varepsilon} = (\mathscr{F}_{iu+\varepsilon})(\mathscr{F}_{iv+\varepsilon}) = \mathscr{F}_{iu}\mathscr{F}_\varepsilon\mathscr{F}_{iv}\mathscr{F}_\varepsilon$$
$$= \mathscr{F}_{iu}\mathscr{F}_{iv}\mathscr{F}_{2\varepsilon}.$$

Hence in this case, equation (5.4.4.4) may be taken as demonstrating the existence of $\mathscr{F}_{i(u+v)}$. Thus $\mathscr{F}_{iy}$ exists for all $y \in \mathbf{R}$, and (5.4.4.4) is valid. Moreover, the semigroup law (5.2.3) and the law for adjoints clearly extend to the whole closed half-plane. Further, from Lemma 5.4.3.2 and formula (5.4.4.2) we can see that $\mathscr{F}_z$ is continuous in z whenever $z \ne in\pi$. But the same argument used to establish the existence of $\mathscr{F}_{iy}$ for all $y \in \mathbf{R}$ now shows $\mathscr{F}_z$ is continuous in z for $\mathrm{Re}\,z \ge 0$. Thus we have extended the semigroup $\mathscr{F}_z$ continuously from the open half-plane to the closed half-plane.

5.5. For z real, we have the factorization (5.3.1). It is trivial (and a direct consequence of Lemma 5.4.3.1) that as $\varepsilon \to 0$ through positive values, the operators $M_{\gamma_{\tanh(\varepsilon/2)}}$ converge BS* to I, the identity operator. Further, we may observe that as $\varepsilon \to 0$ the operators $\Gamma_{\sinh \varepsilon}$ form a "Dirac sequence" or "approximate identity" in $L^1(\mathbf{R})$ [**HR, L2**]. It is then a standard result (again see [**HR, L**]) that the $\Gamma_{\sinh \varepsilon}$ converge BS* to I. Because BS* limits respect products, we conclude

$$(5.5.1) \qquad \mathscr{F}_0 = \mathrm{BS}^* \lim_{\varepsilon \to 0} \mathscr{F}_\varepsilon = I$$

is the identity operator.

The formula (5.2.4) extends by continuity to the imaginary axis, where it says

$$(5.5.2) \qquad (\mathscr{F}_{iy})^* = \mathscr{F}_{-iy}, \qquad y \in \mathbf{R}.$$

An application of (5.2.3), plus (5.4.5.1), gives us

$$(5.5.3) \qquad (\mathscr{F}_{iy})^* \mathscr{F}_{iy} = \mathscr{F}_{-iy} \mathscr{F}_{iy} = \mathscr{F}_0 = I = \mathscr{F}_{iy}(\mathscr{F}_{iy})^*, \qquad y \in \mathbf{R}.$$

In other words, the operators $\mathscr{F}_{iy}$ on the edge of our extended semigroup are unitary.

5.6. We summarize what we have found out about our extended semigroup.

PROPOSITION. *The semigroup $\mathscr{F}_z$, initially defined by formula (5.2.1) for $\mathrm{Re}\, z > 0$, extends to a self-adjoint BS* continuous semigroup of contractions on the closed right complex half-plane $\{z \in \mathbf{C} : \mathrm{Re}\, z \geq 0\}$. For $z = iy$ pure imaginary, the operators $\mathscr{F}_{iy}$ are unitary, and so constitute a continuous one-parameter semigroup of unitary operators.*

5.7. Consider the value of $\mathscr{F}_{iy}$ at $y = \pi/2$. According to formula (5.4.4.2) we have

$$(5.7.1) \qquad \mathscr{F}_{i\pi/2}(f)(t) = i^{1/2} \int_{-\infty}^{\infty} f(s) e^{-2\pi i s t}\, dt = i^{1/2} \hat{f}(t)$$

where $\hat{}$ denotes the Fourier transform. In other words,

$$(5.7.2) \qquad \hat{} = i^{-1/2} \mathscr{F}_{i\pi/2}.$$

Here

$$i^{1/2} = \frac{\sqrt{2}(1+i)}{2}.$$

From our knowledge (Proposition 5.6) of the $\mathscr{F}_{iy}$ we can deduce the standard classical results about the L^2-theory of the Fourier transform. (We have worked only with one variable; but the n-variable analogs can be proved in exactly the same way, so we state them.)

COROLLARY 5.7.3 (PLANCHEREL THEOREM). *The Fourier transform defines a unitary operator on $L^2(\mathbf{R}^n)$.*

COROLLARY 5.7.4 (FOURIER INVERSION).

$$(\hat{})^{-1}(f) = (\hat{})^*(f) = \hat{}(\omega(-1)f),$$

where -1 is minus the identity in $\mathrm{GL}_n(\mathbf{R})$, and $\omega(-1)$ is given by formula (4.1.1). In other words

$$(5.7.5) \qquad \hat{}^2 = \omega(-1).$$

6. Description of infinitesimal generators. If $t \to U_t$, $t \geq 0$, is a strongly continuous one-parameter semigroup of operators on a Banach space V, then U_t has an *infinitesimal generator* G, defined by

$$(6.1) \qquad\qquad G(v) = \lim_{t \to 0} \frac{U_t(v) - v}{t}$$

for all v in V for which the limit exists. The relation between the semigroup and its infinitesimal generator is a major concern of the theory of semigroups [**HP**]. For that reason, and because they are interesting operators, we describe the infinitesimal generators of one-parameter subgroups of Ω. However, because we make no use of it elsewhere in the paper, we simply state the result without proof. The tools for the proof can be found in §§12, 14, 23 and 24. In §12, the problem is reduced to a Lie-theoretic question, and §23, plus §§14 and 24, provides information for the explicit computation.

Consider the operators

$$(6.2) \qquad ix_j x_k, \quad \frac{1}{2}\left(x_j \frac{\partial}{\partial x_k} + \frac{\partial}{\partial x_k} x_j\right) = x_j \frac{\partial}{\partial x_k} + \frac{\delta_{jk}}{2}, \quad i\frac{\partial^2}{\partial x_j \partial x_k}.$$

These operators, it is easy to check, are formally skew-adjoint, and their real span is a Lie algebra, which we shall denote by $\omega(\mathfrak{sp})$. The complex span of the operators (6.2) is denoted $\omega(\mathfrak{sp})_{\mathbf{C}}$. We have a decomposition

$$\omega(\mathfrak{sp})_{\mathbf{C}} = \omega(\mathfrak{sp}) \oplus i\omega(\mathfrak{sp}).$$

PROPOSITION. *The set of infinitesimal generators of one-parameter subgroups of Ω is a cone in $\omega(\mathfrak{sp})_{\mathbf{C}} \oplus \mathbf{C}$, where $\mathbf{C}$ here denotes the scalar operators. It has the form*

$$(6.3) \qquad\qquad \omega(\mathfrak{sp}) \oplus \mathbf{C} \oplus i\mathscr{K}$$

where $\mathscr{K}$ is a cone in $\omega(\mathfrak{sp})$. The cone $\mathscr{K}$ consists of operators $E \in \omega(\mathfrak{sp})$ such that

$$(6.4) \qquad (iEf, f) = \int_{\mathbf{R}^n} (iEf)(x)\overline{f(x)}\,dx < 0, \qquad f \in \mathscr{S}(\mathbf{R}^n), \ f \neq 0.$$

Concretely,

$$(6.5a) \qquad i\mathscr{K} = \{-x^T A x + i(x^T B \nabla + \nabla^T B^T x) + \nabla^T D \nabla\}$$

such that

$$(6.5b) \qquad\qquad \begin{bmatrix} A & B \\ B^T & D \end{bmatrix} \text{ is positive definite.}$$

Here the notation with ∇ follows §4.3.

Examples. (a) We have already remarked in §4.3 that the generators of Gaussian convolution semigroups have the form $\nabla^T D \nabla$ with the real part of D positive definite; these operators lie on the boundary of $\omega(\mathfrak{sp}) \oplus i\mathscr{K}$.

(b) It may easily be calculated that the generator of a semigroup $M_{\gamma_{tA}}$ of multiplication operators is just the quadratic function $-\pi x^T A x$. Here A should have positive definite real part. These operators likewise lie on the boundary of the cone (6.3).

(c) It is standard that the infinitesimal generators of $\omega(\mathrm{GL}_n(\mathbf{R}))$ are the real span of the operators $x_j(\partial/\partial x_k) + (\partial/\partial x_k)x_j$; these form a subspace of $\omega(\mathfrak{sp})$.

(d) From the factorization (5.3.1) and §4.3 it is easy to compute that the infinitesimal generator of the semigroup $\{\mathscr{F}_z\}$ of equation (5.2.1) is

$$-\pi x^2 + \frac{1}{4\pi}\frac{d^2}{dx^2}.$$

7. Twisted convolution. To study more thoroughly the group law of the oscillator semigroup Ω, it is convenient to work in the twisted convolution algebra associated to the Heisenberg group. This algebra is discussed in detail in [**Hw**] among other places, and we will refer to [**Hw**] for the basic formulas we will use. Note that the essential fact needed to develop the theory of [**Hw**] is the Plancherel Theorem for $\mathbf{R}^n$, which we have established in §5 (see §5.7). Thus the union of [**Hw**] and this paper provide a more or less self-contained account of some aspects of Fourier analysis on $\mathbf{R}^n$.

7.1. We establish notation and recall a few essential facts from [**Hw**]. Let $W = \mathbf{R}^n \oplus (\mathbf{R}^n)^*$. By $\mathbf{R}^n$ we understand the space of column vectors of length n with entries in $\mathbf{R}$, and by $(\mathbf{R}^n)^*$ the space of row vectors of length n with entries in $\mathbf{R}$. We identify $(\mathbf{R}^n)^*$ with $\mathbf{R}^n$ by means of the standard transpose map, and so identify W with $\mathbf{R}^{2n}$, the first n coordinates of $w \in \mathbf{R}^{2n}$ being assigned to $\mathbf{R}^n$ and the second n coordinates being assigned to $(\mathbf{R}^n)^*$. We define a symplectic form (the standard symplectic form) $\langle\ ,\ \rangle$ on W by the recipe

(7.1.1a) $\qquad \langle (x, \lambda), (x', \lambda') \rangle = \lambda(x') - \lambda'(x), \qquad x, x' \in \mathbf{R}^n;\ \lambda, \lambda' \in (\mathbf{R}_n)^*.$

Or if $(x, \lambda) = w = (x_1, x_2, \ldots, x_n, y_1, \ldots, y_n)^T$, where $x = (x_1, \ldots, x_n)^T \in \mathbf{R}^n$, and $\lambda = (y_1, \ldots, y_n) \in (\mathbf{R}_n)^*$, then

(7.1.1b) $$\langle w, w' \rangle = \sum_{i=1}^{n} y_i x_i' - x_i y_i' = w'^T J w,$$

where

(7.1.1c) $$J = \begin{bmatrix} 0 & I \\ -I & 0 \end{bmatrix},$$

I being the $n \times n$ identity matrix, is the matrix defining $\langle\ ,\ \rangle$.

For $f_1, f_2 \in \mathscr{S}(W)$, define the *twisted convolution* of f_1 and f_2 by the recipe

(7.1.2)
$$f_1 \natural f_2(w') = \int_W f_1(w) f_2(w' - w) e^{\pi i \langle w, w' \rangle}\, dw$$
$$= \int_W f_1(w' - w) f_2(w) e^{-\pi i \langle w, w' \rangle}\, dw.$$

Twisted convolution defines a structure of associative algebra on $\mathscr{S}(W)$. There is a natural involution $f \to f^*$ defined by

(7.1.3) $$f^*(w) = \overline{f(-w)}$$

where ‾‾, as usual, denotes complex conjugate. So under twisted convolution, $\mathscr{S}(W)$ is an algebra with involution.

There is an isomorphism of algebras-with-involution

(7.1.4) $$\rho \colon \mathscr{S}(W) \xrightarrow{\sim} \mathscr{S}(\mathbf{R}^n \times \mathbf{R}^n)$$

which takes twisted convolution to composition of integral operators, as given by formula (2.2). The mapping ρ is called the *Weyl transform* [**Sg**]. It is essentially a partial Fourier transform; it is given explicitly in [**Hw**], formula (1.3.17). The Weyl transform ρ extends to a unitary isomorphism of $L^2(W)$ to $L^2(\mathbf{R}^n \times \mathbf{R}^n)$; the kernel operators associated with these are the Hilbert-Schmidt operators [**L2**].

We will work in the twisted convolution algebra for much of the remainder of this paper.

8. The oscillator semigroup and twisted convolution. Since the Weyl transform ρ ((7.1.4)) is essentially a partial Fourier transform, it will take Gaussians on W to Gaussians on $\mathbf{R}^n \times \mathbf{R}^n$. Thus the oscillator semigroup Ω (cf. (3.2.1)) can also be realized as the set of (multiples of) Gaussians on W, with semigroup law given by twisted convolution. We will compute this explicitly.

Since we have identified W with $\mathbf{R}^{2n}$, we can use the notation of §§1, 3. Thus a typical $2n \times 2n$ complex symmetric matrix with positive definite real part will be denoted $\mathscr{A}$; the set of all such $\mathscr{A}$ is $\mathfrak{S}_{2n}$; and the Gaussian function, given by the analog of formula (1.2) is $\gamma_{\mathscr{A}}$.

Combining formulas (7.1.2), (7.1.1b, c), and (1.3.4) yields the semigroup law

(8.1a) $$\gamma_{\mathscr{A}_1} \natural \gamma_{\mathscr{A}_2} = \det(\mathscr{A}_1 + \mathscr{A}_2)^{-1/2} \gamma_{\mathscr{A}_3},$$

where

(8.1b) $$\mathscr{A}_3 = \mathscr{A}_2 - (\mathscr{A}_2 + iJ/2)(\mathscr{A}_1 + \mathscr{A}_2)^{-1}(\mathscr{A}_2 - iJ/2).$$

Some elementary manipulations show this formula can be rewritten

(8.2a) $$\mathscr{A}_3 + iJ/2 = (\mathscr{A}_2 + iJ/2)(\mathscr{A}_1 + \mathscr{A}_2)^{-1}(\mathscr{A}_1 + iJ/2)$$

or

(8.2b) $$\mathscr{A}_3 - iJ/2 = (\mathscr{A}_1 - iJ/2)(\mathscr{A}_1 + \mathscr{A}_2)^{-1}(\mathscr{A}_2 - iJ/2).$$

Observe also, from formula (7.1.3), taking adjoints simply corresponds to complex conjugation in $\mathfrak{S}_{2n}$:

(8.3) $$(\gamma_{\mathscr{A}})^* = \gamma_{\overline{\mathscr{A}}}.$$

9. The symplectic group. Let $\mathrm{Sp}(W) = \mathrm{Sp}_{2n}(\mathbf{R})$ ($= \mathrm{Sp}$ when more specific notation is unnecessary) denote the group of isometries of the symplectic form $\langle\,,\,\rangle$ on W (cf. (7.1.1)). Specifically,

(9.1a) $$\mathrm{Sp}(W) = \{g \in \mathrm{GL}(W); \langle g(w), g(w')\rangle = \langle w, w'\rangle, \ w, w' \in W\}.$$

Or, in terms of coordinates,

(9.1b) $$\mathrm{Sp}_{2n}(\mathbf{R}) = \{g \in \mathrm{GL}_{2n}(\mathbf{R}); \ g^T J g = J\}.$$

We can let $\mathrm{Sp}(W)$ act on $\mathscr{S}(W)$ or $L^2(W)$ in analogy with formula (4.1.1). However, for reasons which will appear in §12, we wish to use a slightly different notation for this action. Thus we set

(9.2) $\omega_{1,1}(g)(f)(w) = f(g^{-1}(w)), \qquad g \in \mathrm{Sp}(W), \; f \in \mathscr{S}(W), \; w \in W.$

An easy calculation verifies that $g \to \omega_{1,1}(g)$ defines an action of $\mathrm{Sp}(W)$ via *-automorphisms of the twisted convolution algebra structure. Furthermore, one sees that $\omega_{1,1}(\mathrm{Sp})$ normalizes the oscillator semigroup Ω, viewed as set in $\mathscr{S}(W)$. Thus in speaking of operator-theoretic properties of elements of Ω, we may conjugate elements of Ω by Sp whenever convenient. The existence of this large group $\omega_{1,1}(\mathrm{Sp})$ of symmetries of Ω facilitates the study of structural properties of Ω inside the twisted convolution algebra.

10. The symplectic Lie algebra and Cayley transform.

10.1. Denote the Lie algebra of $\mathrm{Sp}(W)$ by $\mathfrak{sp}(W)$ or $\mathfrak{sp}_{2n}(\mathbf{R})$ or $\mathfrak{sp}$, in parallel with our notations for Sp. Concretely, $\mathfrak{sp}$ is the space of "infinitesimal isometries" of the form $\langle \, , \, \rangle$:

(10.1.1a) $\mathfrak{sp}(W) = \{S \in \mathrm{End}(W); \langle Sw, w' \rangle + \langle w, Sw' \rangle = 0\}$

or

(10.1.1b) $\mathfrak{sp}_{2n}(\mathbf{R}) = \{S \in M_{2n}(\mathbf{R}); \; JS + S^T J = 0\}.$

The conditions (10.1.1) say S is skew-adjoint with respect to the form $\langle \, , \, \rangle$. But since the matrix J is itself skew-adjoint, this condition may also be expressed

(10.1.2) $JS = -(S^T J) = (JS)^T.$

That is, JS is a symmetric matrix.

10.2. There is a well-known rational map, the *Cayley transform* [**Wy**1]:

(10.2.1a) $c: \mathfrak{sp} \leftrightarrow \mathrm{Sp}$

given by the formula

(10.2.1b) $c(x) = \dfrac{x+1}{x-1}.$

Here 1 is understood to be the identity of whatever algebra x belongs to. The Cayley transform is defined for all matrices x for which 1 is not an eigenvalue. The set of such x is invariant under c, and on this set c is an involution:

(10.2.2) $c(c(x)) = x.$

Via the Cayley transform we can transport the group structure on Sp to $\mathfrak{sp}$, outside the set where c blows up. We compute this transferred product. Given

$x, y \in \mathfrak{sp}$, we have

(10.2.3)

$$c(c(x)c(y)) = (c(x)c(y) + 1)(c(x)c(y) - 1)^{-1}$$
$$= ((x-1)^{-1}(x+1)(y+1)(y-1)^{-1} + 1)$$
$$\times ((x-1)^{-1}(x+1)(y+1)(y-1)^{-1} - 1)^{-1}$$
$$= (x-1)^{-1}((x+1)(y+1) + (x-1)(y-1))(y-1)^{-1}$$
$$\times ((x-1)^{-1}((x+1)(y+1) - (x-1)(y-1))(y-1)^{-1})^{-1}$$
$$= (x-1)^{-1}(2(xy+1))(y-1)^{-1}(y-1)(2(x+y))^{-1}(x-1)$$
$$= (x-1)^{-1}(xy+1)(x+y)^{-1}(x-1)$$
$$= (x-1)^{-1}((xy+1)(x+y)^{-1} - 1)(x-1) + 1$$
$$= (x-1)^{-1}(xy+1 - x - y)(x+y)^{-1}(x-1) + 1$$
$$= (y-1)(x+y)^{-1}(x-1) + 1.$$

11. Normalization in Ω. For $\mathscr{A} \in \mathfrak{S}_{2n}$, set

(11.1)
$$\gamma_{\mathscr{A}}^0 = (\det(\mathscr{A} + iJ/2))^{1/2} \gamma_{\mathscr{A}}.$$

Implicit in the square root of $\det(\mathscr{A} + iJ/2)$ there is an ambiguity of sign, and we will abuse notation somewhat to consider $\gamma_{\mathscr{A}}^0$ to be either the two-element set defined unambiguously by equation (11.1), or any convenient member of it. Observe that $\det(\mathscr{A} + iJ/2)$ does have zeros in $\mathfrak{S}_{2n}$ (take $\mathscr{A} = I/2$), so it is not possible to select a single branch of $\det(\mathscr{A} + iJ/2)^{1/2}$ to use in equation (11.1).

From formulas (8.1) and (8.2) we can verify that introduction of the normalization factor $\det(\mathscr{A} + iJ/2)^{1/2}$ eliminates the scalar factor involved in (8.1), up to the inherent ambiguity of sign:

(11.2)
$$\gamma_{\mathscr{A}_1}^0 \, \natural \, \gamma_{\mathscr{A}_2}^0 = \gamma_{\mathscr{A}_3}^0.$$

The price to be paid for this normalization is that some of the $\gamma_{\mathscr{A}}^0$ are simply zero. Let us define

(11.3)
$$\Omega^0 = \{\gamma_{\mathscr{A}}^0 : \mathscr{A} \in \mathfrak{S}_{2n}\} - \{0\}$$

where $\{0\}$ is the (set containing the) zero operator. Let $\mathscr{V} \subseteq \mathfrak{S}_{2n}$ be the variety of zeros of $\det(\mathscr{A} + iJ/2)$. Then formula (11.2) shows Ω^0 is a subsemigroup of Ω, and the map

(11.4)
$$\Omega^0 \overset{\gamma^{-1}}{\to} \mathfrak{S}_{2n} - \mathscr{V}$$
$$\gamma_{\mathscr{A}}^0 \overset{\gamma^{-1}}{\to} \mathscr{A}$$

is a 2-to-1 covering map. We will call Ω^0 the *normalized oscillator semigroup*.

12. Homomorphism to the complex symplectic group. Let $\text{Sp}_{2n}(\mathbf{C})$

be the complex symplectic group, and $\mathfrak{sp}_{2n}(\mathbf{C})$ its Lie algebra. These are defined just as $\text{Sp}_{2n}(\mathbf{R})$ and $\mathfrak{sp}_{2n}(\mathbf{R})$ are defined (in §9 and §10 respectively), except the coefficients of the matrices can be complex numbers. Of course $\text{Sp}_{2n}(\mathbf{R})$ is the subgroup of $\mathrm{Sp}_{2n}(\mathbf{C})$ consisting of matrices with real entries, and $\mathfrak{sp}_{2n}(\mathbf{R})$ is likewise a (real) Lie subalgebra of $\mathfrak{sp}_{2n}(\mathbf{C})$. In fact we have the decomposition

$$(12.1) \qquad \mathfrak{sp}_{2n}(\mathbf{C}) = \mathfrak{sp}_{2n}(\mathbf{R}) \oplus i\mathfrak{sp}_{2n}(\mathbf{R}).$$

We have a similar decomposition

$$(12.2) \qquad \begin{array}{ccc} \varsigma^2(\mathbf{C}^{2n}) & \simeq & \varsigma^2(\mathbf{R}^{2n}) \oplus i\varsigma^2(\mathbf{R}^{2n}) \\ \cup| & & \cup| \\ \mathfrak{S}_{2n} & \simeq & \varsigma^{2+}(\mathbf{R}^{2n}) \oplus i\varsigma^2(\mathbf{R}^{2n}) \end{array}$$

of spaces of symmetric matrices. (See §1 for notation.) According to formula (10.1.2) the map β, defined by

$$(12.3) \qquad \mathscr{A} \to \beta(\mathscr{A}) = -2iJ\mathscr{A}, \qquad \mathscr{A} \in \varsigma^2(\mathbf{C}^n),$$

defines a linear isomorphism between $\varsigma^2(\mathbf{C}^{2n})$ and $\mathfrak{sp}_{2n}(\mathbf{C})$, and takes $i\varsigma^2(\mathbf{R})$ to $\mathfrak{sp}_{2n}(\mathbf{R})$. Thus $\beta(\mathfrak{S}_{2n})$ will be an open cone in $\mathfrak{sp}_{2n}(\mathbf{C})$ invariant under translations by $\mathfrak{sp}_{2n}(\mathbf{R})$.

Observe that the open subset $\mathfrak{S}_{2n} - \mathscr{V}$ (cf. (11.4)) of $\mathfrak{S}_{2n}$ is precisely the inverse image under β of the domain of definition of the Cayley transform (formula (10.2.1)). Thus we may apply c to elements of $\beta(\mathfrak{S}_{2n} - \mathscr{V})$ to obtain elements of $\mathrm{Sp}_{2n}(\mathbf{C})$. Comparison of formulas (8.2) and (12.3) with formula (10.2.3) yields the following conclusion.

THEOREM. *The mapping*

$$(12.4a) \qquad \omega^{-1} \colon \Omega^0 \to \mathrm{Sp}_{2n}(\mathbf{C})$$

defined by

$$(12.4b) \qquad \omega^{-1} = c \circ \beta \circ \gamma^{-1}, \omega^{-1}(\gamma^0_{\mathscr{A}}) = c(-2iJ\mathscr{A}) = \frac{2iJ\mathscr{A} - 1}{2iJ\mathscr{A} + 1}$$

defines a 2-to-1 homomorphism from Ω^0 to an open subsemigroup of $\mathrm{Sp}_{2n}(\mathbf{C})$.

REMARK. We will sometimes abuse notation and write

$$\gamma^0_{\mathscr{A}} = \omega(c(-2iJ\mathscr{A})).$$

13. Comparison of twisted convolution and integral operator realizations. We have now learned something about Ω from each of the two realizations we have given it, the original realization as integral operators, and the twisted convolution realization. Comparison of the two realizations will allow us to pass information between them, and so further sharpen our understanding of Ω.

According to [Hw], formulas (1.3.17), and (1.6.3), the Weyl transform ρ of (7.1.4) is given as follows.

$$(13.1a) \qquad \rho(f) = T_{K_\rho}(f), \qquad f \in \mathscr{S}(W),$$

$$(13.1b) \qquad K_\rho(f)(x, u) = \int_{\mathbf{R}^n} f(x - u, y) e^{\pi i(x+u)^T y} \, dy.$$

Combining this with formula (1.3.4), we can compute that if

$$\mathscr{A} = \begin{bmatrix} A & B \\ B^T & D \end{bmatrix}$$

is in $\mathfrak{S}_{2n}$ then

(13.2a) $K_\rho(\gamma_\mathscr{A}) = (\det D)^{-1/2}\gamma_{\tilde{\mathscr{A}}},$

where

(13.2b)
$$\tilde{\mathscr{A}} = \begin{bmatrix} A - (B-i/2)D^{-1}(B^T - i/2) & -A + (B-i/2)D^{-1}(B^T + i/2) \\ -A + (B+i/2)D^{-1}(B^T - i/2) & A - (B+i/2)D^{-1}(B^T + i/2) \end{bmatrix}.$$

Let us specialize this formula to the case $n = 1$ and

$$\mathscr{A} = \begin{bmatrix} s & 0 \\ 0 & s \end{bmatrix}, \qquad s > 0.$$

We find

(13.3a) $\rho(\gamma_{\begin{bmatrix} s & 0 \\ 0 & s \end{bmatrix}}) = (s^2 - \tfrac{1}{4})^{1/2}s^{-1/2}T_{\begin{bmatrix} \alpha & -\beta \\ -\beta & \alpha \end{bmatrix}},$

where

(13.3b) $\alpha = s + 1/4s, \qquad \beta = s - 1/4s.$

If we set

(13.4a) $s = \dfrac{1}{2}\coth\left(\dfrac{z}{2}\right) = \left(\dfrac{1}{2}\right)\dfrac{e^z + 1}{e^z - 1},$

then quick computations show

(13.4b) $s + 1/4s = \coth z, \quad s - 1/4s = \operatorname{csch} z.$

Note also that if $z > 0$, then $s > \tfrac{1}{2}$, so $s^2 - \tfrac{1}{4} > 0$. Comparing (13.3) and (13.4) with (5.2.1), we see that

(13.5) $\rho(\gamma_\Psi^0) = \mathscr{F}_z, \qquad z > 0,$

where

$$\Psi = \begin{bmatrix} \dfrac{\coth(z/2)}{2} & 0 \\ 0 & \dfrac{\coth(z/2)}{2} \end{bmatrix}.$$

Here we have taken the positive root of $\coth^2 z - 1$ in defining γ^0. Thus the properties of the $\mathscr{F}_z$ developed in §5 apply likewise to the

$$\gamma^0_{\begin{bmatrix} s & 0 \\ 0 & s \end{bmatrix}}$$

with $s > \tfrac{1}{2}$. In particular, we know these operators are contractions.

It is convenient also to record here the specialization of the semigroup law (8.1) to the

$$\gamma^0_{\begin{bmatrix} s & 0 \\ 0 & s \end{bmatrix}}.$$

It is

$$(13.6) \qquad \gamma^0 \begin{bmatrix} s & 0 \\ 0 & s \end{bmatrix} \natural \gamma^0 \begin{bmatrix} t & 0 \\ 0 & t \end{bmatrix} = \gamma^0 \begin{bmatrix} u & 0 \\ 0 & u \end{bmatrix} \qquad \text{where } u = \frac{4st + 1}{4(s + t)}.$$

14. Positive definite conjugacy classes in $\mathfrak{sp}_{2n}(\mathbf{R})$. We now need an analog for $\mathfrak{sp}_{2n}(\mathbf{R})$ of the principal axis theorem [**L1**, p. 525]. This fact is well known, but we lack a convenient reference, so provide a proof to maintain continuity. (A closely related result is in [**Hr2**, Theorem 21.5.3].) In §10 (see (10.1.2)) we have observed that the map

$$S \to JS, \qquad S \in \mathfrak{sp}_{2n}(\mathbf{R})$$

induces an isomorphism between $\mathfrak{sp}_{2n}(\mathbf{R})$ and $\varsigma^{2n}(\mathbf{R})$. More intrinsically, the bilinear form

$$(14.1) \qquad B_S(w, w') = \langle Sw, w' \rangle = w'^T J S w$$

is a symmetric bilinear form on W.

A basis $\{e_i, f_i\}_{1 \le i \le n}$ is called a *symplectic basis* for W if

$$(14.2) \qquad \langle e_i, e_j \rangle = 0 = \langle f_i, f_j \rangle, \qquad \langle f_i, e_j \rangle = \delta_{ij},$$

where δ_{ij} is Kronecker's delta. Referring to formula (7.1.1b) we see that, in our identification of W with $\mathbf{R}^{2n}$, the standard basis of $\mathbf{R}^{2n}$ is a symplectic basis. Thus a symplectic basis may also be described as a basis for $W \simeq \mathbf{R}^{2n}$ which may be transformed to the standard basis by an element of $\mathrm{Sp}(W)$.

PROPOSITION. *Suppose $S \in \mathfrak{sp}_{2n}(\mathbf{R})$ is such that the form B_S of (14.1) is positive definite on W. Then there is a symplectic basis $\{e_i, f_j\}$ of W, and numbers $a_i > 0$ such that*

$$(14.3) \qquad S(e_i) = a_i f_i, \qquad S(f_i) = -a_i e_i.$$

Equivalently, S is conjugate by some $g \in \mathrm{Sp}(W)$ to

$$(14.4) \qquad g S g^{-1} = \begin{bmatrix} 0 & -D \\ D & 0 \end{bmatrix},$$

where D is a diagonal $n \times n$ matrix with diagonal entries $a_i > 0$.

PROOF. The equivalence of the two formulations follows from the remarks just above concerning standard bases of W. We will prove the version stated in terms of symplectic bases.

Suppose $W_1 \subseteq W$ is a subspace invariant by S. If $x \in W_1$ is nonzero, then $Sx \in W_1$, and $\langle Sx, x \rangle = B_S(x, x) > 0$. Hence the restriction of $\langle \ , \ \rangle$ to W_1 is nondegenerate, and we have a direct sum decomposition

$$W = W_1 \oplus W_1^\perp$$

where $\perp$ here indicates orthogonality with respect to $\langle \ , \ \rangle$. Choose $x \in W_1$, $y \in W_1^\perp$. Then

$$\langle Sy, x \rangle = -\langle y, Sx \rangle = 0,$$

since $S \in \mathfrak{sp}(W)$ and W_1 is invariant by S. Hence $W_1^\perp$ is also invariant by S. We have shown that any S-invariant subspace $W_1 \subseteq W$ is part of an S-invariant orthogonal direct sum decomposition of W. Iterating this argument, we can find a decomposition $W = \sum_i U_i$ such that:

(i) Each U_i is invariant by S.

(ii) The U_i are mutually orthogonal with respect to $\langle \, , \, \rangle$.

(iii) The U_i are minimal for property (i), i.e., are irreducible for S.

It follows from condition (ii) that $\langle \, , \, \rangle$ must be nondegenerate when restricted to any U_j. From (iii) and standard eigenvalue theory, we know that either U_i has dimension 1, and consists of eigenvectors for S, or U_j has dimension 2 and on complexification splits into two eigenlines with mutually conjugate complex eigenvalues. The first case cannot arise, because the restriction of the antisymmetric form $\langle \, , \, \rangle$ to a line must be identically zero. Thus $\dim U_j = 2$ and the restriction of S to U_j has eigenvalues $b_j \pm ia_j$, $a_j, b_j \in \mathbf{R}$. We can find a basis e_j, f_j for U_j such that the matrix of $S_{|U_j}$ with respect to e_j, f_j is

$$\begin{bmatrix} b_j & -a_j \\ a_j & b_j \end{bmatrix}.$$

(Take for e_j, f_j the real and imaginary parts of the complex eigenvector.) Scaling e_j, f_j, and multiplying f_j by -1 if necessary, we can arrange that $\langle f_j, e_j \rangle = 1$; and of course $\langle e_j, e_j \rangle = \langle f_j, f_j \rangle = 0$. It is now very easy to verify that for B_S to be symmetric on U_j, we must have $b_j = 0$, and that B_S positive definite implies $a_j > 0$. This concludes the proof of the proposition.

15. The contraction property.

THEOREM. *The normalized operators $\gamma_{\mathscr{A}}^0$, $\mathscr{A} \in \mathfrak{S}_{2n}$, defined in formula (11.1) are contractions, i.e.*

$$(15.1) \qquad\qquad\qquad \|\gamma_{\mathscr{A}}^0\|_{\mathrm{op}} \le 1.$$

Thus the normalized oscillator semigroup Ω^0 (cf. (11.3)) is a self-adjoint, holomorphic semigroup of contractions.

PROOF. We have

$$\|\gamma_{\mathscr{A}}^0\|_{\mathrm{op}}^2 = \|(\gamma_{\mathscr{A}}^0)^* \natural \gamma_{\mathscr{A}}^0\|_{\mathrm{op}} = \|\gamma_{\overline{\mathscr{A}}}^0 \natural \gamma_{\mathscr{A}}^0\|_{\mathrm{op}} = \|\gamma_{\tilde{\mathscr{A}}}^0\|_{\mathrm{op}},$$

where $\tilde{\mathscr{A}}$ can be deduced from formulas (8.1), (8.2), but in any case $\tilde{\mathscr{A}}$ will be real since $\gamma_{\tilde{\mathscr{A}}}^0$ is self-adjoint. Hence it suffices to prove the theorem for $\mathscr{A}$ real. Also, we may conjugate $\mathscr{A}$ by an arbitrary element of $\mathrm{Sp}(W)$, via the action $\omega_{1,1}$ of $\mathrm{Sp}(W)$ as *-automorphisms of the twisted convolution algebra (cf. formula (9.2)). By the proposition of §14, we can arrange that

$$\mathscr{A} = \begin{bmatrix} D & 0 \\ 0 & D \end{bmatrix},$$

where D is a diagonal matrix with positive entries a_i. But for this $\mathscr{A}$, the element $\gamma^0_{\mathscr{A}}$ ia just a (tensor) product of operators

$$\gamma^0 \begin{bmatrix} a_i & 0 \\ 0 & a_i \end{bmatrix}$$

acting on the coordinates x_i, y_i, and its norm will clearly be the product of the norms of the operators. But we know these are contractions from §13 (see formula (13.5)). This establishes the theorem.

16. Closure: The metaplectic group. If Γ is any self-adjoint semigroup of contraction operators on a Hilbert space $\mathscr{H}$, then $\overline{\Gamma}$, its closure in the BS* topology (see §5.4) will again be a self-adjoint semigroup of contraction operators on $\mathscr{H}$. Even if Γ consisted of strict contractions (i.e., operators A such that $\|Ax\| < \|x\|$ for all $x \in \mathscr{H}$), it may happen that $\overline{\Gamma}$ contains unitary operators. Because of the self-adjointness of $\overline{\Gamma}$, the set of unitary operators in $\overline{\Gamma}$ will be a group, not merely a semigroup. We want to consider the group of unitary operators in $\overline{(\Omega^0)}$, the closure of the normalized oscillator semigroup. It turns out this is exactly the metaplectic group, the twofold cover of $\mathrm{Sp}_{2n}(\mathbf{R})$ constructed by Shale [**Sh**] and Weil [**Wi**]. In this section we will show $\overline{\Omega^0}$ contains the metaplectic group.

16.1. Our arguments for dealing with $\overline{(\Omega^0)}$ are very similar to those used for the example of the Hermite semigroup studied in §5. Here we note an analog of Lemma 5.4.3.2. Again the simple proof is left as an exercise.

We recall (see the remarks at the end of §7) that if φ is a function on the symplectic vector space W, then $\|\varphi\|_{\mathrm{op}}$ denotes the norm of φ acting by twisted convolution on the left on $L^2(W)$.

LEMMA. *Let φ_i be a sequence of continuous functions on W. Suppose $\|\varphi_i\|_{\mathrm{op}} \leq M$ for some number M independent of i and that for $f \in \mathscr{S}(W)$, the set $\{\varphi_i \natural f\}$ is bounded in $\mathscr{S}(W)$. Suppose $\varphi_i \to \varphi$ uniformly on compacta. Then $\|\varphi\|_{\mathrm{op}} \leq M$, and φ_i converges BS* to φ.*

16.2. Consider $\mathscr{A} \in \varsigma^2(\mathbf{R}^{2n})$. Let J be the matrix (see (7.1.1c)) of the symplectic form on $W \simeq \mathbf{R}^{2n}$. Suppose that $\det(\mathscr{A} + J/2) \neq 0$, that is, $\mathscr{A} + J/2$ is nonsingular. Then the function $\gamma^0_{i\mathscr{A}}$ is well defined and nonzero. Approaching $\gamma^0_{i\mathscr{A}}$ via functions $\gamma^0_{\tilde{\mathscr{A}}+i\mathscr{A}}$, where $\tilde{\mathscr{A}} \in \varsigma^{2+}(\mathbf{R}^{2n})$ and $\tilde{\mathscr{A}} \to 0$, and applying the lemma of §16.1, we see that $\gamma^0_{i\mathscr{A}}$ is the BS* limit of the $\gamma^0_{\tilde{\mathscr{A}}+i\mathscr{A}}$ as $\tilde{\mathscr{A}} \to 0$; in particular $\gamma^0_{\tilde{\mathscr{A}}}$ is a contraction. (We need the formulas (19.2.5) to verify that the conditions of Lemma 16.1 are satisfied. Although we do not state these formulas until §19, they can be easily verified from the basic definitions. With the use of (19.2.5), checking the hypotheses of Lemma 16.1 is routine.) Also, the formula (11.2) (see also (8.2)) will remain valid by continuity, when either one or both of $\mathscr{A}_1$ and $\mathscr{A}_2$ is pure imaginary, so long as $\det(\mathscr{A}_i + iJ/2) \neq 0$ for $i = 1, 2, 3$. Further we will have

(16.2.1) $\qquad (\gamma^0_{i\mathscr{A}})^* = \gamma^0_{-i\mathscr{A}}, \qquad \mathscr{A} \in \varsigma^2(\mathbf{R}^{2n}); \ \det(\mathscr{A} + J/2) \neq 0.$

(Note that by taking transposes, $\det(\mathscr{A} - J/2) = \det(\mathscr{A} + J/2)$.)

For given $\mathscr{A} \in \varsigma^2(\mathbf{R}^{2n})$ with $\mathscr{A} \pm J/2$ nonsingular, consider the product

(16.2.2a) $\qquad \gamma^0_{\varepsilon I + i\mathscr{A}} \natural \gamma^0_{\varepsilon I - i\mathscr{A}} = \gamma^0_{\mathscr{B}(\varepsilon)}, \qquad \varepsilon > 0.$

Here I is the $2n \times 2n$ identity matrix. We can compute $\mathscr{B}$ via formula (8.2). It is

(16.2.2b) $\qquad \mathscr{B}(\varepsilon) = \frac{1}{2}(\varepsilon I + \varepsilon^{-1}(\mathscr{A} - J/2)(\mathscr{A} + J/2)).$

Evidently, as $\varepsilon \to 0$, $\mathscr{B}(\varepsilon) \to \infty$ in $\varsigma^{2+}(\mathbf{R}^{2n})$. I claim that as $\varepsilon \to 0$, $\gamma^0_{\mathscr{B}(\varepsilon)}$ approaches the identity. This may be seen easily in at least two ways: (i) by appealing to the proposition of §14, and formula (13.5), the claim is reduced to the result that $\mathscr{F}_z$ approaches the identity operator, which fact was established in §5.5; or (ii) one can argue directly in the twisted convolution algebra, by observing that as $\varepsilon \to 0$, integration against $\gamma^0_{\mathscr{B}(\varepsilon)}$ approaches the Dirac delta at the origin in W (here note that the ratio

$$\frac{\det(\mathscr{B}(\varepsilon) + iJ/2)}{\det(\mathscr{B}(\varepsilon))}$$

approaches 1 as $\varepsilon \to 0$), hence the $\gamma^0_{\mathscr{B}(\varepsilon)}$ constitute a Dirac sequence or approximate identity in the twisted convolution algebra. We conclude that

(16.2.3)
$$(\gamma^0_{i\mathscr{A}})\natural(\gamma^0_{i\mathscr{A}})^* = \gamma^0_{i\mathscr{A}}\natural\gamma^0_{-i\mathscr{A}} = \delta_0, \qquad \mathscr{A} \in \varsigma^2(\mathbf{R}^{2n}), \ \det(\mathscr{A} + J/2) \neq 0,$$

where δ_0 is the Dirac delta at the origin in W. Hence $\gamma^0_{i\mathscr{A}}$ is unitary.

16.3. The unitary operators $\gamma^0_{i\mathscr{A}}$ just constructed will generate a group which will belong to $\overline{(\Omega^0)}$. We want to identify this group. The mapping ω^{-1} from Ω^0 to $\mathrm{Sp}_{2n}(\mathbf{C})$ defined in formula (12.4) clearly extends continuously to all $\gamma^0_{\mathscr{A}}$, $\mathscr{A} \in \varsigma^2(\mathbf{C}^{2n})$, such that $\mathrm{Re}\,\mathscr{A}$ is positive semidefinite and $\det(\mathscr{A} + iJ/2) \neq 0$. In particular, it is defined for the unitary operators $\gamma^0_{i\mathscr{A}}$, $\mathscr{A} \in \varsigma^2(\mathbf{R}^{2n})$, $\det(\mathscr{A} + J/2) \neq 0$. Further, we see from formula (12.4) that for real $\mathscr{A}$, the $\omega^{-1}(\gamma^0_{i\mathscr{A}})$ belong to $\mathrm{Sp}_{2n}(\mathbf{R})$, and they fill up a (Zariski) open and dense set in $\mathrm{Sp}_{2n}(\mathbf{R})$. Furthermore, as long as both the factors and the product belong to the image of ω^{-1}, we know, again by continuity, that ω^{-1} takes products to products. If now we apply the result of §17, we can make the following conclusion.

THEOREM. *Denote by* $\widetilde{\mathrm{Sp}}(W)$ *the group of unitary operators generated by the operators* $\gamma^0_{i\mathscr{A}}$, $\mathscr{A} \in \varsigma^2(\mathbf{R}^{2n})$, $\det(\mathscr{A} + J/2) \neq 0$. *The map*

(16.3.1a) $\qquad \omega^{-1}: \gamma^0_{i\mathscr{A}} \to \dfrac{-(1 + 2J\mathscr{A})}{1 - 2J\mathscr{A}} = (\mathscr{A} + J/2)^{-1}(\mathscr{A} - J/2)$

obtained by continuity from (12.4) extends uniquely to an isomorphism of groups

(16.3.1b) $\qquad \omega^{-1}: \widetilde{\mathrm{Sp}}(W)/\{\pm 1\} \simeq \mathrm{Sp}_{2n}(\mathbf{R}) = \mathrm{Sp}(W).$

NOTE. In (16.3.1a), the 1 indicates the identity operator on $W \simeq \mathbf{R}^{2n}$. In (16.3.1b) the 1 indicates the identity of the twisted convolution algebra, or, essentially equivalently, the identity operator on $L^2(\mathbf{R}^n)$.

16.4. EXAMPLE. In the above considerations, we may take $\mathscr{A} = 0$, the zero symmetric matrix. We learn that twisted convolution with $\gamma_0^0 = 2^{-n}$ is unitary, and corresponds via ω^{-1} to the element $-1 \in \mathrm{Sp}(W)$. Explicitly,

$$(16.4.1) \qquad f \natural \gamma_0^0(w') = 2^{-n} \int_W f(w) e^{\pi i \langle w', w \rangle} \, dw.$$

Thus, up to multiples, convolution with γ_0^0 amounts to "symplectic Fourier transform". In [**Hw**], it was used to define the "isotropic symbol" which is also known as the "Weyl symbol" [**Hr**, **Wy2**].

17. Group laws are determined by three-quarters majority.

PROPOSITION. *Let G be a group, and $U \subseteq G$ a subset such that*

$$(17.1a) \qquad U = U^{-1}.$$

If g_i, $1 \leq i \leq 4$, are any 4 elements of G, then

$$(17.1b) \qquad \bigcap_{i=1}^{4} g_i U \neq \varnothing.$$

Let $\tilde{U}$ be a copy of U: $\tilde{U} = \{\tilde{u} : u \in U\}$. Let $\tilde{G}$ be the group generated by $\tilde{U}$ subject to the relations

$$(17.2(\mathrm{i})) \qquad (\tilde{u})^{-1} = (u^{-1})^{\sim}, \qquad u \in U,$$

$$(17.2(\mathrm{ii})) \qquad \tilde{u}_1 \tilde{u}_2 = \tilde{u}_3 \quad \text{if} \quad u_1 u_2 = u_3, \qquad u_i \in U.$$

Then $\tilde{G}$ is isomorphic to G. More precisely, the identification mapping

$$j : \tilde{U} \to U, \qquad \tilde{u} \to u,$$

induces an isomorphism of groups $j : \tilde{G} \overset{\sim}{\to} G$.

REMARKS. (a) In a finite group G condition (17.1b) is guaranteed by

$$\#(U) > \tfrac{3}{4}\#(G).$$

(b) Examples of candidate sets U in locally compact groups are: open dense sets; more generally, the complement of a (Baire) category I set; or the complement of a set of Haar measure zero. Note that if U_1 is a subset of one of these types, then $U = U_1 \cap U_1^{-1}$ satisfies both (17.1a) and (17.1b).

(c) In any infinite group, the complement of $F \cup F^{-1}$, for a finite set F, will satisfy (17.1a) and (17.1b).

PROOF. It is clear that the identification map $j : \tilde{U} \to U$ will induce a homomorphism of groups. If $g \in G$, then $U \cap gU$ is nonempty by condition (17.1b). Hence we can find u_1, u_2 in U so that $u_1 = gu_2$, or $g = u_1 u_2^{-1}$. Since $u_2^{-1} \in U$, we see U generates G. Hence $j : \tilde{G} \to G$ will be surjective. Therefore, to prove the proposition, it suffices to show the homomorphism j is one-to-one.

Since $\tilde{U} = (\tilde{U})^{-1}$ by relation (17.2(i)), we will have

$$\tilde{G} = \tilde{U} \cup \tilde{U}^2 \cup \tilde{U}^3 \cup \cdots.$$

We will show that, in fact, $\tilde{G} = \tilde{U} \cup \tilde{U}^2$. To do this it suffices to show that $\tilde{U}^3 \subseteq \tilde{U} \cup \tilde{U}^2$. Consider a product $\tilde{u}_1 \tilde{u}_2 \tilde{u}_3$, $u_i \in U$. If $u_2 u_3 \in U$, then $\tilde{u}_2 \tilde{u}_3 = (u_2 u_3)^\sim$, so $\tilde{u}_1 \tilde{u}_2 \tilde{u}_3 = \tilde{u}_1 (u_2 u_3)^\sim \in \tilde{U}^2$. In general, property (17.1b) tells us we can find an element u_4 such that

$$u_4 \in U \cap u_2^{-1} U \cap u_3 U \cap (u_1 u_2)^{-1} U.$$

Then we see

$$\tilde{u}_1 \tilde{u}_2 \tilde{u}_3 = \tilde{u}_1 \tilde{u}_2 \tilde{u}_4 \tilde{u}_4^{-1} \tilde{u}_3 = \tilde{u}_1 \tilde{u}_2 \tilde{u}_4 (u_4^{-1})^\sim \tilde{u}_3 = \tilde{u}_1 (u_2 u_4)^\sim (u_4^{-1} u_3)^\sim$$

$$= (u_1 (u_2 u_4))^\sim (u_4^{-1} u_3)^\sim \in \tilde{U}^2.$$

To finish the proof, it suffices to show that if $u_1 u_2 = u_3 u_4$, $u_i \in U$, then $\tilde{u}_1 \tilde{u}_2 = \tilde{u}_3 \tilde{u}_4$. For then the homomorphism j is one-to-one on $\tilde{U}^2$, hence one-to-one on $\tilde{U} \cup \tilde{U}^2$ by relation (17.2(ii)), hence one-to-one by the argument of the previous paragraph.

If $u_1 u_2 \in U$, then relation (17.2(ii)) implies $\tilde{u}_1 \tilde{u}_2 = (u_1 u_2)^\sim = (u_3 u_4)^\sim = \tilde{u}_3 \tilde{u}_4$. In general, we can find $u \in U$ such that

$$u \in U \cap u_2^{-1} U \cap (u_1 u_2)^{-1} U \cap (u_4)^{-1} U.$$

Note then that also $u \in (u_3 u_4)^{-1} U$. Thus

$$\tilde{u}_1 \tilde{u}_2 = \tilde{u}_1 \tilde{u}_2 \tilde{u} \tilde{u}^{-1} = \tilde{u}_1 (u_2 u)^\sim \tilde{u}^{-1} = (u_1 (u_2 u))^\sim \tilde{u}^{-1}$$

$$= ((u_1 u_2) u)^\sim \tilde{u}^{-1} = ((u_3 u_4) u)^\sim \tilde{u}^{-1} = (u_3 (u_4 u))^\sim \tilde{u}^{-1}$$

$$= \tilde{u}_3 (u_4 u)^\sim \tilde{u}^{-1} = \tilde{u}_3 \tilde{u}_4 \tilde{u} \tilde{u}^{-1} = \tilde{u}_3 \tilde{u}_4.$$

18. Extending by Twisted Translations. As explained in [**Hw**], twisted convolution on our symplectic vector space W is derived from the standard convolution on the Heisenberg group associated to W. Convolution in a group algebra is often thought of as an integrated form of translation in the group. Similarly, twisted convolution may be regarded as an integrated form of a "twisted translation", which is derived from ordinary translation on the Heisenberg group. The twisted translations will generate a group isomorphic to the Heisenberg group. Just as usual translations in a group G may be regarded as convolutions with the unit masses concentrated at points of G, so also the twisted translations may be regarded as twisted convolutions by the point masses concentrated at points of W, and we will so treat them here.

18.1. The formulas for twisted translation (see [**Hw**], formulas (1.7.28)) are

(18.1.1a) $\qquad \delta_w \natural f(w') = e^{\pi i \langle w, w' \rangle} f(w' - w) = e^{\pi i w'^T J w} f(w' - w),$

$$w, w' \in W, \ f \in \mathscr{S}(W),$$

(18.1.1b) $\qquad (f \natural \delta_w)(w') = e^{\pi i \langle w', w \rangle} f(w' - w) = e^{-\pi i w'^T J w} f(w' - w).$

Comparison of formulas (18.1.1) and (7.1.2) reveals that, indeed, twisted convolution is obtained by averaging twisted translations:

(18.1.2) $\qquad f_1 \natural f_2(w') = \int_W f_1(w) (\delta_w \natural f_2)(w') \, dw, \qquad f_1, f_2 \in \mathscr{S}(W).$

In connection with these formulas it is useful to note that

(18.1.3a) $$(\delta_w \natural f \natural \delta_w)(w') = f(w - 2w'),$$

(18.1.3b) $$(\delta_w \natural f \natural \delta_{-w})(w') = e^{2\pi i \langle w, w' \rangle} f(w).$$

Thus by appropriately combining twisted translations we can construct ordinary translation in W, and also multiplication by unitary characters of W.

18.2. If we take twisted translations of elements of the oscillator semigroup Ω, we obtain a semigroup combining both types of objects.

PROPOSITION. *The set of elements*

(18.2.1) $$c\delta_{w_1} \natural \gamma_{\mathscr{A}} \natural \delta_{w_2}, \qquad c \in \mathbf{C}, \ w_1, w_2 \in W, \ \mathscr{A} \in \mathfrak{S}_{2n},$$

of $\mathscr{S}(W)$ form a semigroup under twisted convolution.

REMARKS. (a) It is necessary to take twisted translates of the $\gamma_{\mathscr{A}}$ on both sides; twisted translations of the $\gamma_{\mathscr{A}}$ only to the left or only to the right will not yield a semigroup.

(b) Similarly, it is necessary here to have the scalar multiple c; there is no way to normalize so as to eliminate it, as in §11 for the oscillator semigroup by itself. Indeed the behavior of the scalar factor c in the semigroup law controls fundamental properties of pseudodifferential and Fourier integral operators (see §§26–29).

(c) We will refer to the semigroup formed by elements (18.2.1) as the *affine oscillator semigroup*, and denote it $A\Omega$.

PROOF. We will prove the proposition by explicit calculations, which will yield some formulas for the semigroup law in $A\Omega$.

First, observe that, since the elements $c\delta_w$, $|c| = 1$, simply form the Heisenberg group (explicitly

(18.2.2) $$\delta_{w_1} \natural \delta_{w_2} = e^{\pi i \langle w_1, w_2 \rangle} \delta_{w_1 + w_2})$$

and Ω is a semigroup, to prove the proposition it suffices to show that a function

(18.2.3) $$\gamma_{\mathscr{A}_1} \natural \delta_w \natural \gamma_{\mathscr{A}_2}, \qquad \mathscr{A}_1, \mathscr{A}_2 \in \mathfrak{S}_{2n}, \ w \in W,$$

can be expressed in the form (18.2.1). Thus we want to establish an equation

(18.2.4) $$\gamma_{\mathscr{A}_1} \natural \delta_w \natural \gamma_{\mathscr{A}_2} = \kappa \delta_u \natural \gamma_{\mathscr{A}_3} \natural \delta_v,$$

where $\kappa \in \mathbf{C}$, $u, v \in W$, and $\mathscr{A}_3 \in \mathfrak{S}_{2n}$ are functions of $\mathscr{A}_1, \mathscr{A}_2$ and w. (The precise values of u, v are given in formula (18.3.4), and of κ in (18.3.6) to (18.3.8), culminating computations which begin now.)

Specializing formulas (18.1.1) to $f = \gamma_{\mathscr{A}}$, $\mathscr{A} \in \mathfrak{S}_{2n}$, we find

(18.2.5a) $$(\delta_w \natural \gamma_{\mathscr{A}})(w') = e^{\pi i w'^T J w} \gamma_{\mathscr{A}}(w' - w)$$
$$= \gamma_{\mathscr{A}}(w) e^{\pi w'^T (2\mathscr{A} + iJ)w} \gamma_{\mathscr{A}}(w'),$$

(18.2.5b) $$(\gamma_{\mathscr{A}} \natural \delta_w)(w') = e^{-\pi i w'^T J w} \gamma_{\mathscr{A}}(w' - w)$$
$$= \gamma_{\mathscr{A}}(w) e^{\pi i w'^T (2\mathscr{A} - iJ)w} \gamma_{\mathscr{A}}(w').$$

Combining the two formulas gives

$$(18.2.6) \quad \begin{aligned} &(\delta_u \natural \gamma_{\mathscr{A}} \natural \delta_v)(w') \\ &= \gamma_{\mathscr{A}}(u+v)e^{\pi i u^T J v}e^{\pi w'^T(2\mathscr{A}(u+v)+iJ(u-v))}\gamma_{\mathscr{A}}(w'). \end{aligned}$$

Thus a two-sided twisted translate of $\gamma_{\mathscr{A}}$ is a multiple of $\gamma_{\mathscr{A}}$ times an exponential function (a nonunitary character of W). Conversely, suppose that we have such a function, and let us try to express it as a multiple of a twisted translate of $\gamma_{\mathscr{A}}$: thus we seek an equation

$$(18.2.7a) \qquad e^{2\pi w'^T \lambda}\gamma_{\mathscr{A}} = c\delta_u \natural \gamma_{\mathscr{A}} \natural \delta_v.$$

Here $u, v \in W$, c is a complex number, and

$$\lambda = \operatorname{Re}\lambda + i\operatorname{Im}\lambda, \qquad \operatorname{Re}\lambda, \operatorname{Im}\lambda \in W,$$

belongs to $W_{\mathbf{C}} = W \oplus iW$, the complexification of W. Comparing (18.2.6) with (18.2.7a) we see the latter equation is equivalent to the equations

$$(18.2.7b) \qquad \begin{aligned} c &= (\gamma_{\mathscr{A}}(u+v)e^{\pi i u^T J v})^{-1} = e^{-\pi i u^T J v}e^{\pi(u+v)^T \mathscr{A}(u+v)}, \\ \lambda &= \mathscr{A}(u+v) + \tfrac{1}{2}iJ(u-v). \end{aligned}$$

These equations may indeed be solved, uniquely, for u, v, and c. Indeed, c is already expressed in terms of u and v. To find u, v in terms of λ, break both sides of the equation for λ into real and imaginary parts. This gives

$$(18.2.8a) \qquad \operatorname{Re}\lambda = \operatorname{Re}\mathscr{A}(u+v), \qquad 2\operatorname{Im}\lambda = 2\operatorname{Im}\mathscr{A}(u+v) + J(u-v),$$

which may be solved to yield
$$(18.2.8b)$$
$$u+v = (\operatorname{Re}\mathscr{A})^{-1}(\operatorname{Re}\lambda), \qquad u-v = -2J(\operatorname{Im}\lambda - \operatorname{Im}\mathscr{A}((\operatorname{Re}\mathscr{A})^{-1}(\operatorname{Re}\lambda))),$$

or

$$(18.2.8c) \qquad \begin{aligned} u &= (\tfrac{1}{2} + J\operatorname{Im}\mathscr{A})(\operatorname{Re}\mathscr{A})^{-1}(\operatorname{Re}\lambda) - J(\operatorname{Im}\lambda), \\ v &= (\tfrac{1}{2} - J\operatorname{Im}\mathscr{A})(\operatorname{Re}\mathscr{A})^{-1}(\operatorname{Re}\lambda) + J(\operatorname{Im}\lambda). \end{aligned}$$

Further, observing that $2u^T J v = -(u+v)^T J(u-v)$, and looking at the equation (18.2.7b) which expresses λ in terms of u and v, we see we can write

$$(18.2.8d) \qquad c = e^{\pi(\operatorname{Re}\lambda)^T(\operatorname{Re}\mathscr{A})^{-1}\lambda}.$$

Now let us compute the product (18.2.3). This calculation again uses the basic formula (1.3.4), along with (7.1.2), (18.2.2), and (18.2.5). We find

$$
(\gamma_{\mathscr{A}_1} \natural \delta_w \natural \gamma_{\mathscr{A}_2})(w') = (\gamma_{\mathscr{A}_1} \natural (\delta_w \natural \gamma_{\mathscr{A}_2}))(w')
$$

$$
= \int_W \gamma_{\mathscr{A}_1}(y)(\delta_y \natural (\delta_w \natural \gamma_{\mathscr{A}_2}))(w') \, dy
$$

$$
= \int_W \gamma_{\mathscr{A}_1}(y) e^{\pi i w^T J y}(\delta_{y+w} \natural \gamma_{\mathscr{A}_2})(w') \, dy
$$

$$
= \int_W \gamma_{\mathscr{A}_1}(y) \gamma_{\mathscr{A}_2}(y+w) \gamma_{\mathscr{A}_2}(w') e^{\pi(i w^T J y + w'^T(2\mathscr{A}_2 + iJ)(y+w))} \, dy
$$

$$
= \gamma_{\mathscr{A}_2}(w) \gamma_{\mathscr{A}_2}(w') e^{\pi w'^T(2\mathscr{A}_2 + iJ)w}
$$

$$
\times \int_W \gamma_{(\mathscr{A}_1 + \mathscr{A}_2)}(y) e^{\pi w'^T(2\mathscr{A}_2 + iJ)y} e^{-\pi w^T(2\mathscr{A}_2 - iJ)y} \, dy
$$

$$
= \det(\mathscr{A}_1 + \mathscr{A}_2)^{-1/2} \gamma_{\mathscr{A}_2}(w) \gamma_{\mathscr{A}_2}(w') e^{\pi w'^T(2\mathscr{A}_2 + iJ)w} e^{\pi Q},
$$

where

$$
Q = ((\mathscr{A}_2 - iJ/2)w' - (\mathscr{A}_2 + iJ/2)w)^T (\mathscr{A}_1 + \mathscr{A}_2)^{-1}((\mathscr{A}_2 - iJ/2)w' - (\mathscr{A}_2 + iJ/2)w).
$$

Continuing the calculation, separating the terms involving only w, only w', and both, we get

(18.2.9a) $$(\gamma_{\mathscr{A}_1} \natural \delta_w \natural \gamma_{\mathscr{A}_2})(w') = \det(\mathscr{A}_1 + \mathscr{A}_2)^{-1/2} \gamma_{\tilde{\mathscr{A}}_3}(w) \gamma_{\mathscr{A}_3}(w') e^{\pi R}$$

where $\mathscr{A}_3$ is as in the semigroup law (8.1), and

(18.2.9b)
$$
\begin{aligned}
\tilde{\mathscr{A}}_3 &= \mathscr{A}_2 - (\mathscr{A}_2 - iJ/2)(\mathscr{A}_1 + \mathscr{A}_2)^{-1}(\mathscr{A}_2 + iJ/2) \\
&= (\mathscr{A}_2 - iJ/2)(\mathscr{A}_1 + \mathscr{A}_2)^{-1}(\mathscr{A}_1 - iJ/2) + iJ/2 \\
&= (\mathscr{A}_1' + iJ/2)(\mathscr{A}_1 + \mathscr{A}_2)^{-1}(\mathscr{A}_2 + iJ/2) - iJ/2.
\end{aligned}
$$

<u>(A comparison with (8.2) shows that if $\mathscr{A}_1, \mathscr{A}_2$ are real then $\tilde{\mathscr{A}}_3 = \overline{\mathscr{A}}_3$, where</u> denotes complex conjugation.) Finally, we have

(18.2.9c)
$$
\begin{aligned}
R &= 2w'^T(\mathscr{A}_2 + iJ/2)w - 2w'^T(\mathscr{A}_2 + iJ/2)(\mathscr{A}_1 + \mathscr{A}_2)^{-1}(\mathscr{A}_2 + iJ/2)w \\
&= 2w'^T(\mathscr{A}_2 + iJ/2)(\mathscr{A}_1 + \mathscr{A}_2)^{-1}(\mathscr{A}_1 - iJ/2)w.
\end{aligned}
$$

From formula (18.2.9) we can recognize that, as a function of w', the product (18.2.3) has the form of a multiple of the left-hand side of equation (18.2.7a). Hence, it has the form of the right-hand side of (18.2.7a); so equation (18.2.4) holds, and the proposition is proved.

18.3. We would like to combine formulas (18.2.8) and (18.2.9) to get a fairly explicit version of equation (18.2.4). We will do this under the restriction

(18.3.1) $\mathscr{A}_1$ and $\mathscr{A}_2$ are *real*.

REMARK. In view of formula (19.1.1), and the decomposition theorem of §25, the case of $\mathscr{A}_1, \mathscr{A}_2$ real is the essential case to consider.

When the $\mathscr{A}_i$ are real, we can compute

(18.3.2a)
$$\operatorname{Re}\mathscr{A}_3 = \operatorname{Re}((\mathscr{A}_2 + iJ/2)(\mathscr{A}_1 + \mathscr{A}_2)^{-1}(\mathscr{A}_1 + iJ/2) - iJ/2)$$

$$= \mathscr{A}_2(\mathscr{A}_1 + \mathscr{A}_2)^{-1}\mathscr{A}_1 - \frac{J(\mathscr{A}_1 + \mathscr{A}_2)^{-1}J}{4}$$

$$= \mathscr{A}_2(\mathscr{A}_1 + \mathscr{A}_2)^{-1}\mathscr{A}_1 \left(J(\mathscr{A}_1 + \mathscr{A}_2)J + \frac{\mathscr{A}_1^{-1}(\mathscr{A}_1 + \mathscr{A}_2)\mathscr{A}_2^{-1}}{4}\right)J(\mathscr{A}_1 + \mathscr{A}_2)^{-1}J$$

$$= -\mathscr{A}_2(\mathscr{A}_1 + \mathscr{A}_2)^{-1}\mathscr{A}_1(\mathscr{T}(\mathscr{A}_1) + \mathscr{T}(\mathscr{A}_2))J(\mathscr{A}_1 + \mathscr{A}_2)^{-1}J,$$

where we have set

(18.3.3)
$$\mathscr{T}(\mathscr{A}) = (4\mathscr{A})^{-1} - J\mathscr{A}J;$$

(18.3.2b)
$$\operatorname{Im}\mathscr{A}_3 = \operatorname{Im}((\mathscr{A}_2 + iJ/2)(\mathscr{A}_1 + \mathscr{A}_2)^{-1}(\mathscr{A}_1 + iJ/2) - iJ/2)$$
$$= \tfrac{1}{2}(J(\mathscr{A}_1 + \mathscr{A}_2)^{-1}\mathscr{A}_1 + \mathscr{A}_2(\mathscr{A}_1 + \mathscr{A}_2)^{-1}J - J)$$
$$= \tfrac{1}{2}J(\mathscr{A}_1 + \mathscr{A}_2)^{-1}(\mathscr{A}_2 J\mathscr{A}_1 - \mathscr{A}_1 J\mathscr{A}_2)(\mathscr{A}_1 + \mathscr{A}_2)^{-1}J;$$

(18.3.2c)
$$\operatorname{Re}(\mathscr{A}_2 + iJ/2)(\mathscr{A}_1 + \mathscr{A}_2)^{-1}(\mathscr{A}_1 - iJ/2) = \mathscr{A}_2(\mathscr{A}_1 + \mathscr{A}_2)^{-1}\mathscr{A}_1 + \frac{J(\mathscr{A}_1 + \mathscr{A}_2)^{-1}J}{4}$$

$$= \mathscr{A}_2(\mathscr{A}_1 + \mathscr{A}_2)^{-1}\mathscr{A}_1 \left(J(\mathscr{A}_1 + \mathscr{A}_2)J + \frac{\mathscr{A}_1^{-1}(\mathscr{A}_1 + \mathscr{A}_2)\mathscr{A}_2^{-1}}{4}\right)J(\mathscr{A}_1 + \mathscr{A}_2)^{-1}J$$

$$= \mathscr{A}_2(\mathscr{A}_1 + \mathscr{A}_2)^{-1}\mathscr{A}_1(\mathscr{T}(\mathscr{A}_1) + \mathscr{T}(\mathscr{A}_2) + 2J(\mathscr{A}_1 + \mathscr{A}_2)J)J(\mathscr{A}_1 + \mathscr{A}_2)^{-1}J;$$

(18.3.2d)
$$\operatorname{Im}((\mathscr{A}_2 + iJ/2)(\mathscr{A}_1 + \mathscr{A}_2)^{-1}(\mathscr{A}_1 - iJ/2))$$
$$= \tfrac{1}{2}(J(\mathscr{A}_1 + \mathscr{A}_2)^{-1}\mathscr{A}_1 - \mathscr{A}_2(\mathscr{A}_1 + \mathscr{A}_2)^{-1}J)$$
$$= \tfrac{1}{2}J(\mathscr{A}_1 + \mathscr{A}_2)^{-1}(\mathscr{A}_2 J\mathscr{A}_2 - \mathscr{A}_1 J\mathscr{A}_1)(\mathscr{A}_1 + \mathscr{A}_2)^{-1}J.$$

Plugging these expressions in equations (18.2.8c) yields, after some calculation, the formulas

(18.3.4)
$$u = -\left(\frac{\mathscr{A}_1^{-1}}{4} + J\mathscr{A}_1 J\right)(\mathscr{T}(\mathscr{A}_1) + \mathscr{T}(\mathscr{A}_2))^{-1}(w)$$

$$= -\left(\frac{\mathscr{A}_1^{-1}}{4} + J\mathscr{A}_1 J\right)\mathscr{C}(\mathscr{A}_1, \mathscr{A}_2)(w),$$

$$v = -\left(\frac{\mathscr{A}_2^{-1}}{4} + J\mathscr{A}_2 J\right)(\mathscr{T}(\mathscr{A}_1) + \mathscr{T}(\mathscr{A}_2))^{-1}(w)$$

$$= -\left(\frac{\mathscr{A}_2^{-1}}{4} + J\mathscr{A}_2 J\right)\mathscr{C}(\mathscr{A}_1, \mathscr{A}_2)(w)$$

for u, v in (18.2.4). Here we have set

(18.3.5)
$$\mathscr{C}(\mathscr{A}_1, \mathscr{A}_2) = (\mathscr{T}(\mathscr{A}_1) + \mathscr{T}(\mathscr{A}_2))^{-1}.$$

We continue these calculations. Combining (18.2.9a) with (18.2.8d), we obtain for the constant κ in equation (18.2.4) the expression

(18.3.6a)
$$\kappa = \gamma_{\tilde{\mathscr{A}}_3}(w)e^{\pi(\operatorname{Re}\lambda)^T(\operatorname{Re}\mathscr{A}_3)^{-1}\lambda}\det(\mathscr{A}_1 + \mathscr{A}_2)^{-1/2},$$

where $\tilde{\mathscr{A}_3}$ is from (18.2.9b), $\mathscr{A}_3$ is from (8.1) and

(18.3.6b) $\qquad \lambda = (\mathscr{A}_2 + iJ/2)(\mathscr{A}_1 + \mathscr{A}_2)^{-1}(\mathscr{A}_2 - iJ/2)w.$

(This last expression is from (18.2.9c).) We compute separately the absolute value and phase of κ. From (18.3.6) we have

(18.3.7a) $\qquad |\kappa| = \gamma_{\mathrm{Re}\,\tilde{\mathscr{A}_3}}(w)\gamma_{-(\mathrm{Re}\,\mathscr{A}_3)^{-1}}(\lambda)|\det(\mathscr{A}_1 + \mathscr{A}_2)|^{-1/2},$

(18.3.7b)

$$\frac{\kappa}{|\kappa|} = \gamma_{i\,\mathrm{Im}\,\tilde{\mathscr{A}_3}}(w)\exp(\pi i (\mathrm{Re}\,\lambda)^T(\mathrm{Re}\,\mathscr{A}_3)^{-1}(\mathrm{Im}\,\lambda))\left(\frac{\det(\mathscr{A}_1 + \mathscr{A}_2)}{|\det(\mathscr{A}_1 + \mathscr{A}_2)|}\right)^{-1/2}$$

with λ as just above ((18.3.6b)). Plugging in the appropriate quantities in formulas (18.3.7) and reducing the resulting expressions gives formulas

(18.3.8a) $\qquad |\kappa| = e^{-\pi w \mathscr{C} w}$

where $\mathscr{C} = \mathscr{C}(\mathscr{A}_1, \mathscr{A}_2)$ is as in (18.3.5). A similar computation shows that

(18.3.8b) $\qquad \dfrac{\kappa}{|\kappa|} = e^{\pi i w^T \mathscr{V} w}\left(\dfrac{\det(\mathscr{A}_1 + \mathscr{A}_2)}{|\det(\mathscr{A}_1 + \mathscr{A}_2)|}\right)^{-1/2}$

where

(18.3.8c) $\qquad \mathscr{V} = \tfrac{1}{2}J(\mathscr{A}_1 + \mathscr{A}_2)^{-1}((\mathscr{A}_1 - \mathscr{A}_2) + 2(\mathscr{A}_1 J \mathscr{A}_1 - \mathscr{A}_2 J \mathscr{A}_2)J\mathscr{C}).$

18.4. We give some additional formulas complementary to the basic ones (8.1), (8.2), (18.2.4), (18.3.4)–(18.3.8) expressing the composition law in $A\Omega$.

First, note that in terms of $\beta(\mathscr{A}) = -2iJ\mathscr{A}$ (see (12.3)), we can write

(18.4.1) $\qquad \mathscr{T}(\mathscr{A}) = (\beta(\mathscr{A}) + \beta(\mathscr{A})^{-1})(-iJ/2), \qquad \mathscr{A} \in \mathfrak{S}_{2n},$

where $\mathscr{T}(\mathscr{A})$ is given by (18.3.3). Here also it seems worth noting that if c is the Cayley transform (see (10.2.1)) then $\beta(\mathscr{A})^{-1} = c(-c(\beta(\mathscr{A})))$.

Second, we note some special cases of the general formulas. Recall $\mathscr{C} = \mathscr{C}(\mathscr{A}_1, \mathscr{A}_2)$ from formula (18.3.5).

If $\mathscr{A}$ commutes with J, then

(18.4.2a) $\qquad \mathscr{T}(\mathscr{A}) = \tfrac{1}{2}((2\mathscr{A}) + (2\mathscr{A})^{-1}).$

Suppose $i\beta(\mathscr{A}) = 2J\mathscr{A}$ is a multiple of a complex structure on W: $i\beta(\mathscr{A}) = s\mathscr{J}$ where $\mathscr{J}^2 = -1$. Then

(18.4.2b) $\quad \mathscr{T}(\mathscr{A}) = -\mathscr{J}J\left(\dfrac{s + s^{-1}}{2}\right) = -J\mathscr{A}J(1 + s^{-2}) = \mathscr{A}^{-1}\left(\dfrac{s^2 + 1}{4}\right).$

If both $\mathscr{A}_1$ and $\mathscr{A}_2$ satisfy (b), i.e., $\mathscr{A}_i = (-s/2)J\mathscr{J}i$ for complex structures $\mathscr{J}i$, then

(18.4.2c) $\qquad \mathscr{C} = \dfrac{2}{s + s^{-1}}J(\mathscr{J}_1 + \mathscr{J}_2)^{-1}.$

If $\mathscr{A}_1 = \mathscr{A}_2 = \mathscr{A}$, then

(18.4.2d) $\qquad \mathscr{C} = \tfrac{1}{2}\mathscr{T}(\mathscr{A})^{-1} = -iJ(\beta(\mathscr{A}) + \beta(\mathscr{A})^{-1})^{-1}.$

If $\mathscr{A}_1 = \mathscr{A}_2 = \mathscr{A}$, and the condition of (b) holds for $\mathscr{A}$, then

(18.4.2e) $$\mathscr{C} = -\frac{J\mathscr{J}}{s + s^{-1}} = \frac{2\mathscr{A}}{s^2 + 1}.$$

If $\mathscr{A}_1 = \mathscr{A}_2 = \mathscr{A}$, then

(18.4.2f)
$$u = v = \tfrac{1}{2}(\beta(\mathscr{A}) - \beta(\mathscr{A})^{-1})/(\beta(\mathscr{A}) + \beta(\mathscr{A})^{-1})(w)$$
$$= \tfrac{1}{2}(\beta(\mathscr{A})^2 - 1)/(\beta(\mathscr{A})^2 + 1)(w) = \tfrac{1}{2}c(\beta(\mathscr{A})^2)^{-1}(w).$$

If $\mathscr{A}_1 = \mathscr{A}_2$ and the condition of (b) holds for $\mathscr{A}$, then

(18.4.2g) $$u = v = \left(\frac{1}{2}\right)\left(\frac{s^2 - 1}{s^2 + 1}\right)w.$$

If $\mathscr{A}_1 = \mathscr{A}_2 = \mathscr{A}$, then

(18.4.2h) $$\mathscr{A}_3 = -\tfrac{1}{2}J\mathscr{J}(\mathscr{A})J.$$

If $\mathscr{A}_1 = \mathscr{A}_2 = \mathscr{A}$ and the condition of (b) holds for $\mathscr{A}$, then

(18.4.2i) $$\mathscr{A}_3 = -J\mathscr{J}\frac{(s + s^{-1})}{4} = \mathscr{A}\frac{(1 + s^{-2})}{2}.$$

18.5. REMARK. Formula (18.3.8) says, roughly, that the influence of elements of the oscillator semigroup Ω is highly localized in the "phase space" W: if two elements of Ω are separated by translating one of them in phase space, then their product becomes very small. However, there is a limitation to this localization. Since the formula (18.3.3) for $\mathscr{J}(\mathscr{A})$ involves both $\mathscr{A}$ and $\mathscr{A}^{-1}$, the matrix $\mathscr{J}(\mathscr{A})$ can never get too small; hence the matrix $\mathscr{C}(\mathscr{A}_1, \mathscr{A}_2)$ (formula (18.3.5)), which intervenes in formula (18.3.8), can never get too big. Thus the rate of drop-off in w of the size of the product $\gamma_{\mathscr{A}_1}\natural\delta_w\natural\gamma_{\mathscr{A}_2}$ is limited. This limitation on phase-space localization of products in $A\Omega$ is a reflection of the Uncertainty Principle.

19. The Conjugation Property.

19.1. The equations (18.2.5) and (18.2.6) are of course valid for $\mathscr{A} \in \mathfrak{S}_{2n}$. If $\det(\operatorname{Im}\mathscr{A} + J/2) \neq 0$, then we can replace $\gamma_{\mathscr{A}}$ by $\gamma^0_{\mathscr{A}}$, and take a limit as the real part of $\mathscr{A}$ shrinks to zero, and the resulting functions on the right-hand sides of (18.2.5), (18.2.6) will describe unitary operators.

A calculation using (18.2.6) shows that if $\mathscr{A}$ is pure imaginary, say $\mathscr{A} = i\mathscr{A}'$, $\mathscr{A}' \in \varsigma^{2n}(\mathbf{R})$, and if

(19.1.1a)
$$(\mathscr{A}' + J/2)(u) + (\mathscr{A}' - J/2)(v) = 0; \quad \mathscr{A}' \in \varsigma^{2n}(\mathbf{R}), \ \det(\mathscr{A}' \pm J/2) \neq 0,$$

then

(19.1.1b) $$\delta_u\natural\gamma^0_{i\mathscr{A}'}\natural\delta_v = \gamma^0_{i\mathscr{A}'}.$$

This may be rewritten

(19.1.2a) $$\gamma^0_{i\mathscr{A}'}\natural\delta_v\natural(\gamma^0_{i\mathscr{A}'})^{-1} = \delta_{g(v)},$$

where (see (16.3.1))

(19.1.2b) $$g = (\mathscr{A}' + J/2)^{-1}(\mathscr{A}' - J/2) = c(\beta(i\mathscr{A}')) = \omega^{-1}(\gamma^0_{i\mathscr{A}'}).$$

This formula allows us to make the following statement.

THEOREM. *The action (9.2) of the symplectic group by automorphisms of the twisted convolution algebra is induced by conjugation by the metaplectic group $\widetilde{\mathrm{Sp}}(W)$. Precisely, in the notation of formulas (9.2) and (16.3.1), we have*

(19.1.3)
$$\gamma_{i\mathscr{A}}^0 \natural f \natural (\gamma_{i\mathscr{A}}^0)^{-1} = \omega_{1,1}(\omega^{-1}(i\mathscr{A}))(f),$$
$$\mathscr{A} \in \varsigma^{2n}(\mathbf{R}), \ \det(\mathscr{A} + iJ/2) \neq 0, \ f \in \mathscr{S}(W).$$

REMARK. The usual procedure in other constructions of the metaplectic group is to use general theorems about group representations to work backward from the existence of the action $\omega_{1,1}$ of Sp by automorphisms to the existence of the metaplectic group $\widetilde{\mathrm{Sp}}$. In this approach the determination of the cocycle determining the 2-fold cover of Sp defined by $\widetilde{\mathrm{Sp}}$ is somewhat elusive. Our explicit computational approach "from inside" using the oscillator semigroup Ω allows us to reverse the process: we can construct $\widetilde{\mathrm{Sp}}$ directly as a 2-fold cover of Sp, then identify conjugation by $\widetilde{\mathrm{Sp}}$ as the obvious action $\omega_{1,1}$ of Sp.

19.2. The conjugation relations (19.1.1), (19.1.2) have infinitesimal versions. For $w \in W$, we define the distribution of *twisted differentiation in the direction w* by

(19.2.1)
$$\partial_w = \lim_{t \to 0} \frac{\delta_{tw} - \delta_0}{t}, \qquad t \in \mathbf{R}.$$

Or more concretely, in terms of functions

(19.2.2a)
$$\partial_w \natural f(w') = \lim_{t \to 0} \frac{\delta_{tw} \natural f(w') - f(w)}{t}$$
$$= \lim_{t \to 0} \frac{e^{\pi i t \langle w, w' \rangle} f(w' - w) - f(w')}{t}$$
$$= \pi i \langle w, w' \rangle f(w') + (\partial_w * f)(w'),$$

where $\partial_w^* f$ indicates the usual abelian directional derivative of f in the direction of $-w$. Similarly

(19.2.2b)
$$(f \natural \partial_w)(w') = -\pi i \langle w, w' \rangle f(w') + (\partial_w * f)(w').$$

If, in relation (19.1.1b) we convolve on the right with δ_{-v}, then replace u by tu, $t \in \mathbf{R}$, and differentiate as $t \to 0$, we obtain

(19.2.3a) $\quad \gamma_{\mathscr{A}}^0 \natural \partial_v = \partial_{g(u)} \natural \gamma_{\mathscr{A}}^0, \qquad \mathscr{A} = i\mathscr{A}' \in i\varsigma^2(\mathbf{R}^n), \ \det(\mathscr{A}' + J/2) \neq 0$

where

(19.2.3b)
$$g = c(-2iJ\mathscr{A}) = \frac{2iJ\mathscr{A} - 1}{2iJ\mathscr{A} + 1} \in \mathrm{Sp}_{2n}(\mathbf{R}).$$

We know that the map

(19.2.4)
$$\partial : v \to \partial_v, \qquad v \in W,$$

is a linear map of real vector spaces from W to $\mathscr{S}^*(W)$, the space of tempered distributions on W, so that the ∂_v fill out a real $2n$-dimensional subspace of $\mathscr{S}^*(W)$. We denote this space by ∂W. Equation (19.2.3) says that the map ∂ of (19.2.4) is equivariant for the standard action of $\mathrm{Sp}(W) \simeq \mathrm{Sp}_{2n}(\mathbf{R})$ on W and the action, via conjugation by $\widetilde{\mathrm{Sp}}(W)$, on $\mathscr{S}^*(W)$.

By taking complex linear combinations of the distributions ∂_v, we get a complex linear subspace of $\mathscr{S}^*(W)$. This subspace will be isomorphic to the complexification $W_{\mathbf{C}}$ of W, by the unique complex-linear extension of the map ∂ of (19.2.4). We denote this space by $\partial W_{\mathbf{C}}$.

Equation (19.2.3) can be interpreted as saying that by conjugation, the metaplectic group $\widetilde{\mathrm{Sp}}(W)$ acts on $\partial W_{\mathbf{C}}$ via the complexification of the standard action of $\mathrm{Sp}(W)$ on $W \simeq \partial W$. Theorem 12 and Theorem 16.3 show that the normalized oscillator semigroup Ω^0 is a holomorphic extension of the metaplectic group. Thus the formula (19.2.3) will extend to arbitrary $\mathscr{A} \in \mathfrak{S}_{2n}^+ - \mathscr{V}$ (see equation (11.4) for notation). Thus we may state

PROPOSITION. *For $w \in W_{\mathbf{C}}$, $w = u + iv$, $u, v \in W$, set $\partial_w = \partial_u + i\partial_v$. Then for $\mathscr{A} \in \mathfrak{S}_{2n}^+ - \mathscr{V}$, we have*

$$(19.2.5) \qquad \gamma_{\mathscr{A}}^0 \natural \partial_w = \partial_{\omega^{-1}(\gamma_{\mathscr{A}}^0)(w)} \natural \gamma_{\mathscr{A}}^0,$$

where $\omega^{-1}(\mathscr{A}) \in \mathrm{Sp}_{2n}(\mathbf{C})$ is as in equation (12.4).

More generally, given any $\mathscr{A} \in \mathfrak{S}_{2n}$, and $w_1, w_2 \in W_{\mathbf{C}}$ satisfying

$$(19.2.6\mathrm{a}) \qquad (2iJ\mathscr{A} + 1)w_1 = (2iJ\mathscr{A} - 1)w_2,$$

we will have the relation

$$(19.2.6\mathrm{b}) \qquad \partial_{w_1} \natural \gamma_{\mathscr{A}} = \gamma_{\mathscr{A}} \natural \partial_{w_2}.$$

Readers who do not like the above argument by analytic continuation are welcome to verify the proposition by explicit calculation. The analog of (19.2.5) for the integral-operator realization of Ω has been already recorded in §4.4.

20. Action on Gaussian densities/coherent states/wave packets.

20.1. Recall from §1.2 that we denote by $\gamma \mathfrak{S}_n$ the set of multiples of Gaussian functions. It will come as no surprise if we observe that the oscillator semigroup, in its standard action on $L^2(\mathbf{R}^n)$ via the operators $T_{\mathscr{A}}$ of §3, preserves $\gamma \mathfrak{S}_n$. Indeed, using the formulas of §§1–3 we can easily compute, for $\mathscr{A} \in \mathfrak{S}_{2n}$ as in (3.1.1), and $C \in \mathfrak{S}_n$, that

$$(20.1.1\mathrm{a}) \qquad T_{\mathscr{A}} \gamma_C = \det(D + C)^{-1/2} \gamma_{C'}$$

where

$$(20.1.1\mathrm{b}) \qquad C' = A - B(D + C)^{-1} B^T.$$

From formula (20.1.1) we see that by taking $B = 0$ and choosing A to be any desired element of $\mathfrak{S}_n$, $T_{\mathscr{A}}$ may be selected so as to transform γ_C into (some multiple of) an arbitrary element of $\gamma \mathfrak{S}_n$. This gives us the first statement of the following result.

PROPOSITION. (i) *The natural action (20.1.1) of the oscillator semigroup Ω on $\gamma \mathfrak{S}_n$ is transitive: any point may be transformed to any other.*

(ii) *More precisely, let*

$$(20.1.2\mathrm{a}) \qquad P = M \cdot N$$

where

(20.1.2b) $M = \omega(\mathrm{GL}_n(\mathbf{R}))$, *as described in* §4.1, *and*

(20.1.2c) $N = \{M_{\gamma_{iA}} : A \in \varsigma^2(\mathbf{R}^n)\}$.

Set

(20.1.3) $\mathbf{P}\gamma\mathfrak{S}_n = \gamma\mathfrak{S}_n/\mathbf{C}^x = \{\{t\gamma_A : t \in \mathbf{C}^x\} : A \in \mathfrak{S}_n\}$.

Note that $\mathbf{P}\gamma\mathfrak{S}_n$ *is in a natural way a subset of* $\mathbf{P}L^2(\mathbf{R}^n)$, *the projective space associated to* $L^2(\mathbf{R}^n)$. *Thus the action of* P *on* $L^2(\mathbf{R}^n)$ *restricts and factors to define an action of* P *on* $\mathbf{P}\gamma\mathfrak{S}_n$. *This action is also transitive. Thus* $\mathbf{P}\gamma\mathfrak{S}_n$ *is a homogeneous space for* P.

PROOF. Statement (i) of the proposition is already proved. For statement (ii), consider $A \in \mathfrak{S}_n$, and decompose A into its real and imaginary parts, as in equation (1.1). A trivial calculation shows that

(20.1.4) $$M_{\gamma_{-i\,\mathrm{Im}\,A}}(\gamma_A) = \gamma_{\mathrm{Re}\,A}$$

(see (4.2.1)). We further see from (4.1.1) that if $g \in \mathrm{GL}_n(\mathbf{R})$, then

(20.1.5) $$\omega(g^{-1})\gamma_{\mathrm{Re}\,A} = |\det g|^{1/2}\gamma_{g^T\,\mathrm{Re}\,A\,g}.$$

Taking $g = (\mathrm{Re}\,A)^{-1/2}$, we see that via P we can transform γ_A to a multiple of γ_{I_n}, where I_n is the $n \times n$ identity matrix. This proves part (ii) of the proposition.

REMARK. The map

(20.1.6) $\mathfrak{S}_n \to \mathbf{P}\gamma\mathfrak{S}_n$, $A \to \{t\gamma_A : t \in \mathbf{C}^x\}$

is a bijection. It is well known classically [Si] that $\mathfrak{S}_n$ is the symmetric space for $\mathrm{Sp}_{2n}(\mathbf{R})$: a homogeneous space whose isotropy group is the maximal compact subgroup. The action of $\mathrm{Sp}_{2n}(\mathbf{R})$ on $\mathfrak{S}_n$ is typically given in terms of linear fractional transformations. In $\mathrm{Sp}_{2n}(\mathbf{R})$, there is a maximal parabolic subgroup

(20.1.7) $P \simeq M \cdot N = \left\{ \begin{bmatrix} g & 0 \\ 0 & g^{t-1} \end{bmatrix} \begin{bmatrix} I_n & A \\ 0 & I_n \end{bmatrix} = g \in \mathrm{GL}_n(\mathbf{R}),\ A \in \varsigma^2(\mathbf{R}^n) \right\}$

which may be identified to the group P of (20.1.2) in an obvious way. If this is done the standard action of P on $\mathfrak{S}_n$ coincides, again after the straightforward identifications, with the action described in formulas (20.1.4), (20.1.5). One expects, then, that if we extend action (20.1.1) by continuity to obtain an action of $\widetilde{\mathrm{Sp}}_{2n}(\mathbf{R})$ on $\gamma\mathfrak{S}_n$, then the map (20.1.6) will identify $\mathbf{P}\gamma\mathfrak{S}_n$ with $\mathfrak{S}_n$ not just as sets, but as homogeneous spaces for $\mathrm{Sp}_{2n}(\mathbf{R})$. To see this is so, and to gain further insight into Ω and $\gamma\mathfrak{S}_n$, we again appeal to the framework provided by the Heisenberg group and twisted convolution.

21. The Grassmannian of polarizations. In this section we treat a topic in symplectic geometry. This material is not new, but again no convenient reference is known to me. The paper [W] gives a general theory of which this section provides one very special example. To keep this treatment reasonably brief, we leave some of the more elementary lemmas as exercises for the reader, or give brief sketches of proofs.

21.1. We start with some purely algebraic considerations. To emphasize their general nature we work over a base field F which is arbitrary, except that it should not be of characteristic 2.

Let W be a finite-dimensional vector space over F, equipped with a symplectic form $\langle\,,\,\rangle$. Thus $\langle\,,\,\rangle$ is antisymmetric and nondegenerate. The nondegeneracy of $\langle\,,\,\rangle$ is equivalent to the condition that the map

$$(21.1.1a) \qquad\qquad \alpha\colon W \to W^*$$

from W to its dual space W^*, given by

$$(21.1.1b) \qquad\qquad \alpha(w)(w') = \langle w, w'\rangle$$

be an isomorphism.

Let $X \subseteq W$ be a subspace. Define $X^\perp$, the orthogonal complement of X with respect to $\langle\,,\,\rangle$, by

$$(21.1.2) \qquad\qquad X^\perp = \{w \in W : \langle w, x\rangle = 0,\ \text{all } x \text{ in } X\}.$$

If we form the diagram

$$(21.1.3) \qquad\qquad
\begin{array}{ccc}
X & \xrightarrow{\ i\ } & W \\
{\scriptstyle \alpha_X}\downarrow & & \downarrow{\scriptstyle \alpha} \\
X^* & \xleftarrow{\ i^*\ } & W^*
\end{array}$$

where i is the inclusion of X in W and $\alpha_X\colon X \to X^*$ is defined just as in (21.1.1b), then it is clear that

$$X^\perp = \ker(i^* \circ \alpha).$$

Standard facts about duality in linear algebra [**L1**, p. 500] plus antisymmetry of $\langle\,,\,\rangle$ tell us that

(21.1.4a) $\dim X + \dim X^\perp = \dim W$;

(21.1.4b) $(X^\perp)^\perp = X$;

(21.1.4c) for two subspaces $X, Y \subseteq W$, we have $(X + Y)^\perp = X^\perp \cap Y^\perp$.

We observe that $X \cap X^\perp$ is precisely the radical of the restriction of the form $\langle\,,\,\rangle$ to X, i.e., the kernel of α_X. Thus the restriction of $\langle\,,\,\rangle$ to X will be nondegenerate precisely when $X \cap X^\perp = \{0\}$; whence, by (21.1.4a), we will have a direct sum decomposition $W = X \oplus X^\perp$. At the other extreme the restriction of $\langle\,,\,\rangle$ to X will be identically zero if and only if $X \subseteq X^\perp$. We call such an X *isotropic*.

If $X \subseteq W$ is isotropic, then (21.1.4a) implies $\dim X \leq \frac{1}{2}\dim W$. On the other hand, if $\{e_j, f_j\}$ is a symplectic basis for W (cf. (14.2)) and if X is the span of the e_i's, then X is isotropic and of dimension $\frac{1}{2}\dim W$. We call an isotropic subspace of dimension $\frac{1}{2}\dim W$ a *polarization* of W. (Other common terms are 'Lagrangian subspace' [**GS**] and 'maximal isotropic subspace'.) Evidently a subspace $X \subseteq W$ will be a polarization of W if and only if $X^\perp = X$.

21.2. Again, if $\{e_j, f_j\}$ is a symplectic basis for W, then X, the span of the e_j's, and Y, the span of the f_j's, are both maximal isotropic. Further, we have $W = X \oplus Y$. We call such a pair, of complementary maximal isotropic subspaces, a *complete polarization* of W.

21.2. LEMMA. (a) *Let* $X \subseteq W$ *be a polarization, and let* $\{e_j\}_{i=1}^n$ *be a basis for* X. *Then there exists a symplectic basis* $\{e_j, f_j\}$ *for* W *containing the* e_j.

(b) *If* Y *is another polarization of* W *complementary to* X, *so that* $\{X, Y\}$ *make a complete polarization, then the* f_i *may be chosen to belong to* Y; *and, so chosen, they are uniquely determined.*

PROOF. Since X is a polarization, $X = X^\perp$. Referring to diagram (21.1.3), we see that $i^* \circ \alpha$ defines an isomorphism from W/X to X^*. Let $\{e_j^*\}$ be the basis of X^* to $\{e_j\}$. Choose $f_1 \in W$ such that $i^* \circ \alpha(f_1) = e_1$. Then $\{e_1, f_1\}$ spans a two-dimensional subspace $W_1 \subseteq W$, on which $\langle \, , \, \rangle$ is nondegenerate. Thus we have a decomposition

$$W = W_1 \oplus W_1^\perp.$$

The space $\tilde{X} = X \cap W_1^\perp$ is a polarization of $W_1^\perp$, and has $\{e_j\}_{j \geq 2}$ as basis. By induction on $\dim W$, we can extend $\{e_j\}_{j \geq 2}$ to a symplectic basis $\{e_j, f_j\}_{j \geq 2}$ of $W_1^\perp$. Then adjoining $\{e_1, f_1\}$ to this gives a symplectic basis for W. This proves part (a).

If $Y \subseteq W$ is complementary to X, then $i^* \circ \alpha$ defines an isomorphism from Y to X^*. Let the f_j be the uniquely defined elements of Y such that $i^* \circ \alpha(f_j) = e_j^*$ (with $\{e_j^*\}$ the basis dual to $\{e_j\}$, as above.) Then if Y is isotropic $\{e_j, f_j\}$ is easily checked to be a symplectic basis for W.

21.3. LEMMA. *Let* X, Z *be two polarizations of* W, *not necessarily complementary.*

(i) $(X + Z)^\perp = X \cap Z$.

Let U *be any vector space complement to* $X \cap Z$ *in* $X + Z$.

(ii) *The restriction of* $\langle \, , \, \rangle$ *to* U *is nondegenerate, and so defines a symplectic form on* U.

(iii) $\{X \cap U, Z \cap U\}$ *defines a complete polarization for* U.

(iv) $X \cap Z$ *is a polarization of* $U^\perp$.

PROOF. Left to the reader. With regard to (ii), we note that if $V \subseteq W$ is any subspace, and $U \subseteq V$ is a complement to $V \cap V^\perp$, then $\langle \, , \, \rangle$ restricted to U is nondegenerate.

21.4. Let $\Pi(W)$ denote the set of all polarizations of W. We will give a description of $\Pi(W)$.

Let $\mathrm{Sp}(W)$ be the subgroup of $\mathrm{GL}(W)$ consisting of transformations which preserve the form $\langle \, , \, \rangle$. Clearly if $X \subseteq W$ is a polarization and $g \in \mathrm{Sp}(W)$, then $g(X)$ is again a polarization. Thus $\Pi(W)$ allows in a natural way an action of $\mathrm{Sp}(W)$. Also $\mathrm{Sp}(W)$ will act, in a simply transitive manner, on the set of symplectic bases for W. It follows from part (a) of Lemma 21.2 that $\mathrm{Sp}(W)$ acts transitively on $\Pi(W)$; thus $\Pi(W)$ is a homogeneous space for $\mathrm{Sp}(W)$.

For $X \in \Pi(W)$, let $P_X \subseteq \mathrm{Sp}(W)$ denote the subgroup stabilizing X. There is a restriction map

$$(21.4.1) \qquad\qquad r\colon P_X \to \mathrm{GL}(X), \qquad g \to g_{|X}.$$

Let $N_X = N$ denote the kernel of r. Thus elements of N leave X pointwise fixed.

Let $Y \in \Pi(W)$ be complementary to X, so that $\{X, Y\}$ is a complete polarization of W. From Lemma 21.2(b) we can see that

$$r\colon P_X \cap P_Y \xrightarrow{\sim} \mathrm{GL}(X)$$

is an isomorphism of groups. Thus if we set $P_X \cap P_Y = M_{X,Y} = M$, we have

$$(21.4.2) \qquad\qquad P_X = M \cdot N, \qquad M \simeq \mathrm{GL}(X).$$

LEMMA. (i) *The group* $N_X = N$ *acts simply transitively on the set* $\Pi(W)_X = \{Y \in \Pi(W)\colon Y \cap X = \{0\}\}$.

(ii) *Given* $Y \in \Pi(W)_X$, *and* $n \in N$, *we define a bilinear form* β_n *on* Y *by the formula*

$$(21.4.3) \qquad\qquad \beta_n(y_1, y_2) = \langle n(y_1), y_2 \rangle = \langle (n-1)y_1, y_2 \rangle.$$

Then β_n *is a symmetric bilinear form.*

(iii) *The map*

$$(21.4.4) \qquad\qquad \beta\colon N \to \varsigma^{2^*}(Y), \qquad n \to \beta_n,$$

where $\varsigma^{2^*}(Y)$ *is the space of symmetric bilinear forms on* Y, *is an isomorphism of groups.*

PROOF. Left to the reader. Part (i) follows from Lemma 21.2. Parts (ii) and (iii) involve straightforward checking.

COROLLARY. *The orbits of* N *in* $\Pi(W)$ *are the sets*

$$(21.4.5) \qquad\qquad Q_{X_1} = \{Z \in \Pi(W)\colon Z \cap X = X_1\}$$

for X_1 *ranging over all subspaces of* X. *The orbits of* P_X *are the sets*

$$(21.4.6) \qquad\qquad Q_l = \bigcup_{\dim X_1 = l} Q_{X_1}, \qquad 0 \le l \le \dim X.$$

PROOF. Left to the reader.

21.5. We now take W to be a symplectic vector space over $\mathbf{R}$. Let

$$(21.5.1) \qquad\qquad W_{\mathbf{C}} \simeq W \otimes_{\mathbf{R}} \mathbf{C}$$

be the complexification of W. It is considered as a vector space over $\mathbf{C}$. The symplectic form $\langle\,,\,\rangle$ on W extends complex-bilinearly to a symplectic form, still denoted $\langle\,,\,\rangle$, on $W_{\mathbf{C}}$. Let $\mathrm{Sp}(W)$ and $\mathrm{Sp}(W_{\mathbf{C}})$ be the symplectic groups of W and $W_{\mathbf{C}}$ respectively. ($\mathrm{Sp}(W_{\mathbf{C}})$ consists of *complex* linear endomorphisms of $W_{\mathbf{C}}$). An element $g \in \mathrm{Sp}(W)$ may be extended to a complex linear endomorphism, still denoted g, of $W_{\mathbf{C}}$. The extended g will belong to $\mathrm{Sp}(W_{\mathbf{C}})$. Thus the extension process yields an injection

$$(21.5.2) \qquad\qquad \mathrm{Sp}(W) \hookrightarrow \mathrm{Sp}(W_{\mathbf{C}}).$$

Consider the variety $\Pi(W_{\mathbf{C}})$ of polarizations of $W_{\mathbf{C}}$. It is a homogeneous space for $\mathrm{Sp}(W_{\mathbf{C}})$. Via the embedding (21.5.2), we can let $\mathrm{Sp}(W)$ act on $\Pi(W_{\mathbf{C}})$; we want to describe this action. We will see $\Pi(W_{\mathbf{C}})$ is not a homogeneous space for $\mathrm{Sp}(W)$, but consists of only finitely many $\mathrm{Sp}(W)$ orbits.

Complex conjugation on $\mathbf{C}$ induces a (complex *anti*-linear) endomorphism of $W_{\mathbf{C}}$:

(21.5.3)
$$\overline{} : w \otimes z \to w \otimes \bar{z}, \qquad w \in W,\ z \in \mathbf{C},$$

where $z \to \bar{z}$ is complex conjugation. A (complex linear) subspace $X \subseteq W_{\mathbf{C}}$ is of the form $X_0 \otimes \mathbf{C}$ for some (real linear) $X_0 \subseteq W$ if and only if $X = \overline{X}$; then $X_0 = \{x + \bar{x} : x \in X\}$.

Consider $X \in \Pi(W_{\mathbf{C}})$. Then also $\overline{X} \in \Pi(W)$. The spaces $X \cap \overline{X}$ and $X + \overline{X}$ are clearly conjugation-invariant, so

$$X \cap \overline{X} = ((X \cap \overline{X}) \cap W)_{\mathbf{C}} \quad \text{and} \quad X + \overline{X} = ((X + \overline{X}) \cap W)_{\mathbf{C}}.$$

We call X *totally complex* if $X \cap \overline{X} = \{0\}$. In this case X and $\overline{X}$ constitute a complete polarization in $W_{\mathbf{C}}$. If $X = \overline{X}$ (so that $X = (X \cap W)_{\mathbf{C}}$), we call X *real*. The real elements of $\Pi(W_{\mathbf{C}})$ are evidently identifiable with $\Pi(W)$, and constitute one $\mathrm{Sp}(W)$ orbit. Lemma 21.3 allows us to decompose a general $X \in \Pi(W_{\mathbf{C}})$ into real and totally complex parts.

LEMMA. *Consider* $X \in \Pi(W_{\mathbf{C}})$. *Let* U *be a complement to* $(X \cap \overline{X}) \cap W$ *in* $(X + \overline{X}) \cap W$. *Then*

(i) $X \cap U_{\mathbf{C}}$ *is a totally complex polarization of* $U_{\mathbf{C}}$.

(ii) $X \cap (U_{\mathbf{C}}^{\perp})$ *is a real polarization of* $(U^{\perp})_{\mathbf{C}}$.

PROOF. Left to the reader.

It is clear that $\dim_{\mathbf{C}}(X \cap \overline{X}) = \dim_{\mathbf{R}}(X \cap W)$ is an invariant of X under the action of $\mathrm{Sp}(W)$. And conversely, if $\dim_{\mathbf{R}}(X_1 \cap W) = \dim_{\mathbf{R}}(X_2 \cap W)$, then by transforming X_2 by $\mathrm{Sp}(W)$ if necessary, we can arrange that $X_2 \cap \mathbf{R} = X_1 \cap \mathbf{R}$. Thus the lemma tells us that in the study of $\mathrm{Sp}(W)$ orbits in $\Pi(W_{\mathbf{C}})$, greatest interest accrues to the totally complex polarizations. So we consider these.

21.6. By a *complex structure* on W we mean an operator $J \in \mathrm{End}\,W$ such that $J^2 = -1$. The complex structure J is said to be *compatible with* $\langle\ ,\ \rangle$ if $J \in \mathrm{Sp}(W)$.

REMARK. A simple calculation shows that if c is the Cayley transform (see formula (10.2.1)), then J is a complex structure if and only if $c(J) = -J$. Thus we may conclude that $J \in \mathrm{Sp}(W)$ if and only if $J \in \mathfrak{sp}_W$.

A complex structure can have eigenvalues equal only to $\pm i$, and these eigenvalues must occur with equal multiplicity, which will be $\frac{1}{2}\dim W$ for each. Let $W_J^+ \subseteq W_{\mathbf{C}}$ be the $(+i)$-eigenspace of the complex structure J, and let W_J^- be the $(-i)$-eigenspace. Then

(21.6.1a)
$$W_J^- = (W_J^+)^-$$

where the $^-$ outside the parentheses indicates complex conjugation, and

(21.6.1b)
$$W_{\mathbf{C}} = W_J^+ \oplus W_J^-.$$

Statement (21.6.1b) just says W_J^+ is totally complex, which is obvious. The operator

$$(21.6.1c) \qquad\qquad p_J^+ = \frac{1 - iJ}{2}$$

projects $W_{\mathbf{C}}$ onto W_J^+ with kernel W_J^-, and

$$(21.6.1d) \qquad\qquad p_J^- = \frac{1 + iJ}{2} \left(= \overline{p_J^+} \right)$$

projects $W_{\mathbf{C}}$ onto W_J^- with kernel W_J^+. Note that p_J^+ is a real linear isomorphism from W to W_J^+, and that

$$(21.6.1e) \qquad\qquad p_J^+(Jw) = ip_J^+(w), \qquad w \in W.$$

LEMMA. (i) *The correspondence $J \leftrightarrow W_J^+$ is a bijection from the set of complex structures on W to the set of totally complex subspaces of $W_{\mathbf{C}}$ of dimension $\frac{1}{2} \dim W$.*

(ii) *The complex structure J on W is compatible with $\langle\ ,\ \rangle$ if and only if W_J^+ is isotropic for $\langle\ ,\ \rangle$, in which case $W_J^+ \in \Pi(W_{\mathbf{C}})$.*

PROOF. Left to the reader.

21.7. On $W_{\mathbf{C}}$, consider the pairing

$$(21.7.1a) \qquad\qquad H(w, w') = i\langle w, \overline{w'} \rangle, \qquad w, w' \in W_{\mathbf{C}}.$$

This pairing is clearly complex linear in the first variable, and complex antilinear in the second variable. Antisymmetry of $\langle\ ,\ \rangle$ implies that H is Hermitian symmetric:

$$(21.7.1b) \qquad\qquad H(w', w) = \overline{H(w, w')}.$$

(The factor i is inserted to make H Hermitian symmetric rather than Hermitian skew symmetric.) Clearly $\mathrm{Sp}(W)$, in its action on $W_{\mathbf{C}}$, will preserve the form H.

Consider a totally complex $X \in \Pi(W_{\mathbf{C}})$. Since $\overline{X} \cap X^{\perp} = \overline{X} \cap X = \{0\}$, we see that the restriction $H_{|X}$ of the form H to X will be nondegenerate.

Let X' be another totally complex polarization of $W_{\mathbf{C}}$. If $X' = g(X)$ for some $g \in \mathrm{Sp}(W)$, then since $\mathrm{Sp}(W)$ preserves H, the map g must define an isometry from X equipped with $H_{|X}$ to X' equipped with $H_{|X'}$. Thus for X and X' to belong to the same $\mathrm{Sp}(W)$-orbit, the forms $H_{|X}$ and $H'_{|X}$ must be isometric.

Conversely, suppose

$$(21.7.2) \qquad\qquad a: X \to X'$$

is an isometry from $H_{|X}$ to $H_{|X'}$. Write

$$W_{\mathbf{C}} = X \oplus \overline{X} = X' \oplus \overline{X'}$$

and define $g \in \mathrm{End}(W_{\mathbf{C}})$ by

$$(21.7.3) \qquad g(x_1 + \bar{x}_2) = a(x_1) + \overline{a(x_2)}, \qquad x_1, x_2 \in X.$$

Then an easy computation shows $g \in \mathrm{Sp}(W)$. Thus X and X' belong to the same $\mathrm{Sp}(W)$ orbit.

The possible isometry classes of Hermitian forms are well known, being described by Sylvester's Inertia Theorem [**L1**, p. 509]. For our X as above, we can find a basis $\{e_j\}$, and an integer p, $0 \leq p \leq \dim X$, such that

$$(21.7.4) \qquad H(e_j, e_k) = \varepsilon_j \delta_{jk}$$

where $\varepsilon_j = 1$ for $1 \leq j \leq p$, and $\varepsilon_j = -1$ for $p < j$. Set $q = \dim X - p$. The pair (p, q) is called the signature of $H_{|X}$, and it determines the isometry class of $H_{|X}$. Note that if we set

$$(21.7.5) \qquad f_j = -i\varepsilon_j \bar{e}_j$$

then $\{e_j, f_j\}$ is a symplectic basis for $W_{\mathbf{C}}$.

Combining the discussion of this subsection with Lemma 21.5 yields the following conclusion.

PROPOSITION. *The* $\mathrm{Sp}(W)$ *orbit of* $X \in \Pi(W_{\mathbf{C}})$ *is determined by the nonnegative integers* p, q, r, *where*

(i) $r = \dim_{\mathbf{R}}(X \cap W) = \dim_{\mathbf{C}}(X \cap \overline{X})$, *and*

(ii) (p, q) *is the signature of* $H_{|(X \cap U)_{\mathbf{C}}}$, *where* U *is any complement to* $X \cap W$ *in* $(X + \overline{X}) \cap W$.

The triple (p, q, r) *satisfies* $p + q + r = \frac{1}{2} \dim W$ *and otherwise is arbitrary.*

NOTATION. We denote by $\mathscr{O}^{(p,q,r)}$ the orbit whose invariants, described by this proposition, are (p, q, r).

21.8. Let $X \in \Pi(W_{\mathbf{C}})$ be totally complex, and let J be the associated complex structure on W (see §21.6). We have noted in §21.6 that $p_J^+ = (1 - iJ)/2$ defines an isomorphism from W to X. On W define

$$(21.8.1) \qquad \begin{aligned} H_J(u, v) &= H_{|X}(p_J^+(w), p_J^+(v)) = i\langle p_J^+(u), p_J^-(v)\rangle \\ &= \tfrac{1}{2}(\langle Ju, v\rangle + i\langle u, v\rangle). \end{aligned}$$

From formula (21.6.1e), we can see that if W is endowed with the complex structure J—i.e., if it is considered as a complex vector space in which multiplication by i is accomplished by applying J—then H_J is a symmetric Hermitian-bilinear form.

PROPOSITION. (i) *An element of* $\mathrm{GL}(W)$ *stabilizes* X *if and only if it commutes with* J *if and only if it is complex linear with respect to the complex structure defined by* J.

(ii) *For* $g \in \mathrm{Sp}(W)$, *these conditions imply that* g *is an isometry of the Hermitian form* H_J, *or equivalently, of its real part*

$$(21.8.2) \qquad \mathrm{Re}\, H_J = B_J = \langle J\cdot, \cdot\rangle.$$

(iii) *Thus, in particular, the isometry group in* $\mathrm{Sp}(W)$ *of* $X \in \Pi(W_{\mathbf{C}})$ *is isomorphic to the unitary group of the form* H_J, *acting on* W *considered as complex vector space via the complex structure* J *associated to* X.

PROOF. Left to the reader.

21.9. We call $X \in \Pi(W_{\mathbf{C}})$ *positive* if the Hermitian form $H_{|X}$ is positive definite. In this case X must be totally complex. Let J be the associated complex structure on W. Then positivity of X is equivalent to the positive definiteness of the symmetric bilinear form B_J (see (21.8.2)), the real part of H_J. Denote the set of positive polarizations of W by $\Pi^+(W_{\mathbf{C}})$. We also refer to J as a positive complex structure, and via Lemma 21.6 we may identify $\Pi^+(W_{\mathbf{C}})$ to the set of positive complex structures.

We know from Propositions 21.7 and 21.8 that $\Pi^+(W_{\mathbf{C}})$ is a homogeneous space for $\mathrm{Sp}(W)$, of the form

$$(21.9.1) \qquad \Pi^+(W_{\mathbf{C}}) \simeq \mathrm{Sp}(W)/U_{H_J}$$

where U_{H_J} is the isometry group of the Hermitian form H_J (see (21.8.1)) associated to a given $X \in \Pi^+(W)$. We note that since H_J is positive definite, U_{H_J} is a compact group.

We will give a concrete description of $\Pi^+(W_{\mathbf{C}})$. Let $\{X, Y\}$ be a complete polarization of W. Then $\{X_{\mathbf{C}}, Y_{\mathbf{C}}\}$ is a complete polarization of $W_{\mathbf{C}}$. Since $H(x, x) = 0$ for $x \in X_{\mathbf{C}}$, it follows that $X_{\mathbf{C}} \cap Z = \{0\}$ for $Z \in \Pi^+(W_{\mathbf{C}})$. Thus in the notation of Lemma 21.4 we have

$$(21.9.2) \qquad \Pi^+(W_{\mathbf{C}}) \subseteq \Pi(W_{\mathbf{C}})_{X_{\mathbf{C}}} \simeq \varsigma^{2^*}(Y_{\mathbf{C}}).$$

We will describe $\Pi^+(W_{\mathbf{C}})$ as a subset of $\varsigma^{2^*}(Y_{\mathbf{C}})$.

We will perform this calculation in coordinates in order to achieve a maximally explicit result. Thus let $\{e_j\}$ be a basis for X, and let $\{f_i\}$ be the basis for Y such that $\{e_j, f_j\}$ is a symplectic basis for W. Let

$$(21.9.3) \qquad \begin{bmatrix} x \\ y \end{bmatrix} \sim \sum x_j e_j + y_j f_j, \qquad x, y \in \mathbf{R}^n,$$

be coordinates on W with respect to this basis. The matrix of $\langle\ ,\ \rangle$ with respect to this basis is

$$(21.9.4) \qquad J_0 = \begin{bmatrix} 0 & I \\ -I & 0 \end{bmatrix}.$$

(This is the matrix J of (7.1.1). Interpreted as a linear transformation, it is a complex structure; we are now labeling it with a subscript to distinguish it from the general complex structure.)

Let $J_1 \in \mathrm{End}(W)$ be a positive compatible complex structure. Let the matrix of J_1 in coordinates (21.9.3) be

$$(21.9.5) \qquad J_1 = \begin{bmatrix} A & B \\ C & D \end{bmatrix}.$$

The condition that J_1 be compatible is that

$$(21.9.6) \qquad J_0 J_1 = \begin{bmatrix} -C & -D \\ A & B \end{bmatrix}$$

be symmetric, or

$$(21.9.7) \qquad C = C^T, \qquad B = B^T, \qquad D = -A^T.$$

The condition that J_1 be positive is the condition that $J_0 J_1$ be a positive definite matrix. The condition that J_1 be a complex structure is that

$$(21.9.8) \qquad J_1^2 = \begin{bmatrix} A^2 + BC & AB + BD \\ CA + DC & CB + D^2 \end{bmatrix} = \begin{bmatrix} -I & 0 \\ 0 & -I \end{bmatrix}.$$

Following Lemma 21.4, the bilinear form $\beta_{J_1} \in \varsigma^{2^*}(Y) \simeq \varsigma^2(\mathbf{C}^n)$ is defined as follows. Let Z_1 be the totally complex polarization associated to J_1. Then for $y_1, y_2 \in Y$, we have

$$(21.9.9) \qquad \beta_{J_1}(y_1, y_2) = \langle y_1 + x + ix', y_2 \rangle = \langle x + ix', y_2 \rangle,$$

where $x, x' \in X$ are such that $y_1 + x + ix'$ belongs to Z. This condition amounts to the requirement that

$$0 = p_{J_1}^-(y_1 + x + ix') = \frac{(1 + iJ_1)}{2}(y_1 + x + ix')$$
$$= \tfrac{1}{2}((y_1 + x - J_1 x') + i(J_1(y_1 + x) + x')).$$

This amounts to the equation

$$y_1 = J_1 x' - x.$$

In terms of the decomposition (21.9.5), this equation amounts to

$$(21.9.10) \qquad x' = C^{-1} y_1, \qquad x = AC^{-1} y_1.$$

Thus we compute

$$(21.9.11) \qquad \begin{aligned} \beta_{J_1}(y_1, y_2) &= -y_2^T (AC^{-1} y_1 + iC^{-1} y_1) \\ &= -i(y_2^T (I - iA) C^{-1} y_1). \end{aligned}$$

Note that relations (21.9.7) and the lower left corner of equation (21.9.8) imply C and CA are symmetric. Hence $AC^{-1} = C^{-1}(CA)C^{-1}$ is also symmetric. Thus we see that the matrices $-i\beta_{J_1} \in \varsigma^2(\mathbf{C}^n)$, belong to the Siegel upper half-space $\mathfrak{S}_n$. Since the subgroup P_X of $\mathrm{Sp}(W)$ acts transitively on $i\mathfrak{S}_n$, as described in Proposition 20.1, we conclude that the map

$$(21.9.12) \qquad \Pi^+(W_{\mathbf{C}}) \to \mathfrak{S}_n, \qquad J_1 \to -i\beta_{J_1},$$

is a bijection.

REMARK. If one replaces the complete polarization $\{X_{\mathbf{C}}, Y_{\mathbf{C}}\}$ with another one, of the form $\{\overline{Z}, Z\}$ where $Z \in \Pi^+(W_{\mathbf{C}})$, then one obtains a description of $\Pi^+(W_{\mathbf{C}})$ as a bounded domain in $\varsigma^{2^*}(Z)$. This is the "bounded realization" of $\mathfrak{S}_n$ [Si]. Other choices of complete polarizations will lead to concrete realizations of $\Pi^+(W_{\mathbf{C}})$ intermediate between $\mathfrak{S}_n$ and the bounded domain.

21.10. Since the set $\Pi(W_{\mathbf{C}})$ is a homogeneous space for the complex Lie group $\mathrm{Sp}(W_{\mathbf{C}})$, and the isotropy group of a point is a complex Lie subgroup, we see $\Pi(W_{\mathbf{C}})$ naturally has the structure of a complex manifold. In particular it is a topological space. We recall from Proposition 21.7 the notation $\mathcal{O}^{(p,q,r)}$ for the $\mathrm{Sp}(W)$ orbits in $\Pi(W_{\mathbf{C}})$.

PROPOSITION. (a) *The open* $\mathrm{Sp}(W)$-*orbits in* $\Pi(W_{\mathbf{C}})$ *are the orbits* $\mathscr{O}^{(p,q,0)}$ *consisting of totally complex elements* $X \in \Pi(W_{\mathbf{C}})$ *such that the Hermitian form* $H_{|X}$ *on* X *has signature* (p,q).

(b) *The* $\mathrm{Sp}(W)$-*orbits that are in the closure of* $\mathscr{O}^{(p,q,0)}$ *are the orbits* $\mathscr{O}^{(p',q',r')}$ *with* $p' \le p$ *and* $q' \le q$.

PROOF. The Hermitian form $H_{|X}$, being a restriction to $X \in \Pi(W_{\mathbf{C}})$ of a form on the ambient space $W_{\mathbf{C}}$, clearly varies continuously with X. Thus the numbers (p,q,r) attached to X in Proposition 21.7 are semicontinuous: if a sequence $\{X_n\}_{n \in \mathbf{Z}^+} \subseteq \mathscr{O}^{(p,q,r)}$ approaches $X_\infty \in \mathscr{O}^{(p',q',r')}$ we must have $p' \le p$, $q' \le q$, and $r' \ge r$. On the other hand, if these inequalities do hold, it is easy to construct such sequences. For example, let $\{e_j, f_j\}$ be a symplectic basis for W. Choose real numbers t_j, and let $X = X(t_1, t_2, \ldots)$ be the span of $\{it_j e_j + f_j\}$. Then the $it_j e_j + f_j$ are an orthogonal basis for X, and the p, q, and r for X are easily seen to be equal to the number of t_j which are positive, negative, and zero respectively. This concludes the proof of the proposition.

REMARK. Since $\Pi^+(W_{\mathbf{C}}) = \mathscr{O}^{(n,0,0)}$, where $n = \frac{1}{2}\dim W$, we see from the proposition that the closure $\Pi^+(W_{\mathbf{C}})^-$ consists of the $n+1$ $\mathrm{Sp}(W)$-orbits $\mathscr{O}^{(p,0,n-p)}$. These are linearly ordered by inclusion of closures. Since $\Pi(W_{\mathbf{C}})$ is a compact manifold, the closed subset $\Pi^+(W_{\mathbf{C}})^-$ will be compact. It is the (minimal) *Satake compactification* of $\Pi^+(W_{\mathbf{C}})$ [**Sa**].

22. Quantization of semipositive polarizations.

22.1. Return now to the real symplectic vector space $W = \mathbf{R}^n \oplus (\mathbf{R}^n)^*$ which plays host to twisted convolution, as described in §7. In §19.2 we have described the twisted differentiations $f \to \partial_w \natural f$, $w \in W$, which fill out a real linear subspace ∂W of $\mathscr{S}^*(W)$. We saw there that ∂W is invariant under conjugation by the metaplectic group, and that the map $w \to \partial_w$ intertwines the two natural actions of the symplectic group.

The symplectic form on W has a natural interpretation on ∂W. As is very well known, and easy to compute from formula (19.2.2), twisted convolutions of the ∂_w satisfy the Canonical Commutation Relations

$$(22.1.1) \qquad [\partial_w, \partial_{w'}] = \partial_w \natural \partial_{w'} - \partial_{w'} \natural \partial_w = 2\pi i \langle w, w' \rangle \delta_0,$$

where δ_0 is the unit mass at the origin in W (and the identity element of the twisted convolution algebra).

Consider the complexification $\partial W_{\mathbf{C}}$ of ∂W. The CCR (22.1.1) of course extend complex-bilinearly to $\partial W_{\mathbf{C}}$. It is obvious from the CCR that, given a subspace $X \subseteq W_{\mathbf{C}}$, the elements of ∂X commute with each other if and only if X is isotropic.

22.2. The twisted convolution algebra acts on the Schwartz space $\mathscr{S}(\mathbf{R}^n)$ via the Weyl transform ρ (see (13.1)). The first part of the following result is in Cartier [**Ct**].

THEOREM. (a) *Let $X \in \Pi^+(W_{\mathbf{C}})$. Then there is a function $q_X \in \mathscr{S}(\mathbf{R}^n)$, unique up to multiples, such that*

$$(22.2.1) \qquad\qquad \rho(\partial \bar{x})q_X = 0, \qquad x \in X.$$

(b) *In fact, $q_X \in \gamma \mathfrak{S}_n$, and the map*

$$(22.2.2) \qquad\qquad \mathscr{Q}: \Pi^+(W_{\mathbf{C}}) \to \mathbf{P}\gamma \mathfrak{S}_n \subseteq \mathscr{S}(\mathbf{R}^n), \qquad X \to \mathbf{C}^x \cdot q_X,$$

is bijective and equivariant for the natural action of $\mathrm{Sp}(W)$ on $\Pi^+(W_{\mathbf{C}})$ and the action induced on $\mathbf{P}\gamma\mathfrak{S}_n$ by the action ω of $\widetilde{\mathrm{Sp}}(W)$.

PROOF. Suppose we know q_X exists and is unique up to multiples. Then the equivariance for $\mathrm{Sp}(W)$ follows from formulas (19.2.3) or (19.2.5). Furthermore, given those formulas, and the transitivity of $\mathrm{Sp}(W)$ on $\Pi^+(W_{\mathbf{C}})$, it suffices to verify existence and uniqueness of q_X for a single X. Further, given that this $q_X \in \gamma\mathfrak{S}_n$, the bijectivity follows from (20.1.6) and (21.9.12). Thus all we need to do is verify existence and uniqueness for one X. But this is very easy. Note that $\rho(\partial W_{\mathbf{C}})$ is the real linear span of the operators $\partial/\partial x_j$ and ix_j. Consider the subspace $Z \subseteq W$ such that $\rho(\partial Z)$ is spanned by the operators

$$(22.2.3a) \qquad\qquad a_j^+ = -\frac{1}{2}\left(\frac{\partial}{\partial x_j} - 2\pi x_j\right) = -\frac{1}{2}\left(\frac{\partial}{\partial x_j} + 2\pi i(ix_j)\right).$$

Then we have

$$(22.2.3b) \qquad \overline{(a_j^+)} = a_j^- = -\frac{1}{2}\left(\frac{\partial}{\partial x_j} - 2\pi i(ix_j)\right) = -\frac{1}{2}\left(\frac{\partial}{\partial x_j} + 2\pi x_j\right)$$

and we compute that

$$(22.2.3c) \qquad\qquad [a_j^+, \overline{(a_j^+)}] = \pi\delta_{jk}, \qquad [a_j^+, a_k^+] = 0.$$

Comparison of (22.2.3c) with (22.1.1) and (21.7.1) shows that Z is in $\Pi^+(W_{\mathbf{C}})$. On the other hand, the Existence and Uniqueness Theorem for O.D.E., which in this case simply amounts to the Fundamental Theorem of Calculus, since

$$(22.2.4) \qquad\qquad a_j^- = -\frac{1}{2}e^{-\pi x_j^2}\left(\frac{\partial}{\partial x_j}\right)e^{\pi x_j^2}$$

or the calculations of [**Hw**, §1.7], or various other arguments show the unique Schwartz function annihilated by the a_j^- is

$$(22.2.5) \qquad\qquad q_Z(x) = e^{-\pi x^T x}, \qquad x \in \mathbf{R}^n.$$

This concludes the proof of the theorem.

REMARK. For the general element $\gamma_A \in \gamma\mathfrak{S}_n$, the subspace of $\rho(\partial W_{\mathbf{C}})$ annihilating γ_A consists of the operators

$$(22.2.6) \qquad\qquad \partial_u - 2\pi u^T Ax, \qquad u \in \mathbf{R}^n.$$

Here ∂_u indicates the negative of the directional derivative in the direction of u; this is consistent with formula (19.2.1).

22.3. The operators $\rho(\partial W_{\mathbf{C}})$ act on the space $\mathscr{S}^*(\mathbf{R}^n)$ of tempered distributions by duality. Via the usual inner product on $L^2(\mathbf{R}^n)$, we can embed $\mathscr{S}(\mathbf{R}^n)$ in $\mathscr{S}^*(\mathbf{R}^n)$. If we regard the functions q_X of Theorem 22.2 as tempered distributions, then the map $\mathscr{Q}$ of (22.2.2) extends to the compactification $\Pi^+(W_{\mathbf{C}})^-$ of semipositive polarizations of $W_{\mathbf{C}}$.

THEOREM. (a) *For $X \in \Pi^+(W_{\mathbf{C}})^-$, there is a distribution Q_X, unique up to multiples, such that*

$$(22.3.1) \qquad Q_X(\rho(\partial x)f) = 0, \qquad x \in X, \ f \in \mathscr{S}(\mathbf{R}^n).$$

(b) *For $X \in \Pi^+(W_{\mathbf{C}})$, we have*

$$(22.3.2) \qquad Q_X(f) = \int_{\mathbf{R}^n} f(u) q_X(u)\, du, \qquad f \in \mathscr{S}(\mathbf{R}^n).$$

(c) *The map*

$$\mathscr{Q}: \Pi^+(W_{\mathbf{C}})^- \to \mathbf{P}\mathscr{S}^*(\mathbf{R}^n)$$

is injective, continuous, and equivariant for the natural action of $\mathrm{Sp}(W)$ on $\Pi^+(W_{\mathbf{C}})^-$ and the action induced on $\mathbf{P}\mathscr{S}^*(\mathbf{R}^n)$ by the contragredient of the representation ω of $\widetilde{\mathrm{Sp}}(W)$ on $\mathscr{S}(\mathbf{R}^n)$.

REMARK. Again, this result is a slight extension of results in Cartier [Ct].

PROOF. The proof of this result is very much like the proof of Theorem 22.2. We only need to verify the existence and uniqueness of Q_X for one X in each $\mathrm{Sp}(W)$-orbit in $\Pi^+(W_{\mathbf{C}})^-$, and then verify continuity at the boundary (the nonopen orbits).

For $0 \le p \le n = \frac{1}{2}\dim W$, let X_p be the polarization of $W_{\mathbf{C}}$ whose image $\rho(\partial X_p)$ is spanned by the operators

$$(22.3.3) \qquad a_j^+, \quad 1 \le j \le p, \qquad \frac{\partial}{\partial x_j}, \quad p < j \le n.$$

Here the a_j^+ are as in formula (22.2.3). It is a standard elementary lemma in distribution theory that the only distribution Q_0 on $\mathbf{R}^n$ such that

$$(22.3.4a) \qquad Q_0\left(\frac{\partial f}{\partial x_j}\right) = 0, \qquad f \in \mathscr{S}(\mathbf{R}^n), \ 1 \le j \le n$$

is Lebesgue measure.

$$(22.3.4b) \qquad Q_0(f) = \int_{\mathbf{R}^n} f(x)\, dx.$$

It follows from this and formula (22.2.4) that the only distribution Q_p such that

$$Q_p(a_j^+ f) = 0, \qquad 1 \le j \le p, \ f \in \mathscr{S}(\mathbf{R}^n),$$

$$(22.3.5a) \qquad Q_p\left(\frac{\partial f}{\partial x_j}\right) = 0, \qquad p < j \le n,$$

is

$$(22.3.5b) \qquad \exp(-\pi(x_1^2 + x_2^2 + \cdots + x_p^2))\, dx$$

where dx indicates Lebesgue measure. This takes care of existence and uniqueness.

For continuity, we refer to the remark at the end of §22.2. From that remark we can see that if $A \in \varsigma^2(\mathbf{C}^n)$ has positive semidefinite real part, then the operators

(22.3.6a) $$\partial_u + 2\pi u^T A x$$

are the image of $\rho(\partial X_A)$ for $X_A \in \Pi^+(W_{\mathbf{C}})^-$, and that

(22.3.6b) $$Q_{X_A} = e^{-\pi x^T A x} \, dx.$$

This is clearly continuous as a function of A, which parametrizes an open set in $\Pi^+(W_{\mathbf{C}})^-$, by Lemma 21.4. This concludes the proof.

23. Characterization of $\operatorname{im} \omega^{-1}$.

23.1. Let $\operatorname{Sp}_{2n}(\mathbf{C})^+$ denote the set of elements g in $\operatorname{Sp}_{2n}(\mathbf{C})$ such that

(23.1.1) $$g(\Pi^+(W_{\mathbf{C}})) \subseteq \Pi^+(W_{\mathbf{C}}).$$

Let $\operatorname{Sp}_{2n}(\mathbf{C})^{++}$ denote the set of g such that

(23.1.2) $$g(\Pi^+(W_{\mathbf{C}})^-) \subseteq \Pi^+(W_{\mathbf{C}}).$$

It is clear that

(23.1.3(i)) $\operatorname{Sp}_{2n}(\mathbf{C})^+$ and $\operatorname{Sp}_{2n}(\mathbf{C})^{++}$ are subsemigroups of $\operatorname{Sp}_{2n}(\mathbf{C})$.

(23.1.3(ii)) $\operatorname{Sp}_{2n}(\mathbf{C})^{++} \subseteq \operatorname{Sp}_{2n}(\mathbf{C})^+$.

(23.1.3(iii)) $\operatorname{Sp}_{2n}(\mathbf{C})^+$ is closed in $\operatorname{Sp}_{2n}(\mathbf{C})$, and $\operatorname{SP}_{2n}(\mathbf{C})^{++}$ is open.

(23.1.3(iv)) $\operatorname{Sp}_{2n}(\mathbf{R}) \subseteq \operatorname{Sp}_{2n}(\mathbf{C})^+$.

From §12 (see formula 12.4), we have a two-to-one homomorphism ω^{-1} of Ω^0, the normalized oscillator semigroup, into $\operatorname{Sp}_{2n}(\mathbf{C})$. From Theorem 22.3, we know that $\omega^{-1}(\Omega^0) \subseteq \operatorname{Sp}_{2n}(\mathbf{C})^{++}$.

THEOREM. *We have*

(23.1.4) $$\omega^{-1}(\Omega^0) = \operatorname{Sp}_{2n}(\mathbf{C})^{++}.$$

The proof of this equality is spread over the next several subsections.

23.2. Recall the Hermitian form H defined in equation (21.7.1). Set

(23.2.1a) $$W_{\mathbf{C}}^+ = \{w \in W_{\mathbf{C}} : H(w, w) > 0\}.$$

Then

(23.2.1b) $$\overline{W_{\mathbf{C}}^+} = \{w \in W_{\mathbf{C}} : H(w, w) \geq 0\}$$

is the closure of $W_{\mathbf{C}}^+$.

LEMMA. *An element $g \in \operatorname{Sp}(W_{\mathbf{C}})$ belongs to $\operatorname{Sp}(W_{\mathbf{C}})^+$ if and only if $g(W_{\mathbf{C}}^+) \subseteq W_{\mathbf{C}}^+$; and g belongs to $\operatorname{Sp}(W_{\mathbf{C}})^{++}$ if and only if $g(\overline{W_{\mathbf{C}}^+}) \subseteq W_{\mathbf{C}}^+$.*

PROOF. Consider $w \in W_{\mathbf{C}}$. Let $U \subseteq W_{\mathbf{C}}$ be the subspace spanned by w and its complex conjugate $\bar{w}$. The space U is defined over $\mathbf{R}$:

$$U = (U \cap W_{\mathbf{R}}) \otimes \mathbf{C}.$$

If $H(w, w) = i\langle w, \bar{w}\rangle > 0$, then U is nondegenerate for $\langle \, , \, \rangle$, and we can write

$$W_{\mathbf{C}} = U \oplus U^{\perp}.$$

The space $U^{\perp}$ is also defined over $\mathbf{R}$. Let $X_1 \subseteq U$ be a positive polarization. Then if X is the span of w and X_1, we see that X is a positive polarization W.

If $H(w, w) = 0$, then U is isotropic. We can find another isotropic space V, defined over $\mathbf{R}$, such that $U \oplus V$ is nondegenerate. Then if $X \subseteq (U \oplus V)^{\perp}$ is any semipositive polarization, the space $X \oplus U$ will be a semipositive polarization of W.

Thus we have shown

$$W_{\mathbf{C}}^+ \subseteq \bigcup_{X \in \Pi^+(W_{\mathbf{C}})} X, \qquad \overline{W_{\mathbf{C}}^+} \subseteq \bigcup_{X \in \Pi^+(W_{\mathbf{C}})^-} X.$$

On the other hand, it is clear from the definitions that if $X \in \Pi(W_{\mathbf{C}})$, then $X \in \Pi^+(W_{\mathbf{C}})$ if and only if $X - \{0\} \subseteq W_{\mathbf{C}}^+$. And similarly $X \in \Pi^+(W_{\mathbf{C}})^-$ if and only if $X \subseteq \overline{W_{\mathbf{C}}^+}$. Hence we have

$$W_{\mathbf{C}}^+ = \bigcup_{X \in \Pi^+(W_{\mathbf{C}})} X - \{0\}, \qquad \overline{W_{\mathbf{C}}^+} = \bigcup_{X \in \Pi^+(W_{\mathbf{C}})^-} X.$$

From these equalities, and the criteria for membership in $\Pi^+(W_{\mathbf{C}})$ and $\Pi^+(W_{\mathbf{C}})^-$ just given, the lemma follows.

23.3. Let $w \to \bar{w}$ be complex conjugation in $W_{\mathbf{C}}$ (see (21.5.3)). Define an involution $g \to \bar{g}$ on $\mathrm{Sp}(W_{\mathbf{C}})$ by

(23.3.1) $\bar{g}(w) = (g(\bar{w}))^-.$

One checks easily that $g = \bar{g}$ if and only if $g \in \mathrm{Sp}(W)$.

We may use (23.3.1) also to define conjugation on the Lie algebra

$$\mathfrak{sp}(W_{\mathbf{C}}) \simeq \mathfrak{sp}(W) \oplus i\mathfrak{sp}(W).$$

The $(+1)$-eigenspace of $\overline{}$ on $\mathfrak{sp}(W_{\mathbf{C}})$ is $\mathfrak{sp}(W)$, and the (-1)-eigenspace is $i\mathfrak{sp}(W)$, and $\mathfrak{sp}(W_{\mathbf{C}})$ is the sum of these spaces.

Consider the Cayley transform, as given by formula (10.2.1). Of course c takes $\mathfrak{sp}(W)$ to $\mathrm{Sp}(W)$. Thus $g = \bar{g}$ if and only if $c(g) = \overline{c(g)}$ (providing both are defined). More generally,

(23.3.2a) $c(\bar{g}) = \overline{c(g)}.$

Thus we also have

(23.3.2b) $\bar{g} = g^{-1}$ if and only if $c(g) \in i\mathfrak{sp}(W).$

For general $g \in \mathrm{Sp}(W_{\mathbf{C}})$, consider the product

(23.3.3a) $h = h(g) = \bar{g}^{-1}g.$

Check that

(23.3.3b) $\bar{h} = h^{-1}.$

Suppose we can find p such that

(23.3.3c) $$\bar{p} = p^{-1} \quad \text{and} \quad p^2 = h.$$

Set

(23.3.3d) $$u = gp^{-1}.$$

Then we check

$$\bar{u}^{-1}u = 1 \quad \text{or} \quad u = \bar{u}.$$

Hence we have a factorization

(23.3.3e) $$g = up, \quad u \in \mathrm{Sp}(W), \quad c(p) \in i\mathfrak{sp}(W).$$

As a consequence of Lemma 23.2, we see that

(23.3.4) The semigroups $\mathrm{Sp}(W_{\mathbf{C}})^+$ and $\mathrm{Sp}(W_{\mathbf{C}})^{++}$ are invariant under the map $g \to \bar{g}^{-1}$.

23.4. LEMMA. (a) *Consider* $g \in \mathrm{Sp}(W_{\mathbf{C}})^+$. *Let* $w \in \overline{W_{\mathbf{C}}^+}$ *and* $v \notin \overline{W_{\mathbf{C}}^+}$ *be eigenvectors for* g, *with corresponding eigenvalues* λ *and* μ *respectively. Thus*

$$g(w) = \lambda w, \quad gv = \mu v.$$

Then

(23.4.1) $$|\lambda| \geq |\mu|.$$

(b) *Suppose further that* $g \in \mathrm{Sp}(W_{\mathbf{C}})^{++}$. *Then* $w \in W_{\mathbf{C}}^+$, *and* $|\lambda| > |\mu|$.

PROOF. Let U be the plane spanned by w and v. Then g defines a transformation $\mathbf{P}g$ on the projective space $\mathbf{P}U$. (This is essentially the Riemann sphere.) The lines $\mathbf{C}w$ and $\mathbf{C}v$ are the fixed points of $\mathbf{P}g$. If $|\lambda| > |\mu|$, then $\mathbf{C}w$ is an attractive fixed point of $\mathbf{P}g$, and every compact $K \subseteq \mathbf{P}U - \{\mathbf{C}v\}$ is eventually drawn toward (i.e., into an arbitrarily small neighborhood of) $\mathbf{C}w$. If, on the other hand, $|\lambda| < |\mu|$, then $\mathbf{C}v$ is the attractive fixed point. If $g \in \mathrm{Sp}(W_{\mathbf{C}})^+$, obviously v cannot be the attractive fixed point, so (23.4.1) must hold. If $g \in \mathrm{Sp}(W_{\mathbf{C}})^{++}$, then $g(w) = \lambda w \in W_{\mathbf{C}}^+$, so $w \in W_{\mathbf{C}}^+$. Also if $|\lambda| = |\mu|$, then $\mathbf{P}g$ is a measure-preserving transformation for a suitable smooth measure on $\mathbf{P}U$, so that $g(W_{\mathbf{C}}^+ \cap U) \subseteq W_{\mathbf{C}}^+ \cap U$ implies $g(W^+ \cap U) = W_{\mathbf{C}}^+ \cap U$; hence we must have $|\lambda| > |\mu|$ for $g \in \mathrm{Sp}(W_{\mathbf{C}})^{++}$.

23.5. LEMMA. *If* $g \in \mathrm{Sp}(W_{\mathbf{C}})^{++}$, *then we can write*

(23.5.1) $$W_{\mathbf{C}} = X_g \oplus Y_g,$$

where X_g *is the sum of the generalized eigenspaces of* g *for eigenvalues of absolute value greater than* 1, *and* Y_g *is the sum of generalized eigenspaces of* g *of absolute value less than* 1. *Further* $X_g \in \Pi^+(W_{\mathbf{C}})$, *and* $\overline{Y}_g$, *the conjugate of* Y, *also belongs to* $\Pi^+(W_{\mathbf{C}})$.

PROOF. Suppose for now that g is diagonalizable, and that no two eigenvalues of g have the same absolute value. Each eigenvector of g either belongs to $\overline{W_{\mathbf{C}}^+}$, or does not. Let X be the span of the eigenvectors which belong to $\overline{W}_{\mathbf{C}}^+$, and let Y be the span of the remaining eigenvectors. Then certainly $W_{\mathbf{C}} = X \oplus Y$. According to Lemma 23.4, the eigenvalues of g acting on X are all greater in absolute value than the eigenvalues of g acting on Y.

It is well known (and easy to show, by reasoning similar to that used for establishing the basic spectral properties of Hermitian operators) that if λ is an eigenvalue of $g \in \mathrm{Sp}(W_{\mathbf{C}})$, then $1/\lambda$ is also an eigenvalue (and with the same multiplicity). And if V_λ is the generalized λ-eigenspace for g, then V_λ and $V_{1/\lambda}$ are paired nondegenerately by $\langle\ ,\ \rangle$, and V_λ is orthogonal to V_μ if $\mu \neq 1/\lambda$. Thus our g has n eigenvalues of absolute value greater than 1, and n eigenvalues of absolute value less than one. Evidently X will be the sum of the V_λ with $|\lambda| > 1$ if and only if $\dim X = n$; and then Y will be the sum of the V_λ with $|\lambda| < 1$.

Let us show $\dim Y \leq n$. Suppose on the contrary that $\dim Y > n$. Since the Hermitian form H defined in formula (21.7.1) allows isotropic subspaces of dimension n (where $2n = \dim_{\mathbf{C}} W_{\mathbf{C}}$), it has signature (n, n). It follows that any subspace of $W_{\mathbf{C}}$ of dimension larger than n has a nonempty intersection with $W_{\mathbf{C}}^+$. In particular $Y \cap W_{\mathbf{C}}^+ \neq \varnothing$.

Among the eigenvalues for g acting on Y, let λ have the largest absolute value; let v be the eigenvector of g corresponding to λ. Then for the transformation $\mathbf{P}g$ induced by g on the projective space $\mathbf{P}Y$, the line $\mathbf{C}v$ is an attractive fixed point: under the successive powers of $\mathbf{P}g$, every point outside the hyperplane spanned by the other eigenvectors of g will eventually be drawn into an arbitrarily small neighborhood of $\mathbf{C}v$. Since $Y \cap W_{\mathbf{C}}^+$ must be invariant by g, we conclude $v \in \overline{W}_{\mathbf{C}}^+$; and from Lemma 23.4, we get $v \in W_{\mathbf{C}}^+$. But this contradicts the definition of Y. Hence we must have $\dim Y \leq n$.

Applying the same argument to $\bar{g}^{-1}$ (which is also in $\mathrm{Sp}(W_{\mathbf{C}})^{++}$, by (23.3.4)), we conclude $\dim X \leq n$. Hence both have dimension n, and X is the span of the eigenvectors with eigenvalues of absolute value greater than 1, and Y the span of eigenvectors for eigenvalues of absolute value less than 1. It also follows, from our remarks above on the structure of symplectic transformations, that X and Y are isotropic.

Finally, let us show X is a positive polarization. The same argument applied to $\bar{g}^{-1}$ will prove Y is negative, i.e., $\overline{Y}$ is positive. Among the eigenvalues of g acting on X, let μ be the one of smallest absolute value. Let X be the corresponding eigenvector. Then for the transformation $\mathbf{P}g$ induced by g on $\mathbf{P}X$, the line $\mathbf{C}x$ defines a repelling fixed point. Under iteration of $\mathbf{P}g$, an arbitrarily small neighborhood of $\mathbf{C}x$ will fill all of $\mathbf{P}X$, except for the hyperplane spanned by the other eigenvectors. Since a suitably small neighborhood of $\mathbf{C}x$ will belong to $\overline{W}_{\mathbf{C}}^+$, we conclude $X \subseteq \overline{W}_{\mathbf{C}}^+$. Then since X is g-invariant and $g \in \mathrm{Sp}(W_{\mathbf{C}})^{++}$, it follows that $X - \{0\} \subseteq W_{\mathbf{C}}^+$, i.e., $X \in \Pi^+(W_{\mathbf{C}})$.

The lemma is now proved for g with eigenvalues of distinct absolute values. But $\mathrm{Sp}(W_{\mathbf{C}})^{++}$ is an open set in $\mathrm{Sp}(W_{\mathbf{C}})$. Thus given arbitrary $g \in \mathrm{Sp}(W_{\mathbf{C}})^{++}$, we can find arbitrarily small perturbations g' of g such that g' is still in $\mathrm{Sp}(W_{\mathbf{C}})^{++}$, and such that the above argument applies to g'. Taking a limit as g' approaches g implies the lemma for g. Note that Lemma 23.4 prevents g from having any eigenvalues of absolute value 1.

23.6. Consider $g \in \mathrm{Sp}(W_{\mathbf{C}})^{++}$, and suppose $g = \bar{g}^{-1}$. Lemma 23.5 guarantees that the Cayley transform $c(g)$ (see formula (10.2.1)) is defined. According to (23.3.2) we will have $c(g) = a \in i\mathfrak{sp}(W)$.

Consider the space X_g, defined by (23.5.1). This will be invariant by a. Consider the form $H(ax, x')$ on X_g. Using the fact that $a \in i\mathfrak{sp}(W)$ and formula (21.7.1), we compute

$$(23.6.1) \qquad \begin{aligned} H(ax, x') &= i\langle ax, \overline{x'} \rangle = i\langle x, -a\overline{x'} \rangle = i\langle x, -\bar{a}\overline{x'} \rangle \\ &= i\langle x, \overline{ax'} \rangle = H(x, ax'). \end{aligned}$$

Thus on X_g, the operator a is self-adjoint with respect to the Hermitian form H. Since H is positive definite on X_g, we see that a must be diagonalizable, with real eigenvalues. Thus $g = c(a)$ will also have real eigenvalues.

If λ is an eigenvalue of a on X_g, then $(\lambda + 1)/(\lambda - 1)$ is the corresponding eigenvalue of $g = c(a)$. We must have $|(\lambda + 1)/(\lambda - 1)| > 1$. This is equivalent to requiring $\lambda > 0$.

More precisely, we check that

$$(23.6.2) \qquad \begin{aligned} (\lambda + 1)/(\lambda - 1) &> 1 \quad \text{when } \lambda > 1, \\ (\lambda + 1)/(\lambda - 1) &< -1 \quad \text{when } 0 < \lambda < 1. \end{aligned}$$

We leave as an exercise to the reader to verify a converse to this analysis. Suppose λ_i are some positive numbers, none equal to 1. Suppose we have an orthogonal direct sum decomposition

$$W_{\mathbf{C}} = \sum_i (V_{\lambda_i} \oplus \overline{V}_{\lambda_i})$$

where $X = \sum V_{\lambda_i}$ is a positive polarization of $W_{\mathbf{C}}$. Define g by letting V_{λ_i} be the $(\lambda_i + 1)/(\lambda_i - 1)$ eigenspace for g, and $\overline{V}_{\lambda_i}$ the $(\lambda_i - 1)/(\lambda_i + 1)$ eigenspace for g. Then $g = \bar{g}^{-1}$, and $g \in \mathrm{Sp}(W_{\mathbf{C}})^{++}$.

Return to our element $g = \bar{g}^{-1}$ of $\mathrm{Sp}(W_{\mathbf{C}})^{++}$. Suppose g has all eigenvalues positive. Then g has a unique square root with positive eigenvalues; denote it by $g^{1/2}$. The exercise just above guarantees $g^{1/2}$ also belongs to $\mathrm{Sp}(W_{\mathbf{C}})^{++}$.

23.7. Consider now a general element of g of $\mathrm{Sp}(W_{\mathbf{C}})^{++}$. The element $\bar{g}^{-1}g = \tilde{g}$ is again in $\mathrm{Sp}(W_{\mathbf{C}})^{++}$, by (23.3.4). Also $\bar{\tilde{g}} = \tilde{g}^{-1}$, so the analysis of §23.6 applies to $\tilde{g}$. If $w \in W_{\mathbf{C}}^+$, we can compute that

$$(23.7.1) \qquad \begin{aligned} H(\tilde{g}w, w) &= i\langle \tilde{g}w, \bar{w} \rangle = i\langle \bar{g}^{-1}gw, \bar{w} \rangle = i\langle gw, \bar{g}\bar{w} \rangle \\ &= i\langle gw, \overline{gw} \rangle > 0 \end{aligned}$$

since $gw \in W_{\mathbf{C}}^+$ also. Letting w be an eigenvector of $\tilde{g}$, we find $\tilde{g}$ has positive eigenvalues. Thus §23.6 guarantees us a square root $\tilde{g}^{1/2}$ of $\tilde{g}$, such that $(\tilde{g}^{1/2})^- = (\tilde{g}^{1/2})^{-1}$, and $\tilde{g}^{1/2} \in \mathrm{Sp}(W_{\mathbf{C}})^{++}$. If now we apply the computations (23.3.3), we obtain the following result.

PROPOSITION. *Given $g \in \mathrm{Sp}(W_{\mathbf{C}})^{++}$, we have a factorization*

(23.7.2) $$g = up,$$

where $u = \bar{u} \in \mathrm{Sp}(W)$ and $p = \bar{p}^{-1} \in \mathrm{Sp}(W_{\mathbf{C}})^{++}$, and p has positive eigenvalues.

23.8. It is now easy to prove Theorem 23.1. Since $\omega^{-1}(\Omega^0)$ is invariant under multiplication on the right and left by $\mathrm{Sp}(W)$, Proposition 23.7 shows it is enough to check that it contains any $p \in \mathrm{Sp}(W_{\mathbf{C}})^{++}$ satisfying $p = \bar{p}^{-1}$. But using §23.6, we compute directly that $(iJ/2)c(p)$ belongs to $\varsigma^{2+}(W) \subset \mathfrak{S}_{2n}$. Thus, denoting $(iJ/2)c(p) = \mathscr{A}$, we see

$$p = \omega^{-1}(\gamma_{\mathscr{A}}^0),$$

which proves Theorem 23.1.

24. Spectrum and positivity. If $\mathscr{A} \in \mathfrak{S}_{2n}$ is real, i.e., $\mathscr{A} \in \varsigma^{2+}(\mathbf{R}^{2n})$, then the associated element $\gamma_{\mathscr{A}}^0$ of Ω^0 is self-adjoint. Accordingly, it will have a spectrum of real eigenvalues $\{\lambda_j\}$, converging to zero, and a corresponding orthonormal eigenbasis. We will compute the spectrum and eigenbasis. This will in particular allow us to give a criterion for $\gamma_{\mathscr{A}}^0$ to be a positive operator. It will also give an operator-theoretic interpretation to the normalization factor $\det(\mathscr{A} + iJ/2)^{1/2}$.

24.1. The spectrum of $\gamma_{\mathscr{A}}^0$ is clearly invariant under conjugation by $\mathrm{Sp}_{2n}(\mathbf{R})$. Thus, using the arguments of §§13–15, we know that to analyze the spectrum of $\gamma_{\mathscr{A}}^0$, it is enough to treat the one-variable case of

$$\mathscr{A} = \begin{bmatrix} s & 0 \\ 0 & s \end{bmatrix}, \qquad s \in \mathbf{R}^+.$$

According to formulas (13.3), (13.5), the elements

$$\gamma^0 \begin{bmatrix} s & 0 \\ 0 & s \end{bmatrix}$$

are sent via the Weyl transform ρ essentially to the operators $\mathscr{F}_z$ of §5. Hence we will study the operators $\mathscr{F}_z$. According to the notation of §1.2, we have

$$\gamma_1(x) = e^{-\pi x^2}, \qquad x \in \mathbf{R}.$$

Combining formulas (5.1.1), (5.1.3), (5.2.1) and (20.1.1) gives us the relation

(24.1.1) $$\mathscr{F}_z(\gamma_1) = e^{-z/2}\gamma_1.$$

Also formulas (19.2.7) specialize to yield

(24.1.2a) $$\left(\frac{d}{dx} + 2\pi x\right)\mathscr{F}_z = e^{-z}\mathscr{F}_z\left(\frac{d}{dx} + 2\pi x\right),$$

(24.1.2b) $$\left(\frac{d}{dx} - 2\pi x\right)\mathscr{F}_z = e^{z}\mathscr{F}_z\left(\frac{d}{dx} - 2\pi x\right).$$

As explained in [**Hw**, §1.7], the functions

(24.1.3) $$\varphi_m = (\pi^m m!)^{-1/2}2^{1/4}(a^+)^m(\gamma_1), \qquad m = 0, 1, 2, \ldots,$$

form an orthogonal basis for $L^2(\mathbf{R})$. (They are essentially Hermite functions.) Here a^+ is from formula (22.2.3). Using relations (24.1.2b) and (24.1.1) we can easily verify that

(24.1.4) $\qquad J_z(\varphi_m) = e^{-(m+1/2)z}\varphi_m, \qquad m = 0, 1, 2, \ldots$.

Via the relation (13.5), and the identity

$$e^{-z} = \frac{\coth(z/2) - 1}{\coth(z/2) + 1},$$

we can see that the eigenvalues of

$$\gamma^0\begin{bmatrix} s & 0 \\ 0 & s \end{bmatrix}, \qquad s > 1/2,$$

are

$$\left(\frac{2s-1}{2s+1}\right)^{m+1/2} \qquad m \in \mathbf{Z}^+, \ s > \frac{1}{2}.$$

On the other hand, the eigenvalues of $\omega^{-1}(\gamma^0_{\begin{bmatrix} s & 0 \\ 0 & s \end{bmatrix}})$, described in equation (12.4), are easily computed to be

$$\left(\frac{2s-1}{2s+1}\right)^{\pm 1}, \qquad s \in \mathbf{R} - \left\{\pm\frac{1}{2}\right\}.$$

For $s < 1/2$, formula (13.5) no longer applies, but we can still use (13.3). Combining (13.3) with (4.1.2), we can compute that

(24.1.5) $\qquad \rho(\gamma^0\begin{bmatrix} s & 0 \\ 0 & s \end{bmatrix}) = \pm i\omega(-1)\rho(\gamma^0\begin{bmatrix} t & 0 \\ 0 & t \end{bmatrix}), \qquad 0 < s < \frac{1}{2},$

where $t = 1/4s$. We can easily verify that

(24.1.6) $\qquad \omega(-1)\varphi_m = (-1)^m\varphi_m$

with φ_m as in (24.1.3). Since if $t = 1/4s$, we have

$$\frac{2t-1}{2t+1} = -\left(\frac{2s-1}{2s+1}\right).$$

Hence, again in this case, we can say the eigenvalues of

$$\rho(\gamma^0\begin{bmatrix} s & 0 \\ 0 & s \end{bmatrix})$$

are

(24.1.7) $\qquad \lambda_m = \left(\frac{2s-1}{2s+1}\right)^{m+1/2}, \qquad m \in \mathbf{Z}^+.$

24.2. The computations above can be interpreted and generalized as follows. Consider $\gamma^0_{\mathscr{A}} \in \Omega^0$, and let $\omega^{-1}(\gamma^0_{\mathscr{A}}) = g \in \mathrm{Sp}(W_{\mathbf{C}})^{++}$ be the image of $\gamma^0_{\mathscr{A}}$ in the complex symplectic group. According to Lemma 23.5, g stabilizes a positive polarization X_g, and a complementary negative polarization Y_g. From the formula (19.2.5) we can conclude that the quantization map $\mathcal{Q}$ of Theorem 22.3 is in fact equivariant, not simply for $\mathrm{Sp}(W)$, but for the full semigroup $\mathrm{Sp}(W_{\mathbf{C}})^+$. In particular we can conclude that the line $\mathcal{Q}(X_g)$ is invariant under $\rho(\gamma^0_{\mathscr{A}})$.

Since Y_g is isotropic, the algebra generated by ∂Y_g is commutative, isomorphic to the symmetric algebra $\varsigma(\partial Y_g)$ on ∂Y_g. Similar remarks apply to X_g. The line $\mathcal{Q}(X_g)$ is invariant under $\rho(\partial X_g)$, hence under the full algebra $\rho(\varsigma(\partial X_g))$. As explained for example in [**Hw**, §1.7], the operators $\rho(\varsigma(\partial \overline{X}_g))$ applied to $\mathcal{Q}(X_g)$ generate a dense subspace of $L^2(\mathbf{R}^n)$. More precisely, the evaluation map

$$\rho(\varsigma(\partial \overline{X}_g)) \otimes \mathcal{Q}(X_g) \to \mathscr{S}(\mathbf{R}^n),$$
$$u \otimes \varphi \to \rho(u)(\varphi), \qquad u \in \varsigma(\partial \overline{X}_g), \ \varphi \in \mathcal{Q}(X_g),$$

is injective with dense image. Since $X_g \oplus \overline{X}_g = W_{\mathbf{C}} = X_g \oplus Y_g$, we may conclude (by a calculation or by application of the Poincaré-Birkhoff-Witt Theorem) that

$$\rho(\varsigma(\partial Y_g))(\mathcal{Q}(X_g)) = \rho(\varsigma(\overline{X}_g))(\mathcal{Q}(X_g)).$$

Hence the evaluation map

(24.2.1)
$$\rho(\varsigma(\partial Y_g)) \otimes \mathcal{Q}(X_g) \xrightarrow{m} \mathscr{S}(\mathbf{R}^n),$$
$$\partial u \otimes \varphi \xrightarrow{m} \rho(\partial u)\varphi, \qquad u \in \varsigma(Y_g), \ \varphi \in \mathcal{Q}(X_g),$$

is an injection, with dense image. Appealing again to formula (19.2.5), we can make the following conclusion.

PROPOSITION. *The map (24.2.1) is a map of g-modules. Precisely, with u, φ as in (24.2.1), we have*

(24.2.2) $$\rho(\gamma^0_{\mathscr{A}})(m(\partial u \otimes \varphi)) = m(\partial(g(u)) \otimes \rho(\gamma^0_{\mathscr{A}})(\varphi)),$$

where $g = \omega^{-1}(\gamma^0_{\mathscr{A}})$. Also

(24.2.3) $$\rho(\gamma^0_{\mathscr{A}})(\varphi) = (\det g_{|Y_g})^{1/2}\varphi.$$

(*The two possible square roots of $\det g_{|Y}$ are taken by the two possible values of $\rho(\gamma^0_{\mathscr{A}})$.*) *Thus if $u \in \varsigma(Y_g)$ is an eigenvector for g:*

(24.2.4a) $$g(u) = \lambda(u),$$

then

(24.2.4b) $$\rho(\gamma^0_{\mathscr{A}})(\rho(\partial u)(\varphi)) = \lambda(\det g_{|Y_g})^{1/2}\rho(\partial u)(\varphi).$$

In particular, the eigenvalues of $\rho(\gamma^0_{\mathscr{A}})$ are of the form

(24.2.5) $$(\det g_{|Y_g})^{1/2}\lambda_1^{a_1}\lambda_2^{a_2}\cdots\lambda_n^{a_n} = \lambda_1^{a_1+1/2}\cdots\lambda_n^{a_n+1/2}, \qquad a_i \in \mathbf{Z}^+,$$

where the λ_i are the eigenvalues of g acting on Y_g.

PROOF. The formula (24.2.3) is the only part of the proposition that does not follow from the discussion just preceding it; it is implied by formulas (24.1.7) and (20.1.1).

24.3. As a corollary of Proposition 24.2, and the calculations of §24.1, we have

POSITIVITY CRITERION. *If $\mathscr{A} \in \varsigma^{2+}(\mathbf{R}^n) \subseteq \mathfrak{S}_{2n}$, then the appropriate choice of sign for $\rho(\gamma^0_{\mathscr{A}})$ will make $\rho(\gamma^0_{\mathscr{A}})$ a positive operator if and only if $\omega^{-1}(\gamma^0_{\mathscr{A}}) \in \mathrm{Sp}_{2n}(W_{\mathbf{C}})$ has positive eigenvalues. This is equivalent to requiring that the eigenvalues of $\mathscr{A}$ be greater than $1/2$.*

REMARK. Qualitatively this says that if $\rho(\gamma^0_{\mathscr{A}})$ is positive, then $\gamma^0_{\mathscr{A}}$, as a function on W, cannot be too spread out—it must be concentrated near the origin. Since the symbol, in the sense of isotropic symbol, as in [**Hw**, §2], of $\gamma^0_{\mathscr{A}}$ is just $\gamma^0_{\mathscr{A}-1}$, this is the same as requiring the symbol of $\gamma^0_{\mathscr{A}}$ not to be highly localized in W, i.e., phase space. This may be regarded as a reflection of the Uncertainty Principle.

25. Polar decomposition. A standard basic fact about operators on Hilbert space is that they have polar decompositions. An operator T on a Hilbert space can be factored uniquely as

(25.1) $$T = UP,$$

where U is a partial isometry from $(\ker T)^{\perp}$ to $(\mathrm{im}\, T)^{\perp}$, and P is a positive operator. In this section, we wish to observe that the polar decomposition of elements of $\rho(\Omega^0)$ is consistent with the structure of Ω^0, as determined in §23.

By virtue of §23, we may regard the oscillator semigroup as the image of a representation of a twofold cover of the semigroup $\mathrm{Sp}(W_{\mathbf{C}})^{++}$. This leads us to use, when convenient, the following slightly abused notation. Given $\gamma^0_{\mathscr{A}}$ in Ω^0, let $\omega^{-1}(\gamma^0_{\mathscr{A}}) = g \in \mathrm{Sp}(W_{\mathbf{C}})^{++}$ be its image under the 2-fold covering map (12.4). Then we will write

(25.2) $$\rho(\gamma^0_{\mathscr{A}}) = \omega(g).$$

It should be kept in mind that $\omega(g)$ is defined only up to a factor of ± 1.

Recall from §23.3 that the semigroup $\mathrm{Sp}(W_{\mathbf{C}})^{++}$ is stabilized by the (anti-) involution $g \to \bar{g}^{-1}$. Let $T \to T^*$ be the usual adjoint map for operators on Hilbert space. Then comparison of formulas (23.3.2) and (8.3) shows that, in the notation (25.2), we have

(25.3) $$\omega(g)^* = \omega(\bar{g}^{-1}), \qquad g \in \mathrm{Sp}(W_{\mathbf{C}})^{++}.$$

Combining this with the factorization (23.3.3e) gives us the polar decomposition of $\omega(g)$.

PROPOSITION. *Let*

$$g = up, \qquad g \in \mathrm{Sp}(W_{\mathbf{C}})^{++},$$

where $u \in \mathrm{Sp}(W)$, and $p \in \mathrm{Sp}(W_{\mathbf{C}})^{++}$ satisfies $\bar{p}^{-1} = p$ and has positive spectrum, be the factorization (23.3.3e) of g. Then the corresponding factorization

(25.4) $$\omega(g) = \omega(u)\omega(p)$$

is the polar decomposition of $\omega(g)$.

PROOF. Theorem 16.3 says $\omega(u)$ is unitary, and the Positivity Criterion 24.3 says, with appropriate choice of sign, the operator $\omega(p)$ is positive. Since the polar decomposition of an operator is unique, for $\omega(g)$ it must be given by (25.4).

26. Estimates I: The Calderon-Vaillancourt $(0,0)$**-estimate.** The structure of the oscillator semigroup as developed above—most particularly, the form of the group law in the affine oscillator semigroup $A\Omega$, described in §18—has fairly direct applications to L^2-estimates for pseudodifferential and Fourier integral operators. The idea is to use the Cotlar-Stein Lemma [**KS**] to bound the norm of superpositions of elements of $A\Omega$. Because of the rapid decay of products in $A\Omega^0$, as the phase space separation between them increases (precisely, because of the formula (18.3.8a), expressing the size of the product (18.2.3)), this produces L^2-estimates for broad classes of operators. The method seems to be roughly equivalent in power to the techniques of Beals [**Be**], Hörmander [**Hr**], and Unterberger [**U1**]. It does not seem, at the moment, to yield estimates of the type of Fefferman-Phong [**FP1, FP2**]. The method is quite flexible. In particular, it deals with pseudodifferential and Fourier integral operators on more or less the same footing. Loosely speaking, we can say that the general element of Ω^0 is a mixture of Fourier integral and pseudodifferential pieces. This is expressed in the factorization (25.4): the factor $\omega(u)$ is the Fourier integral part of $\omega(g)$, and $\omega(p)$ is the pseudodifferential part. Depending on which part you emphasize, you can get results about either type of operator.

In this and the next three sections, we demonstrate these methods. We do not strive for maximal sharpness or generality, but only try to illustrate how the procedure works. In this section we give a proof, using these ideas, of the standard Calderon-Vaillancourt $(0,0)$-estimate [**CV**]. In the next section we discuss how our superpositions behave under composition—a sort of "symbolic calculus". Then in §§28 and 29 we discuss more general superpositions, yielding "spatially inhomogeneous" pseudodifferential operators in §28, and Fourier integral operators in §29.

26.1. Our basic tool for producing estimates will be the Cotlar-Stein Lemma [**KS**], which we will use in the following form.

LEMMA (COTLAR-STEIN). *Let $T(w)$ be a continuous function on W with values in the bounded operators on Hilbert space $\mathscr{H}$. Then*

(26.1.1)
$$\left\| \int_W T(w)\, dw \right\| \leq$$
$$\left\{ \max_w \left(\int_W \|T^*(u)T(w)\|^{1/2}\, du \right), \max_w \left(\int_W \|T(u)T^*(w)\|^{1/2}\, du \right) \right\}^{1/2}$$

26.2. For $f \in \mathscr{S}(W)$, let

(26.2.1a)
$$L_w(f)(w') = f(w - w')$$

be the (abelian) translation of f by w. Using formulas (18.1.1) we can write

(26.2.1b)
$$L_w(f) = \delta_{w/2} \natural f \natural \delta_{w/2}.$$

It is obvious that the superposition

(26.2.2)
$$\int_W L_w(f)\, dw = \left(\int_W f(w)\, dw \right) 1$$

of translates of f equals the constant function on W with value equal to the integral of f. In particular, taking f equal to

$$\left(\frac{\det \mathscr{A}}{\det(\mathscr{A} + iJ/2)}\right)^{1/2} \gamma_{\mathscr{A}}^{0}$$

for $\mathscr{A} \in \mathfrak{S}_{2n}$, we see from formulas (11.1) and (13.3) that this f has total integral equal to 1. Hence

$$(26.2.3a) \qquad \eta(\mathscr{A}) \int_{W} \delta_{w/2} \natural \gamma_{\mathscr{A}}^{0} \natural \delta_{w/2} \, dw = 1,$$

where we have set

$$(26.2.3b) \qquad \eta(\mathscr{A}) = \left(\frac{\det \mathscr{A}}{\det(\mathscr{A} + iJ/2)}\right)^{1/2}$$

26.3. Consider $\phi \in L^{\infty}(W)$. According to the previous paragraph, we may regard

$$(26.3.1) \qquad T_{\mathscr{A}}(\phi) = \phi * (\eta(\mathscr{A})\gamma_{\mathscr{A}}^{0}) = \eta(\mathscr{A}) \int_{W} \phi(w)\delta_{w/2} \natural \gamma_{\mathscr{A}}^{0} \natural \delta_{w/2} \, dw$$

as a smoothed-out version of ϕ. We want to give an estimate for the norm of $T_{\mathscr{A}}(\phi)$ considered as an operator on $L^{2}(\mathbf{R}^{n})$. Using the Cotlar-Stein Lemma (§26.1), we see we will get an estimate for $\|T_{\mathscr{A}}(\phi)\|_{\mathrm{op}}$, the operator norm of $\rho(T_{\mathscr{A}}(\phi))$, in terms of the maximum over w of integrals of the form

$$|\eta(\mathscr{A})| \int_{W} |\phi(u)\phi(w)|^{1/2} \|\delta_{u/2} \natural \gamma_{\mathscr{A}}^{0} \natural \delta_{u/2} \natural \delta_{-w/2} \natural \gamma_{\mathscr{A}}^{0} \natural \delta_{-w/2}\|_{\mathrm{op}}^{1/2} \, du$$

and the similar one with the order of u and w reversed. Since

$$\|\delta_{u/2} \natural \gamma_{\mathscr{A}}^{0} \natural \delta_{u/2} \natural \delta_{-w/2} \natural \gamma_{\mathscr{A}}^{0} \natural \delta_{-w/2}\|_{\mathrm{op}} = \|\gamma_{\mathscr{A}}^{0} \natural \delta_{(u-w)/2} \natural \gamma_{\mathscr{A}}^{0}\|_{\mathrm{op}}$$

we see that, if *we take $\mathscr{A}$ real*, we can use formula (18.3.8) to conclude

$$\|\gamma_{\mathscr{A}}^{0} \natural \delta_{y} \natural \gamma_{\mathscr{A}}^{0}\|_{\mathrm{op}} \leq e^{-\pi y^{T} \mathscr{C} y/8}$$

where $\mathscr{C}$ is given by formulas (18.3.3), (18.3.5) as

$$(26.3.2) \qquad \mathscr{C} = \mathscr{C}(\mathscr{A}) = \mathscr{T}(\mathscr{A})^{-1} = ((4\mathscr{A})^{-1} - J\mathscr{A}J)^{-1}.$$

Plugging this into the integral above and using formula (1.3.3) again, we find the estimate

$$(26.3.3) \qquad \|T_{\mathscr{A}}(\phi)\|_{\mathrm{op}} \leq 2^{4n} \|\phi\|_{\infty} \eta(\mathscr{A})(\det \mathscr{T}(\mathscr{A}))^{1/2}.$$

REMARK. Suppose $\mathscr{A}$ is very large. Then $\eta(\mathscr{A}) \sim 1$, and $\det \mathscr{T}(\mathscr{A}) \sim \det \mathscr{A}$, so we have

$$\|T_{\mathscr{A}}(\phi)\|_{\mathrm{op}} \leq 2^{4n} \|\phi\|_{\infty} \det \mathscr{A}^{1/2}.$$

On the other hand, γ_{0}^{0} is the constant function $\alpha 2^{-n}$, where α is an eighth root of unity, and we know γ_{0}^{0} is unitary and thus has norm 1. Our estimate gives

$$1 = \|\gamma_{0}^{0}\|_{\mathrm{op}} = \|T_{\mathscr{A}}(\alpha 2^{-n})\|_{\mathrm{op}} \leq 2^{3n} \det \mathscr{A}^{1/2}.$$

So for γ_0^0 our estimate (26.3.3) loses a factor depending on $\mathscr{A}$ and n.

26.4. Estimate (26.3.3) is a sort of fuzzy version of the Calderon-Vaillancourt $(0,0)$-estimate [**CV**]. To turn it into the standard version is purely a matter of abelian harmonic analysis: one shows that if a function has bounded derivatives, it can be approximated efficiently by functions $T_{\mathscr{A}}(\phi)$.

We will work with a fixed but arbitrary positive definite symmetric $n \times n$ matrix A. We will need some auxiliary functions for certain constructions. Let $\varphi \in C_c^\infty(\mathbf{R})$ be a nonnegative function which is monotone decreasing on $\mathbf{R}^+$, and satisfies

$$(26.4.1a) \qquad \varphi(t) = 1, \quad 0 \le t \le 1, \qquad \varphi(t) = 0, \quad t > 2.$$

Define $\varphi_1 \in C_c^\infty(\mathbf{R}^n)$ by

$$(26.4.1b) \qquad \varphi_1(x) = \varphi(x^T A x).$$

For $a \in \mathbf{R}^x$, define dilation operators

$$(26.4.2) \qquad \omega_p(a)f(x) = |a|^{-n/p} f(x/a), \qquad 1 \le p \le \infty, \ f \in \mathscr{S}(\mathbf{R}^n).$$

Note that $\omega_p(a)$ is an isometry on $L^p(\mathbf{R}^n)$. Fix an $a > 1$. We can construct a decomposition of the constant function 1 as follows

$(26.4.3a)$

$$1 = \lim_{m \to \infty} \omega_\infty(a^m)(\varphi_1) = \varphi_1 + \sum_{m > 1} [\omega_\infty(a^m)(\varphi_1) - \omega_\infty(a^{m-1})(\varphi_1)]$$

$$= (\varphi_1 e^{\pi x^T A x})\gamma_A(x)$$

$$\quad + \sum_{m \ge 1} [\omega_\infty(a^m)(\varphi_1) - \omega_\infty(a^{m-1})(\varphi_1)]\omega_\infty(a^{m-1})(e^{\pi x^T A x})\gamma_{a^{-2(m-1)}A}(x)$$

$$= \psi_1 \gamma_A(x) + \sum_{m \ge 0} \omega_\infty(a^m)(\psi_2)\gamma_{a^{-2m}A}(x)$$

where

$$(26.4.3b) \qquad \psi_1 = \varphi_1 e^{\pi x^T A x}, \qquad \psi_2 = (\omega_\infty(a)(\varphi_1) - \varphi_1)e^{\pi x^T A x}.$$

Here and below, we are abusing notation, and using $x^T A x$ to denote the function $x \to x^T A x$.

Consider a tempered distribution $D \in \mathscr{S}^*(\mathbf{R}^n)$. If we multiply D by the above decomposition of the constant function 1, we obtain a decomposition of D:

$$(26.4.4) \qquad D = (\psi_1 D)\gamma_A + \sum_{m \ge 0} (\omega_\infty(a^m)(\psi_2)D)\gamma_{a^{-2m}A}.$$

We work with the Fourier transform

$$(26.4.5) \qquad \hat{f}(y) = \int_{\mathbf{R}^n} f(x)e^{-2\pi i y^T x}\, dx, \qquad f \in \mathscr{S}(\mathbf{R}^n)$$

extended to $\mathscr{S}^*(\mathbf{R}^n)$ by duality. Applying the Fourier transform to equation (26.4.4) yields a decomposition

$(26.4.6)$

$$\hat{D} = (\hat{\psi}_1 * \hat{D}) * \det B^{1/2}\gamma_B + \sum_{m \ge 0} (\omega_1(a^{-m})(\hat{\psi}_2) * \hat{D}) * (a^m \det B^{1/2}\gamma_{a^{2m}B}),$$

where $B = A^{-1}$.

Suppose that $\hat{D}$ is a bounded function, and that some of the derivatives of $\hat{D}$ are also bounded. Let

$$(26.4.7) \qquad \nabla^T A \nabla = \sum_{i,j=1}^n a_{ij} \frac{\partial^2}{\partial x_i \partial x_j} = \Delta_A$$

be the "A-Laplacian" (see formula (4.3.4.4)). Standard formulas for the Fourier transform imply that

$$(26.4.8) \qquad \Delta_A \hat{D} = -4\pi^2 (x^T A x D)\hat{\ }.$$

We may write, for any integer $k \geq 0$,

$$\omega_\infty(a^m)(\psi_2)D = \frac{\omega_\infty(a^m)(\psi_2)}{(x^T A x)^k}((x^T A x)^k D)$$

$$= a^{-2mk} \omega_\infty(a^m) \left(\frac{\psi_2}{(x^T A x)^k} \right)((x^T A x)^k D).$$

Thus if we set

$$(26.4.9) \qquad \psi_{2,k} = \frac{\psi_2}{(x^T A x)^k}$$

we can modify formula (26.4.6) to the decomposition
(26.4.10)

$$\hat{D} = (\hat{\psi}_1 * \hat{D}) * \det B^{1/2} \gamma_B$$
$$+ (-4\pi^2)^{-k} \sum_m a^{-2mk} \omega_1(a^{-m})(\hat{\psi}_{2,k}) * (\Delta_A^k \hat{D}) * (a^m \det B^{1/2} \gamma_{a^{2m}B}).$$

The functions $\omega_1(a^{-m})(\hat{\psi}_{2,k})$ are all in $L^1(\mathbf{R}^n)$, with L^1-norm independent of m. Precisely,

$$(26.4.11a) \qquad \|\omega_1(a^{-m})(\hat{\psi}_{2,k})\|_1 = \|\hat{\psi}_{2,k}\|_1 \leq c^k \|\hat{\psi}_2\|_1,$$

where c is a constant which can be bounded by

$$(26.4.11b) \qquad c \leq \|\hat{\xi}\|_1$$

where ξ is any smooth function which equals $(x^T A x)^{-1}$ on the support of ψ_2. Thus we have the standard estimate

$$(26.4.12) \qquad \|\omega_1(a^{-m})(\hat{\psi}_{2,k}) * (\Delta_A^k \hat{D})\|_\infty \leq \|\hat{\psi}_{2,k}\|_1 \|\Delta_A^k \hat{D}\|_\infty.$$

Thus if $\Delta_A^k \hat{D}$ is bounded for some $k \geq 1$, the series (26.4.6) converges uniformly to $\hat{D}$.

26.5. Now let us replace $\mathbf{R}^n$ by W, B by $\mathscr{B} \in \mathfrak{S}_{2n}$ (but keep $\mathscr{B}$ real), and $\hat{D}$ by a bounded smooth function ϕ on W. Using the notation of (26.3.1), we may rewrite equation (26.4.10) in the form

$$(26.5.1) \quad \phi = T_{\mathscr{B}}(\hat{\psi}_1 * \phi) + (-4\pi^2)^{-k} \sum_{m \geq 0} a^{-2mk} T_{a^{2m}\mathscr{B}}(\omega_1(a^{-m})(\hat{\psi}_{2,k}) * (\Delta_{\mathscr{B}}^k \phi)).$$

If we use estimate (26.3.3) to bound the operator norms of the individual terms in the summation, we find
(26.5.2)
$$\|\phi\|_{op} \leq \|\phi\|_{\infty}[2^{4n}\|\hat{\psi}_1\|_1 \eta(\mathscr{B})(\det \mathscr{T}(\mathscr{B}))^{1/2}]$$
$$+ \|\Delta_{\mathscr{B}}^k \phi\|_{\infty}\left[2^{4n}\|\hat{\psi}_{2,k}\|_1(4\pi)^{-k}\left(\sum_{m \geq 0} a^{-2mk}\eta(a^m\mathscr{B})(\det(\mathscr{T}(a^m\mathscr{B}))^{1/2}\right)\right].$$

For large s, we have
$$\eta(s\mathscr{B}) \sim 1, \qquad (\det \mathscr{T}(s\mathscr{B}))^{1/2} \sim s^n(\det \mathscr{B})^{1/2}.$$

Hence we can estimate

(26.5.3) $$\eta(a^m\mathscr{B})(\det \mathscr{T}(a^m\mathscr{B}))^{1/2} \leq v(\mathscr{B})a^{mn}, \qquad m \geq 0,$$

for a suitable number $v(\mathscr{B})$ depending on $\mathscr{B}$. Plugging this is (26.5.2), we obtain

$$\|\phi\|_{op} \leq \|\phi\|_{\infty}[2^{4n}\|\psi_1\|_1 v(\mathscr{B})]$$
(26.5.4)
$$+ \|\Delta_{\mathscr{B}}^k \phi\|_{\infty}[2^{4n}\|\psi_{2,k}\|_1(4\pi)^{-k}v(\mathscr{B})\sum_{m \geq 0} a^{2m(n-k)}].$$

If $k > n$, the summation is finite. Thus we obtain

THEOREM (CALDERON-VAILLANCOURT). *For ϕ a bounded smooth function on W, and real $\mathscr{B} \in \mathfrak{S}_{2n}$, we have*

(26.5.5) $$\|\phi\|_{op} \leq \beta(\mathscr{B})(\|\phi\|_{\infty} + \|\Delta_{\mathscr{B}}^{n+1}\phi\|_{\infty})$$

for an appropriate number $\beta(\mathscr{B})$.

REMARKS. (a) Study of the argument will show that we do not need $2(n+1)$ derivatives in estimate (26.5.5), but only $2n + \varepsilon$ for any $\varepsilon > 0$.

(b) In this argument, the role smoothness of ϕ plays is simply that it permits ϕ to be approximated closely by the smoothed functions $T_{a^{2m}\mathscr{B}}(\omega_1(a^{-m})(\hat{\psi}_{2,k}) * \Delta_{\mathscr{B}}^k(\phi))$. One can show that, conversely, if ϕ can be closely approximated by functions $T_{\mathscr{A}}(\xi)$ for $\xi \in L^{\infty}(W)$, then ϕ is smooth.

(c) In the sections following, we will consider more general superpositions of elements of $A\Omega$. We will concentrate there on the phenomenology of these superpositions, and ignore the question of what functions on W can be well approximated by them.

27. Symbolic composition.

27.1. In [Hw, §2.1], the isotropic symbol (almost the same as the symbol Weyl defined in [Wy2]) of a distribution $D \in \mathscr{S}^*(W)$ is just its symplectic Fourier transform:

(27.1.1) $$\sigma(D) = \hat{D} = D \natural \gamma_0^0.$$

The "symbolic calculus" or "Weyl calculus" expresses $\sigma(D_1 \natural D_2)$ in terms of $\sigma(D_1)$ and $\sigma(D_2)$. We will denote it by $\mathbb{S}$. It is given by

$$(27.1.2) \qquad \begin{aligned} \sigma(D_1)\mathbb{S}\sigma(D_2) &= \sigma(D_1 \natural D_2) = D_1 \natural D_2 \natural \gamma_0^0 \\ &= D_1 \natural \gamma_0^0 \natural \gamma_0^0 \natural D_2 \natural \gamma_0^0 = \sigma(D_1) \natural \gamma_0^0 \natural \sigma(D_2). \end{aligned}$$

Since the oscillator semigroup is invariant under twisted convolution by γ_0^0, the distinction between an element $\gamma_0^0 \in \Omega^0$ and its symbol is somewhat blurred. However, the superpositions $T_{\mathscr{A}}(\phi)$ of §26.2 look like the kind of distributions which have traditionally been considered as symbols. In this section, we will discuss their symbolic composition.

27.2. Since the $T_{\mathscr{A}}(\phi)$ are defined as superpositions, our basic problem is to compute

$$(\delta_{u/2} \natural \gamma_{\mathscr{A}_1}^0 \natural \delta_{u/2}) \mathbb{S} (\delta_{v/2} \natural \gamma_{\mathscr{A}_2}^0 \natural \delta_{v/2}), \qquad \mathscr{A}_1, \mathscr{A}_2 \in \mathfrak{S}_{2n};\ u, v \in W.$$

By formulas (27.1.2), (11.2), and (19.1.3), this product is equal to

$$\delta_{u/2} \natural \gamma_{\mathscr{A}_1'}^0 \natural \delta_{-u/2} \natural \delta_{v/2} \natural \gamma_{\mathscr{A}_2}^0 \natural \delta_{v/2}$$

$$= e^{\pi i (u^T J v)/4} \delta_{u/2} \natural \gamma_{\mathscr{A}_1'}^0 \natural \delta_{(v-u)/2} \natural \gamma_{\mathscr{A}_2}^0 \natural \delta_{v/2},$$

where $\mathscr{A}_1' = -(J\mathscr{A}_1 J)/4$. This may be computed explicitly by means of the formulas of §18. (Here, as there, we take $\mathscr{A}_1$, $\mathscr{A}_2$ real.) The result is

$$(27.2.1a) \qquad (\delta_{u/2} \natural \gamma_{\mathscr{A}_1}^0 \natural \delta_{u/2}) \mathbb{S} (\delta_{v/2} \natural \gamma_{\mathscr{A}_2}^0 \natural \delta_{v/2}) = v \delta_c \natural \gamma_{\mathscr{A}_3'}^0 \natural \delta_d,$$

where

(27.2.1b) $\mathscr{A}_3' = -(J\mathscr{A}_3/J)/4$, with $\mathscr{A}_3$ as in formula (8.2),

$$(27.2.1c) \qquad c = \left(\frac{u+v}{4}\right) - (\mathscr{T}(\mathscr{A}_1) - \mathscr{T}(\mathscr{A}_2) + 4J\mathscr{A}_1 J)\mathscr{C}\left(\frac{u-v}{4}\right),$$

$$(27.2.1d) \qquad d = \left(\frac{u+v}{4}\right) - (\mathscr{T}(\mathscr{A}_1) - \mathscr{T}(\mathscr{A}_2) - 4J\mathscr{A}_2 J)\mathscr{C}\left(\frac{u-v}{4}\right).$$

Here $\mathscr{T}(\mathscr{A}_i)$ is given in formula (18.3.3), and $\mathscr{C} = \mathscr{C}(\mathscr{A}_1, \mathscr{A}_2)$ is from formula (18.3.5). We note that

$$\mathscr{T}(-(J\mathscr{A}_1^{-1}J)/4) = \mathscr{T}(\mathscr{A}_1).$$

Also

(27.2.1e) $\qquad |v| = \exp(-\pi(u-v)^T \mathscr{C}(u-v)/4),$

(27.2.1f) $\qquad \dfrac{v}{|v|} = \exp(\pi i(u^T \mathscr{A}_1 + v^T \mathscr{A}_2)J\mathscr{C}(v-u)/2).$

We will record some specializations of these formulas. First, suppose

$$(27.2.2a) \qquad \mathscr{A}_1 = -s\frac{J\mathscr{J}}{2}, \qquad \mathscr{A}_2 = -t\frac{J\mathscr{J}}{2},$$

where $\mathscr{J}$ is a compatible complex structure (see formula (18.4.2), or §21.6). Then the formulas (27.2.1) reduce to

$$(27.2.2\text{b}) \qquad \mathscr{A}_3' = -\left(\frac{s+t}{1+st}\right)\frac{J\mathscr{J}}{2},$$

$$(27.2.2\text{c}) \qquad c = \frac{u+v}{4} - \left(\frac{(s-t)(1-st)+4s^2t}{(s+t)(1+st)}\right)\left(\frac{v-u}{4}\right),$$

$$(27.2.2\text{d}) \qquad d = \frac{u+v}{4} - \left(\frac{(s-t)(1-st)-4s^2t}{(s+t)(1+st)}\right)\left(\frac{v-u}{4}\right),$$

(27.2.2e)

$$|v| = \exp(\pi(u-v)^T rJ\mathscr{J}(u-v)), \quad \text{where } r = \frac{st}{2(s+t)(1+st)},$$

$$(27.2.2\text{f}) \qquad \frac{v}{|v|} = \exp(\pi i(s^{-1}u^T + tv^T)rJ(u-v)).$$

For a second special case, take

$$(27.2.3\text{a}) \qquad \mathscr{A}_1 = \mathscr{A}_2 = \mathscr{A}.$$

Then the formulas (27.2.1) become

$$(27.2.3\text{b}) \qquad \tilde{\mathscr{A}_3} = \mathscr{C}(\mathscr{A},\mathscr{A}) = \frac{1}{2}\mathscr{T}(\mathscr{A})^{-1},$$

$$(27.2.3\text{c}) \qquad \left(\frac{u+v}{4}\right) + 2J\mathscr{A}\,J\mathscr{T}(\mathscr{A})^{-1}\left(\frac{v-u}{4}\right),$$

$$(27.2.3\text{d}) \qquad \left(\frac{u+v}{4}\right) - 2J\mathscr{A}\,J\mathscr{T}(\mathscr{A})^{-1}\left(\frac{v-u}{4}\right),$$

$$(27.2.3\text{e}) \qquad |v| = \exp(-\pi(u-v)^T\mathscr{T}(\mathscr{A})^{-1}(u-v)/8),$$

$$(27.2.3\text{f}) \qquad \frac{v}{|v|} = \exp(\pi i(u+v)^T J\mathscr{T}(\mathscr{A})^{-1}(v-u)/4),$$

27.3. Consider two functions ϕ_1, ϕ_2 in $L^\infty(W)$. Using formulas (27.2.1) and (26.3.1) we can compute the symbolic product
(27.3.1)
$$T_{\mathscr{A}_1}(\phi_1)\circledS T_{\mathscr{A}_2}(\phi_2)$$

$$= \eta(\mathscr{A}_1)\eta(\mathscr{A}_2)\int_{W\times W}\phi_1(u)\phi_2(v)(\delta_{u/2}\natural\gamma^0_{\mathscr{A}_1}\natural\delta_{u/2})\circledS(\delta_{v/2}\natural\gamma^0_{\mathscr{A}_2}\natural\delta_{v/2})\,du\,dv$$

$$= \eta(\mathscr{A}_1)\eta(\mathscr{A}_2)\int_{W\times W}\phi_1(u)\phi_2(v)\exp(-\pi(u-v)^T\mathscr{C}(u-v)/4)$$

$$\times e^{\pi i\alpha}(\delta_c\natural\gamma^0_{\mathscr{A}_3'}\natural\delta_d)\,du\,dv$$

where c, d are as in (27.2.1c and d), and $e^{\pi i\alpha} = v/|v|$ is as in (27.2.1f). Set

$$y = (u+v)/2, \qquad z = u-v,$$

$$-(\mathscr{T}(\mathscr{A}_1) - \mathscr{T}(\mathscr{A}_2) + 4J\mathscr{A}_1 J)\mathscr{C}/4 = \Lambda_1,$$

$$-(\mathscr{T}(\mathscr{A}_1) - \mathscr{T}(\mathscr{A}_2) - 4J\mathscr{A}_1 J)\mathscr{C}/4 = \Lambda_2.$$

Referring to formulas (27.2.1c,d), we see that this notation permits us to rewrite the integral just above as follows:

(27.3.2)
$$T_{\mathscr{A}_1}(\phi_1) \circledS T_{\mathscr{A}_2}(\phi_2) = \eta(\mathscr{A}_1)\eta(\mathscr{A}_2)$$

$$\times \int_{W\times W} \phi_1(y+z/2)\phi_2(y-z/2)e^{-\pi z^T \mathscr{C} z/4}e^{\pi i \alpha'}\delta_{\Lambda_1(z)}\natural(\delta_{y/2}\natural\gamma^0_{\mathscr{A}_3'}\natural\delta_{y/2})\natural\delta_{\Lambda_2(z)}\,dy\,dz$$

$$= \frac{\eta(\mathscr{A}_1)\eta(\mathscr{A}_2)}{\eta(\mathscr{A}_3')}\int_W e^{-\pi z^T \mathscr{C} z/4}\delta_{\Lambda_1(z)}\natural T_{\mathscr{A}_3'}(e^{\pi i \alpha'}L_{-z/2}(\phi_1)L_{z/2}(\phi_2))\natural\delta_{\Lambda_2(z)}\,dz.$$

Here L_z denotes translation in W by z (cf. formula (26.2.1)), and

$$e^{\pi i \alpha'} = e^{\pi i \alpha}\exp(-\pi i y^T J(\Lambda_1(z) - \Lambda_2(z))/2) = e^{\pi i \alpha}\exp(-\pi i y^T(\mathscr{A}_1 + \mathscr{A}_2)J\mathscr{C} z)$$

REMARKS. (a) Formula (27.3.2) exhibits the symbolic composition of two $T_{\mathscr{A}_i}(\phi_i)$ as a rapidly decreasing superposition of (left and right) twisted translates of other $T_{\mathscr{A}}(\phi)$'s. Thus the space of rapidly decreasing superpositions of twisted translates of the $T_{\mathscr{A}}(\phi)$'s will be an algebra with a reasonably explicit (symbolic) product.

(b) When $z = 0$, the function $e^{\pi i \alpha'}$ is identically 1 as a function of y. Thus, modulo the factor $\eta(\mathscr{A}_3')$ (which is near 1 if $\mathscr{A}_3'$ is large), formula (27.3.2) exhibits $T_{\mathscr{A}_1}(\phi_1)\circledS T_{\mathscr{A}_2}(\phi_2)$ as a perturbation of the pointwise product $\phi_1\phi_2$. Particularly, if ϕ_1 and ϕ_2 are slowly varying, then because $e^{-\pi z^T C z/4}$ dies off rapidly in z, the difference $\phi_1\phi_2 - \eta(\mathscr{A}_3')T_{\mathscr{A}_1}(\phi_1)\circledS T_{\mathscr{A}_2}(\phi_2)$ will be small.

(c) It is useful to consider the case when, say, ϕ_2 is the Dirac δ at some point $w \in W$. Then (27.3.2) reduces to

(27.3.3)
$$T_{\mathscr{A}_1}(\phi_1)\circledS(\delta_{w/2}\natural\gamma^0_{\mathscr{A}_2}\natural\delta_{w/2})$$

$$= \eta(\mathscr{A}_1)\int_W e^{-\pi z^T \mathscr{C} z/4}e^{\pi i \alpha'}\phi_1(w+z)\delta_{\Lambda_1(z)}\natural(\delta_{w+z/2}\natural\gamma^0_{\mathscr{A}_3'}\natural\delta_{w+z/2})\natural\delta_{\Lambda_2(z)}\,dz.$$

One sees from this that $T_{\mathscr{A}_1}(\phi)\circledS(\delta_{w/2}\natural\gamma^0_{\mathscr{A}_2}\natural\delta_{w/2})$ is strongly localized near w. More precisely, we can observe that if ϕ_1 is constant, then $T_{\mathscr{A}}(\phi_1) = (2i)^n\phi_1\gamma^0_0$, and

$$((2i)^n\phi_1\gamma^0_0)\circledS(\delta_{w/2}\natural\gamma^0_{\mathscr{A}_2}\natural\delta_{w/2}) = (2i)^n\phi_1(\delta_{w/2}\natural\gamma^0_{\mathscr{A}_2}\natural\delta_{w/2}).$$

Therefore we may think of $T_{\mathscr{A}_1}(\phi_1)\circledS(\delta_{w/2}\natural\gamma^0_{\mathscr{A}_2}\natural\delta_{w/2})$ as a perturbation of $(2i)^n\phi_1(w)(\delta_{w/2}\natural\gamma^0_{\mathscr{A}_2}\natural\delta_{w/2})$. The more slowly ϕ varies, the smaller the perturbation.

28. Estimates II: Symbols of Beals-Hörmander-Unterberger (ρ, ρ) type.
It is clear that the operators $T_{\mathscr{A}}(\phi)$ of formula (26.3.1) are a very restricted class of superpositions of elements of $A\Omega$. More general superpositions can be handled by essentially the same techniques, and yield correspondingly more general estimates. In this section and the next we will consider as examples some classes of superpositions that produce symbols of types considered in the recent literature. Here we consider superpositions that will permit synthesis of symbols of the type considered by Beals [Be] and later by Hörmander

[**Hr**] and Unterberger [**U1, U2**], which in turn generalize symbols of Calderon-Vaillancourt (ρ, ρ) type [**CV**].

28.1. We study superpositions

$$(28.1.1) \qquad T(\phi, \mathscr{A}) = \int_W \phi(w)(\delta_{w/2} \natural \gamma^0_{\mathscr{A}(w)} \natural \delta_{w/2}) \, dw, \qquad \phi \in L^\infty(W),$$

which generalize (26.3.1) by allowing $\mathscr{A}$ to depend on w. We continue to require that $\mathscr{A}(w)$ be real. By means of formula (18.3.8) and the Cotlar-Stein Lemma (§26.1), we see we can estimate

$$(28.1.2a) \qquad \begin{aligned} \|T(\phi, \mathscr{A})\|_{\mathrm{op}} &\leq \max_u \int_W |\phi(u)\phi(w)| e^{-\pi(u-w)^T \mathscr{C}(u-w)/8} \, dw \\ &\leq \|\phi\|_\infty \max_u \int_W e^{-\pi(u-w)^T \mathscr{C}(u-w)/8} \, dw, \end{aligned}$$

where

$$(28.1.2b) \qquad \mathscr{C} = \mathscr{C}(u,w) = \mathscr{C}(\mathscr{A}(u), \mathscr{A}(w)) = (\mathscr{T}(\mathscr{A}(u)) + \mathscr{T}(\mathscr{A}(w))^{-1}.$$

(See formulas (18.3.3), (18.3.5).)

As long as $(u - w)^T \mathscr{C}(u,w)(u - w)$ grows reasonably fast as $w \to \infty$, the integral in (28.1.2a) will yield a finite bound for $\|T(\phi, \mathscr{A})\|_{\mathrm{op}}$. Suppose we have an estimate like

$$(28.1.3)$$
$$y^T \mathscr{C}(u,w)y \geq M((u - w)^T \mathscr{T}(\mathscr{A}(u))^{-1}(u - w) + 1)^{-\alpha}(y^T \mathscr{T}(\mathscr{A}(u))^{-1}y)$$

where M is a positive constant, and $0 \leq \alpha < 1$. Then the final integral in (28.1.2) can be estimated as follows. In the calculation below, we use the notation

$$(28.1.4) \qquad\qquad e^x = \exp(x).$$

Now we estimate

$$\begin{aligned} \int_W &\exp(-\pi(u - w)^T \mathscr{C}(u,w)(u - w)/8) \, dw \\ &\leq \int_W \exp(-\pi M((u - w)^T \mathscr{T}(\mathscr{A}(u))^{-1}(u - w) + 1)^{-\alpha} \\ &\quad \times ((u - w)^T \mathscr{T}(\mathscr{A}(u))^{-1}(u - w)/8) \, dw \\ &= \det(\mathscr{T}(\mathscr{A}(u)))^{1/2} \int_W \exp(-\pi M(\|w\|^2 + 1)^{-\alpha}\|w\|^2/8) \, dw \\ &= \beta(M, \alpha) \det(\mathscr{T}(\mathscr{A}(u)))^{1/2}. \end{aligned}$$

Here $\|w\|$ is the standard Euclidean norm on $W \simeq \mathbf{R}^{2n}$, and $\beta(M, \alpha)$ is given by

$$(28.1.5) \qquad \begin{aligned} \beta(M, \alpha) &= \int_W \exp(-\pi M(\|w\|^2 + 1)^{-\alpha}\|w\|^2/8) \, dw \\ &= \frac{(2\pi)^n}{\Gamma(n)} \int_0^\infty \exp(-\pi M(t^2/8(t^2 + 1)^\alpha))t^{2n-1} \, dt. \end{aligned}$$

Hence under assumption (28.1.3), we get an estimate

$$(28.1.6) \qquad \|T(\phi, \mathscr{A})\|_{\mathrm{op}} \leq \beta(M, \alpha)\|\phi\|_\infty \max_u \det(\mathscr{T}(\mathscr{A}(u)))^{1/2}.$$

28.2. As an example of a function $\mathscr{A}(w)$ which satisfies condition (28.1.3), we offer

(28.2.1a)
$$\mathscr{A}(w) = \frac{s}{2} \begin{bmatrix} (\|x\|^2 + 1)^\alpha I & 0 \\ 0 & (\|x^2\| + 1)^{-\alpha} I \end{bmatrix}.$$

In this formula we have written $w = \begin{bmatrix} x \\ y \end{bmatrix}$, $x, y \in \mathbf{R}^n$, and $\|x\|$ is the usual Euclidean norm of x. Also s is a positive number, and I is the $n \times n$ identity matrix.

With this choice for $\mathscr{A}(w)$ we have

(28.2.1b)
$$\mathscr{T}(w) = \frac{s + s^{-1}}{2} \begin{bmatrix} (\|x\|^2 + 1)^{-\alpha} I & 0 \\ 0 & (\|x\|^2 + 1)^\alpha I \end{bmatrix}$$

and

(28.2.1c) $\mathscr{C}(w, w') =$
$$\frac{2}{s + s^{-1}} \begin{bmatrix} ((\|x\|^2 + 1)^{-\alpha} + (\|x'\|^2 + 1)^{-\alpha})^{-1} I & 0 \\ 0 & ((\|x^2\| + 1)^\alpha + (\|x'\|^2 + 1)^\alpha)^{-1} I \end{bmatrix}.$$

This choice of $\mathscr{A}(w)$ satisfies condition (28.1.3) with $M = (1 + s + s^{-1})^{-1}$, and the same α. To see this, set $\|x\| = t$, $\|x'\| = t'$. Then condition (28.1.3) follows from the inequalities

(28.2.2a) $\quad ((t^2 + 1)^{-\alpha} + (t'^2 + 1)^{-\alpha})^{-1}$
$$\geq \frac{1}{(1 + s + s^{-1})} (1 + 2(t - t')^2 (t^2 + 1)^\alpha / (s + s^{-1}))^{-\alpha} (t^2 + 1)^\alpha,$$

(28.2.2b) $\quad ((t^2 + 1)^\alpha + (t'^2 + 1)^\alpha)^{-1}$
$$\geq \frac{1}{(1 + s + s^{-1})} (1 + 2(t - t')^2 (t^2 + 1)^\alpha / (s - s^{-1}))^{-\alpha} (t^2 + 1)^{-\alpha}.$$

Taking reciprocals of each side converts the first inequality to

$$(t^2 + 1)^{-\alpha} + (t'^2 + 1)^{-\alpha} \leq (1 + s + s^{-1})(1 + 2(t - t')^2 (t^2 + 1)^\alpha / (s + s^{-1}))^\alpha (t^2 + 1)^{-\alpha}.$$

Since $1 + 2(t - t')^2 (t^2 + 1)^\alpha / (s + s^{-1}) \geq 1$, this inequality is implied by

$$(t'^2 + 1)^{-\alpha} \leq (s + s^{-1})(1 + 2(t - t')^2 (t^2 + 1)^\alpha / (s + s^{-1}))^\alpha (t^2 + 1)^{-\alpha}.$$

Raising this to the α^{-1} power gives

$$(t'^2 + 1)^{-1} \leq (s + s^{-1})^{1/\alpha} (1 + 2(t - t')^2 (t^2 + 1)^\alpha / (s + s^{-1}))(t^2 + 1)^{-1}.$$

Since $\alpha < 1$, and $s + s^{-1} \geq 2 > 1$, this follows from

$$(t'^2 + 1)^{-1} \leq (s + s^{-1})(1 + 2(t - t')^2 (t^2 + 1)^\alpha / (s + s^{-1}))(t^2 + 1)^{-1}.$$

In turn this inequality is equivalent to

$$1 + t^2 \leq (s + s^{-1})(1 + 2(t - t')^2 (1 + t^2)^\alpha / (s + s^{-1}))(1 + t'^2).$$

Since $s + s^{-1} \geq 2$, this follows from

$$1 + t^2 \leq 2(1 + t'^2 + (t - t')^2 (1 + t^2)^\alpha (1 + t'^2)).$$

This is implied by

$$1 + t^2 \leq 2(1 + t'^2 + (t - t')^2)$$

which is true. Hence inequality (28.2.2a) holds. The procedure to establish (28.2.2b) is very similar.

28.3. REMARKS. (a) The matrices $\mathscr{A}(w)$ of equation (28.2.1) are all conjugate to the identity matrix times $s/2$. In particular $\det(\mathscr{A}(w)) = (s/2)^{2n}$, independent of w.

(b) If one looks at the behavior under differentiation of the functions $e^{-\pi w^T \mathscr{A}(u)w}$, one sees that the superpositions $T(\phi, \mathscr{A})$, with $\mathscr{A}(w)$ as in (28.2.1), will satisfy estimates

$$(28.2.3) \qquad \left\| \frac{\partial^{|m|}}{\partial y^m} \frac{\partial^{|m'|}}{\partial x^{m'}} T(\phi, \mathscr{A}) \right\|_\infty \leq C_{m,m'}(1 + \|x\|^2)^{(|m'| - |m|)\alpha/2}$$

for multi-indices α, α'. These are the estimates for symbols of type (ρ, ρ), with $\rho = \alpha$, in the sense of Calderon-Vaillancourt [CV]. An approximation procedure of the sort described in §26.4 would turn estimates (28.1.6), for $\mathscr{A}(w)$ given by (28.2.1), into the C-V (ρ, ρ)-estimates, $0 \leq \rho < 1$.

(c) Since the $\mathscr{A}(w)$ are positive definite real matrices, they may be regarded as defining a Riemannian metric on W. This point of view is espoused by Hörmander [Hr] and Unterberger [U2]. From this point of view, condition (28.1.3) is a requirement that the metric $\mathscr{A}(w)$ changes slowly, with respect to distance as measured from a given point w_0 with respect to $\mathscr{A}(w_0)$. This is the sort of condition imposed by Beals, Hörmander, and Unterberger.

(d) Condition (28.1.3) has both a local and a global aspect. Locally, it requires $\mathscr{A}(w)$ not to change too fast. This permits approximation schemes such as described in §26.3. Globally, it guarantees that the "spheres of influence", the sets $\{w + y : y^T \mathscr{A}(w)y \leq 1\}$, recede to ∞ as w does. This prevents contributions from the elements $\delta_{w/2} \natural \gamma^0_{\mathscr{A}(w)} \natural \delta_{w/2}$, of which $T(\phi, \mathscr{A})$ is a superposition, from piling up in one place and creating an unbounded operator.

29. Fourier integral operators. So far we have only considered superpositions of (twisted translates of) $\gamma^0_{\mathscr{A}}$'s with $\mathscr{A}$ real. Essentially as a consequence, the operators we have constructed are of the type usually called pseudodifferential. If we allow $\mathscr{A}$ to have imaginary part, then the operators will resemble Fourier integral operators. (Indeed, elements of the metaplectic group are the "constant coefficient" Fourier integral operators—according to formula (19.1.3), they induce the linear "canonical transformations" of W.)

29.1. In dealing with syntheses involving $\gamma^0_{\mathscr{A}}$ for $\mathscr{A}$ nonreal, we will use the polar decomposition described in §25. We will express elements of the metaplectic group directly in terms of the corresponding element of the symplectic group, rather than in terms of the symmetric matrix obtained from it by the Cayley transform. Thus we write elements of Ω^0 in the form

$$\omega(g)\gamma^0_{\mathscr{A}}$$

where g belongs to the metaplectic group, and $\mathscr{A}$ is real.

We consider integrals of the form

$$(29.1.1) \quad T(\phi, \mathscr{A}, g_1, g_2) = \int_W \phi(w)\omega(g_1(w))\natural(\delta_{w/2}\natural\gamma^0_{\mathscr{A}(w)}\natural\delta_{w/2})\natural\omega(g_2(w))\, dw.$$

To estimate this using Cotlar-Stein, we need to know

$$\|\omega(g_1(w))\natural(\delta_{w/2}\natural\gamma^0_{\mathscr{A}(w)}\natural\delta_{w/2})\natural\omega(g_2(w))\natural\omega(g_2(u))^{-1}$$

$$\natural(\delta_{-u/2}\natural\gamma^0_{\mathscr{A}(u)}\natural\delta_{-u/2})\natural\omega(g_1(u))^{-1}\|_{\mathrm{op}}$$

$$(29.1.2a) \qquad = \|\gamma^0_{\mathscr{A}(w)}\natural\delta_{w/2}\natural\omega(g_2(w))\natural\omega(g_2(u))^{-1}\natural\delta_{-u/2}\natural\gamma^0_{\mathscr{A}(u)}\|_{\mathrm{op}}$$

$$= \|\gamma^0_{\mathscr{A}_2(w)}\natural\delta_{(w''-u'')/2}\natural\gamma^0_{\mathscr{A}_2(u)}\|$$

$$\le \exp(-\pi(w''-u'')^T\mathscr{C}(\mathscr{A}_2(w), \mathscr{A}_2(u))(w''-u'')/8),$$

where

$$(29.1.2b) \qquad\qquad \mathscr{A}_2(w) = g_2(w)^T\mathscr{A}(w)g_2(w)$$

and

$$(29.1.2c) \qquad\qquad w'' = g_2(w)^{-1}(w).$$

We also need to know the norm of the product (29.1.2a) with the factors involving u preceding the factors involving w. The same procedure gives a bound

$$(29.1.3a) \qquad\qquad \exp(-\pi(w'-u')^T\mathscr{C}(\mathscr{A}_1(w), \mathscr{A}_1(u))(w'-u')/4),$$

where

$$(29.1.3b) \qquad\qquad \mathscr{A}_1(w) = (g_1(w)^{-1})^T\mathscr{A}(w)g_1(w)^{-1},$$

$$(29.1.3c) \qquad\qquad w' = g_1(w)(w).$$

For purposes of estimation we would like to know that the quantity

$$(w'-u')^T\mathscr{C}(\mathscr{A}_1(w), \mathscr{A}_1(u))(w'-u')$$

grows reasonably rapidly as w recedes from u; and likewise for $\mathscr{A}_2$ and w''. For example, we could suppose that

(29.1.4a)
Each of the $\mathscr{A}_i(w)$ satisfies condition (29.1.3).

(29.1.4b)
$$1 + (w'-u')^T\mathscr{T}(\mathscr{A}_1(u))^{-1}(w'-u')$$

$$\ge \mu(1 + (w-u)^T\mathscr{T}(\mathscr{A}_1(u))^{-1}(w-u))^\beta, \quad \mu, \beta > 0 \quad \text{and similarly for}$$

$$w'' \text{ and } \mathscr{A}_2(u).$$

By an argument analogous to that given for estimate (28.1.6), we could conclude if $\mathscr{A}$ and the g_i satisfy conditions (29.1.4), then the operator defined by $T(\phi, \mathscr{A}, g_1, g_2)$ is bounded by $\|\phi\|_\infty$ times a factor depending on the parameters (M, α, μ, β) in conditions (29.1.4), times $\sup_u \det(\mathscr{T}(\mathscr{A}_i(u)))^{1/2}$.

29.2. It is perhaps not immediately transparent what sorts of functions $\mathscr{A}$ and g_i satisfy conditions (29.1.4). In fact, the standard Fourier integral operators corresponding to homogeneous canonical transformations will satisfy much more stringent conditions than (29.1.4). We will discuss the phase space version of these transformations.

Suppose $F: W - \{0\} \to W - \{0\}$ is a symplectic diffeomorphism which commutes with dilations:

$$(29.2.1) \qquad\qquad F(\lambda w) = \lambda F(w), \qquad \lambda \in \mathbf{R}^x.$$

Then the derivative

$$(29.2.2) \qquad\qquad DF: W \to \mathrm{Sp}(W), \qquad w \to DF_w$$

is invariant under dilations:

$$DF_{\lambda w} = DF_w,$$

and F may be recaptured from DF by the simple relation

$$(29.2.3) \qquad\qquad F(w) = DF_w(w).$$

By virtue of the fact that F is a diffeomorphism and is homogeneous, we obviously have

$$(29.2.4) \qquad\qquad \|DF_w(w) - DF_u(u)\| = \|F(w) - F(u)\| \geq \mu\|w - u\|$$

for a suitable $\mu > 0$. Here $\| \ \|$ again indicates the Euclidean norm on $W \simeq \mathbf{R}^{2n}$.

To the mapping F, and any real $\mathscr{A} \in \mathfrak{S}_{2n}$, we associate the operator

$$(29.2.5) \quad T_{F,\mathscr{A}} = T_F = T(1, \mathscr{A}, F, 1) = \eta(\mathscr{A}) \int_W \omega(DF_w) \natural (\delta_{w/2} \natural \gamma^0_{\mathscr{A}} \natural \delta_{w/2})\, dw.$$

Here we abuse notation, and indicate by DF_w one of its two inverse images in the metaplectic group. We insist $\omega(DF_w)$ vary continuously with w. Thus T_F, like the $\gamma^0_{\mathscr{A}}$, is only defined up to sign. Clearly conditions (29.1.4) apply to T_F; hence T_F is bounded. If $F = g \in \mathrm{Sp}(W)$, one can check $T_F = \omega(g) \natural \gamma^0_0 = \omega(g)\omega(-1)$.

29.3. The idea behind Fourier integral operators, of course, is that they are supposed to approximately realize symplectic diffeomorphisms on the level of symbolic composition of operators. More precisely, under symbolic composition they are supposed to be unitary modulo pseudodifferential and compact operators, and conjugation by them is supposed to transform symbols approximately according to a canonical transformation. We will check these properties for the operators T_F defined just above. Similar messier computations will hold for the operators $T(\phi, \mathscr{A}, g_1, g_2)$ of equation (29.1.1).

If T_F is regarded as a symbol, then the operator of which it is a symbol is $T_F \natural \gamma^0_0$. The adjoint of this is $\gamma^0_0 \natural T_F^*$, which has as symbol $\gamma^0_0 \natural T_F^* \natural \gamma^0_0$. The symbolic composition of these two operators is
(29.3.1)
$$T_F \circledS (\gamma^0_0 \natural T_F^* \natural \gamma^0_0) = T_F \natural \gamma^0_0 \natural (\gamma^0_0 \natural T_F^* \natural \gamma^0_0) = T_F \natural T_F^* \natural \gamma^0_0 = \eta(\mathscr{A})^2$$

$$\times \int_{W \times W} \omega(DF_u) \natural (\delta_{u/2} \natural \gamma^0_{\mathscr{A}} \natural \delta_{u/2}) \natural (\delta_{-v/2} \natural \gamma^0_{\mathscr{A}} \natural \delta_{-v/2}) \natural \omega(DF_v)^{-1} \natural \gamma^0_0\, du\, dv.$$

Computing again with the formulas of §18.3, we find

(29.3.2a)
$$(\delta_{u/2}\natural\gamma^0_{\mathscr{A}}\natural\delta_{u/2})\natural(\delta_{-v/2}\natural\gamma^0_{\mathscr{A}}\natural\delta_{-v/2})$$
$$= (\exp(-\pi(u-v)^T\mathscr{T}(\mathscr{A})^{-1}(u-v)/8))$$
$$\times \delta_{L(v-u)}\natural(\delta_{(u+v)/4}\natural\gamma^0_{\mathscr{A}'}\natural\delta_{-(u+v)/4})\natural\delta_{L(v-u)},$$

where

(29.3.2b)
$$\mathscr{A}' = -(J\mathscr{T}(\mathscr{A})J)/2$$

and

(29.3.2c)
$$L = -(J\mathscr{A}J)\mathscr{T}(\mathscr{A})^{-1}/4.$$

Using this formula to continue the calculation of (29.3.1), we obtain
(29.3.3)
$$T_F\natural T_F^*\natural\gamma^0$$
$$= \eta(\mathscr{A})^2\int_{W\times W}\exp(-\pi(u-v)^T\mathscr{T}(\mathscr{A})^{-1}(u-v)/8)\omega(DF_u)^{-1}$$
$$\natural(\delta_{L(u-v)}\natural\delta_{(u+v)/4}\natural\gamma^0_{\mathscr{A}'}\natural\delta_{-(u+v)/4}\natural\delta_{L(u-v)})\natural\omega(DF_v)^{-1}\natural\gamma^0_0\,du\,dv.$$

Now set $(u+v)/2 = y$, $u - v = z$, and continue.

$$= \eta(\mathscr{A})^2\int_{W\times W}\exp(-\pi z^T\mathscr{T}(\mathscr{A})^{-1}z/8)\omega(DF_{y+z/2})$$
$$\natural(\delta_{L(z)}\natural\delta_{y/2}\natural\gamma^0_{\mathscr{A}'}\natural\delta_{-y/2}\natural\delta_{L(z)})\natural\omega(DF_{y-z/2})^{-1}\natural\gamma^0_0\,dy\,dz$$
$$= \eta(\mathscr{A})^2\int_W\exp(-\pi z^T\mathscr{T}(\mathscr{A})^{-1}z/8)\left(\int_W\omega(DF_{y+z/2}\right.$$
$$\left.\natural(\delta_{L(z)}\natural\delta_{y/2}\natural\gamma^0_{\mathscr{T}(\mathscr{A})^{-1}/2}\natural\delta_{y/2}\natural\delta_{-L(z)})\natural\omega(DF_{y-z/2})^{-1}\,dy\right)dz.$$

When $z = 0$, the integral in y is

(29.3.4)
$$\int_W\omega(DF_y)\natural\delta_{y/2}\natural\gamma^0_{\mathscr{T}(\mathscr{A})^{-1}/2}\natural\delta_{y/2}\natural\omega(DF_y)^{-1}\,dy$$
$$= \int_W\delta_{F(y)/2}\natural\gamma^0_{\mathscr{B}(y)}\natural\delta_{F(y)/2}\,dy = \int_W\delta_{w/2}\natural\gamma^0_{\mathscr{B}'(w)}\natural\delta_{w/2}\,dw$$

where
(29.3.4b)
$$\mathscr{B}(y) = \tfrac{1}{2}((DF_y)^T)^{-1}\mathscr{T}(\mathscr{A})^{-1}(DF_y)^{-1}, \qquad \mathscr{B}'(w) = \mathscr{B}(F^{-1}(y)).$$

Note there is no Jacobian factor involved in passing from the integral in y to the integral in $w = F(y)$ since F is symplectic, hence volume-preserving.

The final expression in (29.3.4) is clearly of the type (28.1.1). In fact, $\tilde{\mathscr{B}}(w)$ is constant on rays and so in particular, only varies in a compact set, so (29.3.4) barely misses being of the type (26.3.1). Further, since DF is constant on rays, we see that for fixed $z \neq 0$, as $y \to \infty$, the elements $DF_{y\pm z/2}$ of $\mathrm{Sp}(W)$ differ from DF_y by an amount of the order of $\|w\|^{-1}$. This allows one to show that the difference

$$\int_W \omega(DF_{y+z/2})\natural(\delta_{L(z)}\natural\delta_{y/2}\natural\gamma^0_{\mathscr{F}(\mathscr{A})^{-1}/2}\natural\delta_{y/2}\natural\delta_{-L(z)})\natural\omega(DF_{y-z/2})^{-1}\,dy$$

$$-\int_W \delta_{DF_y(L(z))}\natural\delta_{w/2}\natural\gamma^0_{\mathscr{B}'(w)}\natural\delta_{w/2}\natural\delta_{-DF_y(L(z))}\,dw \quad \text{(where } y = F(w))$$

defines a compact operator. (In fact, it is in some Schatten p-class.) Taking into account the damping factor $\exp(-\pi z^T\mathscr{F}(\mathscr{A})^{-1}z/8)$, we can conclude that the difference

$$T_F\natural T_F^*\natural\gamma^0_0 - \eta(\mathscr{A})^2 \int_W \exp(-\pi z^T\mathscr{F}(\mathscr{A})^{-1}z/8)$$

$$\left(\int_W \delta_{DF_y(L(z))}\natural\delta_{w/2}\natural\gamma^0_{\mathscr{B}(w)}\natural\delta_{w/2}\natural\delta_{-DF_y(L(z))}\,dw\right)\,dz$$

is compact. Again, the inner integral defines an operator which is a mild generalization of the operators $T_{\mathscr{A}}(\phi)$ defined in (26.3.1).

Thus we have verified that T_F is approximately unitary. We now check the effect of "symbolic conjugation" by T_F. Choose w in W and $\mathscr{B} \in \varsigma^{2+}(\mathbf{R}^N)$, and consider the symbolic composition

(29.3.5)
$$T_F\circledS(\delta_{w/2}\natural\gamma^0_{\mathscr{B}}\natural\delta_{w/2})\circledS(\gamma^0_0\natural T_F^*\natural\gamma^0_0)$$
$$= T_F\natural\gamma^0_0\natural(\delta_{w/2}\natural\gamma^0_{\mathscr{B}}\natural\delta_{w/2})\natural T_F^*\natural\gamma^0_0.$$

If $F = g \in \mathrm{Sp}(W)$, then as we have noted, $T_F = \omega(g)\natural\gamma^0_0$. In this case, one computes that expression (29.3.5) is exactly

$$\delta_{g(w)/2}\natural\gamma^0_{(g^T)^{-1}\mathscr{B}(g)^{-1}}\natural\delta_{g(w)/2}.$$

In general, a formula similar to (27.3.3) holds for the product $T_F\circledS\delta_{w/2}\natural\gamma^0_{\mathscr{B}}\natural\delta_{w/2}$:

(29.3.6)
$$T_F\circledS\delta_{w/2}\natural\gamma^0_{\mathscr{B}}\natural\delta_{w/2} = \eta(\mathscr{A}) \int_W e^{-\pi z^T\mathscr{C}z/4}e^{\pi i\alpha'}\omega(DF_{w+z})\natural\delta_{\Lambda_1(z)}$$

$$\natural\delta_{w+z/2}\natural\gamma^0_{\mathscr{A}_3'}\natural\delta_{w+z/2}\natural\delta_{\Lambda_2(z)}\,dz.$$

Here the notation is as in (27.3.3), except that $\mathscr{A}_3'$ is the product of $\mathscr{A}$ and $\mathscr{B}' = -J\mathscr{B}^{-1}J/4$ with respect to the composition law (8.2).

We note the integrand in (29.3.6) is rapidly decreasing in z. Since DF_{w+z} will vary slowly in z—precisely, $DF_{w+z} - DF_w$ dies off like $\|w\|^{-1}$—we see also that the difference

$$(T_F - \omega(DF_w))\circledS(\delta_{w/2}\natural\gamma^0_{\mathscr{B}}\natural\delta_{w/2})$$

dies off like $\|w\|^{-1}$ as $w \to \infty$. Repeating the argument to deal with the further convolution with $T_F^*\natural\gamma^0_0$, we find that the difference

$$T_F\circledS(\delta_{w/2}\natural\gamma^0_{\mathscr{B}}\natural\delta_{w/2})\circledS(\gamma^0_0\natural T_F^*\natural\gamma^0_0) - \delta_{F(w)/2}\natural\delta_{\mathscr{B}(w)}\natural\delta_{F(w)/2},$$

where $\mathscr{B}(w) = (DF_w^T)^{-1}\mathscr{B}DF_w^{-1}$, dies off like $\|w\|^{-1}$ as $w \to \infty$. This is what one means by saying that conjugation by T_F approximately effects the canonical transformation F.

ACKNOWLEDGEMENT. I am grateful to J. Faraut for leading me to the following references relevant to this paper. The papers [**Vb**] and [**Ol**] contain general results on Lie semigroups which imply most of the structural results about Ω proved here. Further, [**Ol**] puts the existence of a complex extension of a unitary representation in the context of holomorphic representations of groups with hermitian structure. Moreover, the existence of Ω, in a Fock model, and a sketch of the extension to the boundary, has appeared in the physics literature [**BK**]. Also, an application to nuclear physics is given in [**JK**]. Also, I have been informed by J. Peetre of the existence of the paper [**Pt**], and that it anticipates some parts of this paper.

REFERENCES

[**Be**] R. Beals, *A general calculus of pseudo-differential operators*, Duke Math. J. **42** (1975), 1–42.

[**Bk**] W. Beckner, *Inequalities in Fourier analysis*, Ann. of Math. (2) **102** (1975), 159–182.

[**BK**] M. Brunet and P. Kramer, *Complex extension of the representation of the symplectic group associated with the canonical commutation relations*, Group Theoretical Methods in Physics, Springer-Verlag, New York, 1976.

[**CV**] A. Calderon and R. Vaillancourt, *A class of bounded pseudo-differential operators*, Proc. Nat. Acad. Sci. U.S.A. **69** (1972), 1185–1187.

[**Ct**] P. Cartier, *Quantum mechanical commutation relations and theta functions*, Proc. Sympos. Pure Math., vol. 9, Amer. Math. Soc., Providence, R.I., 1966, 361–383.

[**CF**] A. Cordoba and C. Fefferman, *Wave packets and Fourier integral operators*, Comm. Partial Diff. Equations **3** (1978), 979–1006.

[**DB1**] N. DeBruijn, *Uncertainty principles in Fourier analysis*, Proc. Sympos. on Inequalities, Academic Press, New York, 1967, 57–71.

[**DB2**] _____, *A theory of generalized functions, with applications to Wigner distribution and Weyl correspondences*, Nieuw Archief vor Wiskunde **21** (1973), 205–280.

[**FP1**] C. Fefferman and D. Phong, *The uncertainty principle and sharp Gårding inequalities*, Comm. Pure Appl. Math. **34** (1981), 285–331.

[**FP2**] _____, *On the asymptotic eigenvalue distribution of a pseudo-differential operator*, Proc. Nat. Acad. Sci. U.S.A. **77** (Oct. 1980), 5622–5625.

[**GS**] V. Guillemin and S. Sternberg, *Geometric asymptotics*, Math. Surveys, No. 14, Amer. Math. Soc., Providence, R.I., 1977.

[**HP**] E. Hille and R. Phillips, *Functional analysis and semi-groups*, rev. ed., Amer. Math. Soc., Providence, 1957.

[**HR**] E. Hewitt and K. Ross, *Abstract harmonic analysis*. I, Grundlehren Math. Wiss., vol. 115, Springer-Verlag, New York, 1963.

[**Hr**] L. Hörmander, *The Weyl calculus of pseudo-differential operators*, Comm. Pure Appl. Math. **23** (1979), 359–443.

[**Hr2**] _____, *The analysis of linear partial differential operators*. III, Grundlehren Math. Wiss., vol. 274, Springer-Verlag, New York, 1984.

[**Hw**] R. Howe, *Quantum mechanics and partial differential equations*, J. Funct. Anal. **38** (1980), 188–254.

[**JK**] G. John and P. Kramer, *Canonical transformations and Gaussian integral kernels in nuclear physics*, Group Theoretical Methods in Physics (J. Ehlers et. al., eds.), Springer Lecture Notes in Physics, vol. 50, Springer-Verlag, New York, 1976.

[**KR**] R. Kadison and J. Ringrose, *Fundamentals of the theory of operator algebras. Vol. I: Elementary theory*, Academic Press, New York, 1983.

[**Kl**] J. Kelley, *General topology*, Van Nostrand, Princeton, N. J., 1955.

[**KN**] J. Kelley and I. Namioka, *Linear topological spaces*, Van Nostrand, Princeton, N. J., 1963.

[KS] A. Knapp and E. Stein, *Intertwining operators for semisimple groups*, Ann. of Math. (2) **93** (1971), 489–578.

[L1] S. Lang, *Algebra*, 2nd ed., Addison-Wesley, Reading, Mass., 1984.

[L2] _____ , *Real analysis*, 2nd ed., Addison-Wesley, Reading, Mass., 1983.

[ML] W. Myller-Lebedeff, *Die Theorie der Integralgleichungen in Anwendung auf einige Reihenentwicklungen*, Math. Ann. **64** (1907), 388–416.

[Ol] G. Ol'shanski, *Invariant cones in Lie algebras, Lie semigroups, and the holomorphic discrete series*, Functional Anal. and Appl. **15** (1981), 275–375.

[Pt] J. Peetre, *A class of kernels related to the kernels of Beckner and Nelson*, A Tribute to Ake Pleijel, University of Uppsala, 1980, pp. 171–210.

[Sa] I. Satake, *On compactifications of the quotient spaces for arithmetically defined discontinuous groups*, Ann. of Math. (2) **72** (1960), 555–580.

[Sg] I. Segal, *Transforms for operators and symplectic automorphisms over a locally compact abelian group*, Math. Scand. **13** (1963), 31–43.

[Sh] D. Shale, *Linear symmetries of free boson fields*, Trans. Amer. Math. Soc. **103** (1962), 149–167.

[Si] C. Siegel, *Symplectic geometry*, Academic Press, New York, 1964.

[U1] A. Unterberger, *Oscillateur harmonique et opérateurs pseudo-differentiels*, Ann. Inst. Fourier (Grenoble) **29** (1979), 201–221.

[U2] _____ , *Les opérateurs metadifférentiels*, C. R. Acad. Sci. Paris Ser. A **288** (1979), 449–451.

[Vb] E. Vinberg, *Invariant convex cones and orderings in Lie groups*, Functional Anal. and Appl. **14** (1980), 1–10.

[Wl] A. Weil, *Sur certains groupes d'opérateurs unitaires*, Acta Math. **111** (1964), 143–211.

[Wy1] H. Weyl, *The classical groups*, Princeton Univ. Press, Princeton, N. J., 1946.

[Wy2] _____ , *The theory of groups and quantum mechanics*, Dover Publications, New York, 1950.

YALE UNIVERSITY

Proceedings of Symposia in Pure Mathematics
Volume 48 (1988)

The Classical Groups and Invariants of Binary Forms

ROGER HOWE

1. One goal Hermann Weyl had in writing *The Classical Groups* was to provide representation-theoretic foundations for invariant theory. In view of this aim, it is interesting that, except for a $1\frac{1}{2}$-page discussion of the binary quadratic form, Weyl made no attempt to apply the extensive theory he developed to elucidating the ur-problem of invariant theory, namely the invariants of binary forms. The purpose of this essay is to fill this gap: I hope to show that Weyl's general results and other standard tools of representation theory go very far toward organizing into a succinct, readily comprehensible picture a substantial amount of the 19th-century work on invariants of binary forms.

There have appeared in recent times several general accounts of invariant theory. We mention in particular Dieudonné [**DC**], Springer [**Sp1**], and Kung and Rota [**KR**]. There is necessarily substantial overlap between those articles and this one (the latter two have been quite useful in the preparation of this one) but the focus here is distinct from any of them. Dieudonné reviews the Schur-Young theory of GL_n-S_m duality (see §4.1) which also figures heavily in Weyl, but he scants binary forms. Springer does the opposite, and his emphasis is modern. Kung and Rota attempt to recapture some flavor of the 19th-century theory, but their treatment is of a heavily combinatorial cast, virtually free of representation-theoretic or geometric tools. Thus there still seems to be room in the literature for an account which makes a direct connection between the representation-theoretic tools of Weyl and others and the classical constructions.

This article is arranged as follows. §2 reviews the basic notions of invariants and covariants, especially as they apply to connected reductive groups. §3 presents elementary facts about SL_2, especially the structure of its representations, and makes some first applications to invariant theory (relation between covariants and invariants, Weitzenböck's Theorem, numbers of covariants of given degree, Hermite reciprocity). In §4 we review the First Fundamental Theorem of classical invariant theory for GL_n, as developed by Weyl. This includes the

---

1980 *Mathematics Subject Classification* (1985 *Revision*). Primary 15A72; Secondary 14D20, 14D25, 14L30, 14M12, 14M15, 15A69, 20G05.

GL_n-symmetric group duality established by Schur, the polynomial version of it (this is the First Fundamental Theorem proper), and a representation-theoretic version, giving a $(\mathrm{GL}_n, \mathrm{GL}_m)$ duality that has been noticed by more recent authors [**GK, Ho1, KV, Se2, Zh**]. It is this last version of the general theory which is most relevant to invariant theory of binary forms. Also in §4, we give an explication of the legendary "symbolic method" in terms of the Schur theory. Finally, §5 presents the applications of the general theory to binary forms. We construct a simple polynomial mapping that translates questions about invariants of binary forms to the context of §4. This was known as "expressing invariants in terms of symbolical determinantal factors" [**GY**, pp. 14–16] or in terms of "homogenized roots" [**KR**, p. 46]. A point to emphasize is that the $(\mathrm{GL}_2, \mathrm{GL}_n)$-duality is controlling the situation very strongly. However, for one key fact, we must still rely on a 19th-century result [**Ke**], virtually unchanged. (Kung and Rota do likewise [**KR**, p. 75].) We use the facts gathered from the representation-theoretic construction to establish finite generation of invariants in an effective manner, compute and estimate various dimensions, and give an efficient calculation of the most tractable examples $(n = 2, 3, 4)$.

There are few if any new facts here. For some of the dimension formulas, and the asymptotic result 5.8.2, I am not aware of references, but in a subject with a literature so large and so old, ignorance should only inspire modesty. As Borel [**B**] notes, even Weyl, when writing his book, was unaware of some relevant papers. However, I do hope the organization of the story presented here will make the old constructions more accessible to readers with a modern training in representation theory.

I thank R. Stanley for some helpful correspondence, and J. Towber for a careful reading of the manuscript.

2. General notions.

2.1. Let V be a (finite-dimensional) complex vector space. Let $\mathscr{P}(V)$ denote the algebra of polynomial functions on V. It is a graded algebra:

$$\mathscr{P}(V) = \sum_{k=0}^{\infty} \mathscr{P}^k(V),$$

where $\mathscr{P}^k(V)$ is the space of polynomials homogeneous of degree k.

2.2. Let G be a group. Let

$$\rho: G \to \mathrm{GL}(V)$$

be a homomorphism, i.e., a representation of G on V. We can extend ρ to an action of G on $\mathscr{P}(V)$ by the formula

$$\rho^*(g)(p)(v) = p(g^{-1}(v)), \qquad g \in G, \ v \in V, \ p \in \mathscr{P}(V).$$

Note that $\rho^*(g)$ is an algebra automorphism of $\mathscr{P}(V)$.

2.3. A polynomial $p \in \mathscr{P}(V)$ is called a G-invariant if

$$\rho^*(g)(p) = p$$

for all $g \in G$. The set of G-invariants in $\mathscr{P}(V)$ is denoted $\mathscr{P}(V)^G$. It is a subalgebra of $\mathscr{P}(V)$.

2.4. Let σ be an irreducible representation of G on a vector space U. A G-intertwining operator

$$T: U \to \mathscr{P}(V)$$

is called a *covariant of type σ* for ρ. Alternatively, a covariant of type σ is an invariant for the natural tensor product G-action on $\text{Hom}(U, \mathscr{P}(V)) \simeq \mathscr{P}(V) \otimes U^*$. Denote the space of such covariants by $\text{Hom}(\sigma, \mathscr{P}(V))^G$.

2.5. Observe that if $p \in \mathscr{P}(V)^G$, and $T \in \text{Hom}(\sigma, \mathscr{P}(V))^G$, then pT, defined by

$$(pT)(ju) = pT(u)$$

(product in $\mathscr{P}(V)$), is again a covariant of type σ. Thus the space $\text{Hom}(\sigma, \mathscr{P}(V))^G$ of covariants of type σ is a module for the algebra $\mathscr{P}(V)^G$ of G-invariants.

2.7. Given a nonzero $T \in \text{Hom}(\sigma, \mathscr{P}(V))^G$, the image $T(U)$ will be a G-invariant subspace of $\mathscr{P}(V)$ on which G acts irreducibly by a representation isomorphic to σ. Conversely, given such a subspace $U' \subseteq \mathscr{P}(V)$, there is a G-intertwining operator $T_{U'}: U \to U'$, unique up to multiples by Schur's Lemma. The span of all images $T(U)$, for $T \in \text{Hom}(\sigma, \mathscr{P}(V))^G$, is called the *$\sigma$-isotypic component* of $\mathscr{P}(V)$, and denoted $\mathscr{P}(V)^{G,\sigma}$, or if G is understood, simply by $\mathscr{P}(V)^\sigma$. Evidently knowledge of $\text{Hom}(\sigma, \mathscr{P}(V))^G$ is essentially equivalent to knowledge of $\mathscr{P}(V)^\sigma$. More precisely, there is a natural isomorphism

$$\mathscr{P}(V)^\sigma \simeq U \otimes \text{Hom}(\sigma, \mathscr{P}(V))^G.$$

Note that $\mathscr{P}(V)^\sigma$ is a $\mathscr{P}(V)^G$ submodule of $\mathscr{P}(V)$.

2.7.1. Suppose G is a reductive connected complex algebraic group. Fix a Borel subgroup $B \subseteq G$, and a Cartan subgroup $A \subseteq B$. Let $\hat{A}$ denote the character group of A, and $\hat{A}^+$ the positive Weyl chamber in $\hat{A}$ determined by B. According to the theory of the highest weight, the irreducible representations of G are parametrized by $\hat{A}^+$. Precisely, let N be the unipotent radical of B. Let σ be an irreducible representation of G on a vector space U. In U there is a unique line L invariant by N; in fact L will be fixed pointwise by N. Because of the uniqueness of L, it will be invariant under B. Thus elements of L are eigenvectors for A, and there is a character $\psi_\sigma = \psi$ of A such that

$$\sigma(a)(v) = \psi_\sigma(a)v, \qquad a \in A, \ v \in L.$$

This character ψ belongs to $\hat{A}^+$, and the correspondence

$$\sigma \to \psi_\sigma \in \hat{A}^+$$

is the stated parametrization. We call the line L of N-invariants in U the *highest weight line*, and ψ_σ is the highest weight of σ.

2.7.2. Again let σ be an irreducible representation of G on the space U. Let $T: U \to \mathscr{P}(V)$ be a covariant of type σ. Let $L \subseteq U$ be the highest weight line. Then $T(L)$ will be in $\mathscr{P}(V)^N$, and more specifically, will be in $\mathscr{P}(V)^{B,\psi_\sigma}$. (When convenient, we regard ψ_σ as a character of B trivial on N.) Further, given a line L' in $\mathscr{P}(V)^{B,\psi_\sigma}$, the G-module generated by L' will be irreducible and isomorphic to σ. Thus there will be a unique (up to multiples) covariant of type σ, $T': U \to \mathscr{P}(V)$, such that $T'(L) = L'$. Hence we have an isomorphism

$$\operatorname{Hom}(\sigma, \mathscr{P}(V))^G \simeq \mathscr{P}(V)^{B,\psi_\sigma} \subseteq \mathscr{P}(V)^\sigma.$$

In other words when G is a reductive connected algebraic group, we can consider the covariants of type σ for ρ to be the subspace $\mathscr{P}(V)^{B,\psi_\sigma}$ of ψ_σ eigenvectors for B in $\mathscr{P}(V)$. Note that $\mathscr{P}^{B,\psi_\sigma}$ is a module for $\mathscr{P}(V)^G$ (via multiplication in $\mathscr{P}(V)$), and that the isomorphism between $\mathscr{P}(V)^{B,\psi_\sigma}$ and $\operatorname{Hom}(\sigma, \mathscr{P}(V))^G$ is an isomorphism of $\mathscr{P}(V)^G$ modules.

2.7.3. We have a direct sum decomposition

$$\mathscr{P}(V)^N = \sum_{\psi \in \hat{A}^+} \mathscr{P}(V)^{B,\psi}.$$

(Some of the summands may be trivial.) Taking 2.7.2 into account, we may regard $\mathscr{P}(V)^N$ as the sum, over all σ, of the spaces of covariants of type σ. We note that $\mathscr{P}(V)^N$ is also a subalgebra of $\mathscr{P}(V)$; it is called the *algebra of covariants*. It is graded by $\mathbf{Z}^+ \times \hat{A}^+$. Evidently, knowledge of $\mathscr{P}(V)^N$ as $\mathbf{Z}^+ \times \hat{A}^+$ graded algebra is equivalent to knowledge of the structure of $\mathscr{P}(V)$ as G-module.

2.8.1. The main problems of invariant theory are:

(i) Describe the space $\mathscr{P}(V)^G$ of G-invariants.

(ii) Describe the spaces $\operatorname{Hom}(\sigma, \mathscr{P}(V))^G$ of covariants of type σ for ρ.

2.8.2. Each of these problems may be understood in several different ways. One can give a basis for each $\mathscr{P}^k(V)^G$, or another explicit description of the homogeneous components of $\mathscr{P}(V)^G$. One can give a formula for $\dim \mathscr{P}^k(V)^G$, or describe the *Hilbert series*

$$H_\rho(t) = \sum_{k=0}^{\infty} (\dim \mathscr{P}^k(V)^G) t^k.$$

Or one can describe $\mathscr{P}(V)^G$ as an algebra: give a set of generators, or a set of generators with their relations, or even describe all the "higher syzygies" these generators give rise to; or describe a homomorphism or isomorphism to another algebra (which one would hope to be of determinable structure).

Similarly, one might describe $\operatorname{Hom}(\sigma, \mathscr{P}(V))^G$ in terms of the homogeneous components $\operatorname{Hom}(\sigma, \mathscr{P}^k(V))^G$, or by their dimensions, or by the generating function

$$H_{\rho,\sigma}(t) = \sum_{k \geq 0} \dim(\operatorname{Hom}(\sigma, \mathscr{P}^k(V))^G) t^k.$$

Or one can try to describe $\operatorname{Hom}(\sigma, \mathscr{P}(V))^G$ as a $\mathscr{P}(V)^G$-module, by describing generators and if possible relations and higher syzygies between the generators.

One could also attempt to give a full description of $\mathscr{P}(V)$ as a G-module, by giving simultaneous descriptions of all the spaces $\operatorname{Hom}(\sigma, \mathscr{P}(V))^G$ of covariants or of the isotypic components $\mathscr{P}(V)^\sigma$. In case G is connected reductive, one could attempt to describe the algebra of covariants $\mathscr{P}(V)^N$. This description could be of the various types noted above for $\mathscr{P}(V)^G$; additionally, one would hope the description would give information about the $(\mathbf{Z}^+ \times \hat{A}^+)$-grading on $\mathscr{P}(V)^N$.

2.9. A very general result concerning questions 2.8.1 is the well-known

THEOREM. *If G acts reductively on V, then:*

(a) *The algebra $\mathscr{P}(V)^G$ is finitely generated.*

(b) *For any irreducible rational representation σ of G, the space $\mathrm{Hom}(\sigma, \mathscr{P}(V))^G$ of covariants of type σ is finitely generated as a $\mathscr{P}(V)^G$-module.*

These theorems essentially go back to Hilbert [**Hi1**]. The key fact underlying them is what is now known as the Hilbert Basis Theorem [**J1**], which is a great divide in the history of algebra. It caused consternation among the computationally oriented invariant theorists of the late 19th century, and greatly dampened interest in invariant theory [**Re, DC; W**, p. 27].

3. Representations of SL_2, and elementary consequences for invariant theory.

3.1. Consider the group $G = \mathrm{SL}_2(\mathbf{C}) = \mathrm{SL}_2$. This acts via its defining representation on $\mathbf{C}^2$; as in 2.2 we obtain an action of G on $\mathscr{P}(\mathbf{C}^2)$. A basic fact in representation theory as well as in the theory of invariants of binary forms is that the spaces $\mathscr{P}^k(\mathbf{C}^2)$ define exactly the irreducible representations of G, with no repetitions.

3.1.1. We can connect 3.1 with 2.7.1 as follows. Set

$$a(t) = \begin{bmatrix} t & 0 \\ 0 & t^{-1} \end{bmatrix}, \qquad t \in \mathbf{C}^x.$$

Then $A = \{a(t): t \in \mathbf{C}^x\}$ is a Cartan subgroup of G. Set

$$n(z) = \begin{bmatrix} 1 & z \\ 0 & 1 \end{bmatrix}, \qquad z \in \mathbf{C}.$$

Then $N = \{n(z): z \in \mathbf{C}\}$ is the unipotent radical of the Borel subgroup $B = AN$.

3.1.2. The characters of A are the mappings

$$\psi_l(a(t)) = t^l, \qquad t \in \mathbf{C}^x, \ l \in \mathbf{Z}.$$

The positive Weyl chamber is

$$\hat{A}^+ = \{\psi_l: \ l \geq 0\}$$

The space $\mathscr{P}^l(\mathbf{C}^2)$ has dimension $l + 1$. If x, y are the standard coordinates on $\mathbf{C}^2$, then $\mathscr{P}^l(\mathbf{C}^2)$ has a basis consisting of the monomials

$$q_{ab} = x^a y^b, \qquad 0 \leq a, b; \ a + b = l.$$

These are eigenvectors for A; the eigencharacter of A defined by $q_{a,b}$ is $\psi_{b-a} = \psi_{l-2a}$. The fixed vectors for N are the multiples of y^b. Hence the highest weight of $\mathscr{P}^l(\mathbf{C}^2)$ is ψ_l.

Denote the representation of G on $\mathscr{P}^l(\mathbf{C}^2)$ by σ_l.

3.1.3. Note that the eigencharacters of A acting on $\mathscr{P}^l(\mathbf{C}^2)$ are all ψ_k such that $|k| \leq l$ and $k \equiv l \pmod 2$.

3.2. Let ρ be a representation of G on a vector space V. Consider the space of covariants of type σ_l for ρ. This is the space $\operatorname{Hom}(\mathscr{P}^l(\mathbf{C}^2), \mathscr{P}(V))^G \simeq (\mathscr{P}^l(\mathbf{C}^2)^* \otimes \mathscr{P}(V))^G$. Since σ_l is the only irreducible representation of G of dimension $l + 1$, it must be isomorphic to its own contragredient:

$$\sigma_l \simeq \sigma_l^*$$

as G-modules. Thus

$$(\mathscr{P}^l(\mathbf{C}^2) \otimes \mathscr{P}(V))^G \simeq (\mathscr{P}^l(\mathbf{C}^2)^* \otimes \mathscr{P}(V))^G \simeq \mathscr{P}(V)^{B, \psi_l}.$$

3.2.1. Taking a direct sum over l of the isomorphisms just noted, we obtain

$$\mathscr{P}(V)^N \simeq \sum_{l \geq 0} \mathscr{P}(V)^{B, \psi_l} \simeq \sum_{l \geq 0} (\mathscr{P}^l(\mathbf{C}^2) \otimes \mathscr{P}(V))^G \simeq (\mathscr{P}(\mathbf{C}^2) \otimes \mathscr{P}(V))^G$$

$$\simeq \mathscr{P}(V \oplus \mathbf{C}^2)^G.$$

Thus in the case of our group $G = \mathrm{SL}_2$ we have

PROPOSITION. *The algebra of covariants of G acting on $\mathscr{P}(V)$ is isomorphic (as $(\mathbf{Z}^+)^2$-graded algebra) to the algebra of invariants for G acting on $\mathscr{P}(V \oplus \mathbf{C}^2)$.*

3.2.2. REMARK. Proposition 3.2.1 reduces the problem of computing the covariants of a general representation of G (in classical terms, the covariants of an arbitrary collection of binary quantics), to the problem of computing the invariants (of $\mathbf{C}^2$ plus the given representation, which is just another representation of G).

3.2.3. REMARK. A construction analogous to 3.2.1 is available for other reductive groups. See [**Had, HM**].

3.2.4. REMARK. The isomorphism $\mathscr{P}(V)^N \simeq \mathscr{P}(V \oplus \mathbf{C}^2)^G$ of 3.2.1, combined with the general result 2.9, implies that $\mathscr{P}(V)^N$ is a finitely generated algebra. We can see from (3.1.2) that $\sigma_l(n(z))$ is a unipotent transformation with one Jordan block of length $l + 1$. It follows from Jordan Canonical Form that if $\eta(t)$, $t \in \mathbf{C}$, is a one-parameter group of unipotent transformations on a vector space V, then we may define a representation ρ of $\mathrm{SL}_2(\mathbf{C})$ on V such that $\rho(n(t)) = \eta(t)$, $t \in \mathbf{C}$. Together with the preceding considerations, this proves

WEITZENBÖCK'S THEOREM [**Se1**]. *If $Z = \{\eta(t): t \in \mathbf{C}\}$ is a one-parameter group of unipotent transformations on a vector space V, then the algebra $\mathscr{P}(V)^Z$ of invariants for Z is finitely generated.*

3.3. In this exposition, we focus on describing

$$\mathscr{I}_n = \mathscr{P}(\mathscr{P}^n(\mathbf{C}^2))^{\mathrm{SL}_2}$$

—the algebra of invariants of a binary n-ic. This is known explicitly for small n ($n \leq 6$ and $n = 8$ [**Sy2**]). For larger n, the problem has resisted analysis [**Sp1**,

p. 62]. (See, however, [**D2**]. J. Dixmier has written papers over the last several years clarifying a number of questions about binary invariants. See for example [**D1, DL, DEN**] as well as [**D2**].) We will explain the major 19th-century results on the question.

3.3.1.1. One can give an elementary combinatorial description of the dimension of $\mathscr{P}^k(\mathscr{P}^n(\mathbf{C}^2))^{\mathrm{SL}_2} = \mathscr{I}_n^k$—the number of invariants of a given degree. This is due to Cayley and Sylvester [**C, Sy1**]. (See also [**Sp1**].) In fact, one can give an elementary description of $\dim(\mathrm{Hom}(\sigma_l, \mathscr{P}^k(\mathscr{P}^n(\mathbf{C}^2)))^G)$—the number of covariants of degree k and type l for the binary n-ic. These depend on the following observation.

PROPOSITION. *Let ρ be a representation of $G = \mathrm{SL}_2(\mathbf{C})$ on a vector space V. Consider the multiplicities $\dim \mathrm{Hom}(\sigma_l, V)^G$ with which the irreducible representations σ_l occur in ρ. Similarly, for the torus A (see 3.1.1) of G, consider the multiplicities $\dim \mathrm{Hom}(\psi_m, V)^A = \dim V^{A,\psi_m}$ with which the characters of A appear in V. Then*

$$\dim \mathrm{Hom}(\sigma_l, V)^G = \dim V^{A,\psi_{l+2}} - \dim V^{A,\psi_l}, \qquad l \geq 0$$

3.3.1.2. PROOF. According to observation 3.1.3, the character ψ_m, $m \geq 0$, occurs in $\sigma_l|A$ if and only if $m = l + 2k$ for some integer $k \geq 0$, and it occurs once. Thus

$$\dim V^{A,\psi_m} = \sum_{k \geq 0} \dim \mathrm{Hom}(\sigma_{m+2k}, V)^G.$$

Setting $m = l$ and $m = l + 2$ and subtracting yields Proposition 3.3.1.1.

3.3.1.3. To apply Proposition 3.3.1.1 to $\mathscr{P}^k(\mathscr{P}^n(\mathbf{C}^2))$ we need to count the number of ψ_m-eigenvectors for A in $\mathscr{P}^k(\mathscr{P}^n(\mathbf{C}^2))$. This is easy to do. Let $\alpha_{a,b}$ denote the coordinates on $\mathscr{P}^n(\mathbf{C}^2)$ with respect to the basis $q_{a,b}$ of 3.1.2. Then a basis for $\mathscr{P}^k(\mathscr{P}^n(\mathbf{C}^2))$ consists of monomials

$$\mu(r_0, r_1, \ldots, r_n) = \alpha_{n,0}^{r_0} \alpha_{n-1,1}^{r_1} \cdots \alpha_{0,n}^{r_n}, \qquad \sum_{i=0}^n r_i = k, \ r_i \geq 0.$$

These are eigenvectors for A. Precisely, $\mu(r_0, r_1, \ldots, r_n)$ transforms under A according to the character ψ_m, where

$$m = kn - 2\sum_{i=1}^n ir_i.$$

The sum $\sum_{i=1}^n ir_i$ defines a partition of $\frac{1}{2}(kn - m)$. The number of parts in the partition is $\sum_{i=1}^n r_i = k - r_0$, so is at most k. The size of the parts is i with $1 \leq i \leq n$. Conversely, one sees directly that a partition of $\frac{1}{2}(kn - m)$ satisfying these restrictions arises from some monomial $\mu(r_0, r_1, \ldots, r_n)$.

Define $\pi(n, k, d)$ to be the number of partitions of the number d into at most k parts of size at most n. Then we have shown

$$\dim \mathscr{P}^k(\mathscr{P}^n(\mathbf{C}^2))^{A,\psi_m} = \pi(n, k, \tfrac{1}{2}(kn - m)).$$

3.3.1.4. Combining 3.3.1.1 and 3.3.1.3 gives, for any $l \geq 0$,

$$\dim \operatorname{Hom}(\sigma_l, \mathscr{P}^k(\mathscr{P}^n(\mathbf{C}^2)))^G = \pi(n, k, \tfrac{1}{2}(kn - l)) - \pi(n, k, \tfrac{1}{2}(kn - l) - 1).$$

In particular,

$$\dim \mathscr{I}_n^k = \pi(n, k, \tfrac{1}{2}kn) - \pi(n, k, \tfrac{1}{2}kn - 1).$$

3.3.1.5. REMARK. By considering the effect of restricting parts to be of length $n - 1$ rather than n, one finds that the numbers $\pi(n, k, d)$ satisfy a recursion relation

$$\pi(n, k, d) = \pi(n - 1, k, d) + \pi(n, k - 1, d - n).$$

This relation provides a scheme for rapid calculation of the $\pi(n, k, d)$, and thence of the $\dim(\operatorname{Hom}(\sigma_l, \mathscr{P}^k(\mathscr{P}^n(\mathbf{C}^2)))^G)$.

3.3.2.1. It is common to visualize a partition as defining a Young's diagram—an array of small boxes arranged in horizontal rows, all with the same left-hand edge, and with each row no longer than the one above it. For example, the partition $6 + 5 + 3 = 14$ is pictured as

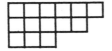

The *size* of the partition, or of the diagram, is the total number of boxes. In terms of this picture, the condition that a partition have at most k parts, and parts of size at most n, is simply the requirement that its Young's diagram fit inside a k by n rectangle. Thus $6 + 5 + 3 = 14$ satisfies these requirements with respect to $k = 4$ and $n = 7$:

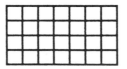

There is a well-known involution on partitions. In terms of Young's diagrams, it amounts to reflecting a diagram across the diagonal line that slants down and to the right from the upper left-hand corner. Thus $(6, 5, 3) \to (3, 3, 3, 2, 2, 1)$:

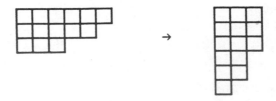

It is obvious that this involution turns a partition of k parts into a partition with parts of size at most k. And it turns a partition with parts of size at most n into a partition with at most n parts. It follows that

$$\pi(n, k, d) = \pi(k, n, d).$$

3.3.2.2. This formula combined with 3.3.1.4 has the following consequence for invariants of binary forms.

HERMITE RECIPROCITY. *For all $l, k, n \geq 0$, and $G = \mathrm{SL}_2(\mathbf{C})$,*

$$\dim \mathrm{Hom}(\sigma_l, \mathscr{P}^k(\mathscr{P}^n(\mathbf{C}^2)))^G = \dim \mathrm{Hom}(\sigma_l, \mathscr{P}^n(\mathscr{P}^k(\mathbf{C}^2)))^G.$$

In words, the number of covariants of type σ_l and degree k of a binary n-ic equals the number of covariants of type σ_l and degree n of a binary k-ic. In particular

$$\dim \mathscr{I}_n^k = \dim \mathscr{I}_k^n$$

The equivalent representation-theoretic statement is: $\mathscr{P}^k(\mathscr{P}^n(\mathbf{C}^2))$ and $\mathscr{P}^n(\mathscr{P}^k(\mathbf{C}^2))$ are isomorphic G-modules.

4. The first fundamental theorem of classical invariant theory for GL_n.

4.0. The fundamental problem Weyl considers in *The Classical Groups* is: Given a group $G \subseteq \mathrm{GL}(V)$, describe the n-fold tensor product

$$V^{\otimes n} = \underbrace{V \otimes V \otimes \cdots \otimes V}_{n \text{ factors}}$$

as a G-module.

4.1.1. A cornerstone of Weyl's analysis in the book is the answer, due to Schur, to this problem for $G = \mathrm{GL}(V)$. Observe that the symmetric group S_n acts on $V^{\otimes n}$ by permuting the factors in a tensor:

$$s^{-1}(v_1 \otimes v_2 \otimes \cdots \otimes v_n) = v_{s(1)} \otimes v_{s(2)} \otimes \cdots \otimes v_{s(n)}, \qquad s \in S_n, \ v_i \in V.$$

SCHUR'S THEOREM [**W**, p. 130]. *The algebras in $\mathrm{End}(V^{\otimes n})$ generated by $\mathrm{GL}(V)$ and by S_n are mutual commutants: each is the full subalgebra of operators commuting with the other.*

4.1.2. Combining Schur's Theorem with the Double Commutant Theorem [**W**, Chapter 3, §5], we conclude:

COROLLARY. *As a $\mathrm{GL}(V) \times S_n$-module, the space $V^{\otimes n}$ decomposes as follows:*

$$V^{\otimes n} \simeq \sum_D \sigma_D \otimes J_D,$$

where σ_D is an irreducible representation of $\mathrm{GL}(V)$ and J_D is an irreducible representation of S_n. The decomposition is such that σ_D determines J_D and vice versa (so it makes sense to label them by a common parameter).

4.1.3.1. We will make Corollary 4.1.2 more explicit by describing how to parametrize the representations involved in the decomposition of $V^{\otimes n}$ under $\mathrm{GL}(V) \times S_n$.

The representations of $G = \mathrm{GL}(V)$ can be understood by the scheme of 2.8.1. Take V to be $\mathbf{C}^m$, so that $\mathrm{GL}(V)$ is $\mathrm{GL}_m(\mathbf{C})$, the group of invertible $m \times m$ matrices. As Cartan subgroup, take the group $A_m = A$ of diagonal matrices. As Borel subgroup, take the group $B_m = B$ of upper triangular matrices. The unipotent radical $N_m = N$ of B is the group of upper triangular matrices with diagonal entries all 1's.

A typical element of A_m can be written

$$a = a(t_1, t_2, \ldots, t_m) = \begin{bmatrix} t_1 & & & 0 \\ & t_2 & & \\ & & \ddots & \\ 0 & & & t_m \end{bmatrix}, \qquad t_i \in \mathbf{C}^x.$$

The map

$$(t_1, t_2, \ldots, t_m) \to a(t_1, t_2, \ldots, t_m)$$

defines a group isomorphism between $(\mathbf{C}^x)^m$ and A_m.

4.1.3.2. Define characters

$$\varphi_i \colon A \to \mathbf{C}^x, \qquad \varphi_i(a(t_1, \ldots, t_m)) = t_i.$$

The φ_i define a basis for the character group $\hat{A}_m \simeq \mathbf{Z}^m$. The general character of A_m is

$$\varphi = \varphi_1^{\alpha_1} \varphi_2^{\alpha_1} \varphi_3^{\alpha_2} \cdots \varphi_n^{\alpha_m} = \prod_i \varphi_i^{\alpha_i}, \qquad \alpha_i \in \mathbf{Z},$$

$$\varphi(a(t_1, t_2, \ldots, t_m)) = t_1^{\alpha_1} t_2^{\alpha_2} \cdots t_m^{\alpha_m}.$$

If $\varphi = \prod_i \varphi_i^{\alpha_i}$, then we write $\alpha_i = \alpha_i(\varphi)$.

We call φ *polynomial* if $\alpha_i(\varphi) \geq 0$ for all i.

4.1.3.3. The positive Weyl chamber $\hat{A}_m^+$ in $\hat{A}_m$ is

$$\hat{A}_m^+ = \{\varphi \in \hat{A} \colon \alpha_i(\varphi) \geq \alpha_{i+1}(\varphi)\}.$$

Observe that $\varphi \in \hat{A}_m^+$ will be polynomial if and only if $\alpha_m(\varphi) \geq 0$. If $\sigma = \sigma_\varphi$ is the irreducible representation $\mathrm{GL}_m(\mathbf{C})$ with highest weight φ, we say σ_φ is polynomial if φ is polynomial. If σ is polynomial, then all weights of σ are polynomial.

4.1.3.4. Write

$$\mathrm{det}_m = \prod_{i=1}^m \varphi_i.$$

This character det_m is of course just the restriction to A of the determinant, which is a character of the whole group $\mathrm{GL}_m(\mathbf{C})$. By abuse of notation, we will also let det_m denote the determinant character of $\mathrm{GL}_m(\mathbf{C})$. In the notation of 4.1.3.3, this says

$$\sigma_{\mathrm{det}_m} = \mathrm{det}_m.$$

Since det_m is one-dimensional, if σ is an irreducible representation of $\mathrm{GL}_m(\mathbf{C})$, the tensor product $\mathrm{det}_m \otimes \sigma$ is also irreducible. If $\sigma = \sigma_\varphi$ has highest weight φ, then $\mathrm{det}_m \otimes \sigma$ has highest weight $(\mathrm{det}_m)\varphi$.

4.1.3.5. Obviously any $\varphi \in \hat{A}$ can be written in the form

$$\varphi = (\det_m)^\beta \varphi_0$$

where φ is a polynomial character of A. This decomposition is not unique. If it holds for $\beta = \beta_0$ then also $\varphi = (\det_m)^{\beta_0 - k}(\det_m^k \varphi_0)$ is another such decomposition for any $k \geq 0$. Thus there is a maximum β_0 for which this decomposition holds, and then it is valid also for all $\beta < \beta_0$. In fact, it is easy to see that $\beta_0 = \min_i a_i(\varphi)$.

Combining these observations with 4.1.3.4, we see that any irreducible representation $\sigma = \sigma_\varphi$, $\varphi \in \hat{A}_m^+$, of $\mathrm{GL}_m(\mathbf{C})$ can be written

$$\sigma \simeq (\det_m)^\beta \otimes \sigma_0,$$

where σ_0 is a polynomial representation, and $\beta \leq \min_i \alpha_i(\varphi)$.

4.1.3.6. Let $\sigma = \sigma_\varphi$ be a polynomial irreducible representation of $\mathrm{GL}_m(\mathbf{C})$. We can associate to σ_φ a Young's diagram $D = D_\varphi$. Indeed, since φ is polynomial, the numbers $\alpha_i(\varphi)$ are a decreasing sequence of nonnegative integers. We let the ith row of D_φ have length $\alpha_i(\varphi)$. Thus if $\varphi = \varphi_1^6 \varphi_2^5 \varphi_3^3$, then D_φ is the Young's diagram used in 3.3.2.1.

4.1.3.7. The irreducible representations of the symmetric group S_n are also parametrized by Young's diagrams. This is explained in [**W**, Chapter 4, §3]. We cannot review the details of the correspondence, but we will briefly recall its essential properties.

We imagine S_n as the group of all permutations of the set $\{1, 2, 3, \ldots, n\}$. Fix a Young's diagram D of size n. Denote by $\alpha_i(D)$ the length of the ith row of D, and $\beta_j(D)$ the length of the jth column of D. Let P be a partition of $\{1, 2, \ldots, n\}$ into subsets of sizes $\alpha_i(D)$, and let P' be a partition into subsets of sizes $\beta_j(D)$. Let $S_P \subseteq S_n$ be the subgroup which leaves each of the sets in P invariant. Similarly let $S_{P'}$ be the subgroup of S_n which leaves invariant each of the sets comprising P'. Then there is a unique irreducible representation of S_n which simultaneously contains a vector which is invariant under S_P and another vector which transforms according to the signum character of $S_{P'}$. This is the representation corresponding to D. Denote it by J_D.

4.1.3.8. As the reader probably expects, the parameter D in Corollary 4.1.2 is indeed a Young's diagram, and the parametrization of the σ_D and J_D is as explained in 4.1.3.6 and 4.1.3.7 respectively. To be precise, the summation in Corollary 4.1.2 runs over all diagrams D of size n and having at most m rows.

4.1.4.0. Corollary 4.1.2 also allows us to characterize the $\mathrm{GL}(V)$ isotypic components of $V^{\otimes n}$ as isotypic components for S_n, and vice versa. This gives rise to a reciprocal relation between the representations of $\mathrm{GL}(V)$ and the representations of the symmetric groups which can be useful in computations. In particular, it can be used to reduce various questions about $\mathrm{GL}_m(\mathbf{C})$ to questions about symmetric groups; these questions are clearly in principle effectively decidable (though they may involve very lengthy computations).

4.1.4.1. Here is an example. We can regard S_n and S_r as subgroups of S_{n+r} in the obvious way: S_n acts as usual on $\{1, 2, \ldots, n\}$ and leaves $n+1, n+2, \ldots, n+r$

all fixed; and S_r fixes $1, 2, \ldots, n$, and permutes $\{n+1, \ldots, n+r\}$ arbitrarily. The product $S_n \times S_r$ of these two subgroups is clearly also a subgroup of S_{n+r}. Given representations J_1 of S_n, J_2 of S_r, we can define the (outer) tensor product $J_1 \otimes J_2$ as a representation of $S_n \times S_r$.

4.1.4.2. PROPOSITION (FROBENIUS-SCHUR RECIPROCITY). *Consider Young's diagrams D_1, D_2. Let $\sigma_{D_1}, \sigma_{D_2}$ be the irreducible polynomial representations of $\mathrm{GL}_m(\mathbf{C})$ corresponding to the D_i. (See 4.1.3.6.) Let the size of D_i be n_i, and let J_{D_i} be the representation of S_{n_i} associated to D_i. (See 4.1.3.7.) Then for each Young's diagram D of size $n_1 + n_2$ and at most m rows, we have*

$$\dim \mathrm{Hom}(\sigma_{D_1} \otimes \sigma_{D_2}, \sigma_D)^{\mathrm{GL}_m(\mathbf{C})} = \dim \mathrm{Hom}(J_{D_1} \otimes J_{D_2}, J_D)^{S_{n_1} \times S_{n_2}}.$$

In words, the multiplicity of the representation σ_D of $\mathrm{GL}_m(\mathbf{C})$ in the tensor product of the σ_{D_i} is equal to the multiplicity of the corresponding representation J_D of $S_{n_1+n_2}$ in the "Zelevinsky product" [Ze1] $J_1 \times J_2$. (See Remark 4.1.4.5.)

4.1.4.3. PROOF. Consider the σ_{D_1} isotypic subspace of $(\mathbf{C}^m)^{\otimes n_1}$ under the action of $\mathrm{GL}_m(\mathbf{C})$. According to Corollary 4.1.2, this is also the J_{D_1}-isotypic subspace of $(\mathbf{C}^m)^{\otimes n_1}$ under the action of S_{n_1}. Similarly the σ_{D_2}-isotypic subspace for $\mathrm{GL}_m(\mathbf{C})$ acting on $(\mathbf{C}^m)^{\otimes n_2}$ is the J_{D_2}-isotypic subspace for S_{n_2}. These spaces have the form $\sigma_{D_1} \otimes J_{D_1}$ and $\sigma_{D_2} \otimes J_{D_2}$ respectively.

The tensor product $Y = (\sigma_{D_1} \otimes J_{D_1}) \otimes (\sigma_{D_2} \otimes J_{D_2})$ is naturally a subspace of $(\mathbf{C}^m)^{\otimes(n_1+n_2)}$. The isomorphism

$$(\sigma_{D_1} \otimes J_{D_1}) \otimes (\sigma_{D_2} \otimes J_{D_2}) \simeq (\sigma_{D_1} \otimes \sigma_{D_2}) \otimes (J_{D_1} \otimes J_{D_2})$$

exhibits Y as the $J_{D_1} \otimes J_{D_2}$-isotypic component of $(\mathbf{C}^m)^{\otimes(n_1+n_2)}$ under the action of $S_{n_1} \times S_{n_2}$. Let us write

$$\sigma_{D_1} \otimes \sigma_{D_2} \simeq \sum m_D \sigma_D,$$

where m_D is the multiplicity of σ_D in $\sigma_{D_1} \otimes \sigma_{D_2}$. Then we have

$$Y \simeq \sum m_D (\sigma_D \otimes J_{D_1} \otimes J_{D_2}).$$

Again by Corollary 4.1.2, the σ_D-isotypic component of $(\mathbf{C}^m)^{\otimes(n_1+n_2)}$ is also the J_D-isotypic component for $S_{n_1+n_2}$, and has the form $\sigma_D \otimes J_D$. It follows that $\sigma_D \otimes (m_D(J_{D_1} \otimes J_{D_2}))$ is the $J_{D_1} \otimes J_{D_2}$-isotypic component of $\sigma_D \otimes J_{D|S_{n_1} \times S_{n_2}}$, and hence m_D is also the multiplicity of $J_{D_1} \otimes J_{D_2}$ in the restriction of J_D to $S_{n_1} \times S_{n_2}$. But this is exactly the statement of Proposition 4.1.4.2.

4.1.4.4. REMARK. A result analogous to Proposition 4.1.4.2 clearly holds for k-fold tensor products.

4.1.4.5. REMARK. Proposition 4.1.4.2 can also be stated using the "Zelevinsky product" [Ze1] of representations of symmetric groups. Indeed, with J_{D_i} as above, define the representation $J_{D_1} \times J_{D_2}$ of $S_{n_1+n_2}$ as an induced representation

$$J_{D_1} \times J_{D_2} = \mathrm{ind}_{S_{n_1} \times S_{n_2}}^{S_{n_1+n_2}} J_{D_1} \otimes J_{D_2}.$$

Then by Frobenius reciprocity, Proposition 4.1.4.2 can be restated

$$\dim \operatorname{Hom}(\sigma_{D_1} \otimes \sigma_{D_2}, \sigma_D)^{\operatorname{GL}_m(\mathbf{C})} = \dim \operatorname{Hom}(J_{D_1} \times J_{D_2}, J_D)^{S_{n_1} + n_2}.$$

4.1.4.6. REMARK. The multiplicities involved in Proposition 4.1.4.2 are explicitly described by the Littlewood-Richardson Rule [**M**, **Ze2**].

4.2. Corollary 4.1.2 and Proposition 4.1.4.2 form the basis for the famous "Symbolic Method" of constructive invariant theory. See [**W**, Chapter 8, §2], and [**DC**]. Although we do not use this method, at least in its pure form, in our main construction, it was such a fixture in the 19th century that to pass over it without mention would be a solecism.

Let σ be a representation of $\operatorname{GL}_m(\mathbf{C})$ on a space V. We would like to compute the invariants or covariants of $\operatorname{SL}_m(\mathbf{C})$ acting on $\mathscr{P}(V)$. To compute the covariants of a given degree k amounts to decomposing the representation of $\operatorname{SL}_m(\mathbf{C})$ on $\mathscr{P}^k(V)$. Since $\operatorname{SL}_m(\mathbf{C})$ and $\operatorname{GL}_m(\mathbf{C})$ are very much alike, to facilitate the use of the results of 4.1 we will discuss the action of $\operatorname{GL}_m(\mathbf{C})$ on $\mathscr{P}^k(V)$. The reader should be able to make the modifications required to deal with $\operatorname{SL}_m(\mathbf{C})$.

Also, to simplify the discussion we will assume $\sigma \simeq \sigma_D^*$ is the contragredient of an irreducible polynomial representation σ_D of $\operatorname{GL}_m(\mathbf{C})$. (In fact, classically σ would usually be the dual of a symmetric power of the standard representation on $\mathbf{C}^m$.) Again, the reader should be able to modify the treatment to cover the general case. Since for any vector space U we have $\mathscr{P}^k(U^*) \simeq \mathscr{S}^k(U)$, where $\mathscr{S}^k(U)$ denotes the kth symmetric power of U, our problem is to decompose $\mathscr{S}^k(\sigma_D)$, the kth symmetric power of an irreducible polynomial representation σ_D of $\operatorname{GL}_m(\mathbf{C})$.

If the Young's diagram D has size n, we may realize σ_D as a subspace of $(\mathbf{C}^m)^{\otimes n}$. Corollary 4.1.2 then tells us that the σ_D-isotypic subspace of $(\mathbf{C}^m)^{\otimes n}$ has the form $\sigma_D \otimes J_D$, where J_D is the representation of S_n associated to D. The kth symmetric power $\mathscr{S}^k(\sigma_D)$ is a subspace of the kth tensor power $\sigma_D^{\otimes k}$ of σ_D. This of course is contained in $(\mathbf{C}^m)^{\otimes nk}$, inside $(\sigma_D \otimes J_D)^{\otimes k} \simeq (\sigma_D)^{\otimes k} \otimes (J_D \otimes J_D \otimes \cdots \otimes J_D)$, where we regard $J_D \otimes \cdots \otimes J_D$ (the outer tensor product of k copies of J_D) as a representation of $(S_n)^k$.

The space $\mathscr{S}^k(\sigma_D)$ is characterized inside $(\sigma_D)^{\otimes k}$ as the subspace invariant under the action of S_k which permutes the various factors σ_D among themselves. Think of $(S_n)^k$ as the subgroup of S_{nk} which leaves invariant the subsets $Y_a = \{(a-1)n + 1, (a-1)n + 2, \ldots, an\}$ for $1 \le a \le k$. Think of S_k as the subgroup of S_{nk} which permutes the sets Y_a "bodily" among themselves; precisely $s \in S_k$ acts by a recipe

$$\delta_n(s)((a-1)n + l) = (s(a) - 1)n + l, \qquad 1 \le a \le k, \ 1 \le l \le n,$$

where $a \to s(a)$ is the usual action of S_k on $\{1, 2, \ldots, k\}$. It is clear that the action $\delta_n(S_k)$ is such that it induces the standard permutation action of S_k on the factors $\sigma_D \otimes J_D$ of $(\sigma_D \otimes J_D)^{\otimes k}$. In particular, let L be a line in (the space of) J_D, so that $\sigma_D \otimes L \subseteq \sigma_D \otimes J_D$ is a copy of (i.e., is isomorphic as $\operatorname{GL}_m(\mathbf{C})$-module to) σ_D. Then

$$(\sigma_D \otimes L)^{\otimes k} \simeq (\sigma_D)^{\otimes k} \otimes (L)^{\otimes k} \subseteq (\sigma_D)^{\otimes k} \otimes (J_D \otimes J_D \otimes \cdots \otimes J_D) \subseteq (\mathbf{C}^m)^{\otimes nk}$$

is a copy of $(\sigma_D)^{\otimes k}$. It is clear that $\delta_n(S_k)$ will preserve $(\sigma_D \otimes L)^{\otimes k}$, and that the $\delta_n(S_k)$ fixed vectors $((\sigma_D \otimes L)^{\otimes k})^{S_k}$ will form a copy of $\mathscr{S}^k(\sigma_D)$.

On the other hand, it is clear that $\delta_n(S_k)$ normalizes $(S_n)^k$ inside S_{nk}. If we represent elements of $(S_n)^k$ as k-tuples $(s_1, s_2, \ldots, s_k)$, $s_i \in S_n$, then the action by conjugation of $\delta_n(S^k)$ on $(S_n)^k$ simply permutes the s_i. Thus the group $\delta_n(S_k) \cdot (S_n)^k$ is the wreath product [**Ha**] of S_n with S_k. In any case the representation $J_D \otimes \cdots \otimes J_D$ of $(S_n)^k$ is clearly left fixed by $\delta_n(S_k)$. Further, we can extend $J_D \otimes \cdots \otimes J_D$ to a representation of $\delta_n(S_k) \cdot (S_n)^k$. Indeed, if J_D is realized on a space U, then $J_D \otimes \cdots \otimes J_D$ is realized on $U^{\otimes k}$; if we let S_k act by its standard action on $U^{\otimes k}$, this action will normalize $J_D \otimes \cdots \otimes J_D((S_n)^k)$, and will give the desired extension. Denote this extension of $J_D \otimes \cdots \otimes J_D$ to $\delta_n(S^k) \cdot (S_n)^k$ by J_D^{wk}.

As noted above, we may realize $\mathscr{S}^k(\sigma_D)$ as the subspace of $\delta_n(S_k)$-invariant vectors in $(\sigma_D)^{\otimes k} \otimes (L \otimes \cdots \otimes L)$, where L is an arbitrarily chosen line in the space of J_D. The salient observation to make is that, if e is the minimal idempotent in the group algebra of $\delta_n(S_k) \cdot (S_n)^k$ which projects to the line $L \otimes \cdots \otimes L$ in the representation J_D^{wk}, then this copy of $\mathscr{S}^k(\sigma_D)$ is precisely the image of this minimal idempotent acting on $(\mathbf{C}^m)^{\otimes nk}$ (via the permutation action of S_{nk} and the embedding described above of $\delta_n(S_k) \cdot (S_n)^k$ in S_{nk}). Thus we can write

$$\mathscr{S}^k(\sigma_D) \simeq e((\mathbf{C}^m)^{\otimes nk}) = e\left(\sum_E \sigma_E \otimes J_E\right) = \sum_E \sigma_E \otimes e(J_E).$$

Here the sum is over Young's diagrams of size nk. On one hand, $\dim(e(J_E))$ is the dimension of the image of e in the representation J_E of S_{nk}. Since e is a minimal idempotent in $\delta_n(S_k) \cdot (S_n)^k$, this dimension is the same as the multiplicity of J_D^{wk} in the restriction of J_E to $\delta_n(S^k) \cdot (S_n)^k$. On the other hand, $\dim(e(J_E))$ is clearly the multiplicity with which σ_E occurs in $\mathscr{S}^k(\sigma_D)$. Thus we have

PROPOSITION (SYMBOLIC METHOD). *For Young's diagrams D of size n and E of size nk (and both with at most m rows), we have*

$$\dim \operatorname{Hom}(\sigma_E, \mathscr{S}^k(\sigma_D))^{\operatorname{GL}_m(\mathbf{C})} = \dim \operatorname{Hom}(J_D^{wk}, J_E)^H$$
$$= \dim \operatorname{Hom}(J_E, \operatorname{ind}_H^{S_{nk}} J_D^{wk}),$$

where we have abbreviated $\delta_n(S_k) \cdot (S_n)^k = H$.

The point of this result is that it reduces questions about invariants and covariants of $\operatorname{GL}_m(\mathbf{C})$ or $\operatorname{SL}_m(\mathbf{C})$ to questions about the finite group S_{nk}, and these latter questions are obviously solvable by finite computations—in fact by straightforward (but lengthy!) algorithms.

4.2.1. REMARK. The above discussion has a straightforward extension to general plethysms in the sense of Littlewood [**M**].

4.3. A drawback of the symbolic method is that, to decompose the polynomials of degree k, it requires computations with S_{nk}, which becomes arbitrarily large (fast) as k increases. Thus for each k the computations are finite, but they increase without bound as k increases. The 19th-century invariant theorists found a method of computing invariants for the binary n-ic which only involves computations with S_n. A description of this method is our goal.

4.3.1. We require some alternative versions, formulated in terms of polynomials rather than tensors, of Schur's Theorem 4.1.1. One version is

THEOREM (FIRST FUNDAMENTAL THEOREM OF CLASSICAL INVARIANT THEORY FOR GL_m) [**W**, Chapter II, §6, or Chapter V, §1]. *Let V be a (complex, finite-dimensional) vector space, with dual V^*. Let $\mathrm{GL}(V)$ act on $V^k \simeq V \otimes \mathbf{C}^k$ by means of the k-fold direct sum of its standard action on V; let $\mathrm{GL}(V)$ act on $V^{*l} \simeq V^* \otimes \mathbf{C}^l$ by means of the l-fold direct sum of its standard action on V^* (dual to its action on V); and let $\mathrm{GL}(V)$ act on $V^k \oplus V^{*l}$ by the sum of these two actions. Then the algebra $\mathscr{P}(V^k \oplus V^{*l})^{\mathrm{GL}(V)}$ of polynomial invariants for this action is generated by the kl quadratic polynomials*

$$\lambda^j(v^i), \qquad 1 \le j \le l, \ 1 \le i \le k,$$

where $(v_1, \ldots, v_k)$, $v_i \in V$, denotes a typical element of V^k, and $(\lambda_1, \ldots, \lambda_l)$, $\lambda_j \in V^*$, denotes a typical element of V^{*l}.

4.3.2. Although Theorem 4.3.1 purportedly describes only invariants, in fact, because it describes invariants for arbitrarily many variables it can be turned into a description of all covariants by a procedure of "doubling the variables" [**Ho1**]. To describe this in detail would take us too far afield, but we sketch the procedure.

4.3.3.1. Consider $\mathscr{PD}(V)$, the space of polynomial coefficient differential operators on $\mathscr{P}(V)$. This is an algebra, generated by the first-order operators

$$\lambda: \text{multiplication by } \lambda \in V^* = \mathscr{P}^1(V),$$
$$\partial_v: \text{directional derivative in the direction } v \in V.$$

4.3.3.2. We may place a "total degree" filtration on $\mathscr{PD}(V)$: let the subspace $\mathscr{PD}^{(j)}(V)$ be the linear span of all products of j or less of the generators 4.3.3.1. We obviously have

$$\mathscr{PD}^{(j)}(V) \cdot \mathscr{PD}^{(k)}(V) \subseteq \mathscr{PD}^{(j+k)}(V).$$

4.3.3.3. The generators 4.3.3.1 satisfy the "canonical commutation relations"

$$[\partial_v, \lambda] = \lambda(v),$$

where $\lambda(v)$ here indicates the operator of multiplication by the number $\lambda(v)$. It follows that

$$\left[\mathscr{PD}^{(j)}(V), \mathscr{PD}^{(k)}(V)\right] \subseteq \mathscr{PD}^{(j+k-2)}(V).$$

That is, the commutator of elements from $\mathscr{PD}^{(j)}$ and $\mathscr{PD}^{(k)}$ belongs to $\mathscr{PD}^{(j+k-2)}(V)$.

4.3.3.4. From the filtered algebra $\mathscr{PD}(V)$ we can form the associated graded algebra [**J**, p. 147]

$$\mathrm{Gr}\,\mathscr{PD}(V) = \sum_j \mathscr{PD}^{(j)}(V)/\mathscr{PD}^{(j-1)}(V) = \sum_j \mathrm{Gr}\,\mathscr{PD}^j(V).$$

The multiplication in $\mathrm{Gr}\,\mathscr{P}\mathscr{D}(V)$ is the direct sum of the quotients by lower-order terms of the maps

$$\mathscr{P}\mathscr{D}^{(j)}(V) \otimes \mathscr{P}\mathscr{D}^{(k)}(V) \to \mathscr{P}\mathscr{D}^{(j+k)}(V), \qquad P \otimes Q \to PQ.$$

4.3.3.5. By virtue of 4.3.3.3, the graded algebra $\mathrm{Gr}\,\mathscr{P}\mathscr{D}(V)$ is a commutative algebra. It is generated by

$$\mathrm{Gr}\,\mathscr{P}\mathscr{D}^{1}(V) = \mathscr{P}\mathscr{D}^{(1)}(V)/\mathscr{P}\mathscr{D}^{(0)}(V) \simeq V \oplus V^*.$$

Since the symmetric algebra $\mathscr{S}(U)$ on a vector space U is the universal commutative algebra generated by U, there is a surjective homomorphism of algebras

$$\mathscr{S}(V \oplus V^*) \to \mathrm{Gr}\,\mathscr{P}\mathscr{D}(V).$$

It is not hard to check that this map is in fact an isomorphism.

4.3.3.6. On $V \oplus V^*$ we may define the symplectic form

$$\langle (v, \lambda), (v', \lambda') \rangle = \lambda(v') - \lambda'(v), \qquad v, v' \in V;\ \lambda, \lambda' \in V^*.$$

This form induces an identification $V \oplus V^* \simeq (V \oplus V^*)^*$, which in turn induces an isomorphism $\mathscr{S}(V \oplus V^*) \simeq \mathscr{P}(V \oplus V^*)$. From 4.3.3.5, we conclude

$$\mathrm{Gr}\,\mathscr{P}\mathscr{D}(V) \simeq \mathscr{P}(V \oplus V^*).$$

4.3.3.7. Since $\mathscr{P}\mathscr{D}(V) \subseteq \mathrm{End}(\mathscr{P}(V))$, and $\mathrm{GL}(V)$ acts by automorphisms on $\mathscr{P}(V)$, we may conjugate an element of $\mathscr{P}\mathscr{D}(V)$ by an element of $\mathrm{GL}(V)$. The result will again be an element of $\mathscr{P}\mathscr{D}(V)$. This yields an action of $\mathrm{GL}(V)$ on $\mathscr{P}\mathscr{D}(V)$ by automorphisms. On the other hand, $\mathrm{GL}(V)$ acts by automorphisms on $\mathscr{P}(V \oplus V^*)$ in the obvious way (see Theorem 4.3.1). It is not hard to check that the isomorphism 4.3.3.6 identifies these two actions.

4.3.3.8. Now replace V by $V^k \oplus V^{*l} \simeq V \otimes \mathbf{C}^k \oplus V^* \otimes \mathbf{C}^l$. Then

$$(V^k \oplus V^{*l}) \oplus (V^k \oplus V^{*l})^* \simeq V \otimes (\mathbf{C}^k \oplus \mathbf{C}^{l*}) \oplus V^* \otimes (\mathbf{C}^l \oplus \mathbf{C}^{k*})$$
$$\simeq V^{(k+l)} \oplus V^{*(k+l)}.$$

We can describe operators in $\mathscr{P}\mathscr{D}^{(2)}(V^k \oplus V^{l*})$ whose images in

$$\mathrm{Gr}\,\mathscr{P}\mathscr{D}(V^k \oplus V^{*l})$$

correspond via isomorphism 4.3.3.6 to the generators of $\mathscr{P}(V^{(k+l)} \oplus V^{*(k+l)})$ described in Theorem 4.3.1.

It is probably simplest to describe these operators in terms of coordinates. Let $\dim V = m$. Choose a basis $\{b_i\}_{i=1}^m$ for V; let $\{\lambda_i\}$ be the dual basis for V^*. Let $\{e_a\}_{i=1}^r$ be the standard basis of $\mathbf{C}^r$. Then $b_i \otimes e_a$, $1 \leq i \leq \dim V$, $1 \leq a \leq k$, constitute a basis for $V \otimes \mathbf{C}^k \simeq V^k$. Let x_{ia} be coordinates on V^k relative to this basis. Similarly let y_{ib}, $1 \leq i \leq m$, $1 \leq b \leq l$, denote coordinates on V^{*l} relative to the basis $\lambda_i \otimes e_b$ of $V^{*l} \simeq V^* \otimes \mathbf{C}^l$. Consider the operators

(a) $$\sum_{i=1}^m x_{ia} y_{ib}, \qquad 1 \leq a \leq k,\ 1 \leq b \leq l,$$

(these are multiplication operators by the polynomials of Theorem 4.3.1);

(b) $\qquad \sum_{i=1}^{m} x_{ia} \dfrac{\partial}{\partial x_{ia'}}, \quad 1 \leq a, a' \leq k, \qquad \sum_{i=1}^{m} y_{ib} \dfrac{\partial}{\partial y_{ib'}}, \quad 1 \leq b, b' \leq l;$

(c) $\qquad \sum_{i=1}^{m} \dfrac{\partial^2}{\partial x_{ia} \partial y_{i\beta}}.$

We can check that, on one hand, these operators correspond via the isomorphism 4.3.3.6 to the generating invariants for $V^{(k+l)} \otimes V^{*(k+l)}$ described in Theorem 4.3.1; and on the other hand that they are invariant under conjugation by G. A short argument comparing the filtered algebra $\mathscr{PD}$ with its associated graded algebra $\mathrm{Gr}\,\mathscr{PD}$ yields the following consequence of Theorem 4.3.1.

THEOREM [**Ho1**]. *The algebra* $\mathscr{PD}(V^k \otimes V^{*l})^{\mathrm{GL}(V)}$ *of operators invariant under conjugation is generated by the operators of types* (a), (b), (c) *above.*

4.3.3.9. Suppose in 4.3.1 and 4.3.3.8 that $l = 0$, so that we are considering $\mathrm{GL}(V)$ acting on $V^k \simeq V \otimes \mathbf{C}^k$. Then the only operators from 4.3.3.8 that are relevant here are the operators $\sum x_{ia} \partial / \partial x_{i\beta}$. Weyl calls these *polarization operators*; they were also known as Aronhold operators [**GY**, p. 12]; [**W**, p. 5]. They have a simple geometric interpretation as follows. The action of $\mathrm{GL}(V)$ on the tensor product $V \otimes \mathbf{C}^k$ is via $g \to g \otimes I_k$, where $g \in \mathrm{GL}(V)$ and I_k is the identity operator on $\mathbf{C}^k$. Clearly $\mathrm{GL}_k(\mathbf{C})$ also acts on $V \otimes \mathbf{C}^k$, via the map $A \to I_V \otimes A$, where $A \in \mathrm{GL}_k(\mathbf{C})$, and I_V is the identity operator on V. This action extends in the standard way to an action of $\mathrm{GL}_k(\mathbf{C})$ on $\mathscr{P}(V \otimes \mathbf{C}^k)$. Clearly the actions of $\mathrm{GL}(V)$ and $\mathrm{GL}_k(\mathbf{C})$ on $V \otimes \mathbf{C}^k$ and on $\mathscr{P}(V \otimes \mathbf{C}^k)$ commute with one another. It is a straightforward matter to check that the polarization operators are precisely the Lie algebra (the infinitesimal generators) of this action of $\mathrm{GL}_k(\mathbf{C})$.

The actions of $\mathrm{GL}(V)$ and $\mathrm{GL}_k(\mathbf{C})$ preserve the spaces $\mathscr{P}^n(V \otimes \mathbf{C}^k)$ of homogeneous polynomials. Since $\mathscr{P}^n(V \otimes \mathbf{C}^k)$ is finite dimensional, the subalgebras of $\mathrm{End}(\mathscr{P}^n(V \otimes \mathbf{C}^k))$ generated by $\mathrm{GL}_k(\mathbf{C})$ and by its infinitesimal generators are the same. It follows from Theorem 4.3.3.8 that the subalgebras of $\mathrm{End}(\mathscr{P}^n(V \otimes \mathbf{C}^k))$ spanned by $\mathrm{GL}(V)$ and by $\mathrm{GL}_k(\mathbf{C})$ are mutual commutants. Thus the same reasoning used to pass from Schur's Theorem 4.1.1 to Corollary 4.1.2, plus a calculation, yields the following statement.

THEOREM [**GK, Ho1, KV, Se2, Zh**]. *Under the joint action of* $\mathrm{GL}(V) \times \mathrm{GL}_k(\mathbf{C})$, *we have a decomposition*

$$\mathscr{P}^n(V \otimes \mathbf{C}^k) \simeq \sum \sigma_D^m \otimes \sigma_D^k.$$

Here the sum is over all Young's diagrams of size n and with at most $\min(m, k)$ rows. (Recall $m = \dim V$.) The superscript on σ_D^k indicates the representation of $\mathrm{GL}_k(\mathbf{C})$ attached to the Young's diagram D.

REMARK. This theorem also has deep roots in the literature. It is closely connected to a famous 19th-century identity (Cauchy's Lemma) for "S-functions"

([**M; W**, p. 202]). Interestingly, while Weyl proves and uses Cauchy's Lemma, he does not seem to geometrize it as in the above Theorem. Some of Weyl's uses of Cauchy's Lemma provide instances of the reciprocity law stated in 4.3.3.10. Seshadri [**Se2**] notes that the case $m = k$ (from which the general case follows) is essentially an example of the Peter-Weyl Theorem. Also Turnbull's theory of "double standard tableaux" [**T**] translates into this result via the theory of Gelfand-Cetlin bases [**Zh**]. A characteristic-free discussion of these results is given in [**DRS**].

4.3.3.10. REMARK. From Theorem 4.3.3.9 there follows a reciprocity law analogous to Frobenius-Schur Reciprocity (Proposition 4.1.4.2). We omit the proof, which is similar to that of Proposition 4.1.4.2.

Observe that the decomposition $\mathbf{C}^{k+l} = \mathbf{C}^k \oplus \mathbf{C}^l$ induces an embedding of $GL_l(\mathbf{C}) \times GL_k(\mathbf{C})$ into $GL_{k+l}(\mathbf{C})$.

PROPOSITION. *Let D_1 and D_2 be Young's diagrams. Let D_i have k_i or fewer nonzero rows. Take $n \geq k_i$. Then for any Young's diagram D with at most $\min(n, k_1 + k_2)$ rows, we have*

$$\dim \operatorname{Hom}(\sigma^n_{D_1} \otimes \sigma^n_{D_2}, \sigma^n_D)^{GL_n} = \dim \operatorname{Hom}(\sigma^{k_1+k_2}_D, \sigma^{k_1}_{D_1} \otimes \sigma^{k_2}_{D_2})^{GL_{k_1} \times GL_{k_2}}.$$

5. Application to invariants of binary forms.

5.1.1. We will study the invariants of a binary n-ic using the theory of §4, especially Theorem 4.3.3.9. The procedure we follow bears a clear analogy with the Symbolic Method (§4.2), but rather than study the homogeneous polynomials one degree at a time, we provide a way of looking at all of them at once.

The idea is simply this. Recall the invariants of the binary n-ic form the algebra $\mathscr{I}_n = \mathscr{P}(\mathscr{P}^n(\mathbf{C}^2))^{SL_2}$. We study it by comparing it with another algebra we understand well. More precisely, we will construct an injection

$$\mu^* : \mathscr{I}_n \to \mathscr{P}(\mathbf{C}^2 \otimes \mathbf{C}^n)^{SL_2}.$$

Our understanding of the second algebra comes from Theorem 4.3.3.9. The map μ^* is in fact the restriction of an SL_2-invariant injection of algebras

$$\mu^* : \mathscr{P}(\mathscr{P}^n(\mathbf{C}^2)) \to \mathscr{P}(\mathbf{C}^2 \otimes \mathbf{C}^n).$$

5.1.2. To construct μ^* we recall there is a natural equivalence between
   (i) polynomial mappings $P : V \to U$ from a vector space V to a vector space U,
   (ii) linear mappings $P^* : U^* \to \mathscr{P}(V)$,
   (iii) algebra homomorphisms $P^* : \mathscr{P}(U) \to \mathscr{P}(V)$.

If both V and U are G-modules for some group G, then one of these maps will be G-equivariant if and only if the others are. The mapping (iii) will be injective if and only if the mapping (i) has rank (in the sense of the maximum rank of its derivative (Jacobian matrix) at points) equal to $\dim U$.

5.1.3. Here is a polynomial map:

$$\mu : (\mathbf{C}^2)^n \to \mathscr{P}^{(n)}(\mathbf{C}^2)$$

that gives rise via 5.1.2 to the homomorphisms μ^* of 5.1. Write a typical element of $(\mathbf{C}^2)^n$ as

$$\begin{bmatrix} a_1 & a_2 & \dots & a_n \\ b_1 & b_2 & \dots & b_n \end{bmatrix}.$$

Write a typical element of $\mathscr{P}^n(\mathbf{C})$ as

$$\sum_{a+b=n} \alpha_{a,b} x^a y^b.$$

Define

$$\mu\left(\begin{bmatrix} a_1 & \dots & a_n \\ b_1 & \dots & b_n \end{bmatrix}\right) = \prod_{i=1}^{n}(b_i x - a_i y).$$

One sees directly from Definition 2.2 that this mapping is equivariant for $\mathrm{GL}_2(\mathbf{C})$. The Fundamental Theorem of algebra tells us it is surjective. Hence the associated algebra homomorphisms are injective.

5.2.1. To use μ^* effectively, we have to be able to describe both the image $\mu^*(\mathscr{P}(\mathscr{P}^n(\mathbf{C}^2)))$ and the invariants $\mathscr{P}(\mathbf{C}^2 \otimes \mathbf{C}^n)^{\mathrm{SL}_2}$.

5.2.2. Consider the invariants first. A homogeneous polynomial p in $\mathscr{P}^l(\mathbf{C}^2 \otimes \mathbf{C}^n)^{\mathrm{SL}_2}$ will be invariant under SL_2 if and only if it is an eigenvector for GL_2; then it will transform by some power of the determinant $\det_2$. The Young's diagram $D(k,k)$ for $(\det_2)^k$ consists of two equal rows of length k. According to Theorem 4.3.3.9 the isotypic subspace for $\det_2^k$ occurs in $\mathscr{P}^{2k}(\mathbf{C}^2 \otimes \mathbf{C}^n)$, and is irreducible for the action of GL_n, supporting the irreducible representation $\sigma^n_{D(k,k)}$.

For $k = 1$, this representation is $\sigma^n_{D(1,1)}$ and

$$D_{(1,1)} = \quad \begin{array}{|c|} \hline \\ \hline \\ \hline \end{array}$$

This is the second fundamental representation of GL_n, on the Grassmann space $\Lambda^2(\mathbf{C}^n)$. It has dimension $\binom{n}{2} = n(n-1)/2$ and has a basis consisting of

$$[ij] = \det\left\| \begin{matrix} a_i & a_j \\ b_i & b_j \end{matrix} \right\| = a_i b_j - a_j b_i.$$

These quantities $[ij]$ are known as homogenized roots [**KR**] or bracket factors [**W**] or symbolical determinantal factors [**GY**], symbols for short.

PROPOSITION. *The algebra* $\mathscr{P}(\mathbf{C}^2 \otimes \mathbf{C}^n)^{\mathrm{SL}_2}$ *is generated by the space* $\mathscr{P}^2(\mathbf{C}^2 \otimes \mathbf{C}^n)^{\mathrm{SL}_2} \simeq \det_2 \otimes \sigma^n_{D(1,1)}$, *spanned by the symbols* $[ij]$ *described just above. The* $[ij]$ *satisfy the relations* ("*the syzygy*")

$$(\mathfrak{P}) \qquad\qquad [ij][kl] + [ik][lj] + [il][jk] = 0.$$

PROOF. The algebra generated by the $[ij]$ will contain nonzero polynomials of degree $2k$ for every $k \geq 1$ (e.g., $[ij]^k$). Since $\mathscr{P}^2(\mathbf{C}^2 \otimes \mathbf{C}^n)^{\mathrm{SL}_2}$ is GL_n-invariant, the algebra it generates will also be GL_n-invariant. Since $\mathscr{P}^{2k}(\mathbf{C}^2 \otimes \mathbf{C}^n)^{\mathrm{SL}_2}$ is irreducible under the action of GL_n, the $[ij]$ must generate the whole algebra of invariants. The identity between the $[ij]$ can be verified by direct computation, or observed to be equivalent to

$$\det \begin{bmatrix} a_1 & a_2 & a_3 & a_4 \\ b_1 & b_2 & b_3 & b_4 \\ a_1 & a_2 & a_3 & a_4 \\ b_1 & b_2 & b_3 & b_4 \end{bmatrix} = 0.$$

5.2.2.1. REMARKS. (a) Proposition 5.2.2. has both representation-theoretic and geometric interpretations. Guided by observation 5.2.2, we see that the map

$$\lambda \colon \begin{bmatrix} a_1 & a_2 & \cdots & a_n \\ b_1 & b_2 & \cdots & b_n \end{bmatrix} \to \left\{ \begin{bmatrix} a_i & a_j \\ b_i & b_j \end{bmatrix} \right\}$$

defines a GL_n-equivariant mapping $\mathbf{C}^2 \otimes \mathbf{C}^n$ to $\Lambda^2(\mathbf{C}^n)$. If we think of an element of $\mathbf{C}^2 \otimes \mathbf{C}^n$ as a pair $\begin{bmatrix} a \\ b \end{bmatrix}$ of vectors $a, b \in \mathbf{C}^n$ (thus reversing the roles of GL_2 and GL_n!), this map simply sends $\begin{bmatrix} a \\ b \end{bmatrix}$ to $a \wedge b$ (exterior product). Thus the image of λ consists of the subvariety of decomposable vectors in $\Lambda^2(\mathbf{C}^n)$. This of course is a homogeneous variety, whose image in the projective space $\mathbf{P}\Lambda^2(\mathbf{C}^n)$ is simply $Gr_{2,n}$, the Grassmann variety of 2-planes in n-space. Thus the relations $(\mathfrak{P})$ of Proposition 5.2.2 are simply the pullback via $\lambda^* \colon \mathscr{P}(\Lambda^2(\mathbf{C}^n)) \to \mathscr{P}(\mathbf{C}^2 \otimes \mathbf{C}^n)$ of the famous Plücker relations [**HP**, Chapter VII, §6], which give the defining equations inside $\Lambda^k(\mathbf{C}^n)$ for the Grassmann varieties $Gr_{k,n}$ of k-planes in n-space.

(b) The Plücker relations also have an interesting representation-theoretic interpretation, by means of which they have been generalized by Kostant. This topic unfortunately is beyond the scope of our discussion.

(c) The syzygies $(\mathfrak{P})$ in fact generate all relations between the homogenized roots. This is proved in [**KR**]. We do not need to know it here. However, if we did, the appropriate result to quote would be the Second Fundamental Theorem of classical invariant theory ([**W**, §6.1]; this is for the symplectic groups, of which SL_2 is one).

(d) From these remarks, we can see that via λ^* the algebra $\mathscr{P}(\mathbf{C}^2 \otimes \mathbf{C}^n)^{SL_2}$ is isomorphic to the "coordinate ring" of (i.e., the restriction of functions on $\Lambda^2(\mathbf{C}^n)$ to) the Grassmannian $Gr_{2,n}$.

5.2.2.2. REMARKS. (a) The dimension of $\mathscr{P}^{2k}(\mathbf{C}^2 \otimes \mathbf{C}^n)^{SL_2}$ is

$$\dim \sigma^n_{D(k,k)} = \frac{(n+k-1)!(n+k-2)!}{(k+1)!\,k!(n-1)!(n-2)!} = \frac{\binom{n+k-1}{n-2}\binom{n+k-2}{n-2}}{(n-1)} = \frac{\binom{n+k-1}{k}\binom{n+k-2}{k}}{(k+1)}.$$

This may be seen from the Weyl Dimension Formula ([**W**, p. 201]; it is due to Schur for GL_n) or from the equation

$$\mathscr{S}^k(\mathbf{C}^n) \otimes \mathscr{S}^k(\mathbf{C}^n) \simeq \mathscr{S}^{k+1}(\mathbf{C}^n) \otimes \mathscr{S}^{k-1}(\mathbf{C}^n) \oplus \sigma^n_{D(k,k)}$$

which is readily deducible from Proposition 4.3.3.10 (see also [**M**, p. 25]).

(b) An explicit basis for $\sigma^n_{D(k,k)}$ is given in [**KR**, Theorem 3.2]. (The cardinality is not given.) This basis is closely related to the Gelfand-Cetlin basis for this representation of GL_n [**Zh**].

(c) As a function of k, dim $\sigma^n_{D(k,k)}$ is a polynomial of degree $2(n-2) =$ dim $\text{Gr}_{2,n}$. The coefficient of $k^{2(n-2)}$ is $((n-1)!(n-2!))^{-1} = C_{n-1}/(2(n-2))!$ where

$$C_n = \frac{1}{n}\binom{2(n-1)}{n-1}$$

is the nth Catalan number. The Catalan numbers are associated with a number of combinatorial problems—polygon division, constrained random walks, and others [**Dr, Kn, F**, III.1]. Also C_{n+1} is the dimension $J_{D(n,n)}$ of the representation of S_{2n} corresponding to the Young's diagram $D(n,n)$. This occurrence of C_n is directly connected with the occurrence of C_n in connection with constrained random walks by means of the branching laws for restrictions of representations of GL_m to GL_{m-1}.

(d) The Hilbert function of $\mathscr{P}((\mathbf{C}^2)^n)^{\text{SL}_2}$ is

$$\sum_{k=0}^{\infty} \dim \sigma^n_{D(k,k)} t^k = \left(\sum_{j=0}^{n-3} \frac{1}{(j+1)}\binom{n-2}{j}\binom{n-3}{j}t^j\right)\Bigg/ (1-t)^{2n-3}.$$

5.2.3. The image $\mu^*(\mathscr{P}(\mathscr{P}^n(\mathbf{C}^2)))$, with μ as in 5.1.3, can be characterized in terms of GL_n. Let $A_n \subseteq \text{GL}_n$ be the group of diagonal matrices. The normalizer of A_n in GL_n is $S_n \cdot A_n$, where the symmetric group S_n is here realized as the group of permutation matrices in GL_n. Denote $S_n \cdot A_n = WA_n$, and let

$$WA_n^1 = WA_n \cap \text{SL}_n$$

denote the subgroup of elements in WA_n with determinant 1.

PROPOSITION. $\mu^*(\mathscr{P}(\mathscr{P}^n(\mathbf{C}^2))) = \mathscr{P}(\mathbf{C}^2 \otimes \mathbf{C}^n)^{WA_n^1}$.

PROOF. The algebra $\mu^*(\mathscr{P}(\mathscr{P}^n(\mathbf{C}^2)))$ is generated by the coefficients of

$$\mu\left(\begin{bmatrix} a_1 & \cdots & a_n \\ b_1 & \cdots & b_n \end{bmatrix}\right).$$

These are homogenized versions of the usual symmetric functions: explicitly they are

$$\gamma_l\left(\begin{bmatrix} a_1 & \cdots & a_n \\ b_1 & \cdots & b_n \end{bmatrix}\right) = \sum_{J \subseteq N}\left(\prod_{j \in J} a_j\right)\left(\prod_{i \notin J} b_i\right), \qquad 0 \le l \le n,$$

where J ranges over all subsets of cardinality l in $N = \{1, 2, \ldots, n\}$.

The condition that some homogeneous polynomial $q \in \mathscr{P}^l((\mathbf{C}^2 \otimes \mathbf{C}^n))$ be invariant under WA_n^1 is equivalent to the two conditions

(i) q should be homogeneous in each pair of variables $[\begin{smallmatrix}a_i\\b_i\end{smallmatrix}]$, and the degree of homogeneity should be the same for each i,

(ii) q should be invariant under permutations of the pairs $[\begin{smallmatrix}a_i\\b_i\end{smallmatrix}]$.

One sees by inspection that the polynomials γ_l satisfy these conditions. Conversely a homogenized version of Langrange's Theorem, that the algebra of symmetric polynomials is generated by the elementary symmetric polynomials, implies that the γ_l will generate the full algebra $\mathscr{P}((\mathbf{C}^2 \otimes \mathbf{C}^n))^{WA_n^1}$.

5.3. To summarize, the results of §5.2 show that we have the square of inclusions

$$\mathscr{P}((\mathbf{C}^2 \otimes \mathbf{C}^n)) \supseteq \mathscr{P}((\mathbf{C}^2 \otimes \mathbf{C}^n))^{WA_n^1} \overset{\mu^\bullet}{\simeq} \mathscr{P}(\mathscr{P}^n(\mathbf{C}^2))$$
$$\cup| \qquad\qquad \cup| \qquad\qquad \cup|$$
$$\mathscr{C}(\mathrm{Gr}_{2,n}) \overset{\lambda^\bullet}{\simeq} \mathscr{P}(\mathbf{C}^2 \otimes \mathbf{C}^n)^{\mathrm{SL}_2} \supseteq \mathscr{P}(\mathbf{C}^2 \otimes \mathbf{C}^n)^{\mathrm{SL}_2 \times WA_n^1} \overset{\mu^\bullet}{\simeq} \mathscr{I}_n$$

Here $\mathscr{C}(\mathrm{Gr}_{2,n})$ indicates the coordinate ring of the Grassmannian $\mathrm{Gr}_{2,n}$. In 5.2.3, the group WA_n^1 was constructed as an extension

$$1 \to A_n^1 \to WA_n^1 \to S_n \to 1.$$

Here A_n^1 is the group of diagonal matrices of determinant 1. A key step in the 19th-century analysis of binary invariants was to use this exact sequence of groups to enlarge the above square of inclusions, yielding the rectangle

$(\mathfrak{R})$

$$\mathscr{P}(\mathbf{C}^2 \otimes \mathbf{C}^n) \quad \supseteq \mathscr{P}(\mathbf{C}^2 \otimes \mathbf{C}^n)^{A_n^1} \quad \supseteq \mathscr{P}(\mathbf{C}^2 \otimes \mathbf{C}^n)^{WA_n^1} \simeq \mathscr{P}(\mathscr{P}^n(\mathbf{C}^2))$$
$$\cup| \qquad\qquad \cup| \qquad\qquad \cup| \qquad\qquad \cup|$$
$$\mathscr{P}(\mathbf{C}^2 \otimes \mathbf{C}^n)^{\mathrm{SL}_2} \supseteq \mathscr{P}(\mathbf{C}^2 \otimes \mathbf{C}^n)^{\mathrm{SL}_2 \times A_n^1} \supseteq \mathscr{P}(\mathbf{C}^2 \otimes \mathbf{C}^n)^{\mathrm{SL}_2 \times WA_n^1} \overset{\mu^\bullet}{\simeq} \mathscr{I}_n$$

It turns out that of the algebras in this rectangle, only the lower right corner, the ultimate object of interest, is intractably mysterious. The algebra just to its left, $\mathscr{P}(\mathbf{C}^2 \otimes \mathbf{C}^n)^{\mathrm{SL}_2 \times A_n^1}$, can be described with some precision. Thus the murkiest aspects of the structure of the algebra of binary invariants occur in the passage from $\mathscr{P}(\mathbf{C}^2 \otimes \mathbf{C}^n)^{\mathrm{SL}_2 \times A_n^1}$ to $\mathscr{P}(\mathbf{C}^2 \otimes \mathbf{C}^n)^{\mathrm{SL}_2 \times WA_n^1} = (\mathscr{P}(\mathbf{C}^2 \otimes \mathbf{C}^n)^{\mathrm{SL}_2 \times A_n^1})^{S_n}$. This passage only involves taking invariants with respect to the finite group S_n. In particular this approach gives effective (though rather bad) estimates on the numbers and degrees of the generators of the algebra of binary invariants.

With these remarks as motivation, we will direct our attention to understanding the middle column of the rectangle $(\mathfrak{R})$.

REMARK. The author has learned that J. Dixmier (unpublished) has also constructed the rectangle $(\mathfrak{R})$.

5.4.1. The space $\mathscr{P}(\mathbf{C}^2 \otimes \mathbf{C}^n)^{A_n^1}$ is very easy to describe. We have the classical decomposition

$$\mathscr{P}^l((\mathbf{C}^2)^n) \simeq \sum_{\sum l_i = n} \mathscr{P}^{l_1}(\mathbf{C}^2) \otimes \mathscr{P}^{l_2}(\mathbf{C}^2) \otimes \cdots \otimes \mathscr{P}^{l_n}(\mathbf{C}^n).$$

The summands in this decomposition are precisely the eigenspaces for A_n. In the notation of 4.1.3.2, the character φ of A_n attached to the typical summand of this decomposition is specified by $\alpha_i(\varphi) = l_i$. In particular, φ will be trivial on A_n^1 if and only if $l_i = l_j$ for all i, j. We conclude:

$$\mathscr{P}^l(\mathbf{C}^2 \otimes \mathbf{C}^n)^{A_n^1} = 0 \quad \text{if } l \text{ is not divisible by } n,$$
$$\mathscr{P}^{nk}(\mathbf{C}^2 \otimes \mathbf{C}^n)^{A_n^1} \simeq (\mathscr{P}^k(\mathbf{C}^2))^{\otimes n}.$$

5.4.1.1. If we now take the S_n-invariants in $\mathscr{P}^{nk}(\mathbf{C}^2 \otimes \mathbf{C}^n)^{A_n^1}$, we see from the isomorphism of 5.4.1 that

$$\mathscr{P}^{nk}(\mathbf{C}^2 \otimes \mathbf{C}^n)^{WA_n^1} \simeq ((\mathscr{P}^k(\mathbf{C}^2))^{\otimes n})^{S_n} \simeq \mathscr{S}^n(\mathscr{P}^k(\mathbf{C}^2)).$$

Combining this with μ^*, we obtain the isomorphism

$$\mathscr{P}^k(\mathscr{P}^n(\mathbf{C}^2)) \overset{\mu^*}{\simeq} \mathscr{S}^n(\mathscr{P}^k(\mathbf{C}^2)).$$

This is an explicit SL_2-isomorphism of the sort implied by Hermite reciprocity (see 3.3.2.2).

5.4.2. Our discussion so far gives us several approaches to $\mathscr{P}(\mathbf{C}^2 \otimes \mathbf{C}^n)^{SL_2 \times A_n^1}$. In particular, with reference to the diagram ($\mathfrak{R}$) of 5.3, we may approach from above, through $\mathscr{P}(\mathbf{C}^2 \otimes \mathbf{C}^n)^{A_n^1}$, or from the left, through $\mathscr{P}(\mathbf{C}^2 \otimes \mathbf{C}^n)^{SL_2}$. Both approaches provide useful information.

5.4.2.1. Approaching from above, and using (5.4.1), we see that

$$\mathscr{P}^l(\mathbf{C}^2 \otimes \mathbf{C}^n)^{SL_2 \times A_n^1} = 0 \quad \text{if } l \text{ is not divisible by } n,$$

$$\mathscr{P}^{nk}(\mathbf{C}^2 \otimes \mathbf{C}^n)^{SL_2 \times A_n^1} \simeq ((\mathscr{P}^k(\mathbf{C}^2))^{\otimes n})^{SL_2}.$$

5.4.2.2. We can compute the dimension of this space by arguments similar to those of §3.3.1 (see especially 3.3.1.4). We find again that it is a difference of partition values, but this time of *ordered* partitions. Precisely, by an *ordered partition* of d into n parts of size at most k we mean an n-tuple $(t_1, t_2, \ldots, t_n)$ of integers t_i, with $0 \le t_i \le k$, such that $\sum_{i=1}^n t_i = d$. Denote the number of such ordered partitions by $\tilde{\pi}(n, k, d)$. In analogy with 3.3.1.4 we have

$$\dim(\mathscr{P}^k(\mathbf{C}^2)^{\otimes n})^{SL_2} = \tilde{\pi}(n, k, \tfrac{1}{2}kn) - \tilde{\pi}(n, k, \tfrac{1}{2}(kn) - 1),$$

with the understanding this is zero if kn is not even.

5.4.2.3. Unlike the case of unordered partitions, we can give a closed form formula for $\tilde{\pi}(n, k, d)$. It is standard that $\tilde{\pi}(n, \infty, d)$, which counts the standard monomials of degree d in $x_1, \ldots, x_n$, is just a binomial coefficient:

$$\tilde{\pi}(n, \infty, d) = \binom{n + d - 1}{n - 1}.$$

Let $(t_1, \ldots, t_n)$ be an ordered partition of d, with no bounds on the t_i. If $t_i > k$, then $(t_1, t_2, \ldots, t_{i-1}, t_i - k - 1, \ldots, t_n)$ is an ordered partition of $d - k - 1$. Thus $\tilde{\pi}(n, \infty, d)$ is the sum of $\tilde{\pi}(n, \infty, d - k - 1)$ and the number of ordered partitions of d in which $t_i \le k$. Summing over all i and using the inclusion-exclusion principle, we find

$$\tilde{\pi}(n, k, d) = \sum_{j=0}^{n} (-1)^j \binom{n}{j} \binom{n + d - 1 - j(k + 1)}{n - 1}.$$

In this formula, it is understood that a binomial coefficient $\binom{a}{b}$ is zero if $a < b$. Combining this formula with 5.4.2.2 gives

$$\dim(\mathscr{P}^{nk}(\mathbf{C}^2 \otimes \mathbf{C}^n))^{SL_2 \times A_n^1} = \sum_{j=0}^{(n-1)/2} (-1)^j \binom{n}{j} \binom{k(n/2 - j) + n - 2 - j}{n - 2}.$$

5.4.2.3.1. If $n = 2m$ is even, then we may take $k = 1$. We can see directly that $\tilde{\pi}(n, 1, d) = \binom{n}{d}$. Hence

$$\dim(\mathscr{P}^{2m}(\mathbf{C}^2 \otimes \mathbf{C}^{2m}))^{SL_2 \times A_{2m}^1} = \binom{2m}{m} - \binom{2m}{m-1} = C_m$$

where C_m is the mth Catalan number (see 5.2.2.2).

5.4.2.4. Approaching from the left, we have $\mathscr{P}^{kn}(\mathbf{C}^2 \otimes \mathbf{C}^n)^{\mathrm{SL}_2 \times A_n^1} \simeq$ $(\sigma^n_{D(k,k)})^{A_n^1}$. This is the zero weight space for $\sigma^n_{D(k,k)}$. To give its dimension, we could use the formulas of Kostant or Freudenthal [**J2**] for dimensions of weight spaces. However, these formulas involve a lot of cancellation, and the formula of 5.4.2.3 is more efficient.

Beyond considerations of dimension, we get explicit representations of the elements of $\mathscr{P}^{nk}(\mathbf{C}^2 \otimes \mathbf{C}^n)^{\mathrm{SL}_2 \times A_n^1}$. We saw in 5.2.2 that the SL_2-invariants are generated by the

$$[ij] = \det \begin{vmatrix} a_i & a_j \\ b_i & b_j \end{vmatrix}.$$

Thus $\mathrm{SL}_2 \times A_n^1$ invariants are sums of products

$$\prod_{i<j} [ij]^{b_{ij}} = M(b_{ij}).$$

To satisfy the condition (see 5.2.3) for A_n^1-invariance, the degree with which each pair of variables $\begin{bmatrix} a_i \\ b_i \end{bmatrix}$ enters into $M(b_{ij})$ must be constant. Thus for all i we have $\sum_{i<j} b_{ij} + \sum_{j<i} b_{ji} = k$, with k independent of i. Summing on i gives

$$2 \sum_{i<j} b_{ij} = 2 \deg M(b_{ij}) = nk.$$

If n is even, this can be satisfied if $\deg M(b_{ij})$ is a half-integer multiple of n; if n is odd, $\deg M(b_{ij})$ must be an integer multiple of n. In particular, the smallest possible positive degree is n if n is odd and $n/2$ if n is even. Here degree means as a polynomial in the $[ij]$. This is $\frac{1}{2}$ the degree in $\mathscr{P}(\mathbf{C}^2 \otimes \mathbf{C}^n)$.

5.4.2.5. The following result is of key importance for the usefulness of this approach to binary invariants. It is due to Kempe [**Ke**]. The proof given here is essentially his original proof. It is almost the only aspect of the classical theory which I have not seen how to reduce to standard constructions of representation theory. In this sense it is the deepest result of the theory.

We will call a monomial $M(b_{ij}) = M$ in the symbols $[ij]$, $1 \le i, j \le n$, *equi-valent* of valence k if

$$\sum_{i<j} b_{ij} + \sum_{j<i} b_{ji} = k.$$

We call an equi-valent monomial *elemental* if it has the smallest possible valence (equivalently, degree) among all nontrivial equi-valent monomials. (This valence will be 1 if n is even, 2 if n is odd.) Observe that if the relation ($\mathfrak{P}$) of 5.2.2 is applied to two symbols appearing in an equi-valent monomial M, there will result an equation $M + M' + M'' = 0$ between 3 equi-valent monomials. An equi-valent monomial M which can be factored as a product of elemental monomials will be called *decomposable*.

THEOREM (KEMPE). *The algebra $\mathscr{P}(\mathbf{C}^2 \otimes \mathbf{C}^n)^{\mathrm{SL}_2 \times A_n^1}$ is generated by its elements of minimum positive degree. More precisely, any equi-valent monomial*

$M(b_{ij}) = M$ *of valence k can be converted via the relation $(\mathfrak{P})$ of 5.2.2 to an integral linear combination of decomposable equi-valent monomials.*

5.4.2.6. PROOF. Consider an equi-valent monomial $M(b_{ij}) = M$. We will say i and j are *connected in M* if $b_{ij} > 0$. Suppose there is some index, say 1, which is connected in M only to two other indices, say 2 and 3. Suppose further that 2 and 3 are not connected to each other in M. Then we can write

$$M = [12]^l [13]^{k-l} \left(\prod_{i>3} [2i]^{b_{2i}} [3i]^{b_{3i}} \right) M',$$

where M' involves only symbols $[ij]$ with $i, j > 3$. Further, k is the valence of M, and

$$\sum_{i>3} b_{2i} = k - l, \qquad \sum_{i>3} b_{3i} = l.$$

Thus we see that the monomial

$$N = \left(\prod_{i>3} [3i]^{b_{2i}+b_{3i}} \right) M'$$

will be equi-valent of valence k on symbols in the indices $3, 4, \ldots, n$. By induction on the number of indices, we can assume the theorem holds for N. Therefore, by means of a finite number of transformations using the relation $(\mathfrak{P})$ of 5.2.2, we can write

$(*)$ $$N = \sum_\alpha c_\alpha E_{\alpha_1} E_{\alpha_2} \cdots E_{\alpha_l},$$

where the E_{α_l} are elemental monomials on symbols in indices 3 to n. If n is even, the number l of factors is k, while if n is odd, the number of factors is $k/2$.

We have noted that equation $(*)$ means that the left-hand side can be transformed to the right-hand side by successively rewriting it using the relations $(\mathfrak{P})$. We can clearly perform a parallel series of operations on

$$N' = \left(\prod_{i>3} [2i]^{b_{2i}} [3i]^{b_{3i}} \right) M'$$

and this will result in an equation

$$N' = \sum c_\alpha E'_{\alpha_1} E'_{\alpha_2} \cdots E'_{\alpha_l},$$

where the E'_{α_p}'s are obtained by replacing 3 by 2 in certain of the symbols of which E_{α_p} is a product. The total number of 2's put in each term is $k - l$, and the number of 3's left is l. Using this equation we may write

$$M = [12]^l [13]^{k-l} N' = \sum c_\alpha [12]^l [13]^{k-l} E'_{\alpha_1} E'_{\alpha_2} \cdots E'_{\alpha_l}$$

$(**)$ $$= \sum c_\alpha E''_{\alpha_1} E''_{\alpha_2} \cdots E''_{\alpha_l},$$

where E''_{α_p} is obtained by the following recipe:

$$E''_{\alpha_p} = [12]^\delta [13]^\epsilon E'_{\alpha_p},$$

where

δ = the number of times 3 occurs in the symbols of which E'_{α_p} is a product;

ϵ = the number of times 2 occurs in the symbols of which E'_{α_p} is a product.

Reviewing this construction, we can see that the equation (∗∗) expresses M in the form required by the theorem.

5.4.2.7. PROOF OF 5.4.2.5 CONTINUED. In the previous subsection, we saw that for M of a certain form the theorem is true. Here we will show that, again by repeated application of the relation (𝔓) of 5.2.2, we can write any equi-valent monomial as a **Z**-linear combination of monomials of the appropriate form. First, consider a monomial M in which 2 and 3 are connected, i.e., $b_{23} > 0$. Suppose $[ij]$ also appears in M, with $i, j > 3$. Then the equation

$$[23][ij] = [2i][3j] - [2j][3i]$$

allows us to write M as a sum of two monomials in which $[23]$ occurs only to the $(b_{23} - 1)$th power. Proceeding in this fashion, as long as there are enough symbols which do not involve 2 or 3, will eventually give an expression for M as a **Z**-linear combination of monomials in which 2 and 3 are not connected. The total number of symbols in which 2 and 3 are involved is $2k - b_{23}$, while the total number of symbols in M is $nk/2$. Hence there will always be symbols of the desired form as long as $nk/2 - (2k - b_{23}) = ((n/2) - 2)k + b_{23} > 0$, thus as long as $n \geq 4$. But it is very easy to verify the theorem if $n \leq 4$.

Thus, if we have a monomial M in which 1 is connected only to 2 and 3, we can write it as a sum of monomials in which 1 is still connected only to 2 and 3, and in addition 2 and 3 are not connected to each other.

To show any monomial can be written as a combination of monomials where some index is joined only to two others, we associate to the monomial $M = M(b_{ij})$ a graph $\Gamma = \Gamma(b_{ij})$. The vertices of Γ are called v_i, $1 \leq i \leq n$, and the vertices v_i and v_j are connected by b_{ij} edges. We imagine the vertices v_i to be arranged at equal intervals around a circle (so they define a regular n-gon), and we imagine the edges to be the chords of the circle connecting the appropriate vertices. (We will label the chord from v_i to v_j with the integer b_{ij} to indicate how many times that edge is repeated.)

Suppose that in the graph Γ associated to M, there are two edges, connecting pairs v_i, v_j and v_k, v_l, which cross inside the circle, as in the figure.

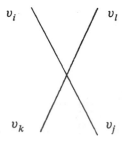

This means the product $[ij][kl]$ divides M. If we apply the syzygy $(\mathfrak{P})$ of 5.2.2 to this product we can replace M by a sum $M' + M''$ such that the graphs Γ', Γ'' associated to M' and M'' are the same as Γ, except that the diagonals of the quadrilateral $v_i v_k v_j v_l$ will be replaced by one or the other of the pairs of opposite sides:

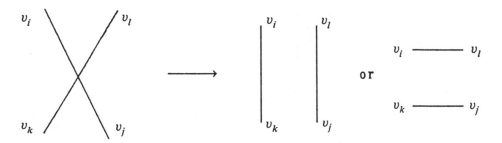

It is easy to see that the sum of the lengths of either pair of opposite sides of a quadrilateral is less than the sum of the lengths of the diagonals. Hence the sum of the lengths of the edges in Γ' or Γ'' will be less than in Γ. Since there are only finitely many possibilities for the sum of the lengths of edges of a graph corresponding to a monomial of a given degree, this process of replacing diagonals of quadrilaterals by pairs of opposite edges must stop after a finite number of steps. We conclude that all monomials are expressible as integral linear combinations of monomials whose associated graphs have no pairs of edges crossing inside the circle; we call such graphs uncrossed.

We claim that an uncrossed graph Γ will have the property we seek: there will be some vertex which is joined only to two other vertices (which will be its immediate neighbors). Indeed, if each vertex is joined only to its immediate neighbors, this is obvious. Otherwise, let v_i, v_j be two non-neighboring vertices which are connected, and for which the length of the chord C_{ij} joining v_i and v_j is minimal subject to the condition that v_i and v_j are not neighbors. Since Γ is uncrossed, deleting the chord C_{ij} from the circle will divide the circle into two regions, R_1 and R_2, and the vertices in one region will be connected only with other vertices in the same regions, or with v_i and v_j. Thus, if, say, R_1 is the smaller of the two regions, and v_k is in R_1, then any chord of Γ emanating from v_k must be strictly shorter than the chord C_{ij}. By choice of C_{ij}, then, we see v_k can only be connected to its nearest neighbors. Finally we note that each of R_1 and R_2 must contain at least one vertex of Γ, since v_i and v_j are not neighbors. This completes the proof of Theorem 5.4.2.5.

5.5. To sum up, we have found that

$$\mathscr{I}_n = \mathscr{P}(\mathscr{P}^n(\mathbf{C}^2))^{\mathrm{SL}_2} \simeq (\mathscr{A})^{S_n}$$

where

$$\mathscr{A} = \mathscr{P}(\mathbf{C}^2 \otimes \mathbf{C}^n)^{\mathrm{SL}_2 \times A_n^1}.$$

We know a fair amount about the structure of $\mathscr{A}$:

(i) $\mathscr{A}$ is a graded subalgebra of $\mathscr{P}(\mathbf{C}^2 \otimes \mathbf{C}^n)$, with homogeneous subspaces of degrees which are nonnegative integral multiples of $m = n$ if n is even, and $m = 2n$ if n is odd;

(ii) $\mathscr{A}$ is generated by $\mathscr{A}^m$, the homogeneous subspace of lowest positive degree (Kempe's Theorem);

(iii) $\mathscr{A}^{kn}$ has dimension 0 if kn is odd, and if k is even, dim $\mathscr{A}^{kn}$ is

$$\sum_{j=0}^{(n/2)-1} (-1)^j \binom{n}{j} \binom{k(n/2) - j) + n - 2 - j}{n - 2}.$$

This is a polynomial in k of degree $n - 3$; although the individual terms in the sum have degree $n - 2$, the sum of the coefficients of k^{n-2} is

$$((n - 2)!)^{-1} \sum_{j=0}^{(n/2)-1} (-1)^j \binom{n}{j} \left(\frac{n}{2} - j\right)^{n-2}$$

and this sum vanishes, as one sees by comparing with the full sum

$$\sum_{j=0}^{n} (-1)^j \binom{n}{j} \left(\frac{n}{2} - j\right)^{n-2},$$

which obviously vanishes since it is the result of the n-fold difference operation applied to a polynomial (namely $(k/2)^{n-2}$) of degree less than n. The coefficient of k^{n-3} is $\frac{1}{2}((n - 3)!)^{-1}$ times

$$c_n = -\sum_{0 \leq j \leq n/2} (-1)^j \binom{n}{j} \left(\frac{n}{2} - j\right)^{n-3}.$$

These facts allow us to make a number of further conclusions about the invariants of a binary n-ic.

5.6. First, they imply an effective finiteness theorem for the generators of $\mathscr{P}(\mathscr{P}^n(\mathbf{C}^2))^{\mathrm{SL}_2}$, with explicit (though not very good) bounds on the orders of the generators.

5.6.1. Indeed, consider $\mathscr{S}(\mathscr{A}^m)$, the symmetric algebra on the lowest-degree subspace $\mathscr{A}^m$ of $\mathscr{A}$. The inclusion $\mathscr{A}^m \hookrightarrow^i \mathscr{A}$ induces an algebra homomorphism

$$i_* : \mathscr{S}(\mathscr{A}^m) \to \mathscr{A}.$$

This homomorphism is surjective, by Kempe's Theorem 5.2.4.5. Thus we have an isomorphism

$$\mathscr{S}(\mathscr{A}^m)/\ker i_* \overset{i_*}{\to} \mathscr{A}.$$

The space $\mathscr{A}^m$ carries a representation of the symmetric group S_n; this action extends uniquely to an action by automorphisms of S_n on $\mathscr{S}(\mathscr{A}^m)$. The homomorphism i_* is clearly a homomorphism of S_n-modules as well as of algebras. Thus we have

$$(\mathscr{S}(\mathscr{A}^m)/\ker i_*)^{S_n} \overset{\sim}{\to} \mathscr{A}^{S_n} \simeq \mathscr{I}_n.$$

Since we are working over $\mathbf{C}$, a field of characteristic zero, Maschke's Theorem guarantees us that the canonical projection

$$\mathscr{S}(\mathscr{A}^m)^{S_n} \to (\mathscr{S}(\mathscr{A}^m)/\ker i_*)^{S_n}$$

is onto. Thus we have maps

$$\mathscr{S}(\mathscr{A}^m)^{S_n} \to \mathscr{S}(\mathscr{A}^m)^{S_n}/(\ker i_*)^{S_n} \simeq (\mathscr{S}(\mathscr{A}^m)/\ker i_*)^{S_n} \simeq \mathscr{I}_n.$$

Consequently, if we find a system of generators for the algebra $\mathscr{S}(\mathscr{A}^m)^{S_n}$, their images under i_* will generate $\mathscr{A}^{S_n}$.

5.6.2. In seeking generators for the algebra $\mathscr{S}(\mathscr{A}^m)^{S_n}$, we can appeal to

NOETHER'S THEOREM (SEE [**W**, §8.15]). *Let the finite G act on the vector space V. Then the algebra $\mathscr{S}(V)^G$ of G-invariants on $\mathscr{S}(V)$ is generated by its elements of degree not more than $\#(G)$.*

This result clearly makes the computation of generators for $\mathscr{I}_n$ a finite job of bounded difficulty. One can find a generating set with degrees no more than $n!$. Good luck!

5.6.3. Noether's Theorem applies to arbitrary representations of a finite group G. For certain representations, there is an alternative result which yields somewhat better bounds on the degrees of a generating set. This may be deduced from Theorem 7.4 of [**GS**].

PROPOSITION. *Let G act by permutations on a set X. Let $\mathbf{C}(X)$ be the vector space with basis X. Then G acts on $\mathscr{P}(\mathbf{C}(X))$ in the obvious way, and $\mathscr{P}(\mathbf{C}(X))^G$ is generated by elements of degree not more than $\#(X)(\#(X)-1)/2$.*

5.6.3.1. REMARK. This result is sharp for the alternating group A_n and its standard action on $\{1, 2, \ldots, n\}$. The invariant of degree $n(n-1)/2$ is the discriminant $\prod_{1 \leq i \leq j \leq n}(x_i - x_j)$.

5.6.3.2. COROLLARY. *If ρ is a representation of G on a vector space V, and if ρ may be embedded in an induced representation*

$$\mathrm{Ind}_H^G 1_H, \qquad 1_H = \text{trivial representation of } H,$$

where $H \subseteq G$ is some subgroup, then $\mathscr{P}(V)^G$ is generated by its elements of order not more than $\#(G/H)(\#(G/H) - 1)/2$. In particular, if ρ is irreducible and V contains a nonzero vector fixed by H, this bound holds.

5.6.3.3. If n is even then $m = n$, and we know from 5.2.2 that $\mathscr{A}^m$ consists of the A_n^1 fixed vectors in the representation $\sigma_{D(l,k)}^n$ where $l = n/2$. From this one sees that, as a representation of S_n, $\mathscr{A}^m$ is irreducible and isomorphic to $\rho_{D(l,l)}$. (Although this is not difficult to prove, we omit the proof, in order not to overburden the discussion.) In particular, $\mathscr{A}^m$ is a subrepresentation

$$\mathrm{Ind}_{S_l \times S_l}^{S_n} 1,$$

which has dimension $\binom{n}{l} = \binom{2l}{l}$ (binomial coefficient). Corollary 5.6.3.2 says in this case the invariants are generated by elements of degree no more than

$\binom{n}{l}\left(\binom{n}{l}-1\right)/2 < \binom{n}{l}^2/2$. Stirling's Formula says that for large n this is roughly $4^n/\pi n$, which is substantially less than $n! \simeq \sqrt{2\pi n}(n/e)^n$. However, it is still dauntingly large.

5.7. Second, we can use 5.5(iii) to give an asymptotic formula as $k \to \infty$ for $\dim \mathscr{I}_n^k$. This formula is due to Hilbert [**Hi2**], and is derived in a purely computational manner in [**Sp1**]. Our description of $\mathscr{I}_n$ in terms of $\mathscr{A}^{S_n}$ gives us a more geometrical approach to this result, by an appeal to the following result.

5.7.1. THEOREM. *Let the finite group G act by degree-preserving automorphisms on the graded algebra A. Suppose A is finitely generated and is an integral domain. Suppose also that the only element of G which acts as a scalar on all homogeneous subspaces A^k of A is the identity. Then*

$$\lim_{k\to\infty} \dim(A^k)^G/\dim A^k = \#(G)^{-1}.$$

This is proved in [**Ho2**].

5.7.2. Combining Theorem 5.7.1 with the computations 5.5(iii), and taking into account the relation between the gradings in $\mathscr{A}$ and $\mathscr{P}(\mathscr{P}^n(\mathbf{C}^2))$, we find

$$\dim \mathscr{I}_n^k = 0 \qquad\qquad\qquad \text{for } nk \text{ odd,}$$
$$\dim \mathscr{I}_n^k \sim \tfrac{1}{2}(n!)^{-1}c_n k^{n-3}/(n-3)! \quad \text{for } nk \text{ even,}$$

with c_n as in 5.5 (iii). Here the $\sim$ means the difference between the two expressions is small in comparison with either, when k is large enough.

5.8. By combining the asymptotic expressions for dimensions with Hermite reciprocity (3.3.2.2), we can see something about the "fundamental invariants" of the binary n-ic, for large n.

5.8.1. Recall the general notion. Let $A = \sum_{k=0}^{\infty} A^k$ be a graded algebra over $\mathbf{C}$, with $A_0 = \mathbf{C}$. Let $A \cdot A$ denote the ideal spanned by all products xy, where x and y are each homogeneous of positive degree. Then $A \cdot A$ is a homogeneous ideal in A. A subspace $B \subseteq A$ will generate A as algebra over $\mathbf{C}$ if and only if B projects onto $A/(\mathbf{C} \oplus A \cdot A)$ under the natural quotient map. If a set $\{x_1, x_2, \ldots, x_m\}$ of homogeneous elements of A projects to a basis of $A/(\mathbf{C} \oplus A \cdot A)$, then the x_i are called *fundamental generators* for A. The integer $\dim A^k/(A \cdot A)^k$ is called the number of fundamental generators of degree k for A.

5.8.2. From the definition, we see that the number of fundamental generators of degree k for A is at least

$$\dim A^k - \sum_{j=1}^{k/2} \dim A^j \dim A^{k-j}.$$

Apply this estimate to $\mathscr{P}(\mathscr{P}^n(\mathbf{C}^2))^{\mathrm{SL}_2}$ for large n. By Hermite reciprocity and 5.7.2, we know that as $n \to \infty$, for fixed k,

$$\dim \mathscr{I}_n^k = \dim \mathscr{I}_k^n \simeq \tfrac{1}{2}(k!)^{-1}c_k n^{k-3}/(k-3)! \quad \text{for } nk \text{ even,}$$
$$\dim \mathscr{I}_n^k = 0 \qquad\qquad\qquad\qquad\qquad\qquad \text{for } nk \text{ odd.}$$

It follows that

$$(\dim \mathscr{I}_n^j)(\dim \mathscr{I}_n^k) \sim \beta_{j,k} n^{j+k-6}, \qquad j, k \geq 3,$$
$$(\dim \mathscr{I}_n^j)(\dim \mathscr{I}_n^2) \simeq \beta_j n^{j-3}.$$

Comparing this with $\dim \mathscr{S}_n^{j+k}$, we can make the following assertion.

THEOREM. *For fixed k, as $n \to \infty$, almost all SL_2-invariants of degree k for the binary n-ic are fundamental. Precisely the ratio of the number of fundamental invariants to $\dim \mathscr{S}_n^k$ approaches 1 as $n \to \infty$. The asymptotic number of fundamental invariants of degree k, $k \geq 4$, is*

$$\begin{array}{ll} \tfrac{1}{2}(k!)^{-1}c_k n^{k-3}/(k-3)! & \textit{for } nk \textit{ even,} \\ 0 & \textit{for } nk \textit{ odd.} \end{array}$$

5.8.3. REMARK. The qualitative part of Theorem 5.8.2, that almost all invariants of fixed degree are eventually fundamental, requires no estimates such as given in §5.7; it requires only that $\mathscr{S}_n$ be an algebra of dimension $n-2$, and Hermite reciprocity. Despite its elementary nature, I am not aware of this fact in the literature; however, it may be there. Theorem 5.8.2 complements a result of Kac [**Kc**] which puts a lower bound on the number of generators of $\mathscr{S}_n$ in terms of the number of partitions of $n-2$, and Springer's observation [**Sp1**] on the degree of the field of fractions of all invariants over the subfield generated by any independent homogeneous set of generators.

5.9. EXAMPLES. The polynomial describing the dimensions of the spaces $\mathscr{A}^{km}$ (cf. 5.5(iii)) can be written more explicitly for small n as follows.

$n = 2$	$\dim \mathscr{A}^{2k} = 1$	$n = 3$	$\dim \mathscr{A}^{6k} = 1$
$n = 4$	$\dim \mathscr{A}^{4k} = k+1$	$n = 5$	$\dim \mathscr{A}^{10k} = 5\binom{k+1}{2} + 1$
$n = 6$	$\dim \mathscr{A}^{6k} = 3\binom{k+2}{3} + k + 1$		

5.9.1. $n = 2$. The only "homogenized root" is $[1,2]$, and it is an invariant. Thus

$$\mathscr{S}_2 \simeq \mathscr{A}^{S_2} = \mathscr{A} = \mathbf{C}[[1,2]].$$

5.9.2. $n = 3$. The space $\mathscr{A}^6$ is spanned by $\gamma = [1,2][2,3][3,1]$. Since $\dim \mathscr{A}^{6k} = 1$, we see that $\mathscr{A} \simeq \mathbf{C}[\gamma]$. It is easily verified that γ transforms according to the signum character of S_3. Thus

$$\mathscr{S}_3 \simeq \mathscr{A}^{S_3} = \mathbf{C}[([1,2][2,3][3,1])^2].$$

5.9.3. $n = 4$. There are three obvious elements of $\mathscr{A}^4$:

$$\gamma_2 = [1,2][3,4], \qquad \gamma_3 = [1,3][4,2], \qquad \gamma_4 = [1,4][2,3].$$

But the syzygy ($\mathfrak{P}$) of 5.2.2 just says $\gamma_2 + \gamma_3 + \gamma_4 = 0$, so $\mathscr{A}$ is generated by γ_2 and γ_3; and since $\dim \mathscr{A}^{4k} = k+1$, we must have $\mathscr{A} \simeq \mathbf{C}[\gamma_2, \gamma_3]$.

One checks very easily that the Sylow 2-subgroup of $A_4 \subseteq S_4$ leaves the γ_i fixed, so that the action of S_4 factors to an effective action of S_3. The action of S_3 is irreducible: the unique irreducible two-dimensional representation, equivalent to the permutation on triples (x_1, x_2, x_3) such that $x_1 + x_2 + x_3 = 0$. The invariants of this representation are well known since Lagrange to be a polynomial algebra on the elementary symmetric polynomials of degrees two and three:

$$\mathscr{S}_4 \simeq \mathscr{A}^{S_4} = \mathbf{C}[J_2, J_3],$$

where

$$J_2 = \gamma_1\gamma_2 + \gamma_2\gamma_3 + \gamma_3\gamma_1 = -(\gamma_1^2 + \gamma_1\gamma_2 + \gamma_2^2) = -\tfrac{1}{2}(\gamma_1^2 + \gamma_2^2 + \gamma_3^2),$$
$$J_3 = \gamma_1\gamma_2\gamma_3 = -\gamma_1\gamma_2(\gamma_1 + \gamma_2) = \tfrac{1}{3}(\gamma_1^3 + \gamma_2^3 + \gamma_3^3).$$

5.9.4. For $n \geq 5$, the computations begin to be some work; we will not burden the reader with the details. Using the recursion relations 3.3.1.5, one can compute dimensions of invariants of different degrees, and by taking differences come up with a candidate for a Hilbert function for the algebra of invariants. Because of the effective bounds given in 5.6 these calculations, if carried far enough, would actually prove the candidate Hilbert function to be correct. Springer [Sp2] has given a closed form expression for the Hilbert function of the invariants of the binary n-ic for all n. It is also some work to evaluate this for a given n. The Hilbert functions for $n = 5, 6$ are

$$\frac{1 + t^{18}}{(1 - t^4)(1 - t^8)(1 - t^{12})}, \qquad n = 5,$$

$$\frac{1 + t^{15}}{(1 - t^2)(1 - t^4)(1 - t^6)(1 - t^{10})}, \qquad n = 6.$$

The Hilbert functions for odd n are substantially more complicated than for even n. For $n = 7$, the denominator is $(1-t^4)(1-t^8)(1-t^{12})^2(1-t^{20})$ and the numerator is a polynomial of degree 48. The Hilbert functions were computed classically for $n \leq 10$ and $n = 12$ [Sy2, Sy3]. Using the recursion relations 3.3.1.5 and a VAX computer, Paul Sally, Jr. [Sa] recently computed the Hilbert functions for $n \leq 22$ and $n = 24$. For $n = 19$, the denominator is $\prod_{j=2}^{18}(1-t^{2j})$ and the numerator is a polynomial of degree 320; the coefficient of k^{160} is $206,054,755,643,582$. For $n = 20$, the denominator is $\prod_{j=2}^{19}(1 - t^j)$, the numerator has degree 168, and the coefficient of k^{84} is $43,665,034,627$. The coefficients of the numerators are symmetric about the middle term. (This reflects the fact that the rings of invariants are Gorenstein rings [St, Ka].) Also, in the form computed by Sally, the coefficients of the numerator are unimodal.

REFERENCES

[BGG] J. Bernstein, I. Gelfand, and S. Gelfand, *Models of representations of Lie groups*, Selecta Math. Sovietica **1** (1981), 121–142.

[B] A. Borel, *Hermann Weyl and Lie groups*, Hermann Weyl, 1885–1895 (K. Chandrasekharan, ed.), Springer-Verlag, Berlin, New York, 1986, pp. 53–82.

[C] A. Cayley, *A second memoir on quantics*, Coll. Math. Papers, vol. 2, Cambridge Univ. Press, Cambridge, 1889, pp. 250–275.

[DC] J. Dieudonné and J. Carrell, *Invariant theory, old and new*, Academic Press, New York, 1971.

[D1] J. Dixmier, *Série de Poincaré et systèmes de paramètres pour les invariants des formes binaires de degré* 7, Bull. Soc. Math. France **110** (1982), 303–318.

[D2] ———, *Série de Poincaré et systèmes de paramètres pour les invariants des formes binaires*, Acta Sci. Math. Univ. Szeged **45** (1983), 151–160.

[DL] J. Dixmier and D. Lazard, *Le nombre minimum d'invariants fondamentaux pour les formes binaires de degré* 7, Portugal. Math. **43** (1985/6), 377–392.

[DEN] J. Dixmier, P. Erdős, and J.-L. Nicolas, *Sur le nombre d'invariants fondamentaux des formes binaires*, preprint, I.H.E.S., April 1987.

[Dr] H. Dorrie, *100 great problems of elementary mathematics* (D. Antin, trans.) Dover Publications, New York, 1965.

[DRS] P. Doubilet, G.-C. Rota, and J. Stein, *On the foundations of combinatorial theory. IX: Combinatorial methods in invariant theory*, Stud. Appl. Math. **53** (1974), 185–216.

[F] W. Feller, *An introduction to probability theory and its applications*, vol. 1, 2nd ed., Wiley, New York, 1957.

[GS] A. Garsia and D. Stanton, *Group actions on Stanley-Reisner rings and invariants of permutation groups*, Adv. in Math. **51** (1984), 107–201.

[GY] J. Grace and A. Young, *Algebra of invariants*, Cambridge Univ. Press, Cambridge, 1903.

[GK] K. Gross and R. Kunze, *Bessel functions and representation theory. II: Holomorphic discrete series and metaplectic representations*, J. Functional Anal. **25** (1977), 1–49.

[Had] D. Hadziev, *Some questions in the theory of vector invariants*, Math. U.S.S.R.-Sbornik **1** (1967), 383–396.

[Ha] M. Hall, *The theory of groups*, Macmillan, New York, 1959.

[Hi1] D. Hilbert, *Über die Theorie der algebraischen Formen*, Math. Ann. **36** (1890), 473–534.

[Hi2] ——, *Über die vollen Invariantensysteme*, Ges. Abh., II², Springer-Verlag, 1970, pp. 287–344.

[HM] G. Hochschild and G. Mostow, *Unipotent groups in invariant theory*, Proc. Nat. Acad. Sci. U.S.A. **70** (1973), 646–648.

[HP] W. Hodge and D. Pedoe, *Methods of algebraic geometry*, vol. I, Cambridge Univ. Press, Cambridge, 1947.

[Ho1] R. Howe, *Remarks on classical invariant theory*, preprint.

[Ho2] ——, *Asymptotic dimensions of spaces of invariants*, in preparation.

[J1] N. Jacobson, *Basic algebra II*, W. H. Freeman, San Francisco, 1980.

[J2] ——, *Lie algebras*, Wiley-Interscience, New York, 1962.

[Kc] V. Kac, *Root systems, representations of quivers and invariant theory*, Invariant Theory (F. Gherardelli, ed.), Lecture Notes in Math., vol. 996, Springer-Verlag, Berlin, New York, 1983, pp. 74–108.

[Ka] I. Kaplansky, *Commutative rings*, Univ. of Chicago Press, Chicago, 1974.

[KV] M. Kashiwara and M. Vergne, *On the Segal-Shale-Weil representations and harmonic polynomials*, Invent. Math. **44** (1978), 1–47.

[Ke] A. Kempe, *On regular difference terms*, Proc. London Math. Soc. **25** (1894), 343–359.

[Kn] D. Knuth, *The toilet paper problem*, Amer. Math. Monthly **91** (1984), 465–470.

[KR] J. Kung and G.-C. Rota, *The invariant theory of binary forms*, Bull. Amer. Math. Soc. (N.S.) **10** (1984), 27–85.

[M] I. MacDonald, *Symmetric functions and Hall polynomials*, Oxford Univ. Press, Oxford, 1979.

[P] R. Proctor, *Solution of two difficult combinatorial problems with linear algebra*, Amer. Math. Monthly **89** (1982), 721–734.

[Re] C. Reid, *Hilbert*, Springer-Verlag, Berlin, New York, 1970.

[Sa] P. Sally, *Hilbert functions for the algebras of invariants of binary forms*, Applied Math. 490, Term Paper, Yale, 1985.

[Se1] C. Seshadri, *On a theorem of Weitzenböck in invariant theory*, J. Math. Kyoto Univ. **1** (1961), 403–409.

[Se2] ——, letter to M. F. Atiyah.

[Sp1] T. Springer, *Invariant theory*, Lecture Notes in Math., vol. 585, Springer-Verlag, Berlin, New York, 1977.

[Sp2] ——, *On the invariant theory of SU_2*, Indag. Math. **42** (1980), 339–345.

[St] R. Stanley, *Hilbert functions of graded algebras*, Adv. in Math. **38** (1978), 57–83.

[Sy1] J. Sylvester, *Proof of the hitherto undemonstrated fundamental theorem of invariants*, Coll. Math. Papers. III, Cambridge Univ. Press, Cambridge, 1912; reprinted by Chelsea, 1973, pp. 117–126.

[Sy2] ——, *Tables of the generating functions and ground forms of the binary quantics of the first 10 orders*, ibid., pp. 283–311.

[Sy3] ——, *Tables of the generating functions and ground forms of the binary duodecemic, with some remarks, and tables of the irreducible syzygies of certain quantics*, ibid., 489–508.

[T] H. Turnbull, *The theory of determinants, matrices, and invariants*, 3rd ed., Dover, New York, 1960.

[W] H. Weyl, *The classical groups*, Princeton Univ. Press, Princeton, 1946.

[Ze1] A. Zelevinsky, *Representations of finite classical groups, a Hopf algebra approach*, Lecture Notes in Math., vol. 869, Springer-Verlag, Berlin, New York, 1981.

[Ze2] ——, *A generalization of the Littlewood-Richardson rule and the Robinson-Schensted-Knuth correspondence*, J. Algebra **69** (1981), 82–94.

[Zh] D. Zhelobenko, *Compact Lie groups and their representations*, Transl. Math. Monographs, vol. 40, Amer. Math. Soc., Providence, R.I., 1973.

YALE UNIVERSITY

Proceedings of Symposia in Pure Mathematics
Volume 48 (1988)

Characters, Harmonic Analysis,
and an L^2-Lefschetz Formula

JAMES ARTHUR

Suppose that U is a locally compact group. It is a fundamental problem to classify the irreducible unitary representations of U. A second basic problem is to decompose the Hilbert space of square integrable functions on U, or on some homogeneous quotient of U, into irreducible U-invariant subspaces. The underlying domain is often attached to a natural Riemannian manifold, and the required decomposition becomes the spectral decomposition of the Laplace-Beltrami operator. Weyl solved both problems in the case of a compact Lie group. His method, which was simple and elegant, was based on the theory of characters.

In this lecture, we shall briefly review Weyl's theory for compact groups. We shall then discuss two newer areas that could claim Weyl's work as a progenitor: the harmonic analysis on noncompact groups, and the analytic theory of automorphic forms. The three areas together form a progression that is natural in several senses; in particular, the underlying algebraic structures of each could be characterized as that of an algebraic group over the field $\mathbf{C}$, $\mathbf{R}$, or $\mathbf{Q}$. However, the latter two areas are vast. Beyond a few general comments, we can attempt nothing like a survey. We shall instead concentrate on a topic that has a particular connection to Weyl's character formula and its later generalizations. We shall describe a Lefschetz formula for the Hecke operators on L^2-cohomology. The formula deals with objects which are highly singular, but turns out to be quite simple nonetheless. It will probably play a role some day in relating the arithmetic objects discussed in Langlands' lecture [8d] to the analytic theory of automorphic forms.

1. Suppose that U is a compact simply connected Lie group. Any representation[1] $\tau \in \hat{U}$ is finite dimensional, and the character

$$\Theta_\tau(x) = \operatorname{tr}\tau(x), \qquad x \in U,$$

1980 *Mathematics Subject Classification* (1985 *Revision*). Primary 11F72, 22E46; Secondary 52T15.

[1]If G is a locally compact group, $\hat{G}$ denotes the set of equivalence classes of irreducible unitary representations of G.

is an invariant function on the conjugacy classes of U. Let T be a maximal torus in U. Weyl introduced the finite group[2]

$$W(U,T) = \mathrm{Norm}_U(T)/\mathrm{Cent}_U(T)$$

to characterize the conjugacy classes in U. Each conjugacy class of U intersects T in a unique $W(U,T)$-orbit. The character Θ_τ, and hence the representation τ itself, is uniquely determined by the $W(U,T)$-invariant function $\Theta_\tau(t)$, $t \in T$.

Weyl [12a] constructed the characters $\{\Theta_\tau\}$ from three simple facts. The first was the observation that the restriction of τ to T is a direct sum of irreducible characters of T. Accordingly, $\Theta_\tau(t)$ is a finite sum of characters of T. The second fact is that the characters $\{\Theta_\tau\}$ form an orthonormal basis of the space of square integrable class functions on U. This is an immediate consequence of the Schur orthogonality relations, established for compact groups by Peter and Weyl in [9]. The third fact is the Weyl integration formula

$$\int_U f(x)\,dx = |W(U,T)|^{-1} \int_T \int_U |\Delta_T^U(t)|^2 f(x^{-1}tx)\,dx\,dt, \qquad f \in C(U),$$

where dx and dt are the normalized Haar measures on U and T, and

$$\Delta_T^U(t) = \prod_{\alpha>0} (\alpha(t)^{1/2} - \alpha(t)^{-1/2}), \qquad t \in T.$$

The product here is taken over the positive roots of (U,T) relative to some ordering, and when multiplied out, the square roots all become well-defined functions of T.

The function Δ_T^U is skew-symmetric. That is,

$$\Delta_T^U(st) = \varepsilon(s)\Delta_T^U(t), \qquad s \in W(U,T),$$

where $\varepsilon(s)$ is the sign of s, regarded as a permutation of the roots. Therefore, the functions

$$\Delta_T^U(t)\Theta_\tau(t), \qquad \tau \in \hat{U},$$

are also skew-symmetric. It follows easily from the second and third facts above that these functions form an orthonormal basis, relative to the measure $|W(U,T)|^{-1}dt$, of the space of square-integrable, skew-symmetric functions on T. Now the group $W(U,T)$ acts by duality on $\hat{T}$. Let $\{\chi\}$ be a set of representatives of those orbits which are regular (in the sense that χ is fixed by only the identity in $W(U,T)$). Then the functions

$$\sum_{s \in W(U,T)} \varepsilon(s)(s\chi)(t), \qquad t \in T,$$

also form an orthonormal basis of the space of skew-symmetric functions. Choose the representatives χ so that their differentials all lie in the chamber defined by the positive roots. It is then a straightforward consequence of the first fact above that the two orthonormal bases are the same.

---

[2]By $\mathrm{Norm}_X(Y)$ and $\mathrm{Cent}_X(Y)$ we of course mean the normalizer and centralizer of Y in X.

This gives the classification of $\hat{U}$. There is a bijection $\tau \leftrightarrow \chi$, with the property that

(1.1) $$\Theta_\tau(t) = \Delta_T^U(t)^{-1} \sum_{s \in W(U,T)} \varepsilon(s)(s\chi)(t),$$

for any point t in

$$T_{\text{reg}} = \{t \in T \colon \Delta_T^U(t) \neq 0\}.$$

This identity is the famous Weyl character formula. It uniquely determines the character Θ_τ and the correspondence $\tau \leftrightarrow \chi$. By taking the limit as t approaches 1, Weyl obtained the simple formula

(1.2) $$\deg(\tau) = \prod_{\alpha > 0} \frac{(d\chi, \alpha)}{(d\chi_1, \alpha)}$$

for the degree of τ. Here, $\chi_1 \in \hat{T}$ corresponds to the trivial representation of U, $d\chi$ is the differential of χ, and $(\cdot, \cdot)$ is a $W(U,T)$-invariant bilinear form on the dual of the Lie algebra of T. The Peter-Weyl theorem asserts that τ occurs in $L^2(U)$ with multiplicity equal to the degree of τ. Therefore, Weyl's classification of $\hat{U}$ also provides a decomposition of $L^2(U)$ into irreducible representations. (See pp. 377–385 of [12b] for a clear elucidation of a special case, and also the survey [2b], in addition to the original papers [12a]).

EXAMPLE. Suppose that $U = \text{SU}(n, \mathbf{R})$, the group of complex unitary matrices of determinant 1. One can take

$$T = \left\{ t = \begin{pmatrix} e^{i\theta_1} & & \\ & \ddots & \\ & & e^{i\theta_n} \end{pmatrix} \colon \det t = 1 \right\}.$$

The Weyl group $W(U,T)$ then becomes the symmetric group S_n. The positive regular characters χ can be identified with the set

$$\{(\lambda_1, \ldots, \lambda_n) \in \mathbf{Z}^n \colon \lambda_1 > \lambda_2 > \cdots > \lambda_n\},$$

taken modulo the diagonal action of $\mathbf{Z}$, and we have

$$\varepsilon(s)(s\chi)(t) = \text{sgn}(s)e^{i\theta_1 \lambda_{s(1)}} \cdots e^{i\theta_n \lambda_{s(n)}}, \qquad s \in S_n,$$

for the summands in the Weyl character formula. The degree of the corresponding representation τ is simply equal to

$$\prod_{1 \leq i < j \leq n} \left(\frac{\lambda_i - \lambda_j}{j - i} \right).$$

There is a unique bi-invariant Riemannian metric on U whose restriction to T is given by the form $(\cdot, \cdot)$. One checks that the Laplacian acts on the subspace of $L^2(U)$ corresponding to τ by the scalar

$$(d\chi, d\chi) - (d\chi_1, d\chi_1).$$

This number represents an eigenvalue of the Laplacian, which occurs with multiplicity

$$\prod_{\alpha > 0} \left(\frac{(d\chi, \alpha)}{(d\chi_1, \alpha)} \right)^2.$$

Weyl's theory thus gives something which is rather rare: an explicit description of the spectrum of the Laplacian on a Riemannian manifold. This is typical of the examples that Lie theory contributes to an increasingly large number of mathematical areas. The examples invariably have an internal structure which is both rich and computable.

2. The compact group U has a complexification. Conversely, a complex semisimple group has a compact real form, which is unique up to conjugation. Weyl exploited this connection with his "unitary trick", in order to study the finite-dimensional representations of a complex group. Now, a complex semisimple group is actually algebraic. Therefore, the choice of the group U in §1 is tantamount to a choice of a semisimple, simply connected algebraic group G over $\mathbf{C}$. In this paragraph, we shall assume that G is in fact defined over $\mathbf{R}$. In other words, we suppose that we are given an arbitrary real form $G(\mathbf{R})$ of $G(\mathbf{C})$. In this context there is again a Riemannian manifold. It is the globally symmetric space

$$X_G = G(\mathbf{R})/K_{\mathbf{R}},$$

in which $K_{\mathbf{R}}$ is a maximal compact subgroup of $G(\mathbf{R})$.

Any representation $\tau \in \hat{U}$ extends to a finite-dimensional representation of $G(\mathbf{C})$. However, the restriction of this representation of $G(\mathbf{R})$, which we shall also denote by τ, need not be unitary. The representations in the unitary dual $\widehat{G(\mathbf{R})}$ are generally infinite-dimensional. The problem of classifying $\widehat{G(\mathbf{R})}$ is much more difficult than in the compact case, and is still not completely solved. (See the lecture [11] of Vogan). However, Harish-Chandra found enough of the representations in $\widehat{G(\mathbf{R})}$ to be able to describe explicitly the decomposition of $L^2(G(\mathbf{R}))$ into irreducible representations.

One of Harish-Chandra's achievements was to establish a theory of characters for infinite-dimensional representations. For a given $\pi \in \widehat{G(\mathbf{R})}$, it was first shown that the operators

$$\pi(f) = \int_{G(\mathbf{R})} f(x)\pi(x)\,dx, \qquad f \in C_c^\infty(G(\mathbf{R})),$$

were of trace class, and that the functional $f \to \mathrm{tr}(\pi(f))$ was a distribution on $G(\mathbf{R})$. Harish-Chandra was able to prove that the distribution was actually a function [5a]. That is, there exists a locally integrable function Θ_π on $G(\mathbf{R})$ such that

$$\mathrm{tr}(\pi(f)) = \int_{G(\mathbf{R})} f(x)\Theta_\pi(x)\,dx,$$

for any $f \in C_c^\infty(G(\mathbf{R}))$. The function Θ_π is invariant on the conjugacy classes of $G(\mathbf{R})$, and is called the *character* of π. It uniquely determines the equivalence class of π.

ing system $R^\vee$

ıeart of the problem of decomposing
ntations in $\widehat{G(\mathbf{R})}$ which behaved very
of a compact group. They are called
:cisely the representations which occur
$\mathbf{R}$)). Harish-Chandra [5b] classified the
e series exist if and only if $G(\mathbf{R})$ has a
ıct. Assume that such a group exists. As
p

$R_\gamma)\cdot(s\chi)'(\gamma),$

$_{(\mathbf{R})}(A_0)/\mathrm{Cent}_{G(\mathbf{R})}(A_0),$

hich depends
)' and γ re-
based on the

$-\ \alpha(t)^{-1/2}), \qquad t \in A_0(\mathbf{R}).$

screte series
entations of
i subgroups
Then M is
enter of M

ıetween $\widehat{G(\mathbf{R})}_{\mathrm{disc}}$ and the $W_{\mathbf{R}}(G, A_0)$-orbits
ıhat

$$\sum_{(G, A_0)} \varepsilon(s)(s\chi)(t), \qquad t \in A_0(\mathbf{R})_{\mathrm{reg}}.$$

ıssification of representations of U. However,
es.

ıneral a proper subgroup of the complex Weyl

$= \mathrm{Norm}_G(A_0)/\mathrm{Cent}_G(A_0).$

:m U with a conjugate, if necessary, we can as-
th the torus T. It is then easy to show that the
(U, T) are the same. Consequently, every regular

$_{(\mathbf{R})_{\mathrm{reg}}},$

contains several $W_{\mathbf{R}}(G, A_0)$-orbits. It follows that
union of finite subsets $\widehat{G(\mathbf{R})}_\tau$, each of order

nooth at
onnected

$= |W(G, A_0)/W_{\mathbf{R}}(G, A_0)|,$

the representations $\tau \in \hat{U}$. The packets $\widehat{G(\mathbf{R})}_\tau$ are
ty that on $T_{\mathrm{reg}} = A_0(\mathbf{R})_{\mathrm{reg}}$, the function

cters of

$-1)^{(1/2)\dim(X_G)} \sum_{\pi \in \widehat{G(\mathbf{R})}_\tau} \Theta_\pi$

Cartan

ıfference from the compact case is that $G(\mathbf{R})$ generally
ıcy classes $\{A(\mathbf{R})\}$ of Cartan subgroups. The function
ı every Cartan subgroup if it is to be specified on all of
gave such formulas for the characters of discrete series.

onding
. The

ıl quote the general character formula for the sum (2.1).
ıary Cartan subgroup of $G(\mathbf{R})$. Then the set R of real

roots of $(G(\mathbf{R}), A(\mathbf{R}))$ is a root system, and it has a correspond
of co-roots. We can choose an isomorphism

$$\mathrm{Ad}(y)\colon A_0(\mathbf{C}) \xrightarrow{\sim} A(\mathbf{C}), \qquad y \in G(\mathbf{C}),$$

over the complex numbers. This allows us to define a character

$$\chi'(\gamma) = \chi(\mathrm{Ad}(y)^{-1}\gamma), \qquad \gamma \in A(\mathbf{C}),$$

on $A(\mathbf{C})$, for every χ in $\hat{T} = \widehat{A_0(\mathbf{R})}$. One then has the formula

$$(-1)^{(1/2)\dim(X_G)} \sum_{\pi \in \widehat{G(\mathbf{R})}_\tau} \Theta_\pi(\gamma) = \Delta_A^G(\gamma)^{-1} \sum_{s \in W(G,A_0)} \varepsilon(s)\cdot\bar{c}(R_{(s\chi)'}^\vee,$$

valid for any point $\gamma \in A(\mathbf{R})_{\mathrm{reg}}$. Here, $\bar{c}(R_{(s\chi)'}^\vee, R_\gamma)$ is an integer w
only on the systems of positive roots in $R^\vee$ and R defined by $(s\chi$
spectively. It can be computed from a simple inductive procedure,
rank of R. (See [6].)

Thus, on the noncompact Cartan subgroups, the characters of di
are slightly different than the characters of finite-dimensional repres
$G(\mathbf{R})$. They are in fact closer to finite-dimensional characters for Lev
of G. Let M be the centralizer in G of the $\mathbf{R}$-split component of A.
a reductive subgroup of G, and the $\mathbf{R}$-split component A_M of the c
is the same as the original $\mathbf{R}$-split component of A. We can write

$$\Delta_A^G(\gamma) = \Delta_M^G(\gamma)\Delta_A^M(\gamma)(-1)^{|R_\gamma \cap (-R^+)|}, \qquad \gamma \in A(\mathbf{R}),$$

where, if $\mathfrak{g}$ and $\mathfrak{m}$ denote the Lie algebras of G and M,

$$\Delta_M^G(\gamma) = |\det(1 - \mathrm{Ad}(\gamma))_{\mathfrak{g}/\mathfrak{m}}|^{1/2}.$$

It can be shown that the function

$$(2.2) \quad \Phi_M(\gamma,\tau) = \Delta_M^G(\gamma)^{-1}(-1)^{(1/2)\dim(X_G)} \sum_{\pi \in \widehat{G(\mathbf{R})}_\tau} \Theta_\pi(\gamma), \qquad \gamma \in$$

extends to a continuous function on $A(\mathbf{R})$. This function is not s
the singular hyperplanes defined by real roots. However, on each c
component of

$$\{\gamma \in A(\mathbf{R})\colon \alpha(\gamma) \neq 1,\ \alpha \in R\},$$

$\Phi_M(\gamma,\tau)$ is an integral linear combination of finite-dimensional chara
$M(\mathbf{R})$.

EXAMPLE. Suppose that $G = \mathrm{SU}(p,q)$. The conjugacy classes of
subgroups are represented by groups

$$\{A_r(\mathbf{R})\colon 0 \leq r \leq \min(p,q)\},$$

in which the split component A_{M_r} of A_r has dimension r. The corresp
Levi subgroup M_r is the product of $\mathrm{SU}(p-r,q-r)$ with an abelian grou

real root system R consists simply of r copies of $\{\pm\alpha\}$, the root system of type A_1. The constant

$$\bar{c}(R_{(s\chi)'}, R_\gamma)$$

can be determined from the case of SL(2). It equals 2^r in chambers where the function

$$(\chi, \gamma) \to (s\chi)'(\gamma)$$

is bounded, and it vanishes otherwise.

The harmonic analysis of $L^2(G(\mathbf{R}))$ is based on the classification of the discrete series. Harish-Chandra [5c] showed that the subrepresentation of $L^2(G(\mathbf{R}))$ which decomposed continuously could be understood in terms of the discrete series on Levi subgroups of G. More precisely, $L^2(G(\mathbf{R}))$ is a direct integral of irreducible representations obtained by inducing discrete series from parabolic subgroups. As in §1, the harmonic analysis is closely related to the spectral decomposition of a Laplacian. If (σ, V_σ) is a finite-dimensional unitary representation on $K_{\mathbf{R}}$, one can form the homogeneous vector bundle

$$G(\mathbf{R}) \times_{K_{\mathbf{R}}} V_\sigma$$

over the Riemannian symmetric space X_G. The Laplacian on X_G then gives an operator on the space of square-integrable sections. A knowledge of the spectral decomposition of this space of sections, for arbitrary σ, is equivalent to the knowledge of the decomposition of $L^2(G(\mathbf{R}))$. Harish-Chandra's theory therefore provides another example of an explicit description of the spectrum of a Laplacian, this time for a noncompact Riemannian manifold.

3. Now suppose that Γ is a discrete subgroup of $G(\mathbf{R})$. We assume that Γ is a congruence subgroup of an arithmetic group. The associated problem in harmonic analysis is to decompose the right regular representation of R of $G(\mathbf{R})$ on $L^2(\Gamma \backslash G(\mathbf{R}))$. The theory has some similarities with that of $L^2(G(\mathbf{R}))$. Write

$$R = R_{\text{disc}} \oplus R_{\text{cont}},$$

where R_{disc} is a direct sum of irreducible representations, and R_{cont} decomposes continuously. Then the decomposition of R_{cont} can be described in terms of the decomposition of the analogues of R_{disc} for Levi subgroups of G. This description is part of the theory of Eisenstein series, initiated by Selberg [10], and established for general groups by Langlands [8c].

The remaining problem, then, is to decompose R_{disc}. This is much harder than the corresponding problem for $L^2(G(\mathbf{R}))$. In fact, one does not really expect ever to obtain a complete description of the decomposition of R_{disc}. Rather, one wants to establish relations or "reciprocity laws" between the decompositions of the representations R_{disc} for different groups. Such relations are summarized in Langlands' functoriality conjecture, and are very deep. (See [8a, 2a, 1a].) The best hope for studying functoriality seems to be through the trace formula. The trace formula is a nonabelian analogue of the Poisson summation formula, which relates the study of R_{disc} to the harmonic analysis on $G(\mathbf{R})$. It provides a reasonably explicit expression for the trace of the operator

$$R_{\mathrm{disc}}(f) = \int_{G(\mathbf{R})} f(x) R_{\mathrm{disc}}(x)\, dx,$$

for a suitable function f on $G(\mathbf{R})$. (Actually, $R_{\mathrm{disc}}(f)$ is not known to be of trace class in general, so one must group the terms in a certain way to insure convergence.) The trace formula was introduced by Selberg [10] in the case of compact quotient and for some groups of rank one. For general rank, we refer the reader to the surveys [1b] and [1d]. We shall not try to summarize the trace formula here, or to discuss the limited applications to functoriality that it has so far yielded. We shall instead discuss an application of the trace formula to L^2-cohomology. This is appropriate in the present symposium, for in the end there is a surprising connection with the character formulas that began with Weyl.

Assume that Γ has no elements of finite order. Then

$$X_\Gamma = \Gamma \backslash G(\mathbf{R})/K_\mathbf{R} = \Gamma \backslash X_G$$

is a locally symmetric Riemannian manifold. Choose an irreducible representation τ of $G(\mathbf{C})$ on a finite-dimensional Hilbert space V_τ. Restricting τ to the subgroup Γ of $G(\mathbf{C})$, we define a locally constant sheaf

$$\mathcal{F}_\tau = V_\tau \times_\Gamma X_G$$

on X_Γ. Let $A^q_{(2)}(X_\Gamma, \mathcal{F}_\tau)$ be the space of smooth q-forms ω on X_Γ with values in $\mathcal{F}_\tau$, such that ω and $d\omega$ are both square integrable. Then $A^*_{(2)}(X_\Gamma, \mathcal{F}_\tau)$ is a differential graded algebra. Its cohomology

$$H^*_{(2)}(X_\Gamma, \mathcal{F}_\tau) = \bigoplus_q H^q_{(2)}(X_\Gamma, \mathcal{F}_\tau)$$

is the L^2-cohomology of X_Γ (with coefficients in $\mathcal{F}_\tau$). Assume that $G(\mathbf{R})$ has a compact Cartan subgroup $A_0(\mathbf{R})$. Then Borel and Casselman [3] have shown that the groups $H^q_{(2)}$ are finite-dimensional. We would like to compute the L^2-Euler characteristic

$$\sum_q (-1)^q \dim(H^q_{(2)}(X_\Gamma, \mathcal{F}_\tau)).$$

More generally, we shall consider the L^2-Lefschetz numbers of Hecke operators.

Hecke operators are best discussed in terms of adèles. As partial motivation for the introduction of adèle groups, consider the question of how one obtains congruence subgroups Γ. The main step is to choose a $\mathbf{Q}$-structure for $G(\mathbf{R})$. In other words, assume that the algebraic group G is defined over $\mathbf{Q}$. Assume also that $G(\mathbf{R})$ has no compact simple factors. Let

$$\mathbf{A} = \mathbf{R} \times \mathbf{A}_0 = \mathbf{R} \times \mathbf{Q}_2 \times \mathbf{Q}_3 \times \cdots$$

be the ring of adèles of $\mathbf{Q}$. Then $G(\mathbf{A})$ is a locally compact group, which contains $G(\mathbf{Q})$ as a discrete subgroup. Suppose that K_0 is an open compact subgroup of the group $G(\mathbf{A}_0)$ of points over the finite adèles. Since G is simply connected, the strong approximation theorem asserts that

$$G(\mathbf{A}) = G(\mathbf{Q}) \cdot G(\mathbf{R})K_0.$$

It follows immediately that there are homeomorphisms

(3.1) $$\Gamma \backslash G(\mathbf{R}) \xrightarrow{\sim} G(\mathbf{Q}) \backslash G(\mathbf{A}) / K_0$$

and

(3.2) $$X_\Gamma = \Gamma \backslash G(\mathbf{R}) / K_{\mathbf{R}} \xrightarrow{\sim} G(\mathbf{Q}) \backslash G(\mathbf{A}) / K_{\mathbf{R}} K_0,$$

where

$$\Gamma = G(\mathbf{Q})K_0 \cap G(\mathbf{R}),$$

a congruence subgroup. Every congruence subgroup is obtained in this way. In other words, given the $\mathbf{Q}$-structure on $G(\mathbf{R})$, a choice of Γ amounts to a choice of an open compact subgroup K_0 of $G(\mathbf{A}_0)$. Suppose that h belongs to the Hecke algebra $\mathcal{H}_{K_0}$ of compactly supported functions on $G(\mathbf{A}_0)$ which are bi-invariant under K_0. If ϕ is any form in $A^q_{(2)}(X_\Gamma, \mathcal{F}_\tau)$, we define

$$h\phi = \int_{G(\mathbf{A}_0)} h(g)(g^*\phi)\, dg,$$

where

$$(g^*\phi)(x) = \phi(xg),$$

for any point x in

$$X_\Gamma \cong G(\mathbf{Q}) \backslash G(\mathbf{A}) / K_{\mathbf{R}} K_0.$$

Then $h\phi$ is also a form in $A^q_{(2)}(X_\Gamma, \mathcal{F}_\tau)$. We obtain an operator

$$H^q_{(2)}(h, \mathcal{F}_\tau) \colon H^q_{(2)}(X_\Gamma, \mathcal{F}_\tau) \to H^q_{(2)}(X_\Gamma, \mathcal{F}_\tau).$$

The problem is to compute the Lefschetz number

$$\mathcal{L}_\tau(h) = \sum_q (-1)^q \mathrm{tr}(H^q_{(2)}(h, \mathcal{F}_\tau)).$$

Of course, if we set h equal to 1_{K_0}, the characteristic function of K_0 divided by the volume of K_0, we obtain the L^2-Euler characteristic.

The homeomorphism (3.1) is compatible with right translation by $G(\mathbf{R})$. It follows that the representations R and R_{disc} of $G(\mathbf{R})$ extend to representations of $G(\mathbf{R}) \times \mathcal{H}_{K_0}$. It turns out that one can find a function $f_\tau \in C^\infty_c(G(\mathbf{R}))$ such that

$$\mathcal{L}_\tau(h) = \mathrm{tr}(R_{\mathrm{disc}}(f_\tau \times h)).$$

One can then evaluate the right-hand side by the trace formula. The result is a rather simple formula for $\mathcal{L}_\tau(h)$. Foregoing the details, which appear in [lc], we shall be content to state the final answer.

We must first describe the main ingredients of the formula. Suppose that M is a Levi component of a parabolic subgroup of G which is defined over $\mathbf{Q}$. If M

contains a Cartan subgroup $A(\mathbf{R})$ which is compact modulo $A_M(\mathbf{R})^0$, we have the function

$$\Phi_M(\gamma, \tau), \qquad \gamma \in A(\mathbf{R})_{\text{reg}},$$

defined by (2.2) in terms of characters of discrete series. Let us extend this to an $M(\mathbf{R})$-invariant function on $M(\mathbf{R})$ which vanishes unless γ is conjugate to an element in $A(\mathbf{R})$. If M does not contain such a Cartan subgroup, we simply set $\Phi_M(\gamma, \tau) = 0$. This function represents the contribution from τ. Now, suppose that γ belongs to $M(\mathbf{Q})$. Let M_γ denote the centralizer of γ in M. We assume that $\Phi_M(\gamma, \tau)$ does not vanish, so in particular, γ is semisimple. The contribution from $h \in \mathcal{H}_{K_0}$ will be of the form

$$h_M(\gamma) = \int_{K_{0,\max}} \int_{N_P(\mathbf{A}_0)} \int_{M_\gamma(\mathbf{A}_0)\backslash M(\mathbf{A}_0)} h(k^{-1}m^{-1}\gamma mnk)\, dm\, dn\, dk,$$

where $P = MN_P$ is a parabolic subgroup over $\mathbf{Q}$ with Levi component M, and $K_{0,\max}$ is a suitable maximal compact subgroup of $G(\mathbf{A}_0)$. This is essentially an invariant orbital integral of h, and is easily seen to be independent of P. The third ingredient in the formula will be a constant $\chi(M_\gamma)$, which is closely related to the (classical) Euler characteristic of the symmetric space of M_γ. Let $\overline{M}_\gamma$ be any reductive group over $\mathbf{Q}$ which is an inner twist of M_γ and such that $\overline{M}_\gamma(\mathbf{R})/A_M(\mathbf{R})^0$ is compact. Then

$$\chi(M_\gamma) = (-1)^{(1/2)\dim(X_{M_\gamma})}\text{vol}(\overline{M}_\gamma(\mathbf{Q})\backslash\overline{M}_\gamma(\mathbf{A}_0))w(M_\gamma),$$

provided that G has no factors of type E_8. This relies on a theorem of Kottwitz [7], which requires the Hasse principle. Otherwise, $\chi(M_\gamma)$ must be given by a slightly more complicated formula.

We can now state the L^2-Lefschetz formula. It is

$$(3.3) \quad \mathcal{L}_\tau(h) = \sum_M (-1)^{\dim(A_M)}|W(G, A_M)|^{-1} \sum_{\gamma \in (M(\mathbf{Q}))} \chi(M_\gamma)\Phi_M(\gamma, \tau)h_M(\gamma),$$

where h is any element in the Hecke algebra $\mathcal{H}_{K_0}$ and

$$W(G, A_M) = \text{Norm}_G(A_M)/\text{Cent}_G(A_M).$$

The outer sum is over the conjugacy classes of Levi subgroups M in G, while the inner sum is over the conjugacy classes of elements γ in $M(\mathbf{Q})$. Since $\Phi_M(\gamma, \tau)$ vanishes unless γ is $\mathbf{R}$-elliptic in M, we can restrict the inner sum to such elements. It can actually be taken over a finite set, which depends only on the support of h. We therefore have a finite closed formula for the Lefschetz number $\mathcal{L}_\tau(h)$.

4. We shall conclude with a few general, and perhaps obvious, remarks. Suppose that the symmetric space X_Γ has complex Hermitian structure. Then one has the Baily-Borel compactification $\overline{X}_\Gamma$ of X_Γ, which is a complex projective algebraic variety with singularities. The Goresky-Macpherson intersection homology is a theory for singular spaces which satisfies Poincaré duality. Zucker's conjecture asserts that the L^2-cohomology of X_Γ is isomorphic to the intersection homology of $\overline{X}_\Gamma$. It can be regarded as an analogue of the de Rham theorem for the noncompact space X_Γ. The L^2-cohomology is of course the analytic ingredient, and the intersection homology represents the geometric ingredient.

Zucker's conjecture opens the possibility of interpreting (3.3) as a fixed point formula. The geometric interpretation of the Hecke operators is easy to describe. For

$$\Gamma = G(\mathbf{Q})K_0 \cap G(\mathbf{R}),$$

as above, there is a bijection

$$\Gamma\backslash G(\mathbf{Q})/\Gamma \xrightarrow{\sim} K_0\backslash G(\mathbf{A}_0)/K_0$$

between double coset spaces. Suppose that $h \in \mathcal{H}_{K_0}$ is given by the characteristic function of a coset

$$\Gamma g \Gamma, \qquad g \in G(\mathbf{Q}).$$

Let Γ' be any subgroup of finite index in $\Gamma \cap g\Gamma g^{-1}$. Then the map

(4.1) $$\Gamma' x \to (\Gamma x, \Gamma g^{-1} x)$$

is an embedding of $X_{\Gamma'}$ into $(X_\Gamma \times X_\Gamma)$. The resulting correspondence gives the Hecke operator on cohomology. Goresky and Macpherson [4] have proved a general Lefschetz fixed point theorem for intersection homology. One could apply it to the correspondence (4.1) and try to duplicate the formula (3.3). However, the correspondence (4.1) does not intersect the diagonal in a nice way. Moreover, the singularities of $\overline{X}_\Gamma$ are quite bad. It therefore seems remarkable that the formula (3.3) is as simple as it is. It is another instance of a general theory which works out nicely in the examples arising from Lie theory.

What one would really like is an interpretation of (3.3) as a fixed point formula in characteristic p, as was done in [8b] for $G = \mathrm{GL}(2)$. Our condition that G be simply connected was purely for simplicity. If we allow G to be an arbitrary reductive group over $\mathbf{Q}$, some of the spaces

$$X_{K_0} = G(\mathbf{Q})\backslash G(\mathbf{A})/K_\mathbf{R} K_0$$

will be associated to Shimura varieties, as in [8d]. They will admit natural definitions over number fields, which are compatible with the action of the Hecke operators. A similar assertion should also apply to the compactifications $\overline{X}_{K_0}$. One could then take the reduction modulo a good prime, and consider the action of the Frobenius. The intersection homology has an l-adic analogue. What is wanted is a formula for the Lefschetz number of the composition of a power of the Frobenius with an arbitrary Hecke correspondence at the invertible primes. The resulting formula could then be compared to (3.3), for suitably chosen h. Of course, one would need to know the structure of the points $\mathrm{mod}\, p$, discussed in [8d], and more elaborate information on the points at infinity. Along the way, one would have to be able to explain the geometric significance of the discrete series characters $\Phi_M(\gamma, \tau)$. Such things are far from known, at least to me. I mention them only to emphasize that the formula (3.3) is just a piece of a larger puzzle.

The ultimate goal is of course to prove reciprocity laws between the arithmetic information conveyed by l-adic representations of Galois groups, and the analytic information wrapped up in the Hecke operators on L^2-cohomology. A comparison of the two Lefschetz formulas would lead to generalizations of the results in [8b] for GL(2). However, as the analytic representative, the formula (3.3) is still somewhat deficient. To get an idea of what more is needed, consider the case that the highest weight of τ is regular. Then one can show that the L^2-cohomology of X_{K_0} is concentrated in the middle dimension. Moreover, the cohomology has a Hodge decomposition

$$H^*_{(2)}(X_{K_0}, \mathcal{F}_\tau) = \bigoplus_{p+q=(1/2)\dim(X_G)} H^{p,q}_{(2)}(X_{K_0}, \mathcal{F}_\tau).$$

Since the Hecke operators commute with the Hodge group $S(\mathbf{R}) \cong \mathbf{C}^*$, there is also a decomposition

$$H^*_{(2)}(h, \mathcal{F}_\tau) = \bigoplus_{p,q} H^{p,q}_{(2)}(h, \mathcal{F}_\tau),$$

for each $h \in \mathcal{H}_{K_0}$. What is required for the comparison is a formula for each number

$$(4.2) \qquad\qquad\qquad \operatorname{tr}(H^{p,q}_{(2)}(h, \mathcal{F}_\tau)),$$

rather than just the sum of traces provided by (3.3). However, this will have to wait until the trace formula has been stabilized. Then one would be able to write (4.2) as a linear combination of Lefschetz numbers attached to endoscopic groups of G. These could then be evaluated by (3.3).

REFERENCES

1. J. ARTHUR
    (a) *Automorphic representations and number theory*, Canadian Math. Soc. Conf. Proc., Vol. 1, 1981, pp. 1–49.
    (b) *The trace formula for noncompact quotient*, Proc. Internat. Congr., Warsaw, 1983, Vol. 2, pp. 849–859.
    (c) *The L^2-Lefschetz numbers of Hecke operators*, preprint.
    (d) *The trace formula and Hecke operators*, Proc. Selberg Symposium (Oslo, 1987), Academic Press (to appear).

2. A. BOREL
    (a) *Automorphic L-functions*, Proc. Sympos. Pure Math., vol. 33, Part 2, Amer. Math. Soc., Providence, R. I., 1979, pp. 27–62.
    (b) *Hermann Weyl and Lie groups*, Hermann Weyl 1885-1985 (K. Chandrasekharan, editor), Springer-Verlag, 1986, pp. 53–82.

3. A. BOREL AND W. CASSELMAN
    L^2-cohomology of locally symmetric manifolds of finite volume, Duke Math. J. **50** (1983), 625–647.

4. M. GORESKY AND R. MACPHERSON
    Lefschetz fixed point theorem for intersection homology, preprint.

5. HARISH-CHANDRA

(a) *Invariant eigendistributions on a semisimple Lie group*, Trans. Amer. Math. Soc. **119** (1965), 457–508; also Collected Papers, Springer-Verlag , Vol. III, pp. 351–402.

(b) *Discrete series for semisimple groups*. I, Acta. Math. **113** (1965), 241–318; II, Acta Math. **116** (1966), 1–111; also Collected Papers, Springer-Verlag, Vol. III, pp. 403–481, and 537–648.

(c) *Harmonic analysis on real reductive groups*. III, *The Maas-Selberg relations and the Plancherel formula*, Ann. of Math. (2) **104** (1976), 117–201; also Collected Works, Springer-Verlag, Vol. IV, pp. 259–343.

6. R. HERB

Characters of averaged discrete series on semisimple real Lie groups, Pacific J. Math. **80** (1979), 169–177.

7. R. KOTTWITZ

Tamagawa numbers, preprint.

8. R P. LANGLANDS

(a) *Problems in the theory of automorphic forms*, Lecture Notes in Math., vol. 170, Springer-Verlag, 1970, pp. 18–86.

(b) *Modular forms and l-adic representations*, Lecture Notes in Math., vol. 349, Springer-Verlag, 1973, pp. 361–500.

(c) *On the functional equations satisfied by Eisenstein series*, Lecture Notes in Math., vol. 544, Springer-Verlag, 1976.

(d) *Representation theory and arithmetic*, these proceedings.

9. F. PETER AND H. WEYL

Der Vollstandigkeit der primitiven Darstellungen einer geschlossen kontinuierlichen Gruppe, Math. Ann. **97** (1927), 737–755; also Ges. Ab. Bd. III, 73, 58–75.

10. A. SELBERG

Harmonic analysis and discontinuous groups in weakly symmetric spaces with applications to Dirichlet series, J. Indian Math. Soc. **20** (1956), 47–87.

11. D. VOGAN

Noncommutative algebras and unitary representations, these proceedings.

12. H. WEYL

(a) *Theorie der Darstellung kontinuierlichen halbeinfach Gruppen durch lineare Transformationen*. I, Math. Z. **23** (1925), 271–309; II, Math. Z. **24** (1926), 328–376; III, Math. Z. **24** (1926), 377–395; Nachtrag, Math. Z. **24** (1926), 789–791; also Ges. Ab. Bd. II, 68, 543–647.

(b) *The theory of groups and quantum mechanics*, Dover Publications, 1950.

UNIVERSITY OF TORONTO, CANADA

Proceedings of Symposia in Pure Mathematics
Volume 48 (1988)

Perspectives on Vertex Operators and the Monster

J. LEPOWSKY

1. Introduction. In the preface to the first edition of his book *The theory of groups and quantum mechanics* [**W2**], Hermann Weyl made these three comments in 1928:

"The importance of the standpoint afforded by the theory of groups for the discovery of the general laws of quantum theory has of late become more and more apparent...."

"[F]rom the purely mathematical viewpoint, it is no longer justifiable to draw such sharp distinctions between finite and continuous groups in discussing the theory of their representations as has been done in the existing texts on the subject...."

"There exists, in my opinion, a plainly discernible parallelism between the more recent developments of mathematics and physics...."

He was referring largely to the physical symmetries induced by classical simple Lie groups and the symmetric groups, the "reciprocity" between representations of the symmetric group and those of the general linear group, and, in the last quoted sentence, to the abstraction required to describe the algebra of observables.

With new and updated meaning, Weyl's words now apply to today's relationships among infinite-dimensional Lie theory, the Monster sporadic group, and string theory (or two-dimensional conformal quantum field theory), as we shall try to point out in this paper. Weyl's work itself plays important roles in several of the historical ideas behind these developments.

The fact that "purely mathematical" Lie algebra theory and fundamental physics have converged is certainly not a coincidence, especially if string theory indeed provides the correct description of nature. Even if this turns out not

1980 *Mathematics Subject Classification* (1985 *Revision*). Primary 17B65; Secondary 05A19, 11F22, 17B67, 20C35, 20C99, 20D08, 20E32, 20F29, 81E99.

Research partially supported by the Guggenheim Foundation and NSF Grant DMS86-03151.

The author would like to thank the Institute for Advanced Study for its hospitality while this paper was completed.

to be the case, string theorists have at the very least taken dramatic steps in the development of an important new physically inspired branch of mathematics, still only partly rigorous and with its boundaries nowhere near visible; see [GG], [GSW], [Wi2] for surveys. String theory incorporates both relativity and quantum mechanics, both of which motivated Weyl. It was problems raised by relativity theory that originally stimulated Weyl to study semisimple Lie theory (see [W4], p. 400; cf. [Bo]); of course, Weyl also built on the work of Frobenius, Schur and mostly Cartan. Meanwhile, always trying to find more symmetry in nature, Weyl undertook to clarify the group-theoretic underpinnings of quantum mechanics, starting from the Heisenberg commutation relations. As we shall sketch, semisimple Lie algebras can be expressed in terms of the Heisenberg commutation relations in a theory of vertex operators expressing (and partly inspired by) some of the important features of string theory.

In this exposition we shall first recall some basic ideas from the theory of affine Kac-Moody algebras, especially some mathematical and physical features of the program to realize their standard modules using variants of the vertex operators of string theory. Then we shall turn to some analogous, but much more subtle, aspects of a recent realization of the Fischer-Griess Monster as a symmetry group of a two-dimensional conformal quantum field theory. Enhanced by a recent contribution of R. Borcherds, this realization is a joint work with I. Frenkel and A. Meurman, whom I warmly thank for their insight over the years.

2. Affine Lie algebras. The Cartan-Weyl theory of semisimple Lie algebras, as continued by Chevalley, Harish-Chandra and Serre, found a natural generalized setting in the theory of Kac-Moody algebras, introduced in [K1], [Kan] and [M1], [M2]; cf. [K3], [F4]. These algebras include the (infinite-dimensional) *untwisted affine Kac-Moody algebras* $\hat{\mathfrak{g}}$, where $\mathfrak{g}$ is a finite-dimensional simple Lie algebra. Here

$$(1) \qquad\qquad \hat{\mathfrak{g}} = \mathfrak{g} \otimes \mathbf{C}[t, t^{-1}] \oplus \mathbf{C}c,$$

where $\mathbf{C}[t, t^{-1}]$ is the algebra of Laurent polynomials in an indeterminate t and where the brackets are given by:

$$(2) \; [x \otimes t^m, y \otimes t^n] = [x, y] \otimes t^{m+n} + \langle x, y \rangle m\delta_{m+n,0}c, \qquad x, y \in \mathfrak{g}, \; m, n \in \mathbf{Z},$$

$$(3) \qquad\qquad [c, \hat{\mathfrak{g}}] = 0,$$

$\langle \cdot, \cdot \rangle$ being an invariant symmetric bilinear form on $\mathfrak{g}$ (such as the Killing form). The algebra $\hat{\mathfrak{g}}$ is $\mathbf{Z}$-graded, with

$$\deg x \otimes t^n = n \quad \text{and} \quad \deg c = 0.$$

The realization (1)–(3) of the *abstract* affine Kac-Moody algebra defined by the extended Cartan matrix of $\mathfrak{g}$ was made in stages: The "loop algebra" $\hat{\mathfrak{g}}/\mathbf{C}c$ (so-called because of its geometric realization when t is replaced by $e^{2\pi i\tau}$, $\tau \in \mathbf{R}$) was described in [K1], [M2] and the central extension term (essentially a cocycle) much later. The algebra (1)–(3) is essentially the same as a "current algebra on the circle," in physical terminology. We shall call (1)–(3) the *affine Lie algebra associated with* $\mathfrak{g}$ (and $\langle \cdot, \cdot \rangle$) even when the Lie algebra $\mathfrak{g}$ is not necessarily simple.

A pinnacle of Weyl's study of semisimple Lie theory was his character formula for finite-dimensional irreducible representations of complex semisimple Lie groups ([**W1**]; see [**Bo**] for an account of Weyl's Lie-theoretic work). He established this by analytic methods, and Freudenthal ([**Freu**]; cf. [**J**]) and Bernstein-Gelfand-Gelfand [**BGG**] later supplied algebraic proofs. A remarkable and at first mysterious link between Weyl's character formula and modular forms arose in Macdonald's identities [**M**] for certain products of powers of Dedekind's η-function. Dyson also found certain of these identities, but "missed the opportunity of discovering a deeper connection between modular forms and Lie algebras, just because the number theorist Dyson and the physicist Dyson were not speaking to each other" (see [**D**]). Macdonald's identities are now known to be one manifestation of a deep symbiosis between Lie theory and string theory.

By adapting the Bernstein-Gelfand-Gelfand proof of Weyl's character formula, Kac [**K2**] generalized the formula to Kac-Moody algebras and the analogues of the finite-dimensional irreducible modules—the "standard" or "integrable highest weight" modules (cf. [**GL**], [**K3**] and [**L1**]), thereby recovering and generalizing Macdonald's identities. These identities are to be understood as specializations of the Weyl-Kac character formula for the trivial one-dimensional module (the denominator formula) for (possibly twisted) affine algebras $\hat{\mathfrak{g}}[\nu]$ with $\mathfrak{g}$ simple and with ν an automorphism of finite order of $\mathfrak{g}$ induced by a Dynkin diagram automorphism. For any automorphism ν of a Lie algebra $\mathfrak{g}$ with $\nu^M = 1$ and with ν also an isometry with respect to an invariant form $\langle \cdot, \cdot \rangle$, the *twisted affine algebra* $\hat{\mathfrak{g}}[\nu]$ is defined by:

$$(4) \qquad \hat{\mathfrak{g}}[\nu] = \coprod_{n \in \mathbf{Z}} \mathfrak{g}_{(n)} \otimes t^{n/M} \oplus \mathbf{C}c$$

where

$$\mathfrak{g}_{(n)} = \{ x \in \mathfrak{g} | \nu x = e^{2\pi i n/M} x \}$$

and where the brackets are given by (2) for $m, n \in (1/M)\mathbf{Z}$, $x \in \mathfrak{g}_{(mM)}$, $y \in \mathfrak{g}_{(nM)}$, and the condition that c again be central. (For this description see [**K1**], [**KKLW**].) The Weyl group of an untwisted affine algebra $\hat{\mathfrak{g}}$ for $\mathfrak{g}$ simple is naturally isomorphic to the affine Weyl group of $\mathfrak{g}$, the semidirect product of the Weyl group of $\mathfrak{g}$ with a translation group, the latter being the source of the theta-functions in Macdonald's identities.

3. Heisenberg algebras. From the standpoint of Lie algebra structure theory, the Lie algebras that least resemble the simple ones are the nilpotent ones, and the most basic of these is the Heisenberg algebra. This has a basis $\{P_i, Q_i, I | 1 \le i \le n\}$ and has structure defined by the quantum-mechanically motivated Heisenberg commutation relations

$$(5) \qquad [P_i, Q_j] = \delta_{ij}I, \quad \text{all other commutators of basis elements} = 0,$$

classically represented on a function space by:

$$Q_i \mapsto x_i \text{ (multiplication by the coordinate function),}$$

(6) $$P_i \mapsto \frac{\partial}{\partial x_i},$$

$$I \mapsto 1 \text{ (the identity operator).}$$

It was Weyl who released these relations from their operator realization by interpreting them as a Lie algebra. He studied the integrated form ("Weyl form") of these relations—the Heisenberg group—and gave heuristic arguments leading to the Stone–von Neumann "uniqueness theorem for the Heisenberg commutation relations." (See [**W2**]; cf. [**Bo**].)

In quantum field theory and string theory, through the process of second quantization, the number n of variables is increased to infinity, and the Heisenberg commutation relations turn into the infinite-dimensional **Z**-graded Lie algebra

(7) $$\hat{\mathfrak{h}} = \mathfrak{h} \otimes \mathbf{C}[t, t^{-1}] \oplus \mathbf{C}c,$$

where $\mathfrak{h}$ is a finite-dimensional vector space, with the brackets in $\hat{\mathfrak{h}}$ given by:

(8) $$[\alpha \otimes t^m, \beta \otimes t^n] = \langle \alpha, \beta \rangle m \delta_{m+n,0} c, \qquad \alpha, \beta \in \mathfrak{h}, \ m, n \in \mathbf{Z},$$

(9) $$[c, \hat{\mathfrak{h}}] = 0,$$

$\langle \cdot, \cdot \rangle$ being a nonsingular symmetric bilinear form on $\mathfrak{h}$. This Lie algebra is exactly the affinization of the abelian Lie algebra $\mathfrak{h}$ in the sense defined above. If we view the $\alpha \otimes t^m$ for $m < 0$ as "Q's" ("creation operators"), then the $\beta \otimes t^n$ for $n > 0$ are essentially the corresponding "P's" ("annihilation operators"), and $\hat{\mathfrak{h}}$ has a canonical realization as differential operators analogous to (6) on the "Fock space" $S(\hat{\mathfrak{h}}^-)$, where $S(\cdot)$ denotes symmetric algebra and

(10) $$\hat{\mathfrak{h}}^- = \coprod_{n<0} \mathfrak{h} \otimes t^n \quad \text{(the span of the ``Q's'');}$$

$S(\hat{\mathfrak{h}}^-)$ is essentially the polynomial algebra on the "Q's". The associative algebra of operators on $S(\hat{\mathfrak{h}}^-)$ generated by the action of $\hat{\mathfrak{h}}$ is sometimes called the Weyl algebra.

Strictly speaking, we should not call the affine algebra $\hat{\mathfrak{h}}$ a Heisenberg algebra because the "zero modes" $\hat{\mathfrak{h}} \otimes t^0 = \mathfrak{h}$ commute with all of $\hat{\mathfrak{h}}$. If we view the elements of $\mathfrak{h}$ as annihilation operators, then we can match them with appropriate creation operators by enlarging $\hat{\mathfrak{h}}$ to the Heisenberg algebra

(11) $$\mathfrak{l} = \hat{\mathfrak{h}} \oplus \mathfrak{h}^*,$$

where we make the identification

(12) $$\mathfrak{h} \simeq \mathfrak{h}^*, \qquad \alpha \mapsto \alpha^*,$$

of $\mathfrak{h}$ with its dual via the form $\langle \cdot, \cdot \rangle$, and we set

$$[\alpha, \beta^*] = \langle \alpha, \beta \rangle c \quad \text{for } \alpha, \beta \in \mathfrak{h}$$

and the other brackets involving $\mathfrak{h}^*$ to be 0. Then $\mathfrak{l}$ has a canonical realization on a larger "Fock space"

$$\text{(13)} \qquad \mathscr{F} = S(\hat{\mathfrak{h}}^-) \otimes \mathscr{H},$$

$\mathscr{H}$ a suitable function space on $\mathfrak{h}$. In bosonic string theory,

$$\mathfrak{h} = \text{momentum space},$$
$$\mathfrak{h}^* = \text{space-time},$$
$$\mathscr{F} = \text{Fock space of a free bosonic string}.$$

Actually, (7)–(9) constitute an algebraic version of an "abelian current algebra on the circle" in the sense of quantum field theory, where t is viewed as $e^{2\pi i \tau}$ ($\tau \in \mathbf{R}$) and the $\alpha \otimes t^n$ are viewed as components of "currents" or "quantum fields"

$$\text{(14)} \qquad \alpha(z) = \sum_{n \in \mathbf{Z}} (\alpha \otimes t^n) z^{-n-1},$$

z a nonzero complex number corresponding in string theory to a coordinate on the world-sheet of the string. The Heisenberg commutation relations (8) become (cf. [**W2**], p. 255)

$$\text{(15)} \qquad [\alpha(z_1), \beta(z_2)] = -\langle \alpha, \beta \rangle z_2^{-1} \frac{\partial}{\partial z_1} \delta\left(\frac{z_1}{z_2}\right)$$

where

$$\text{(16)} \qquad \delta(z) = \sum_{n \in \mathbf{Z}} z^n,$$

which is formally the Fourier expansion of the Dirac δ-function at $z = 1$. To check (15) we simply compare coefficients of monomials in z_1 and z_2 on the two sides. Analogously, the relations (2) for $\hat{\mathfrak{g}}$ correspond to a "nonabelian current algebra" and the corresponding relations for the twisted affine algebra (4) to what might be called a "twisted nonabelian current algebra."

In quantum field theory, equations such as (15) are typically interpreted (not necessarily rigorously) by analytic means related to distribution theory and functional analysis, and the Fock space $\mathscr{F}$ is completed to a Hilbert space (cf. [**W2**]). It is possible, though, and in fact very fruitful, to interpret z as a formal variable and to proceed purely algebraically, freely borrowing from the philosophy of distribution theory or quantum field theory for motivation whenever one wants. We shall adopt this viewpoint.

4. Vertex operators. The "vertex operator for the emission of a tachyon of momentum $\alpha \in \mathfrak{h}$" in bosonic string theory is

$$\text{(17)} \qquad Y(\alpha, z) = {}^{\circ}_{\circ} e^{\int \alpha(z)\, dz} {}^{\circ}_{\circ}.$$

Here

$$\text{(18)} \qquad \int \alpha(z)\, dz = \sum_{n \neq 0} (\alpha \otimes t^n) \frac{z^{-n}}{-n} + \alpha \log z + \alpha^*,$$

where the integration of (14) is carried out formally, and α^* (see (12)) is understood as a constant of integration. This is a formal infinite sum in $\mathfrak{l} = \hat{\mathfrak{h}} \oplus \mathfrak{h}^*$, involving the parameter z. The "e" in (17) refers to the formal exponential series. The open colons designate "normal ordering": all the creation operators are to be placed to the left of all the annihilation operators before the exponential is evaluated. Suitably interpreted, $Y(\alpha, z)$ becomes a rigorous operator on $\mathscr{H}$, or more precisely, if we define the expansion coefficients of $Y(\alpha, z)$ by the "generalized formal Laurent series" expansion

$$(19) \qquad Y(\alpha, z) = \sum_{n \in \mathbf{C}} x_\alpha(n) z^{-n - \langle \alpha, \alpha \rangle / 2}$$

(involving not necessarily integral powers of z), then the $x_\alpha(n)$ can be understood as rigorous operators on $\mathscr{H}$. They would be very cumbersome to define and study directly.

The first important property of the vertex operator (17) is that for $\alpha, \beta \in \mathfrak{h}$,

$$(20) \qquad Y(\alpha, z_1) Y(\beta, z_2) = {}^{\circ}_{\circ} Y(\alpha, z_1) Y(\beta, z_2) {}^{\circ}_{\circ} (z_1 - z_2)^{\langle \alpha, \beta \rangle},$$

where the normal-ordered product is as explained above, and where the last factor is understood as the binomial expansion in nonnegative integral powers of z_2 (not z_1). (When z_1 and z_2 are taken as complex variables, (20) holds in the domain $|z_1| > |z_2|$, after suitable interpretation.) Appropriate integrals of the expressions generalizing $(z_1 - z_2)^{\langle \alpha, \beta \rangle}$ for four vertex operators (vacuum expectation values) lead to the Veneziano amplitude for the scattering of certain particles at a "vertex," the starting point of what later turned into string theory. The desire to explain this amplitude is what originally led physicists to the vertex operator (17). (For background on string theory see e.g. [GSW].)

Formula (20) is actually a straightforward computation, using the Weyl commutation relations, in a formal "Heisenberg group"—the "group" corresponding to $\mathfrak{l}$. For us, it is a rigorous computation in generalized formal Laurent series in z_1 and z_2 with operator coefficients.

So far, we have mentioned two themes in which Weyl's ideas have played historical roles—affine Kac-Moody algebras and vertex operators. In what we have said so far, these themes intersect only in the fact that the algebra $\hat{\mathfrak{h}}$ is an "abelian" *analogue* of the algebra $\hat{\mathfrak{g}}$ with $\mathfrak{g}$ simple. A further obvious connection is that $\hat{\mathfrak{h}}$ is a *subalgebra* of $\hat{\mathfrak{g}}$, where $\mathfrak{h}$ is taken to be a Cartan subalgebra of $\mathfrak{g}$. These observations formed the starting point of a new blending of the two themes.

In 1977, essentially nothing was known about the structure of standard modules for affine Kac-Moody algebras except the abstract Weyl-Kac character formula. Wishing to construct some of these modules "concretely" and partly inspired by Weyl's viewpoint in *The classical groups* [W3] we started with $\hat{A}_1 = \mathfrak{sl}(2, \mathbf{C})\hat{\;}$ and found (see [LM]) that certain level 3 standard modules (the level of a module is the scalar by which the properly normalized central element c acts) suggested a close connection, eventually clarified in [LW2], [LW3], with the classical Rogers-Ramanujan partition identities. From [FL], the character formula also suggested that the "basic" (level 1) standard $\mathfrak{sl}(2, \mathbf{C})\hat{\ }$-modules might involve a Fock space like those described above, and this indeed turned out to be the case for a *twisted* realization of $\mathfrak{sl}(2, \mathbf{C})\hat{\ }$ (see above), using as operator realizations of $\mathfrak{sl}(2, \mathbf{C})\hat{\ }$ some complicated new differential operators which we found could most concisely be described as the expansion coefficients of an exponential expression that we now know to be a suitably twisted analogue of (17) (see [LW1]; it turns out that this operator had already been considered in [CF], but without the connection to a Lie algebra). We learned right away that the best way of studying operators analogous to the $x_\alpha(n)$ in (19) is to embed them first into generating functions like (17).

It was Garland who pointed out the similarity between the operators we had found and the (tachyon) vertex operators of the then-moribund string theory. A certain more general family of twisted vertex operator constructions of $\widehat{A}_n, \widehat{D}_n, \widehat{E}_6, \widehat{E}_7, \widehat{E}_8$ was found in [KKLW], and the constructions of [LW1] and [KKLW] were later related to the soliton solutions of the Korteweg–de Vries equation and certain more general soliton equations, respectively (see [DKM], [DJKM]).

Upon suitably modifying the "untwisted vertex operators" (17) by "discretizing the momentum" to lie in the root lattice of an equal-root-length simple Lie algebra $\mathfrak{g}$ and by working over a central extension of this lattice, Frenkel-Kac [FK] and Segal [S] used the operators described above to realize the untwisted affine algebras $\widehat{A}_n, \widehat{D}_n, \widehat{E}_n$ (cf. also [H], [BHN]). The operators analogous to the $x_\alpha(n)$ in (19) generate the affine algebra. The E_8-case of this construction is incorporated in the 10-dimensional heterotic string ([GHMR1], [GHMR2]), where $E_8 \times E_8$ appears naturally as a gauge group.

Other twisted vertex operator constructions of affine algebras were found in [F2] and [FLM1], and general families of twisted vertex operator constructions were given in [KP3] and [L2]. Much more general commutator results (mentioned below) are described in [B] in the untwisted case and in [FLM5] in the arbitrarily twisted setting. The paper [FLM5] includes expositions of several subjects mentioned here, including the formal algebraic method of computing commutators, initiated by Garland and analogous to standard methods involving "operator product expansions" used in physics. In order for the algebra to work correctly, one needs to construct suitable group extensions in delicate ways. The relevant groups are again closely related to Heisenberg groups. In computing commutators of the (rather subtle in general) vertex operators analogous to (17), one finds it appropriate to combine an infinite family of computations, involving arbitrary numbers of derivatives, into a generating function, a method based ultimately on Taylor's formula, expressed in the form

$$(21) \qquad e^{z_0(d/dz)} f(z) = f(z + z_0)$$

(see [L2], [FLM5]).

Through the mechanism of vertex operators, then, semisimple theory has been fundamentally entwined with Heisenberg-Weyl theory.

We have been discussing vertex operator constructions of (possibly twisted) affine algebras of types $\widehat{A}_n, \widehat{D}_n, \widehat{E}_n$. The constructed modules are level 1 standard modules. Spinor-type constructions, again related to string-theoretic ideas, of several families of affine algebras have also been found ([F1], [F2], [KP1], [FF]). There is also a program to construct $\widehat{B}_n, \widehat{C}_n, \widehat{F}_4, \widehat{G}_2$ and higher level standard modules for untwisted and twisted affine algebras using what we call "Z-algebras," which are "nonlocal" vertex operator algebras (the analogue of the last factor in (20) is not a rational function); this program is complete for the simplest but still highly nontrivial case of $\widehat{A}_1$, in both the twisted and untwisted realizations (see [LW2]–[LW4], [LP], [MP]). It is the level 3 standard module construction for $\widehat{A}_1$ in the twisted setting which "explains" the Rogers-Ramanujan identities. The untwisted $\widehat{A}_1$-case has arisen in "parafermion" conformal quantum field theories [ZF]. An analysis of the equivalence between the constructions of [LP] and of [ZF], together with consequences, is being worked out jointly with C.-Y. Dong. In particular, we show how [LP] gives a "generalized Pauli exclusion principle" for the parafermion currents, thereby perhaps further justifying the use of the term "parafermion," which refers to fractional conformal weight in [ZF]. A number of other cases of Z-algebra constructions of standard modules have been carried out (see the references in [FLM5]); in [C], apparently new combinatorial identities analogous to the Rogers-Ramanujan identities are discovered in the course of the construction of certain modules. The physical implications of such constructions should be investigated further. For other approaches to the construction of affine algebras, see [BT], [Ge], [GQ], [GNORS], [GNOS], [NY] and [Wi1]. Computations of weight multiplicities of standard modules and many related matters are studied in detail in [KP2] using modular transformation properties of the characters.

5. The Monster. The Fischer-Griess Monster or Friendly Giant is the largest of the sporadic finite simple groups, constructed by Griess in [G] as a group of automorphisms of a 196884-dimensional commutative nonassociative algebra now called the Griess algebra or Griess-Norton algebra. Even before it was proved to exist the Monster attracted widespread attention outside the finite group community because of its remarkable (conjectured) "moonshine" properties, described in [CN], relating the Monster to modular functions. Some— but not all—of these properties are "explained" in joint work with Frenkel and Meurman announced in [FLM2] and appearing in detail in a forthcoming book [FLM6]; see also [FLM3]–[FLM5] and [T3], [T4]. We cannot try to review this long-term work in any detail here, but we shall comment briefly on some aspects of the theory.

This work's main thrust is the explicit construction of an infinite-dimensional Z-graded module for the Monster analogous to, but much more subtle than, the untwisted Fock space $\mathscr{F}$ discussed above. This "moonshine module" is equipped with certain algebraic structure preserved by the Monster, which is itself constructed in a self-contained way with the help of vertex operators; by restricting attention to a certain finite-dimensional subspace of the module identified with the Griess algebra we see that the group that we construct can be identified with the group that Griess constructed. The algebraic structure involves a notion of what we call the "cross-bracket" of two vertex operators (see below), and this serves to illuminate the structure of the Griess algebra, which initially seemed to be an ad hoc object. While we were motivated by Griess's construction, our main results are independent of his, except for the identification of the group, and are based on a self-contained development of vertex operator calculus, old and new, including a theory of "triality." The largest part of the work is the use of triality for the realization of the Monster as a natural group of automorphisms of the module preserving its algebraic structure.

While admittedly long, our construction of the Monster itself exhibits the group directly as a natural symmetry group of a conformal quantum field theory. The Griess algebra becomes simply the subspace of the theory of conformal weight two. This is to be contrasted with, for instance, the untwisted vertex operator construction of an equal-root-length affine algebra ([**FK**], [**S**]). This can also be interpreted as a conformal field theory (in an extended sense, not requiring the "partition function" to be a modular function), but in which the underlying finite-dimensional simple Lie algebra, which is identified with the subspace of the theory of conformal weight one, is already a very natural object in isolation, and whose associated Lie group acts in a direct way as an automorphism group of both the simple Lie algebra and the module. The source of the vastly greater difficulty in the Monster case is that while weight one vectors can be exponentiated to continuous symmetries of the theory, the moonshine module has no nonzero weight one vectors. This is related to the lack of elements of square length 2 in the Leech lattice. In order to find a really natural object preserved by the Monster, we must pass to an infinite number of dimensions.

We shall explain something of our comments about conformal field theory. In a recent announcement [**B**], Borcherds defines a family of general vertex operators in the untwisted setting and gives a collection of conditions satisfied by the components (analogous to the $x_\alpha(n)$ in (19)) of these operators. He also states that our moonshine module can be given a structure, again preserved by the Monster, satisfying the same conditions; this amounts to an enlargement of the algebraic structure that we had constructed, such that our structure generates the new one. We prefer to express the algebraic relations in terms of generating functions rather than components as Borcherds does, and we strengthen what amounts to Borcherds' main property to formula (30) below (see [**FLM6**]; Borcherds has recently informed us that he too has found the stronger formula). In [**FLM6**] we include detailed proofs of the corresponding results. The logical relation between Borcherds' statement and ours in [**FLM2**] is that his statement follows from ours together with his announced results in the untwisted setting and known properties of the Monster (see [**FLM6**]); in [**FLM6**] we bypass these properties of the Monster and prove a more explicit (in fact, completely explicit) result by suitably defining general twisted vertex operators (see also [**FLM5**]) and establishing their properties.

The resulting notion of "vertex operator algebra" expresses a new kind of algebraic object which, it turns out, (rigorously) encodes the properties of a holomorphic two-dimensional conformal quantum field theory, in the sense of Belavin-Polyakov-Zamolodchikov [**BPZ**]. Such a theory is based on a suitably axiomatized action of the Virasoro algebra—the canonical central extension of the Lie algebra of "algebraic" vector fields on the circle—and on "operator product expansions" obeying an "associative law" which corresponds to "duality" in the sense of the "dual resonance model"—the precursor of modern string theory. The foundation of the essential equivalence between the mathematical and physical approaches is sketched in [**Wi2**], [**Wi3**]. See also [**FV**], [**CO**], [**ADDF**] for earlier work on general vertex operators, and [**KZ**] and [**TK**] for recent approaches based on gauge and current algebra symmetry. It should be said that the rigorous treatment of these matters is subtle. For instance, the algebraic structure is not an associative algebra, in spite of the terms that we have quoted. Here we shall use the formal variable language to express what amounts to the "associativity of the operator product expansion."

We have already mentioned that one should not study operators like the $x_\alpha(n)$ individually, without combining them into a generating function—a vertex operator. Somewhat analogously, for appropriately general results one should not perform individual commutator computations, each requiring the calculation of a higher derivative, without putting all the computations (and higher derivatives) into a generating function. This is explained in [**FLM5**]; formula (21) is the seed of this idea. Still further, it turns out that one should not calculate just commutators or cross-brackets of vertex operators, but something more general: Given two vertex operators, one should express infinitely many calculations at once by embedding the commutator and cross-bracket processes into a generating function of an infinite family of *products*. All three "levels" of generating function appear in formula (30) below.

We proceed to explain some features of a "vertex operator algebra," partly following [**B**] and partly following [**BPZ**] but noting that there are several natural variants of the notion. Such a structure is based on a **Z**-graded vector space

$$(22) \qquad\qquad V = \coprod_{n \in \mathbf{Z}} V_n$$

such that

$$(23) \qquad\qquad V_n = 0 \quad \text{for } n \text{ sufficiently small}$$

and

$$(24) \qquad\qquad \dim V_n < \infty \quad \text{for all } n.$$

(Sometimes it is useful, for instance in the course of the construction of the moonshine module, to let n range through a larger set than **Z**, and sometimes we

reverse the sign of the grading. Also, conditions (23) and (24) can be relaxed.)
The *graded dimension* $\dim_* V$ of V is the formal Laurent series which is the
generating function of the dimensions of the V_n:

$$(25) \qquad \dim_* V = \sum_{n \in \mathbf{Z}} (\dim V_n) q^n.$$

We have called this the "character" of V before, but we prefer to drop this usage
because it does not characterize anything; in physics language, $\dim_* V$ is called
the "partition function." Sometimes we take $q = e^{2i\pi\tau}$, τ in the upper half-plane.
For a holomorphic conformal field theory, one assumes that $\dim_* V$ is a modular
function with respect to the modular group $\mathrm{SL}(2, \mathbf{Z})$ acting in the standard way
on the upper half-plane.

We are given a linear map

$$(26) \qquad V \to (\mathrm{End}\, V)\{z\}, \qquad v \mapsto Y(v, z),$$

where for a vector space W, $W\{z\}$ denotes the vector space of formal Laurent
series in the indeterminate z (involving integral powers of z) with coefficients in
W. Then, for example,

$$Y(u, z)v \in V\{z\} \quad \text{for } u, v \in V.$$

(It is frequently appropriate to allow nonintegral powers of z.) By an *automorphism* of V we mean a linear automorphism g such that

$$(27) \qquad gY(v, z)g^{-1} = Y(gv, z) \quad \text{for } v \in V.$$

The objects $Y(v, z)$ correspond to "quantum fields" or "(general) vertex operators." We also have a distinguished element $1 \in V$ (the "vacuum") such that

$$(28) \qquad Y(1, z) = 1 \quad \text{(the identity operator, independent of } z\text{)}$$

and such that for $v \in V$,

$$(29) \qquad \lim_{z \to 0} Y(v, z)1 = v,$$

that is, no negative powers of z occur in $Y(v, z)1$ and the constant term is v.
This corresponds to the physical condition that the vertex operator associated
with v creates the state v.

In the setting of the free bosonic string, which we sketched in a mathematically
imprecise way above, the space $\mathscr{F}$ (see (13)) corresponds to V, and $\alpha(z)$ and
$Y(\alpha, z)$ (see (14) and (19)) are examples of quantum fields. The case $\alpha = 0$
corresponds to the vacuum vector 1.

As in (19), we can define component operators. One natural indexing is the
following:

$$Y(v, z) = \sum_{n \in \mathbf{Z}} v_n z^{-n-1},$$

where $v_n \in \mathrm{End}\, V$ (see [B]), and we assume that for $u, v \in V$,

$$u_n v = 0 \quad \text{for } n \text{ sufficiently large.}$$

There is also an action of the Virasoro algebra on V and a number of basic properties, which we shall not describe here. In particular, there is a notion of (conformal) weight, which is the shifting of the degree determined by the condition that the weight of the vacuum vector 1 be 0.

Now we state what we think of as the main property of a vertex operator algebra (see [**FLM6**]): For $u, v \in V$,

$$z_0^{-1}\delta\left(\frac{z_1 - z_2}{z_0}\right) Y(u, z_1)Y(v, z_2) - z_0^{-1}\delta\left(\frac{z_2 - z_1}{-z_0}\right) Y(v, z_2)Y(u, z_1)$$

$$(30) \qquad = z_2^{-1}\delta\left(\frac{z_1 - z_0}{z_2}\right) Y(Y(u, z_0)v, z_2).$$

This is to be understood as follows: The notation $\delta(\cdot)$ is as in (16). The expression $\delta\left(\frac{z_1 - z_2}{z_0}\right)$ is to be expanded as a formal power series in the second term in the numerator, z_2, and analogously for the other δ-function expressions. In particular, while the two δ-function factors on the left-hand side appear equal at first glance, they are not. The linear map (26) is extended in the obvious way from V to such spaces as $V\{z_0\}$ (note the form of the right-hand side). All "convergence" is strictly algebraic: When an expression is applied to any element of V, the coefficient of each monomial in the formal variables is a finite formal sum. The meaning of (30) is that the coefficients of each monomial on the two sides are equal as operators on V.

A very concentrated formula, (30) has many consequences. For instance, if we extract the residue Res_{z_0} with respect to the variable z_0 (the coefficient of z_0^{-1}), we immediately obtain

$$(31) \qquad [Y(u, z_1), Y(v, z_2)] = \mathrm{Res}_{z_0} z_2^{-1}\delta\left(\frac{z_1 - z_0}{z_2}\right) Y(Y(u, z_0)v, z_2),$$

a generating-function version of a formula of Borcherds [**B**] which expresses the commutator of two "quantum fields" in terms of the operators $Y(u, z_0)$. If instead we extract the coefficient of z_0^{-2} we have a formula for

$$(z_1 - z_2)[Y(u, z_1), Y(v, z_2)],$$

which is what we call the *cross-bracket* of $Y(u, z_1)$ and $Y(v, z_2)$ (see [**FLM6**]; this is a slight variant of our notion of cross-bracket introduced in [**FLM1**] and [**FLM2**]). If on the other hand we take Res_{z_2} of (30) we get a formula, which Borcherds uses as an axiom, for $Y(Y(u, z_0)v, z_2)$ in terms of $Y(u, z_1)Y(v, z_2)$ and $Y(v, z_2)Y(u, z_1)$. Note the degenerate case $u = v = 1$ of (30).

Of course, extracting these coefficients is a special case of multiplying by a "test function" and integrating. To use physics language, while the computation of the commutator of $Y(u, z_1)$ and $Y(v, z_2)$ requires only the finitely many singular terms in their operator product expansion (and analogously for the cross-bracket and generalizations of cross-bracket), formula (30) contains all the information in the full operator product expansion, expressing it in terms of $Y(Y(u, z_0)v, z_2)$. Note that knowledge of the algebra structure (commutators and their generalizations) is as explicit as knowledge of the quantum fields $Y(u, z)$ themselves. In [FLM5], [FLM6] and [L3], we construct explicit "twisted vertex operators" $Y(u, z)$ for u in suitable untwisted Fock spaces, acting on suitable twisted Fock spaces, and we prove formula (30) (or more precisely, a generalized form of (30), corresponding to the fact that the operators $Y(u, z)$ involve fractional powers of z) for these operators. Taking the residue recovers the constructions of twisted affine algebras in [KP3] and [L2] as very special cases. In particular, (30) holds in the fully general setting of [L2] and the last section of [FLM5] when we take the element u to be fixed by the underlying automorphism (see [L3]). Note also that Borcherds relates vertex operator algebras to nonaffine Kac-Moody algebras (see [B]; cf. [F3], [GO]).

An especially interesting feature of (30) is a remarkable "hidden $\mathscr{S}_3$-symmetry" property: For a vector $w \in V$, let us call formula (30) applied to w "assertion (*) for the ordered triple (u, v, w)." With the help of the assumed properties of the Virasoro algebra (not stated here), we can show that assertion (*) for (u, v, w) implies assertion (*) for any permutation of (u, v, w) (see [FLM6]). In this sense, formula (30) is completely symmetrical.

With this preparation we can now say that we show in detail in [FLM6] that the moonshine module, which we denote $V^\natural$, has the structure of a vertex operator algebra (or holomorphic conformal field theory) in the sense that we have sketched, with completely explicit expressions given for the $Y(u, z)$; that

$$(32) \qquad \dim_* V^\natural = J(q) = q^{-1} + 0 + 196884q + \cdots,$$

the classical modular function $j(q)$ with its constant term eliminated (that is, $J(q) = j(q) - 744$); and that the Monster acts in a natural and explicit way as a group of grading-preserving automorphisms of $V^\natural$. The identity element of (our presentation of) the Griess algebra gives rise to the (generating function of the) Virasoro algebra under the correspondence $v \mapsto Y(v, z)$, and the 196883-dimensional orthogonal complement of the identity element consists of lowest weight vectors for the Virasoro algebra, giving rise to primary fields.

A theorem of Tits ([T1], [T2]) that the Monster is the *full* automorphism group of the Griess algebra then implies immediately that the Monster is the full group of grading-preserving automorphisms of $V^\natural$.

Thus the Monster is a natural symmetry group of a string theory, or rather, of a holomorphic two-dimensional conformal quantum field theory. The structure $V^\natural$ is the direct sum of an untwisted space and a twisted space, both built canonically from the Leech lattice, and has recently been interpreted as the theory of a string propagating on the 24-dimensional "$\mathbf{Z}_2$-orbifold" obtained as the quotient of $\mathbf{R}^{24}$ by the translation action of the Leech lattice and the negation automorphism; this is potentially interesting because orbifold compactification of the $E_8 \times E_8$ heterotic string is thought to offer some hope in the search for a realistic string theory (see [DHVW1], [DHVW2], [Har]). It is worth noting that there are many variants of the moonshine module construction that can be (and are being) studied (see for instance the discussions of moonshine for $\widehat{E}_8$

and for the Conway sporadic group in [**FLM3**]), and such considerations could help suggest a string theory which has subtle, unexpected symmetries. Indeed, certain nontrivial Monster elements acting on $V^\natural$ can already be thought of as 24-dimensional analogues of the superstring's supersymmetry, which (cf. [**GSW**]) implements the discovery of Gliozzi-Olive-Scherk [**GOS**] (see [**FLM3**], [**FLM4**]). The Monster case exhibits a special "chirality" phenomenon, a kind of intrinsic choice of "parity": While in the "practice" case of $\widehat{E}_8$, either one of the two half-spin modules for $\widehat{D}_8$ can be joined with the basic $\widehat{D}_8$-module to form a copy of the basic $\widehat{E}_8$-module, in the Monster case, only the more subtle (twisted) module can be used in the analogous way to form the moonshine module; the other candidate module gives a "trivial" theory admitting a continuous group of symmetries, which we want to break (see [**FLM3**]). One can say (given what has happened in string theory after 1983) that one of the main points in [**FLM2**] is the explicit construction of hidden symmetries of an orbifold conformal field theory.

More recent studies of such orbifold theories (see [**CH**], [**DFMS**], [**HV**], [**NSV**]) continue to suggest the importance of "twist fields"—vertex operators of the form $Y(v, z)$ with v in a twisted Fock space, in the notation above. Such twist fields, the first cases of which were introduced in [**CF**] in the bosonic setting, are being studied algebraically in relation to (30) in joint work with Y.-Z. Huang and Frenkel.

In this exposition, we have selectively emphasized some representation-theoretic ideas, and we have tried to convey a feeling that certain mathematical and physical streams of thought were destined to converge, sooner or later. Even without reference to the real world, string theory is interesting in great part because it contains so much symmetry. We leave the reader with Weyl's familiar advice ([**W5**], p. 144):

"*Whenever you have to do with a structure-endowed entity* Σ *try to determine its group of automorphisms*, the group of those element-wise transformations which leave all structural relations undisturbed. You can expect to gain a deep insight into the constitution of Σ in this way."

REFERENCES

[**ADDF**] M. Ademollo, E. Del Giudice, P. DiVecchia and S. Fubini, *Couplings of three excited particles in the dual resonance model*, Nuovo Cimento **19A** (1974), 181–203.

[**BHN**] T. Banks, D. Horn and H. Neuberger, *Bosonization of the* SU(N) *Thirring models*, Nuclear Phys. **B108** (1976), 119.

[**BPZ**] A. A. Belavin, A. N. Polyakov and A. B. Zamolodchikov, *Infinite conformal symmetries in two-dimensional quantum field theory*, Nuclear Phys. **B241** (1984), 333–380.

[**BT**] D. Bernard and J. Thierry-Mieg, *Level one representations of the simple affine Kac-Moody algebras in their homogeneous gradations*, Comm. Math. Phys. **111** (1987), 181–246.

[**BGG**] I. N. Bernstein, I. M. Gelfand and S. I. Gelfand, *Structure of representations generated by highest weight vectors*, Functional Anal. i Prilozhen. **5** (1971), 1–9; English transl., Functional Anal. Appl. **5** (1971), 1–8.

[**B**] R. E. Borcherds, *Vertex algebras, Kac-Moody algebras, and the Monster*, Proc. Nat. Acad. Sci. U.S.A. **83** (1986), 3068–3071.

[Bo] A. Borel, *Hermann Weyl and Lie groups*, Hermann Weyl Centenary Lectures, Springer-Verlag, Berlin–Heidelberg–New York, 1986, pp. 53–82.

[C] S. Capparelli, Ph.D. thesis, Rutgers University, 1988.

[CN] J. H. Conway and S. P. Norton, *Monstrous moonshine*, Bull. London Math. Soc. **11** (1979), 308–339.

[CF] E. Corrigan and D. B. Fairlie, *Off-shell states in dual resonance theory*, Nuclear Phys. **B91** (1975), 527–545.

[CH] E. Corrigan and T. J. Hollowood, *Comments on the algebra of straight, twisted and intertwining vertex operators*, to appear.

[CO] E. Corrigan and D. Olive, *Fermion-meson vertices in dual theories*, Nuovo Cimento **11A** (1972), 749–773.

[DJKM] E. Date, M. Jimbo, M. Kashiwara and T. Miwa, *Transformation groups for soliton equations—Euclidean Lie algebras and reduction of the KP hierarchy*, Publ. Research Inst. for Math. Sciences, Kyoto Univ. **18** (1982), 1077–1110.

[DKM] E. Date, M. Kashiwara and T. Miwa, *Vertex operators and τ functions—transformation groups for soliton equations*. II, Proc. Japan Acad. Ser. A Math. Sci. **57** (1981), 387–392.

[DFMS] L. Dixon, D. Friedan, E. Martinec and S. Shenker, *The conformal field theory of orbifolds*, Nuclear Phys. **B282** (1987), 13–73.

[DHVW1] L. Dixon, H. A. Harvey, C. Vafa and E. Witten, *Strings on orbifolds*, Nuclear Phys. **B261** (1985), 651.

[DHVW2] ____, *Strings on orbifolds*. II, Nuclear Phys. **B274** (1986), 285.

[D] F. Dyson, *Missed opportunities*, Bull. Amer. Math. Soc. **78** (1972), 635–652.

[FF] A. Feingold and I. B. Frenkel, *Classical affine algebras*, Adv. in Math. **56** (1985), 117–172.

[FL] A. Feingold and J. Lepowsky, *The Weyl-Kac character formula and power series identities*, Adv. in Math. **29** (1978), 271–309.

[F1] I. B. Frenkel, *Spinor representations of affine Lie algebras*, Proc. Nat. Acad. Sci. U.S.A. **77** (1980), 6303–6306.

[F2] ____, *Two constructions of affine Lie algebra representations and boson-fermion correspondence in quantum field theory*, J. Funct. Analysis **44** (1981), 259–327.

[F3] ____, *Representations of Kac-Moody algebras and dual resonance models*, Applications of Group Theory in Physics and Mathematical Physics (Chicago Summer Seminar, 1982), Lectures in Applied Math., vol. 21, American Math. Soc., Providence, R.I., 1985, pp. 325–353.

[F4] ____, *Beyond affine Lie algebras*, Proc. Internat. Congr. Mathematicians (Berkeley, 1986), Amer. Math. Soc., Providence, R.I., 1987, pp. 821–839.

[FK] I. B. Frenkel and V. G. Kac, *Basic representations of affine Lie algebras and dual resonance models*, Invent. Math. **62** (1980), 23–66.

[FLM1] I. B. Frenkel, J. Lepowsky and A. Meurman, *An E_8-approach to F_1*, Finite Groups—Coming of Age (Proc. Conf., Montreal, 1982), ed. by J. McKay, Contemp. Math., vol. 45, Amer. Math. Soc., Providence, R.I., 1985, pp. 99–120.

[FLM2] ____, *A natural representation of the Fischer-Griess Monster with the modular function J as character*, Proc. Nat. Acad. Sci. U.S.A. **81** (1984), 3256–3260.

[FLM3] ____, *A moonshine module for the Monster*, Vertex Operators in Mathematics and Physics (Proc. 1983 Conf.), ed. by J. Lepowsky, S. Mandelstam and I. M. Singer, Publ. Math. Sciences Res. Inst. # 3, Springer-Verlag, New York, 1985, pp. 231–273.

[FLM4] ____, *An introduction to the Monster*, Workshop on Unified String Theories (Proc. Conf., 1985), M. Green and D. Gross, eds., Inst. for Theoretical Physics, World Scientific, Singapore, 1986, pp. 533–546.

[FLM5] ____, *Vertex operator calculus*, Mathematical Aspects of String Theory (Proc. Conf., San Diego, 1986), ed. by S.-T. Yau, World Scientific, Singapore (1987), pp. 150–188.

[FLM6] ____, *Vertex operator algebras and the Monster*, Pure Appl. Math., Academic Press, to appear.

[Freu] H. Freudenthal, *Zur Berechnung der Charaktere der halbeinfachen Lieschen Gruppen*. I, II, and III.I: Indag. Math. **16** (1954), 369–376. II: ibid., 487–491. III: ibid. **18** (1956), 511–514.

[FV] S. Fubini and G. Veneziano, *Algebraic treatment of subsidiary conditions in dual resonance models*, Ann. Phys. **63** (1971), 12.

[GL] H. Garland and J. Lepowsky, *Lie algebra homology and the Macdonald-Kac formulas*, Invent. Math. **34** (1976), 37–76.

[Ge] D. Gepner, *New conformal field theories associated with Lie algebras and their partition functions*, Nuclear Phys. **B287** (1987), 111.

[GQ] D. Gepner and Z. Qiu, *Modular invariant partition functions and parafermionic field theories*, Nuclear Phys. **B285** (1987), 423.

[GOS] F. Gliozzi, D. Olive and J. Scherk, *Supersymmetry, supergravity theories and the dual spinor model*, Nuclear Phys. **B122** (1977), 253–290.

[GNORS] P. Goddard, W. Nahm, D. Olive, H. Ruegg and A. Schwimmer, *Fermions and octonions*, Comm. Math. Phys. **112** (1987), 385–408.

[GNOS] P. Goddard, W. Nahm, D. Olive and A. Schwimmer, *Vertex operators for non-simply-laced algebras*, Comm. Math. Phys. **107** (1986), 179–212.

[GO] P. Goddard and D. Olive, *Algebras, lattices and strings*, Vertex Operators in Mathematics and Physics (Proc. 1983 Conf.), ed. by J. Lepowsky, S. Mandelstam and I. M. Singer, Publ. Math. Sciences Res. Inst. # 3, Springer-Verlag, New York, 1985, pp. 51–96.

[GG] M. Green and D. Gross, eds., *Workshop on unified string theories* (Proc. Conf., 1985), Inst. for Theoretical Physics, World Scientific, Singapore, 1986.

[GSW] M. B. Green, J. H. Schwarz and E. Witten, *Superstring theory*, Vols. 1 and 2, Cambridge Univ. Press, Cambridge, 1987.

[G] R. L. Griess, Jr., *The Friendly Giant*, Invent. Math. **69** (1982), 1–102.

[GHMR1] D. J. Gross, J. A. Harvey, E. Martinec and R. Rohm, *Heterotic string theory* (I). *The free heterotic string*, Nuclear Phys. **B256** (1985), 253–284.

[GHMR2] ____, *Heterotic string theory* (II). *The interacting heterotic string*, Nuclear Phys. **B267** (1986), 74–124.

[H] M. B. Halpern, *Quantum "solitons" which are* SU(N) *fermions*, Phys. Rev. **D12** (1975), 1684–1699.

[HV] S. Hamidi and C. Vafa, *Interactions on orbifolds*, Nuclear Phys. **B279** (1987), 465–513.

[HAR] J. A. Harvey, *Twisting the heterotic string*, Workshop on Unified String Theories (Proc. Conf., 1985), M. Green and D. Gross, eds., Inst. for Theoretical Physics, World Scientific, Singapore, 1986, pp. 704–718.

[J] N. Jacobson, *Lie algebras*, Wiley-Interscience, New York, 1962.

[K1] V. G. Kac, *Simple irreducible graded Lie algebras of finite growth*, Izv. Akad. Nauk SSSR **32** (1968), 1323–1367; English transl., Math. USSR-Izv. **2** (1968), 1271–1311.

[K2] ____, *Infinite-dimensional Lie algebras and Dedekind's η-function*, Funkcional. Anal. i Prilozhen. **8** (1974), 77–78; English transl., Functional Anal. Appl. **8** (1974), 68–70.

[K3] ____, *Infinite-dimensional Lie algebras*, 2nd ed., Cambridge Univ. Press, Cambridge, 1985.

[KKLW] V. G. Kac, D. A. Kazhdan, J. Lepowsky and R. L. Wilson, *Realization of the basic representations of the Euclidean Lie algebras*, Adv. in Math. **42** (1981), 83–112.

[KP1] V. G. Kac and D. H. Peterson, *Spin and wedge representations of infinite dimensional Lie algebras and groups*, Proc. Nat. Acad. Sci. U.S.A. **78** (1981), 3308–3312.

[KP2] ____, *Infinite-dimensional Lie algebras, theta functions and modular forms*, Adv. in Math. **53** (1984), 125–264.

[KP3] ____, *112 constructions of the basic representation of the loop group of E_8*, Proc. Sympos. Anomalies, Geometry, Topology, 1985, ed. by W. A. Bardeen and A. R. White, World Scientific, Singapore, 1985, pp. 276–298.

[Kan] I. L. Kantor, *Graded Lie algebras*, Trudy Sem. Vect. Tens. Anal., Moscow State University, No. 15 (1970), 227–266. (Russian)

[KZ] V. G. Knizhnik and A. B. Zamolodchikov, *Current algebra and Wess-Zumino models in two dimensions*, Nuclear Phys. **B247** (1984) 83–103.

[L1] J. Lepowsky, *Lectures on Kac-Moody Lie algebras*, Univ. Paris VI, spring, 1978.

[L2] ____, *Calculus of twisted vertex operators*, Proc. Nat. Acad. Sci. U.S.A. **82** (1985), 8295–8299.

[L3] ____, *The algebra of general twisted vertex operators*, to appear.

[LM] J. Lepowsky and S. Milne, *Lie algebraic approaches to classical partition identities*, Adv. in Math. **29** (1978), 15–59.

[LP] J. Lepowsky and M. Primc, *Structure of the standard modules for the affine Lie algebra* $A_1^{(1)}$, Contemp. Math., vol. 46, Amer. Math. Soc., Providence, R.I., 1985.

[LW1] J. Lepowsky and R. L. Wilson, *Construction of the affine Lie algebra* $A_1^{(1)}$, Comm. Math. Phys. **62** (1978), 43–53.

[LW2] ____, *A new family of algebras underlying the Rogers-Ramanujan identities and generalizations*, Proc. Nat. Acad. Sci. U.S.A. **78** (1981), 7254–7258.

[LW3] ____, *The structure of standard modules.* I, *Universal algebras and the Rogers-Ramanujan identities*, Invent. Math. **77** (1984), 199–290.

[LW4] ____, *The structure of standard modules.* II, *The case* $A_1^{(1)}$, *principal gradation*, Invent. Math. **79** (1985), 417–442.

[M] I. G. Macdonald, *Affine root systems and Dedekind's* η*-function*, Invent. Math. **15** (1972), 91–143.

[MP] A. Meurman and M. Primc, *Annihilating ideals of standard modules of* $\mathfrak{sl}(2, \mathbf{C})^\sim$ *and combinatorial identities*, Adv. in Math. **64** (1987), 177–240.

[M1] R. V. Moody, *A new class of Lie algebras*, J. Algebra **10** (1968), 211–230.

[M2] ____, *Euclidean Lie algebras*, Canad. J. Math. **21** (1969), 1432–1454.

[NSV] K. S. Narain, M. H. Sarmadi and C. Vafa, *Asymmetric orbifolds*, Nuclear Phys. **B288** (1987), 551.

[NY] M. Ninomiya and K. Yamagishi, *Nonlocal* SU(3) *current algebra*, Phys. Lett., 1987.

[S] G. Segal, *Unitary representations of some infinite-dimensional groups*, Comm. Math. Phys. **80** (1981), 301–342.

[T1] J. Tits, *Résumé de cours*, Annuaire du Collège de France, 1982–1983, 89–102.

[T2] ____, *On R. Griess' "Friendly Giant,"* Invent. Math. **78** (1984), 491–499.

[T3] ____, *Résumé de cours*, Annuaire du Collège de France, 1985–1986, 101–112.

[T4] ____, *Le module du "moonshine,"* Séminaire Bourbaki, no. 684, juin 1987.

[TK] A. Tsuchiya and Y. Kanie, *Vertex operators and conformal field theory on* $\mathbf{P}^1$ *and monodromy representations of braid groups*, Conformal Field Theory and Solvable Lattice Models, Adv. Studies in Pure Math. **16**.

[W1] H. Weyl, *Theorie der Darstellung kontinuierlicher halbeinfacher Gruppen durch lineare Transformationen.* I (1925), II (1926), III (1926) and Nachtrag (1926), # 68 in Gesammelte Abhandlungen, II, Springer-Verlag, Berlin–Heidelberg–New York, 1968.

[W2] ____, *The theory of groups and quantum mechanics*, S. Hirzel, Leipzig, 1928 and Dover, New York, 1949.

[W3] ____, *The classical groups, their invariants and representations*, Princeton Univ. Press, Princeton, N.J., 1939 and 1946.

[W4] ____, *Relativity theory as a stimulus in mathematical research* (1949), # 147 in Gesammelte Abhandlungen, IV, Springer-Verlag, Berlin–Heidelberg–New York, 1968.

[W5] ____, *Symmetry*, Princeton Univ. Press, Princeton, N.J., 1952.

[Wi1] E. Witten, *Non-abelian bosonization in two dimensions*, Comm. Math. Phys. **92** (1984), 455–472.

[Wi2] ____, *Physics and geometry*, Proc. Internat. Congr. Mathematicians (Berkeley, 1986), Amer. Math. Soc., Providence, R.I., 1987, pp. 267–303.

[Wi3] ____, *Quantum field theory, Grassmannians, and algebraic curves*, Comm. Math. Phys. **113** (1988), 529–600.

[ZF] A. B. Zamolodchikov and V. A. Fateev, *Nonlocal (parafermion) currents in two-dimensional conformal quantum field theory and self-dual critical points in* Z_N*-symmetric statistical systems*, Soviet Phys., JETP **62**(2) (1985), 215–225.

RUTGERS UNIVERSITY

Proceedings of Symposia in Pure Mathematics
Volume 48 (1988)

Some Problems in the Quantization
of Gauge Theories and String Theories

I. M. SINGER

0. Introduction. Although I never met Hermann Weyl, his papers and books affected me deeply. To give a sense of immediacy of Hermann Weyl's work to young mathematicians, it is worth dwelling on his influence.

In his Gibbs lecture [1], Weyl wrote "I feel that these informations about the proper oscillations of a membrane, valuable as they are, are still very incomplete. I have certain conjectures of what a complete analysis of their asymptotic behavior should aim at, but since for more than 35 years I have made no serious effort to prove them, I think I had better keep them to myself."

As background to the quote, one should keep in mind that Hermann Weyl

(1) computed the first term in the asymptotic expansion of $e^{-t\Delta}$ (Δ the Laplacian for a domain in $\mathbf{R}^n$ with smooth boundary and Dirichlet data) [2];

(2) defined the signature of a compact $4k$-manifold [3];

(3) developed singular integral operators in order to quantize their symbols, functions on $\mathbf{R}^n \oplus \mathbf{R}^n \simeq T^*(\mathbf{R}^n)$ [4];

(4) completed the proof of Hodge's theorem by showing that harmonic forms are smooth [5];

(5) computed the volume of a tubular neighborhood of a Riemannian manifold in terms of a curvature and its covariant derivatives [6]. In fact, the Gauss-Bonnet theorem for embedded manifolds is implicit in his computations.

Might Weyl have had in mind in his Gibbs lecture the heat equation proof of the index theorem for the signature or Euler operator? If you think not, you will surely agree that he would feel at home with the current developments in elliptic analysis and index theory.

Weyl's books have been very influential, in my case especially *Group theory and quantum mechanics* [7]. As an undergraduate physicist, courses in quantum mechanics and relativity convinced me that I needed to learn more mathematics. During the second World War, I carried [7] along with me. It convinced me even more and in January 1947 I became a graduate student in mathematics.

1980 *Mathematics Subject Classification* (1985 *Revision*). Primary 81C99.

Thirty years later, I found myself lecturing on gauge theories at Oxford, beginning with the Wu and Yang dictionary [8] and ending with instantons, i.e., self-dual connections. It would be inaccurate to say that after studying mathematics for thirty years, I felt prepared to return to physics. Instead, elementary particle physics turned to modern mathematics, and some of us found the interplay full of promise.

To emphasize the developments of the past decade, we reproduce the dictionary:

Wu & Yang Dictionary, 1975

Gauge field terminology	Bundle terminology
gauge (or global gauge)	principal coordinate bundle
gauge type	principal fibre bundle
gauge potential b_μ^k	connection on a principal fibre bundle
S_{ba}	transition function
phase factor Φ_{QP}	parallel displacement
field strength $f_{\mu\nu}^k$	curvature
source[1] J_μ^k	?[2]
electromagnetism	connection in a $U_1(1)$ bundle
isotopic spin gauge field	connection in a SU_2 bundle
Dirac's monopole quantization	classification of $U_1(1)$ bundle according to first Chern class
electromagnetism without monopole	connection on a trivial $U_1(1)$ bundle
electromagnetism with monopole	connection to a nontrivial $U_1(1)$ bundle

It is remarkable that, in contrast, today's young elementary particle theorist is at home with the following topics: representation theory of affine Lie algebras and the Virasoro algebra (defined in the next section), Riemann surfaces and their moduli spaces, Calabi-Yau spaces (Ricci flat Kähler manifolds), the topology of compact Lie groups, and G-index theory for geometric operators.

Hermann Weyl made many contributions to physics. He was the first to try to interpret electromagnetism as a connection on a bundle. See the interesting discussion in [9]. So, this conference is a natural place to assess the impact of ten years' interplay between mathematics and physics. In fact, M. F. Atiyah, C. Taubes, and E. Witten will be talking about different aspects of mathematical physics. One successful development has been the topological and geometric interpretation of chiral anomalies. But rather than review it, I want to look ahead and discuss some of the basic problems in particle physics which heavily involve mathematics. My view is biased in the direction of geometry and analysis. I am convinced that the quantization of gauge theories and the new/old string theory require analysis on and geometry of special infinite-dimensional manifolds. Many problems can be formulated as the missing infinite-dimensional analogues

---

[1] I.e., electric source; this is the generalization of the concept of electric charges and currents.
[2] *Author's comment*: The ? turns out to be $D_A^* F_A = J$.

of finite-dimensional results. And some of the successes come from ingenious substitutions for the real thing. Yet we lack some fundamental understanding of infinite-dimensional geometries.

I would like to express my gratitude to O. Alvarez, D. Friedan, and E. Witten. They have taught me what little I know of modern physics and have patiently helped me in my cumbersome attempts to merge the new insights with mathematics.

1. Some examples of infinite-dimensional geometries.

(a) For gauge theories the geometric object is $\mathfrak{a}/\mathscr{G}$. Here $\mathfrak{a}$ is the set of connections of a principal G-bundle P over a compact Riemannian 3-manifold M. $\mathscr{G}$ is the group of gauge transformations, the automorphisms of the G-bundle; it acts on $\mathfrak{a}$. G is a compact Lie group. $\mathfrak{a}/\mathscr{G}$ is the orbit space. Since the tangent space $T(\mathfrak{a}, A)$ of $\mathfrak{a}$ at A is the space of equivariant 1-forms on P with values in the Lie algebra of G, there is a natural inner product on $T(\mathfrak{a}, A)$ invariant under $\mathscr{G}$. Therefore, $\mathfrak{a}/\mathscr{G}$ has a Riemannian structure

(b) For the so-called σ-model, the natural geometric object is $\mathscr{L}(M)$, the set of free loops on M; i.e., the smooth maps of S^1 into M, M a Riemannian manifold, usually compact. However, M might be $\mathbf{R}^d$ or Minkowski space $\mathbf{R}^{d-1,1}$. The tangent space of $\mathscr{L}(M)$ at γ, $T(\mathscr{L}(M), \gamma)$, is the set of smooth vector fields along γ (sections of $\gamma^*(T(M))$). This tangent space has an inner product

$$\langle V_1, V_2 \rangle = \int \langle V_1(\gamma(t)), V_2(\gamma(t)) \rangle dt$$

for $V_1, V_2 \in T(\mathscr{L}(M), \gamma)$. Note that the inner product is *not* invariant under the action of Diff S^1, the diffeomorphisms of S^1, on $\mathscr{L}(M)$. Here Diff $S^1 \times \mathscr{L}(M) \to \mathscr{L}(M)$ with $(\phi, \gamma) \to \phi \cdot \gamma$ where $(\phi \cdot \gamma)(t) = \gamma(\phi^{-1}(t))$.

In quantum mechanics, one studies the Schrödinger operator $\Delta/2 + V$ on $L_2(M)$, where Δ is the Laplacian and V is multiplication by a potential function. In quantum field theory, the operators should act on L_2 of certain function spaces or mapping spaces: $\mathfrak{a}$ in (a) and $\mathscr{L}(M)$ in (b). We discuss these operators in the next section. Here we want to emphasize that an alternate to the canonical formalism, studying $\Delta/2 + V$ directly, is to use the Feynman-Kac formula, which expresses the heat kernel $K_T(x, y)$ of $e^{-T(\Delta/2+V)}$ as a path integral over paths from x to y:

$$K_T(x, y) = \int_{\substack{\text{paths } \gamma \\ \gamma(0)=x \\ \gamma(T)=y}} e^{-\int_0^T V(\gamma(t))\, dt} e^{-\dot{\gamma}^2/2} \mathscr{D}t.$$

Here $e^{-\dot{\gamma}^2/2}\mathscr{D}t$ means the Wiener measure of this path space. The path integral approach for operators on $L_2(\mathscr{L}(M))$ requires paths in $\mathscr{L}(M)$; i.e., maps $X: S^1 \times [0, T] \to M$. So the measure space analogous to the space of paths is

(c) $\chi = [X: S^1 \times [0, T] \to M; X(\theta, 0) = \gamma_0(\theta)$ and $X(\theta, T) = \gamma_1(\theta)]$.

For gauge theories, the situation is a little more complicated. Note that a path $t \to f_t(x)$ of functions on M is a function $f(t, x)$ on $[0, T] \times M$. A connection

$A = (A_\mu)$ on $[0, T] \times M$ can be transformed by a gauge transformation on $[0, T] \times M$ so that $A_0 = A(d/dt)$ is 0 (the temporal gauge; integrate the differential equation $dA_0/dt = U(t, x)A_0(t, x)$). Connections on $[0, T] \times M$ become paths of connections on M. Although there are some technical complications, one is led very quickly for path integral purposes to

(d) $\mathfrak{a}/\mathscr{G}$ based on a *four*-dimensional manifold, usually $M \times \mathbf{R}$ (interpreted as paths on $\mathfrak{a}/\mathscr{G}$ based on M).

The last geometric objects we consider are homogeneous spaces of $\mathrm{Diff}_0 S^1$, the orientation preserving diffeomorphisms of S^1. $\mathrm{Diff}\, S^1$ enters string theory (see §2 for a definition) because the theory, involving as it does maps of S^1, should be invariant under reparameterizations of S^1. It is supposed to play a role similar to gauge transformations in gauge theories and $\mathrm{Diff}(M)$ for metrics on M, gravity.

First, some notation: Let $L_n = ie^{in}d/d\theta$. Let $\mathrm{Vect}\, S^1 \otimes \mathbf{C}$ denote the complexification of the Lie algebra of vector fields on S^1; $\{L_n\}$ generate $\mathrm{Vect}\, S^1 \otimes \mathbf{C}$ and $[L_n, L_m] = (n - m)L_{n+m}$. The second cohomology of this Lie algebra [10] is Z and 2-cocycles are

$$\omega(L_m, L_n) = \begin{matrix} am^3 + bm & m + n = 0 \\ 0 & n \neq -m \end{matrix}.$$

The Virasoro algebra $\mathscr{V}$ is the central extension of $\mathrm{Vect}\, S^1 \otimes \mathbf{C}$ by a generator of H^2, say $\frac{1}{2}(m^3 - m)$. In a representation of $\mathscr{V}$, the central charge is the image of the generator of H^2, usually a multiple of I.

(e) The space $\mathrm{Diff}_0 S^1 / S^1$ can be made into a Kähler manifold: The Lie algebra of $\mathrm{Diff}_0 S^1$ is $\mathrm{Vect}(S^1)$. The tangent space of $\mathrm{Diff}_0 S^1 / S^1$ at the identity coset is the set of vector fields whose 0th Fourier coefficient is 0. Thus

$$J = \frac{d/d\theta}{|d/d\theta|}$$

makes $\mathrm{Diff}_0 S^1 / S^1$ into an invariant almost complex structure. It is easy to see that J is integrable and one assumes the Nirenberg-Newlander theorem will hold.

There is a family of Kähler metrics [11] given by the cocycles above with either $a = 0$, $b \neq 0$ or $a \neq 0$, $-b/a \neq n^2$.

Other interesting homogeneous spaces are $\mathrm{Diff}_0 S^1 / K_n$, where K_n is the subgroup with Lie algebra generated by L_0, L_n, and L_{-n}. The case $n = 0$ is (e) above and the case $n = 1$ gives $K_n = \mathrm{S}\ell(2, \mathbf{R}) \subseteq \mathrm{Diff}_0(S^1)$.

Of course homogeneous spaces can be studied by group theoretic means as well as geometric ones. In fact, the unitary highest weight representations of the Virasoro algebra lie at the foundations of string theory.

2. Some basic unsolved problems, canonical formalism. The insights gained from gauge theories are useful in understanding string theory. So we being with some old problems in the quantization of gauge theories. I also want to emphasize that, as successful as perturbative quantized gauge theories have been, one has little understanding of the theory nonperturbatively.

2.1. *The mass gap problem.* Suppose $M = S_R^3$ a 3-sphere with radius R, $G = \mathrm{SU}(N)$, and the principal bundle is (of course) $M \times G$. If $A \in \mathfrak{a}$, let F_A be its curvature 2-form and $\|F_A\|^2$ its L_2 norm $\int_M (F_A, F_A)$. Since $\|F_A\|^2$ is invariant under $\mathscr{G}$, it is a function V on $\mathfrak{a}/\mathscr{G}$. Because $\mathfrak{a}$ and $\mathfrak{a}/\mathscr{G}$ are Riemannian manifolds they have formal Laplacians $\mathscr{L}_\mathfrak{a}$ and $\mathscr{L}_{\mathfrak{a}/\mathscr{G}}$. Consider an analogue of the Schrödinger operator $\mathscr{L}_\mathfrak{a} + V$. The domain of this operator should be $\mathscr{G}$-invariant L_2 functions on $\mathfrak{a}$ relative to the "volume element" of $\mathfrak{a}$. As is well known (check S^1 invariant functions in $\mathbf{R}^2$), $\mathscr{L}_\mathfrak{a}$ on $\mathscr{G}$-invariant functions does not equal $\mathscr{L}_{\mathfrak{a}/\mathscr{G}}$. But with a little manipulation [12] one can show that $\mathscr{L}_\mathfrak{a} + V$ is unitarily equivalent to $H = \mathscr{L}_{\mathfrak{a}/\mathscr{G}} + V + V_c$ on L_2 functions of $\mathfrak{a}/\mathscr{G}$ relative to the "volume element" of $\mathfrak{a}/\mathscr{G}$. The correcting potential function V_c depends on a Jacobian determinant function which we now briefly describe.

Let T_A be the linear transformation from the Lie algebra $\mathscr{g}$ of $\mathscr{G}$ into vector fields on $\mathfrak{a}$ given by the differential of the action of $\mathscr{G}$ on $\mathfrak{a}$. Then $T_A^* T_A$ is a second-order elliptic operator on $\mathscr{g}$ and its determinant is well defined. It represents the volume change going from $\mathscr{G}$ to the orbit $\mathscr{G} \cdot A$ and is the determinant referred to above.

In any event, the problem is to make H well defined (since $\mathrm{vol}\,\mathfrak{a}/\mathscr{G}$ is not). Will H have discrete spectrum? In particular, if λ_0 is the smallest eigenvalue and λ_1 the next, will $\lim_{R \to \infty} \lambda_1(R) - \lambda_0(R)$ exist? Is the limit positive, i.e., is there a mass gap [13]?

The existence of a mass gap is necessary if quark confinement is to hold. By now quark confinement is accepted as an axiom in quantum chromodynamics (QCD) because it is experimentally valid; one does not observe quarks individually, only certain 'colorless' combinations. One still hopes QCD predicts confinement, but to prove it requires a more rigorously defined theory. (See §3 for the path integral formulation.) Perhaps that theory is lattice gauge theory. But there might be a different formulation more consistent with the geometry of $\mathfrak{a}/\mathscr{G}$.

These questions are in the realm of constructive field theory. Answering them, in my view, will lead not only to a mathematical understanding of QCD but to a better physical one as well.

2.2. *The supercharge operator.* Originally the space $\mathscr{L}(M)$ and its associated Schrödinger operator were of interest because they give a quantum field theory for a one-dimensional space (S^1) with properties similar to three-dimensional nonabelian gauge theories—conformal invariance on the classical level and what is called asymptotic freedom. The Schrödinger operator here is the 'Laplacian' on $\mathscr{L}(M)$ plus multiplication by the potential function $V(\gamma) = \int_{S^1} |d\gamma|^2$. The existence of this operator is tantamount to the existence of the quantum field theory. Of particular interest is the behavior of the spectrum as the radius of S^1 goes to ∞.

More recently, $\mathscr{L}(M)$ is of interest because of string theory. For our purposes, I will define string theory as theoretical physics in a new category: points of M are to be replaced by closed curves in M; so M is replaced by $\mathscr{L}(M)$.

To the extent that classical mechanics can be described by the Lagrangian on $T(M)$ or the Hamiltonian function on $T^*(M)$, we would want similar functions on $T(\mathscr{L}(M))$ and $T^*(\mathscr{L}(M))$. Classical paths in M are replaced by paths in $\mathscr{L}(M)$, i.e., cylinders in M. Even at the classical level, a new symmetry group appears: $\mathrm{Diff}(S^1)$ acting on $\mathscr{L}(M)$ as reparameterizations of S^1. (See §1(e).)

Since quantum mechanics normally studies operators (Schrödinger) on $L_2(M)$ we should now study similar operators on $L_2(\mathscr{L}(M))$. That's why the operator in the previous paragraph is of interest in string theory. Note that when $M = \mathbf{R}^d$ or $\mathbf{R}^{d-1,1}$, $\mathscr{L}(M)$ is a vector space and $\mathscr{L}(M) \otimes \mathbf{C}$ is a Hermitian vector space. The Fock space $L^2(\mathscr{L}(M) \otimes \mathbf{C})$ is the free bosonic string, the Veneziano model.

In superstring theory, every field has its superpartner and the theory is invariant under a new symmetry called supersymmetry. From a mathematical point of view, one has to replace the manifold M by a supermanifold and do physics all over again in this new category.

In physics, the first-order Dirac operator, as opposed to the Schrödinger operator, is the supersymmetric charge, the operator implementing supersymmetry. Mathematically, since the Dirac operator is interesting on a spin manifold M, it will also be interesting on $\mathscr{L}(M)$. It is called the Dirac-Ramond operator, or the supercharge Q. We describe it. Let V be the vector field generated by the S^1-action on $\mathscr{L}(M)$, i.e., $V(\gamma) = d\gamma(d/dt)$. Let $\mathscr{V}$ denote Clifford multiplication by V on spinor fields, cross sections of the spin bundle over $\mathscr{L}(M)$. [One needs to know that the spin bundle exists; $\mathscr{L}(M)$ must be a spin manifold. To insure it, we assume that M is a spin manifold and the first Pontrjagin class of M vanishes [14].] Let $\partial\!\!\!/_{\mathscr{L}}$ denote the formal Dirac operator on spinor fields. Then $Q = \partial\!\!\!/_{\mathscr{L}} + \mathscr{V}$. As in the finite even-dimensional case, the spin bundle splits into its $\pm$ chirality subspaces and one has

$$Q = \begin{pmatrix} 0 & Q_- \\ Q_+ & 0 \end{pmatrix}$$

with $Q_\pm$ changing chirality.

Of course these operators exist only formally, and one can't stress enough the importance of establishing the existence of Q on a domain of L_2 sections, and studying its spectrum. ($\mathscr{L}(M)$ does have a measure, Wiener measure, formally denoted by $e^{-\dot\gamma^2/2}\mathscr{D}t$.) One should assume that the Ricci tensor of the Riemannian manifold M is greater than or equal to zero as suggested by renormalization group considerations.

Even though Q doesn't exist (yet), the problem of computing its S^1-index directly has been solved in a beautiful way by E. Witten [15]; and C. Taubes [16] has constructed the weak coupling limit of Q (see below). Their work has significant applications to elliptic genera and elliptic homology [17], which they will describe at this conference. I wish merely to discuss the weak coupling limit and its relation to the fixed point formula for the G-index.

The fixed point formula gives the G-index of an operator in terms of its symbol in an infinitesimal neighborhood of the fixed point set M_g for $g \in G$.

In the case above, $G = S^1$, whose action on $\mathscr{L}(M)$, lifted to the spin bundle, will commute with Q. Since S^1 is generated by an irrational rotation, we need only consider the fixed point set of an irrational rotation. That set is the set of constant maps $M \subset \mathscr{L}(M)$. In the finite-dimensional case, it is not hard to establish that to study $\ker Q$ and $\ker Q^*$, one need only consider an infinitesimal neighborhood of the fixed point set M_g: For consider $Q_u = \emptyset_{\mathscr{L}} + uV$. For u large, and $s \in \ker Q$, most of $\|s\|^2$ will be in a tubular neighborhood U of M_g. As $u \to \infty, \|s\|^2 - \int_U \langle s, s \rangle \to 0$ [18].

Consider then the normal bundle $\mathfrak{n}$ of M in $\mathscr{L}(M)$. Since $T(\mathscr{L}(M), m)$ is the space of smooth maps from S^1 to $T(M, m)$, it is easy to see that $\mathfrak{n}_m$, the fibre of $\mathfrak{n}$ at m, is the subspace of $T(\mathscr{L}(M), m)$ with no constant term in the Fourier expansion.

Let $\widetilde{Q}$ be the operator on spinor fields over $\mathfrak{n}$ given by $\widetilde{Q} = \emptyset_{\mathfrak{n}} + \mathscr{W}$, where $\mathscr{W}$ is Clifford multiplication by the vector field W, the generator of the S^1-action on $\mathfrak{n}$.

In an infinitesimal neighborhood of M in $\mathscr{L}(M)$, Q and $\widetilde{Q}$ have the same symbol. Hence, by analogy with finite dimensions, one expects the S^1-index of Q to equal the S^1-index of $\widetilde{Q}$. Even though Q doesn't exist, $\widetilde{Q}$ does [16] and it has an S^1-index.

In fact, $\widetilde{Q}$ is the weak coupling limit of Q: along M, it is $\emptyset_M$, while along each fibre $\mathfrak{n}_m$ it is the supercharge operator $\widetilde{Q}_m$ for a free theory, namely $\emptyset_{\mathfrak{n}_m} + \mathscr{W}_m$ with $\emptyset_{\mathfrak{n}_m}$ having constant coefficients and W_m varying linearly along $\mathfrak{n}_m$ because the S^1-action on N_m is linear. In fact in one complex dimension when spinors on $\mathbf{C}^1$ are identified with conjugate holomorphic forms, the operator $\widetilde{Q}_m$ becomes $\bar{\partial} + z d\bar{z}$+adjoint.

C. Taubes showed that $\widetilde{Q}$ exists. If $\widetilde{Q}^{(n)}$ is its restriction to H_n, the subspace on which S^1 acts as $e^{in\theta}$, then $\widetilde{Q}^{(n)}$ is Fredholm. Hence the S^1-index of $\widetilde{Q}$ exists.

It is easy to compute. The index formula says take the family of operators $\{\widetilde{Q}_m\}_{m \in M}$. Compute its family's index, an S^1 vector bundle E over M. The desired S^1-index of $\widetilde{Q}$ is obtained by computing the index of $\emptyset_M$ with coefficients in E. Thus the weak coupling limit reduces the problem to a finite-dimensional one coupled to a free normal theory directly solvable. For index theory, the weak coupling limit is exact, whereas in general it is the first approximation to the full theory. In any case, the computations are easy to make [15].

There is another way of computing the index of $\widetilde{Q}$. For the classical case of a geometric elliptic operator, supersymmetric quantum mechanics states that the stationary phase approximation is exact and the index is obtained as a supersymmetric path integral [19]. By analogy, one can study the integral of supersymmetric paths in $\mathscr{L}(M)$. Since a closed path in $\mathscr{L}(M)$ is a map X of a torus into M, the integral will be over the space of maps $\chi = [X: N \to M]$, where N is a flat torus whose complex structure is given by $\tau = \tau_1 + i\tau_2, \tau_2 > 0$. One should think of τ playing the role of T for ordinary paths.

The appropriate path integral [20] is

$$\int_X \int_\Psi \exp\left(\int_N \bar\partial X \wedge \partial X + \psi\bar\partial\psi\right) \mathcal{D}\psi\,\mathcal{D}X$$

where $\Psi = [\psi; \psi$ a vector field along X, i.e., ψ is a cross section of $X^*(T(M))]$. The symbol $\wedge$ means wedge the forms and take the inner product of their values. Also $\bar\partial$ means $\bar\partial$ with coefficients in $X^*(T(M))$ which has a connection coming from $T(M)$. Computing the stationary phase approximation as $q_2 \to 0$ gives the formula for the S^1-index, $\sum_n$ index $\widetilde{Q}^{(n)}q^n$, where $q = e^{2\pi i\tau}$. The path integral point of view explains why the S^1-index is a modular function, i.e., invariant under $S\ell(2, \mathbf{Z})$. Although characteristic class computations give $S\ell(2, \mathbf{Z})$ invariance directly [21], the topological computation doesn't show directly why the S^1-index should be $S\ell(2, \mathbf{Z})$ invariant.

An interesting, different approach to the same issue of modularity, and its relation to anomalies can be found in [22].

The results here are very striking, especially the application to elliptic homology and the constancy of the S^1-index for generalized Rarita-Schwinger operators. It's certainly worth investigating the case where S^1 is replaced by $(S^1)^k$.

One finally intriguing problem about Q or $\widetilde{Q}$: The symbol of the Dirac operator on a spin manifold generates $K(T(M))$ as a module over $K(M)$. The Dirac operator itself, through the families index, gives the periodicity map [23]. As the cognoscenti are well aware, the Dirac operator and Bott periodicity are closely related. What then happens when we stringify, replacing M by $\mathcal{L}(M)$ and study Q on $\mathcal{L}(M)$? Is there a new K-theory and a new periodicity theorem? Perhaps "bundles" of two-dimensional conformally invariant field theories over $\mathcal{L}(M)$, or $\mathfrak{n}$, are involved. And the new index should be an S^1-index with values in the ring of modular functions.

2.3. *Special cases of the supercharge operator.* Despite the success of computing the S^1-index of the nonexistent Q by using the weak coupling limit $\widetilde{Q}$ and the fixed point formula, very little about Q is known analytically, except when M is a Euclidean space, the free bosonic string.

It would be interesting to investigate Q when M is very special, say a compact Lie group G. In the definition of the Dirac operator on a Riemannian manifold, the Riemannian connection is used: $\not{D} = \sigma \circ D$, where D is the covariant differential mapping $C^\infty(S) \to C^\infty(T^* \otimes S)$, S the spin bundle. σ is Clifford multiplication $T^* \otimes S \to S$, with $T \simeq T^*$ via the metric. Another Dirac-like operator is obtained if a different connection is used.

A compact Lie group has left and right invariant flat connections beside the two-sided invariant Riemannian one. Moreover, connections on M induce ones on $\mathcal{L}(M)$ because $T(\mathcal{L}(M), \gamma) = C^\infty(\gamma^*(T(M)))$. So one really has three supercharge operators on $\mathcal{L}(G)$, depending on which orthogonal connection is chosen. We shall denote the other two by Q_L and Q_R.

The loop group $\mathcal{L}(G)$ has been studied extensively [24] by algebraic means: highest weight representations and the Weyl character formula. However, the beautiful methods of Atiyah and Bott [25] obtaining the Weyl character formula via the holomorphic fixed point formula have not been extended to the affine Lie group $\mathscr{L}(G)$. See Pressley and Segal [26] for an elegant exposition of the formal analogy. There, $\mathscr{L}(G)/(\text{constant loops})$ is a complex manifold and Q_L becomes $\bar{\partial}$ + potential function plus the adjoint operator. As in the finite dimensional case, one should couple Q to holomorphic bundles.

It is an open problem to construct Q_L, Q_R, and Q as self-adjoint operators and to find their spectra. More challenging is the construction of Q on $\mathscr{L}(G/K)$.

We close this section by noting some progress in the construction of supercharge operators. Let $M = \mathbf{C}$. Then $\mathscr{L}(M)$ are functions on S^1. Consider $Q_n = \not{\partial}_{\mathscr{L}} + V_n$, where V_n is the gradient of a polynomial function of degree n:

$$p(\phi) = \int_{S^1} \frac{1}{2}m\phi^2 + a_3\phi^3 + \cdots + a_n\phi^n, \qquad a_n \neq 0, \ n \geq 3.$$

Jaffe et al. [27] have shown that the operator Q exists, is Fredholm and has index $n - 1$. They use path integral methods of constructive quantum field theory.

3. Some basic unsolved problems, path integral formalism.
As described in §1, the Feynman-Kac path integral gives a formula for $e^{-tH}(x, y)$ and ultimately for $\langle e^{-tH}u, v\rangle$. The advantage, already noted by Dirac [28], is automatic Lorentz invariance when $M \times \mathbf{R}$ is given the Lorentz metric and imaginary time t is replaced by $\sqrt{-1}\,t$. And a very practical advantage is that frequently the path integral gives a perturbative expansion for e^{-tH} when H in fact is not well defined, as in our examples for $\mathfrak{a}/\mathscr{G}$ and $\mathscr{L}(M)$. Observe that the propagator or Green's function $1/H$ is $\int_0^\infty e^{-tH}\,dt$.

3.1. *Paths in $\mathscr{L}(M)$ and Riemann surfaces.* We have already noted that paths in $\mathscr{L}(M)$ are maps of $S^1 \times [0, T]$ into M and that closed paths are maps of $S^1 \times S^1 \to M$, the latter is particularly useful for index purposes because only traces are used and therefore only the diagonal of e^{-tH} is relevant.

Formally, the Feynman-Kac formula for paths in $\mathscr{L}(M)$ and $H = \Delta_{\mathscr{L}}/2 + \|\dot{\gamma}\|^2$ is $\int_\chi e^{-S(X)}\mathscr{D}X$ where

$$S(X) = \frac{\|dX\|^2}{2} = \frac{1}{2}\int_{S^1 \times [0,T]}\langle dX, dX\rangle, \quad \text{and} \quad \chi = [X; S^1 \times [0, T] \to M].$$

We note that $\|dX\|^2 = \int_{S^1 \times [0,T]} \partial X \wedge \bar{\partial} X$. Hence the action depends only on the conformal structure, not on the metric of $S^1 \times [0, T]$. We also note that when M is Euclidean space, the integration makes sense because the action is a quadratic function on the linear space χ—the free theory once more.

In scalar field theories with polynomial interactions and in gauge theories with its cubic plus quartic interactions, the perturbative expansions can be expressed in terms of Feynman diagrams. Rather than paths from x to y, one has graphs exhibiting annihilation and creation of particles. See Figures (a), (b), and (c). When M is replaced by $\mathscr{L}(M)$, an ordinary path from x to y becomes a cylinder in M. For the perturbative theory, one guesses that (a) should be replaced by

Figure (d). This latter diagram is the basic interaction for a perturbative string theory. One changes (c) to (e), whose thickening is (f).

Only cubic interactions occur. Why quartic interactions should not occur is difficult to explain. Note that (b) becomes (g). Combining the basic three-point interaction with thickening gives Riemann surfaces with ends: Figure (h).

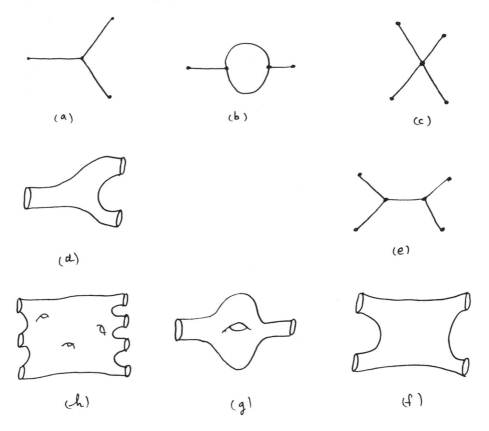

Motivated in part by requiring diffeomorphism invariance of the domain $S^1 \times [0, T]$ (or by now any surface N with ends), Polyakov integrates over all metrics of N [29]. The Polyakov model for string theory becomes

$$\int_{\rho \in \mathfrak{m}_N} \mathscr{D}\rho \int_{X \in \chi} e^{-\int_N |dX|^2} \mathscr{F}(X) \mathscr{D}X$$

where $\mathfrak{m}_N = [\rho, \rho$ a Riemannian metric on $N]$, and $\mathscr{F}$ is an appropriate function of maps $X \in \chi$.

The interpretation of the integral above and its computation has become an industry—an important one because when M is Minkowski space and $\mathscr{F}(X)$ are the functions described below one gets graviton scattering. One obtains a well-defined quantization of gravity, except for some infinities which are understood and dealt with by the introduction of supersymmetry.

However, before discussing the algebraic geometry behind the integral above, it must be emphasized that the expansion in terms of the genus g of N is supposed to represent a perturbative expansion for an unknown action, determined by the basic cubic interaction (d). In scalar field theories and in gauge theories, one begins with natural actions and the perturbative expansion is a consequence. But in string theory, i.e., $\mathcal{L}(M)$, a natural action in closed form has not yet been found even though the cubic interaction is manifest. Finding it and the geometric objects of which it is a function may be the central conceptual problem in string theory.

We return to Polyakov integral and algebraic geometry. See the excellent review by J. Bost [30]. Let us consider the simpler case where the ends are points and N is therefore a closed Riemann surface of genus g. The group of diffeomorphisms of N acts on $\mathfrak{m}_N$ as does the group of conformal weights, positive functions on N. The orbit space of $\mathfrak{m}_N$ under the action of their semidirect product is moduli space M_g. Well-established methods in quantum field theory reduce the measure to one over M_g, at least when there is no conformal anomaly ($M = \mathbf{R}^{26}$). In fact the measure has a beautiful interpretation in the arithmetic of surfaces: the space of holomorphic differentials is a bundle over M_g of complex dimension g. It has a natural Hermitian structure: $\langle \omega, \tau \rangle = \int_N \omega \wedge \bar{\tau}$. Hence its determinant line bundle λ has a Hermitian structure. Mumford observed [31] that λ^{13} is isomorphic to K, the canonical line bundle of M_g. So K inherits a Hermitian structure giving a measure on M_g; turns out to be the Polyakov measure.

Functional integrals can also be interpreted in terms of the arithmetic of curves. For example, consider

$$\int_{\mathfrak{m}_N} \mathscr{D}\rho \int_X \mathscr{D}X \exp\left(-\int_N \bar{\partial}X \wedge \partial X\right)$$

$$\times \int_N \int_N \cdots \int_N \exp\left(-\sum_j k_j X(\sigma_j)\right)((\ ))d\sigma_1 \cdots d\sigma_n.$$

That is, in the previous integral $\mathscr{F}(x) = \iint \cdots \int e^{-\sum_j k_j X(\sigma_j)}((\ ))d\sigma_1 \cdots d\sigma_n$. For tachyon scattering, $((\ )) = 1$. The integral comes down to integrating the following function over moduli space relative to the measure previously described:

$$\int_N \cdots \int_N d\sigma_1 \cdots d\sigma_n \exp\left(-\frac{1}{2}\sum_{i\neq j}\langle k_i, k_j \rangle : G(\sigma_i, \sigma_j): \right)$$

where G is an operator inverse to $\partial\bar{\partial}: \Lambda^0 \to \Lambda^{1,1}$ on forms whose integral is 0, and

$$:G: = \begin{cases} G(p,q) & \text{if } p \neq q; \\ \lim_{p\to q}(G(p,q) - \log r(p,q)) & \text{if } p = q. \end{cases}$$

For graviton-graviton scattering,

$$((\ )) = \prod \bar{\partial}X(\sigma_1) \wedge \partial X(\sigma_1) \wedge \cdots \wedge \bar{\partial}X(\sigma_r) \wedge \partial X(\sigma_r).$$

Now terms of the form $\bar{\partial}_{\sigma_1}\partial_{\sigma_2}(:G:)(\sigma_1, \sigma_2)$ appear.

An inescapable difficulty of this theory is that 1 is not integrable, i.e., the total measure of M_g is infinite. So are the tachyon scattering integrals. This too is reflected by the algebraic geometry. Both λ and K have natural extensions, $\tilde{\lambda}$ and $\tilde{K}$, to the Mumford compactification $\overline{M}_g$. But $\tilde{\lambda}^{-13}\tilde{K}$ is not the trivial bundle on $\overline{M}_g$. Its divisor is $-2(\overline{M}_g - M)$.

Physicists remedy the situation by introducing a superpartner to the map $X: N \rightarrow \mathbf{R}^d$, spinor fields over N. The operator $\bar{\partial}$ coupled to square roots of the canonical bundle appears and one is soon led to supermanifolds, super Riemann surfaces, supermoduli spaces, etc.

We have described how the requisite integrals can be expressed in terms of algebraic geometry. One now expects superintegrals over supermoduli spaces [32]. In fact, the super case should be easier.

There are many approaches. Take your favorite definition of a Riemann surface and favorite line of development of the subject. There is bound to be an extension to the supercategory. But one primary objective is to show that superstring theory is finite.

Q. 1. *Are the integrals of the appropriate functions over moduli spaces of spin structures finite?*

Q. 2. *Is there a principle coming from superalgebraic geometry that insures it?*

Q. 3. *Will the vanishing of the cosmological constant in superstring theory be a consequence?* [33].

3.2. *Curves in $\mathfrak{a}/\mathscr{G}$.* We return to gauge theories and review standard unsolved problems. The exponent in the Feynman-Kac formula for three-dimensional gauge theories is the action $+|\partial A/\partial t|^2 + |F_A|^2$, the curvature norm squared for the corresponding four-dimensional vector potential in the temporal gauge (see §1). The four-manifold here is $M \times [-T, T]$ in the Riemannian metric. Extensions of the Feynman-Kac path integral formalism to quantum fields (see Glimm/Jaffe [34], Part II, for example) lead to consideration of the following integral:

$$\int_{\mathfrak{a}/\mathscr{G}} e^{-S(A)} f(A) \mathscr{D}A$$

where f is a gauge invariant function, i.e., a function on $\mathfrak{a}/\mathscr{G}$ and now $\mathfrak{a} = [A; A$ a connection on a principle $SU(N)$ bundle over a compact *four*-manifold $N]$. One usually takes $N = S^4$ of radius R, or $S^3 \times \mathbf{R}$; ultimately $R \rightarrow \infty$. Here $S(A) = \|F_A\|^2 = \int_N F_A \wedge * F_A$.

Evaluating the integrals for appropriate functions, i.e., defining the measure, amounts to solving the quantum field theory for it gives the vacuum expectation values for operators quantizing functions on the phase space of the example (a) in §1. One does not understand the meaning of these integrals except in the perturbative theory [which means expanding $\|F_A\|^2 = \|F_{A^0}\|^2 +$ cubic in $A - A^0$ + quartic in $A - A^0$ about a fixed connection A^0, and writing $e^{-\|F_A\|^2} = e^{-\|F_{A^0}\|^2}(1 + \cdots)]$. One finite-dimensional approximation is lattice gauge theories. Another approach is through the stochastic integral methods of Itô and Malliavin. There has been some progress recently using this approach. L. Sadun et al. [35] have constructed a nonperturbative gauge invariant regulated measure which agrees with the corresponding perturbative theory.

The gauge invariant functions physicists want to integrate are the Wilson loop functions. If C is a closed curve in N, let $f_C(A)$ equal the trace of parallel transport around C. These functions separate orbits and the algebra they generate are the functions whose integral one would like to compute. L Gross [36] has argued that there is a more natural class, from the viewpoint of constructive field theory.

The Wilson criterion for confinement is easily stated [37]. Suppose $N = \mathbf{R}^4 = \mathbf{R}^3 \times$time. Let C_l be the boundary of a square one of whose sides is in the time direction and length l. Consider

$$-\log \frac{\int_{\mathfrak{a}/\mathscr{G}} e^{-S(A)} f_{C_l}(A)\, \mathscr{D}A}{\int_{\mathfrak{a}/\mathscr{G}} e^{-S(A)}\, \mathscr{D}(A)}$$

as a function of l. It should be quadratic rather than linear as $l \to \infty$.

A less standard problem is to explain spontaneous chiral symmetry breaking. In terms of the integration above, the problem can be formulated as follows:

Let $N_\varepsilon(A)$ be the number of eigenvalues $\leq \varepsilon$ for $\partial_A^2 + m^2$. $\int N_\varepsilon(A) e^{-S(A)} \mathscr{D}A$ is a function $\mathscr{F}(\varepsilon, m, R)$. Is $\lim_{\varepsilon \to 0} \lim_{m \to 0} \lim_{R \to \infty} m\mathscr{F} > 0$? One is computing

$$\int e^{-S(A) + \bar{\psi} i \partial \psi + m \bar{\psi}\psi} \bar{\psi} \cdot \psi\, \mathscr{D}\bar{\psi}\, \mathscr{D}\psi\, \mathscr{D}A.$$

Both of these problems require knowing 'the measure' on $\mathfrak{a}$ or its push-forward on $\mathfrak{a}/\mathscr{G}$. Even if one could make sense of the measure on $\mathfrak{a}/\mathscr{G}$, evaluating integrals is usually not easy. Instead one looks for a new perturbation theory which might give a good approximation for small coupling.

In the large N limit [38], the expansion parameter is N where $G = \mathrm{SU}(N)$. One is interested in the integral

$$\frac{\int e^{-N^2 \|F_A\|^2} f(A)\, \mathscr{D}A}{\int e^{-N^2 \|F_A\|^2}\, \mathscr{D}A}$$

as $N \to \infty$; here $f(A)$ is in the algebra C generated by Wilson loop functionals.

It turns out that the first term in the $(1/N)$-asymptotic expansion of the integral has a multiplicative property. If $\phi(f)$ is this first term, then $\phi(f_1 f_2) = \phi(f_1)\phi(f_2)$. Hence ϕ is a multiplicative linear functional on C. Since C is an algebra of bounded continuous functions on $\mathfrak{a}/\mathscr{G}$ which separates points, ϕ is evaluation at a point in an appropriate compactification of $\mathfrak{a}/\mathscr{G}$.

Such a point is called the master field, but no one has succeeded in describing it explicitly. The simplification of the theory as $N \to \infty$ naturally reminds one of the simplification in the homotopy groups for $\mathrm{SU}(\infty)$. One is tempted to expect the master field to be a connection with gauge group $G = \mathrm{SU}(\infty)$ or the analogous group for a type II factor. It would be useful to find the master field; for $\mathrm{SU}(3)$, the coupling constant in the perturbative expansion about it is $1/N = 1/3$.

We have listed several standard problems. There have been some imaginative attempts at solutions, but they are difficult to describe in purely mathematical terms. For example in [39], G. t'Hooft gets at the confinement phase by computing obstructions to reducing the bundle

$$\mathfrak{a}$$
$$\downarrow \mathscr{G}$$
$$\mathfrak{a}/\mathscr{G}$$

to the subgroup $\mathscr{H} \subset \mathscr{G}$ of gauge transformations with values in a maximal torus T^N in SU(N). In another example, an interesting attempt to characterize the master field involves string theory: integrate over all surfaces that bound the Wilson loop. See Makeenko and Migdal [40] and Gervaise and Neveu [41]. The mathematics involved in these attempts are worth exploring and developing.

4. The diffeomorphism group of S^1. Because reparameterization invariance is necessary, perhaps one should not be surprised at the many ways Diff S^1 enters string theory. Yet the proper role of Diff S^1 remains hidden, at least to me.

(a) The central extension V (see §1) occurs because Diff S^1 appears as automorphisms of a system and the corresponding representation is projective. A prime example is the Fock space $\mathscr{F}_d$ (symmetric algebra) of the Hermitian vector space $\mathscr{L}(R^d) \otimes \mathbf{C}$. The action of S^1 on loops induces a representation of V on $\mathscr{F}_d$ with central charge d.

(b) $\mathrm{Diff}_0 S^1/S^1$ is a moduli space for the Hermitian structures on $\mathscr{L}(R^d) \otimes \mathbf{C}$. As a result one has a holomorphic Fock bundle over $\mathrm{Diff}_0 S^1/S^1$ with fibre $\mathscr{F}_d$ [11].

(c) For Lorentz invariance one considers the Fock space $\mathscr{F}_d$ of $\mathscr{L}(R^{d-1,1}) \otimes \mathbf{C}$, a Hermitian space with nonpositive real form. $\mathscr{H}_{\mathrm{phys}}$ is the subspace of $\mathscr{F}_d$ annihilated by L_n, $n < 0$. The no ghost theorem states that the Hermitian structure on $\mathscr{F}_d$ restricted to $\mathscr{H}_{\mathrm{phys}}$ is positive semidefinite when $d = 26$. Mathematically, this fact is best understood in terms of the relative cohomology of V with coefficients in $\mathscr{F}_d$. In fact this cohomology gives the field content of open-string free field theory. The Witten cubic interaction for open strings can also be explained in cohomological terms [42].

(d) A different approach to understanding the quantization of strings focuses on conformal invariance and hence on the study of two-dimensional conformally invariant field theories [43]. $\mathscr{F}_{26}$ is the simplest example. $\mathrm{Vect}(S^1) \otimes C$ already occurs in the very definition of the subject as infinitesimal holomorphic (and antiholomorphic) transformations of the deleted disc; representations of the Virasoro algebra give the action of these vector fields on the operator-valued distributions of the theory.

Friedan and Shenker obtain the partition function of the field theory from a section of a flat bundle over moduli space M_g. In fact, in their theory, the spaces M_g for all g fit together to form a 'universal' moduli space on which the flat bundle really lives [44].

(e) The foundations of two-dimensional conformally invariant field theory have been axiomatized by Quillen and Segal [45] and, as one might expect, Diff S^1 appears as automorphisms.

We focus on one consequence of their viewpoint communicated to me by G. Segal. It appears that $\mathrm{Vect}\, S^1 \otimes C$ does not have an associated complex group which should be the complexification of $\mathrm{Diff}\, S^1$. However, it does have an associated semigroup

$$\mathscr{E} = [f\colon S^1 \xrightarrow{1-1} \text{ open unit disc}].$$

Let $\mathscr{E}_+ = [f \in \mathscr{E}; f$ is the boundary value of a holomorphic function $\tilde{f}$ with $\tilde{f}(0) = 0]$. We have the following map R from $\mathscr{E}$ to $\mathrm{Diff}\, S^1$. The unbounded component of the complement of $f(S^1)$ in C together with ∞ is holomorphic to the unit disc. Let $\hat{f}$ be the map of this space to the unit disc with $\tilde{f}(\infty) = 0$; define $R(f) = \hat{f} \circ f$. It is easy to see that

$$\mathscr{E}/\mathscr{E}_+ \xrightarrow{\simeq} \mathrm{Diff}_0 S^1/S^1.$$

This isomorphism is perhaps the easiest way of observing that $\mathrm{Diff}_0 S^1/S^1$ is in fact a complex manifold. The isomorphism is analogous to the isomorphism $G/S \simeq K/T$, where G is the complexification of a compact semisimple group K with maximal torus T and S is a maximal solvable subgroup of G containing T.

(f) The 'universal' moduli space of Friedan and Shenker has sparked a search for a more natural representation of such a space. So has the search for the geometric action of closed string field theory. Again, $\mathrm{Diff}\, S^1$ may turn out to be central since it acts transitively (for each genus g!) on the universal curve with local parameter. The geometric action takes the following form: a point on the universal curve with local parameter is a point on a Riemann surface N and a local coordinate function Z. Take the unit disc out and glue it back via an element of $\mathrm{Diff}_0 S^1$, obtaining a new Riemann surface. One can also describe the action infinitesimally using Kodaira-Spencer $H^1(N, T(N))$, in terms of Čech cohomology and the covering $|Z| < 1$ and $N - (0)$ [46]. It remains to be seen whether these actions by $\mathrm{Diff}_0 S^1$ for every genus will lead to the sought for universal moduli space.

(g) The Sato and Sato [47] interpretation of KdV involves the imbedding of any moduli space M_g into $\mathscr{Gr}$, the restricted Grassmannian of half-spaces in a Hilbert space [48]. Hence, one might expect to find the desired universal moduli space imbedded in $\mathscr{Gr}$ [49]. We note here only that $\mathrm{Diff}\, S_0^1$ acts on $\mathscr{Gr}$ and $\mathrm{Diff}\, S_0^1/S^1$ is embedded into an orbit of this action [50].

We have described ways in which the group $\mathrm{Diff}_0 S^1$, its homogeneous space $\mathrm{Diff}_0 S^1/S^1$, and its complexified Lie algebra enter string theory. But, once again, its proper place is unclear, and will remain so, in my opinion, until a 'new geometry' is discovered.

Rather than listing specific geometric problems for $\mathrm{Diff}\, S_0^1/S^1$, or $\mathscr{Gr}$ for that matter, we close on a more philosophical note. A partial analogy illustrates the conceptual problem we have alluded to: Invent Riemannian geometry knowing only $SO(n)$ and its representations. Invent ? knowing Virasoro and its representations, perhaps as they occur in two-dimensional conformally invariant field theories.

Put differently, it is as if we had part of the Taylor's expansion for $g_{\mu\nu}(x)$ about a point, and are beginning to appreciate the importance of certain collections of terms (in the quadratic part, what we call the Riemannian curvature tensor; in the linear part, what we call the Levi-Civita connection). In superstring theory, one has the expansion in terms of the genus of a surface, but the entity analogous to $g_{\mu\nu}(x)$ and its interpretation as differential geometry is missing.

BIBLIOGRAPHY

1. Hermann Weyl, *Ramifications, old and new, of the eigenvalue problem*, Bull. Amer. Math. Soc. **56** (1950), 115–139 .

2. Hermann Weyl, *Gesammelte Abhandlungen*, vol. 1, Springer-Verlag (1968), p. 368 and p. 393.

3. Hermann Weyl, *Gesammelte Abhandlungen*, vol. 2, Springer-Verlag (1968), p. 432.

4. Hermann Weyl, *Quantenmechanik und Gruppen Theorie*, Z. Physik **46** (1927), 1–46 (esp. p. 27).

5. Hermann Weyl, *On Hodge's theory of harmonic integrals*, Ann. of Math. (2) **44** (1943), 1–6.

6. Hermann Weyl, *On the volume of tubes*, Amer. J. Math. **61** (1939), 489–502.

7. Hermann Weyl, *Gruppen Theorie und Quanten mechanik*, Hirzel, Leipzig (1931).

8. T. T. Wu and C. N. Yang, *Concept of nonintegrable phase factors and global formulation of gauge fields*, Phys. Rev. **D12** (1975), 3845–3857.

9. C. N. Yang, *Hermann Weyl's contribution to physics*, Hermann Weyl 1885–1985, Springer-Verlag (1986).

10. I. M. Gelfand and D. B. Fuks, *On cohomologies of a Lie algebra of smooth vector fields*, Soviet Math. Dokl. **11** (1970), 268.

11. M. J. Bowick and S. G. Rajeev, *String theory as the Kähler geometry of loop space*, CTP Paper No. 1414, submitted to Phys. Lett. B.

12. J. Lott, *The Yang-Mills collective-coordinate potential*, Comm. Math. Phys. **95** (1984), 289–300.

13. R. Feynmann, *qualitative discussion of quantum chromodynamics in 2 + 1 dimensions*, Lisbon, Portugal, July 9–15, 1981.

14. E. Witten, *Anomalies, geometry, and topology* (W. Bardeen and A. White, eds.), World Scientific (1985).

15. E. Witten, *Elliptic genera and quantum field theory*, Princeton preprint PUPT-1024; and *The index of the Dirac operator in loop space*, PUPT-1050.

16. C. Taubes, S^1 *actions and elliptic genera*, preprint, Harvard University, Mathematics Department.

17. P. S. Landweber and R. Stong, *Circle actions on spin manifolds and characteristic numbers*, Topology, to appear.

17a. S. Ochanine, *Sur les genres multiplicatifs définis par des intégrales élliptiques*, Topology **26** (1987), 143–152.

18. E. Witten, *Supersymmetry and Morse theory*, J. Diff. Geom. **17** (1982), 661–92.

19. Luis Alvarez-Gaumé, *Supersymmetry and the Atiyah-Singer index theorem*, Comm. Math. Phys. **90** (1983), 161–173.

19a. D. Friedan and P. Windey, *Supersymmetric derivation of the Atiyah-Singer index and the chiral anomaly*, Nuclear Phys. **B235** [FS11] (1984), 395.

20. O. Alvarez, T.-P. Killingback, M. Manzano, and P. Windey, *String theory and loop space index theorems*, Univ. California-Berkeley preprint UCB-PTH 87/12, LBL-23045.

21. D. Zagier, *Note on the Landweber elliptic genus*, preprint.

22. K. Pilch, A. N. Schellekens, and N. P. Warner, *Path integral calculation of string anomalies*, MIT preprint.

23. M. F. Atiyah, *Bott periodicity and the index of elliptic operators*, Quart. J. Math. **19** (1968), 113–140.

24. V. G. Kac, *Infinite dimensional Lie algebras*, Birkhäuser, Boston (1982).

25. M. F. Atiyah and R. Bott, *A Lefshetz fixed point formula for elliptic complexes*: II. *Applications*, Ann. of Math. (2) **88** (1968), 451–491.

26. A. Pressley and G. Segal, *Loop groups*, Oxford Univ. Press, Oxford (1986) (esp. Chapter 14).

27. A. Jaffe, A. Lesniewski, and J. Weitsman, *The two-dimensional, N = 2 Wess Zumino model on a cylinder*, Harvard Preprint HUTMP-87/B203.

28. P. Dirac, *The Lagrangian in quantum mechanics*, Physikalische Zeitschrift der Sovyet-union, Band 3 Heft 1, 1933.

28a. R. P. Feynman, *Space time approach to non-relativistic quantum mechanics*, Rev. Mod. Physics **20** (1948), 367–382.

29. A. M. Polyakov, *Quantum geometry of bosonic strings*, Phys. Lett. **103B** (1981), 207–210.

30. J. B. Bost, *Fibrés déterminants, déterminants régularisés et mesures sur les espaces de modules de courbes complexes*, Séminaire Nicolas Bourbaki (1986–87), 676 .

31. D. Mumford, *Stablity of projective varieties*, Enseign. Math. **23** (1977), 39–110.

32. D. Friedan, *Notes on string theory and two dimensional conformal field theory*, Proc. Workshop on Unified String Theories, I.T.P.

33. D. Arnaudon, C. P. Bachas, V. Rivasseau, and P. Végreville, *On the vanishing of the cosmological constant in four-dimensional superstring models*, preprint, Centre de Physique Théorique, École Polytechnique, preprint A771.0387 (1987).

E. Gava and R. Iengo, *On the cosmological constant in the heterotic string theory*, preprint ISAS Trieste (1987).

G. Moore, J. Harris, P. Nelson, I. M. Singer, *Modular forms and the cosmological constant*, Phys. Lett. **B178** (1986), 167–73.

34. J. Glimm and A. Jaffe, *Quantum physics*, Springer-Verlag, 2nd ed. (1981).

35. L. Sadun, *Schwinger-Dyson renormalization*, Z. Physik (accepted for publication).

Z. Bern, M. B. Halperin, L. Sadun, C. Taubes, *Continuum regularization of QCD*, Phys. Lett. **165B** (1985), 151–156.

——, *Continuum regulariziation of QFT*: I. *Scalar Prototype*, Nuclear Physics **B284** (1987), 1–34.

——, *Continuum regularization of QFT*: II. *Gauge Theory*, Nuclear Physics **B284** (1987), 35–91.

Z. Bern, M. B. Halperin, L. Sadun, *QCD₄β Function*, Nuclear Physics **B284** (1987), 92–.

——, *Langevin renormalization*, Zeitschrift for Physics **C35**, p. 255-.

36. L. Gross, *A Poincaré lemma for connection forms*, J. Funct. Anal. **63**, No. 1 (August 1985).

37. K. Wilson, *Confinement of quarks*, Phys. Rev. **D10** (1974), 2445–2459.

38. G. t'Hooft, *A two dimensional model for mesons*, Nucl. Phys. **B75** (1974), 461–470.

39. ——, *Which topological features of a gauge theory can be responsible for permanent confinement*, Recent Developments in Gauge Theories (G. t'Hooft et al., eds.), Plenum Press (1980).

40. Yu. M. Makeenko and A. A. Migdal, *Exact equation for the loop average in multicolor QCD*, Phys. Lett. **B88** (1979), 135–137.

41. J. L. Gervaise and L. Neveu, *String structure of the master field of U(∞) Yang-Mills*, Nucl. Phys. **B192** (1981), 463–75.

42. I. B. Frenkel, H. Garland, and G. J. Zuckerman, *Semi-infinite cohomology and string theory*, Proc. Nat. Acad. Sci. U.S.A. **83** (1986), 8442–8446.

B. Kostant and S. Sternberg, *Symplectic reduction, BRS cohomology, and infinite dimensional Clifford algebras*, to appear.

43. D. Friedan, *Introduction to Polyakov's string theory*, Ecole d'Eté de Physique Théorique, Les Houches, 1982 (J. B. Zuber and R. Stora, eds.), North-Holland (1984).

44. D. Friedan and S. Shenker, *The integrable analytic geometry of quantum strings*, Phys. Lett. **B175** (1986), 287–296.

45. D Quillen and G. Segal, in preparation.

46. A. A. Beilinson and Yu. I. Manin, *The Mumford form and the Polyakov measure in string theory*, Comm. Math. Physics **107** (1986), 359–376.

47. M. Sato and Y. Sato, *Soliton equations as dynamical systems on infinite dimensional Grassmann manifolds*, preprint.

48. G. Segal and G. Wilson, *Loop groups and equations of KdV type*, Publ. Math. IHES **61** (1985), 5–65.

49. L. Alvarez-Gaumé and C. Gomez, *New methods in string theory*, CERN Preprint.

50. J. Mickelsson, *String quantization on group manifolds and the holomorphic geometry of* Diff S^1/S^1, CTP reprint #1448, M.I.T.

MASSACHUSETTS INSTITUTE OF TECHNOLOGY

Proceedings of Symposia in Pure Mathematics
Volume 48 (1988)

Fully Nonlinear Elliptic Equations

L. NIRENBERG

It is a great honour and pleasure for me to speak in this Symposium on the Mathematical Heritage of Hermann Weyl.

Hermann Weyl did some work on nonlinear elliptic equations. In 1916 he published a paper [27] on the so-called Weyl problem in Riemannian geometry: Given a Riemannian metric $ds^2 = E{dx^1}^2 + 2Fdx^1dx^2 + G{dx^2}^2$ on S^2 having positive Gauss curvature K, can one find an isometric embedding $X: S^2 \to \mathbf{R}^3$ as a closed convex surface? This leads to the problem of solving a nonlinear elliptic equation for a single function

$$u = \tfrac{1}{2}X \cdot X$$

(the origin being inside the convex region bounded by the surface). It is a Monge-Ampère equation, which in terms of local coordinates x^1, x^2 takes the form

$$(u_{x^1x^1} + \sum a^i u_{x^i} - E)(u_{x^2x^2} + \sum c^i u_{x^i} - G) - (u_{x^1x^2} + \sum b^i u_{x^i} - F)^2$$

$$\tag{1} = K(x^1, x^2)f(u, \nabla u).$$

Weyl described a procedure for attacking the problem and carried out a good part of it. He used, essentially, the continuity method. This involved connecting the given metric ds^2 to the usual metric ds_0^2 on the sphere by a one-parameter family of metrics ds_t^2, $0 \le t \le 1$, all having positive Gauss curvatures, with $ds_1^2 = ds^2$. The method then requires proving that the set of t values for which a solution exists is both open and closed.

In particular, towards proving closedness, Weyl derived a priori estimates for the C^2 norm of the function X. Using the maximum principle for second-order elliptic equations, he estimated the maximum of the mean curvature H of the surface—in terms of derivatives up to order 4 of the coefficients E, F, G of the metric.

1980 *Mathematics Subject Classification* (1985 *Revision*). Primary 35J25, 35J60, 35J65, 35B45, 35B60, 53A05, 53A07, 53C45; Secondary 35J20, 35M05, 35A07, 35B50, 35G20.

Weyl's approach was completed in the late 1940s by improvement of elliptic machinery. This in fact was my doctoral thesis [**17**, **18**]. There the following was shown in case $n = 2$:

A. Let u be a scalar solution of an elliptic equation in a domain Ω in $\mathbf{R}^n$,

$$(2) \qquad\qquad F(x, u, Du, D^2 u) = 0,$$

If one has a bound for the C^2 norm of u, $|u|_{C^2}$, then one can estimate the norm $|u|_{C^{2,\mu}}$ for some $0 < \mu < 1$, i.e., the Hölder continuity (exponent μ) of the second derivatives of u.

(This was proved in compact subsets of Ω, and in $\overline{\Omega}$ under various boundary conditions.) Ellipticity should be explained: An equation (2) is called elliptic at a function u if the linearized equation is elliptic there, i.e., if the matrix

$$a^{ij}(x) = \frac{\partial F}{\partial u_{ij}}(x, u(x), Du(x), D^2 u(x))$$

is definite at $x \in \Omega$, say $\{a^{ij}\}$ is positive definite. (Here $u_{ij} = u_{x^i x^j}$.) The proof of this result was based on ideas of C. B. Morrey. Once one has $C^{2,\mu}$ estimates, further regularity and estimates follow from elliptic machinery, in particular the Schauder estimates. Independently, A. V. Pogorelov obtained smooth solutions of Weyl's problem (see the references in [**18**]).

The general question of (even local) isometric embedding into $\mathbf{R}^N$, with N minimal, is still largely open: Consider a Riemannian metric ds^2 in a neighbourhood of the origin in $\mathbf{R}^n$. Can we embed it (or a smaller neighbourhood) isometrically and smoothly in $\mathbf{R}^N$ with $N = n(n+1)/2$ (this is the minimal dimension)? We all know of course the fundamental theorem of Nash on global isometric embedding into $\mathbf{R}^N$ for N sufficiently high. Even for $n = 2$ this local question is not fully settled. It involves solving a nonlinear equation, locally, of Monge-Ampère type; the equation however need not be elliptic or hyperbolic; it may change type. If at the origin the Gauss curvature K of the metric ds^2 is > 0, then the problem is essentially elliptic there, and can be solved locally. If $K(0) < 0$ the problem is hyperbolic and again a local solution can be found. In [**15**] C. S. Lin proved the answer is yes if $K \geq 0$ near the origin. See also [**11**]. Lin also proved [**16**] that the answer is yes in case $K(0) = 0$ and $\nabla K(0) \neq 0$. This involved solving a nonlinear Tricomi-type equation. See also [**10**]. In generic situations the local isometric embedding of a 3-dimensional Riemannian metric into $\mathbf{R}^6$ was proved by R. Bryant, P. A. Griffiths and D. Yang [**2**].

We return to elliptic equations; there has been remarkable progress in recent years, in various directions: 1. variational methods, 2. a priori estimates, 3. study of dimension of sets of singularities for solutions of nonlinear elliptic systems, using techniques of geometric measure theory, 4. Yang-Mills equations and connections with geometry and topology. In this talk I shall confine myself to topic 2; however I would like to mention a few highlights in the use of variational methods.

1. (a) The min-max methods going back to L. A. Lyusternik and L. G. Schnirelman have been developed and refined to yield multiple, sometimes infinitely many, solutions of a wide class of variational problems—in particular for finding time-periodic solutions of Hamiltonian systems. P. Rabinowitz and V. Benci have been central figures in these developments.

(b) In Riemannian geometry, the Yamabe problem, that of finding a Riemannian metric conformal to a given one on a compact manifold, and having constant scalar curvature, has been settled by R. Schoen [21]. He treated all the cases remaining after the initial ones treated by T. Aubin. Their methods are variational.

(c) A related problem is that of finding a positive solution u in a bounded domain Ω in $\mathbf{R}^n$ with smooth boundary of

$$\Delta u + u^{(n+2)/(n-2)} = 0 \quad \text{in } \Omega,$$

$$u = 0 \quad \text{on } \partial\Omega.$$

Here the exponent $(n+2)/(n-2)$ is critical. For smaller exponent > 1 there is always a solution while for this exponent there is no solution in case Ω is star-shaped. A. Bahri and J. M. Coron have proved [1] that the problem has a positive solution in case for some k, $H_k(\Omega, \mathbf{Z}_2) \neq 0$. Nontrivial topological arguments are used in the proof.

(d) Many years ago H. Hopf conjectured that the only compact surface in $\mathbf{R}^3$ with constant mean curvature $H = 1$ is the standard sphere. This was proved to be true for surfaces without self-intersections. However in [26] H. C. Wente, using a variational approach, constructed counterexamples which are topologically equivalent to a torus.

From now on we discuss only topic 2. As I mentioned, much progress has come from technical lemmas refining elliptic machinery. An exception to this, however, is S. T. Yau's proof [28] of Calabi's conjecture which is based on new a priori estimates obtained by sheer strength.

For $n > 2$ the statement A above for solutions of (2), that a C^2 estimate yields a $C^{2,\mu}$ estimate, is not proved, and it is not known whether it is true or not. What has been proved in recent years is:

A′. Statement A holds for $n > 2$ in case F is also *a concave* (or convex) *function in its dependence on the symmetric matrix* $\{D^2 u\}$.

Various people have contributed to this result. First L. C. Evans [8] and N. V. Krylov [12] independently proved it in compact subsets of Ω. Then it was proved in all of $\overline{\Omega}$ (bounded), for the Dirichlet problem, by Krylov [12] and independently by L. Caffarelli, J. J. Kohn, L. Nirenberg, J. Spruck [4]. A crucial ingredient in the proof is the Harnack inequality for second-order elliptic equations of Krylov and M. V. Safonov [14] (see [9] for another proof by Trudinger):

Harnack inequality (Krylov, Safonov): In the unit ball $|x| < 1$ let u be a positive solution of a linear elliptic equation

$$Lu = a^{ij}(x)u_{ij}(x) = 0.$$

Assume L is uniformly elliptic, i.e., for some constant M,

$$M^{-1}|\xi|^2 \leq a^{ij}\xi_i\xi_j \leq M|\xi|^2 \qquad \forall x \text{ in the ball.}$$

Then $\exists C = C(n, M)$ such that

$$\max_{|x| \leq 1/2} u \leq C \min_{|x| \leq 1/2} u.$$

This result is the analogue of the important Harnack inequality (see [9]) of E. DeGiorgi, J. Moser and J. Nash for equations in divergence form $\sum(a^{ij}u_i)_j = 0$.

Recently Caffarelli [3] has developed a unified treatment of the basic estimates for elliptic equations of second order, namely the Schauder theory and L^p theory, based on the maximum principle, the Krylov-Safonov Harnack inequality above, and the inequality of A. D. Alexandroff, I. Bakelman, C. Pucci (see [9, section 9.1]).

Because of A′ we can now treat a variety of nonlinear second-order elliptic equations that were previously intractable. Caffarelli, Nirenberg and Spruck (CNS) have written a series of papers on nonlinear equations of special forms. In addition Krylov has treated classes of nonlinear elliptic problems: see [12] and [13], which also contain further references. The problems (2) that have been treated have been solved with the aid of the continuity method by establishing a priori estimates for the C^2 norms of solutions. These estimates have all been derived using the maximum principle—but often in very tricky ways. The concavity of F in its dependence on D^2u enters at various steps in the proofs though, as I said, it is not known if the hypothesis is truly essential. Let me indicate in a simple case one way in which concavity enters. It is used in estimating second derivatives of u in terms of estimates for them on the boundary (these latter estimates are the hardest to obtain). Consider a special equation (2)

$$F(D^2u) = 0.$$

Let ∂_ξ represent differentiation in any fixed direction ξ; here ξ is a unit vector.

Claim: $\partial_\xi^2 u$ achieves its maximum on the boundary.

Proof. Applying ∂_ξ to the equation we find

$$\frac{\partial F}{\partial u_{jk}}\partial_\xi u_{jk} = 0.$$

Applying it again we find, from the concavity hypothesis,

$$\frac{\partial^2 F}{\partial u_{jk}}(\partial_\xi^2 u)_{jk} = -\frac{\partial^2 F}{\partial u_{jk}\partial u_{rs}}(\partial_\xi u_{jk})(\partial_\xi u_{rs}) \geq 0$$

and the claim follows by the maximum principle.

In the rest of this talk I will mainly discuss some of the results of CNS [5–7]. Further references may be found there and in [12] and [13], including references to earlier work on Monge-Ampère equations by Pogorelov and by S. Y. Cheng and Yau. Some of the results of CNS have been described in the talk [19]. CNS treat problems of essentially two types. To describe these, let Γ be an open convex cone in $\mathbf{R}^n$ with vertex at the origin, $\Gamma \neq \mathbf{R}^n$, containing the positive cone Γ^+, and which is symmetric under permutation, i.e., if $\lambda = (\lambda_1, \ldots, \lambda_n) \in \Gamma$ then any $\tilde{\lambda}$ obtained by permutation of the λ_i is also in Γ. Consider a smooth function $f(\lambda)$ defined in Γ, continuous in $\overline{\Gamma}$ and symmetric (under permutation) in the λ_i, and satisfying:

(i) $f_{\lambda_i} > 0$, and f is concave. For convenience we suppose also $f > 0$ in Γ and $f = 0$ on $\partial \Gamma$.

(ii) For any $\lambda \in \Gamma$, $f(R\lambda) \to \infty$ as $R \to \infty$.

The two types of problems (2) treated by CNS are

(A) [**5**].

$$(3) \qquad F = f(\lambda(u_{jk}(x)) - \psi(x) = 0 \quad \text{in } \Omega,$$

$$(4) \qquad u = \phi \quad \text{in } \partial \Omega,$$

where $\lambda = (\lambda_1, \ldots, \lambda_n)$ are the eigenvalues of the Hessian matrix $\{u_{jk}\}$. Here $\psi > 0$ in $\overline{\Omega}$.

(B) [**6–7**].

$$(5) \qquad F = f(\kappa(\text{graph of } u)) - \psi(x) = 0 \quad \text{in } \Omega;$$

$\psi > 0$ in $\overline{\Omega}$, and (4) is imposed in case Ω is a domain in $\mathbf{R}^n$. Here $\kappa = (\kappa_1, \ldots, \kappa_n)$ are the principal curvatures of the graph of the hypersurface $(x, u(x))$.

For a problem of type (A) we say that $u \in C^2(\overline{\Omega})$ is admissible if $\lambda(u_{jk}(x)) \in \Gamma \; \forall x \in \overline{\Omega}$; we seek an admissible solution. If u is admissible then the condition $f_{\lambda_i} > 0 \; \forall i$ implies that the operator F is elliptic at u, while the concavity of f ensures that F is concave in its dependence on $\{D^2 u\}$. (Similarly for problems of type (B).) In case $\Gamma = \Gamma^+$, u is admissible $\Leftrightarrow u$ is strictly convex. In this case, since we may consider $\phi = 0$, so that $\partial \Omega$ is a level surface of a strictly convex function, we would naturally require Ω to be strictly convex. Thus, part of the problem is to determine in what kind of domains Ω it is possible to solve (3) and (4). This depends on the nature of Γ. It turns out that the results are different depending on whether

(a) Γ is a wide cone, i.e., the positive axes lie inside Γ, or

(b) Γ is a narrow cone: the positive axes lie on $\partial \Gamma$.

We have the following results [**5**] for f satisfying (i) and (ii).

THEOREM 1. *If Γ is a wide cone then there is a unique admissible solution $u \in C^\infty(\overline{\Omega})$ of* (3), (4) $\forall \phi \in C^\infty(\partial \Omega)$ *and for every bounded Ω with smooth boundary.*

THEOREM 2. *If Γ is a narrow cone, assume in addition to* (i), (ii) *that f satisfies*

$$(\text{iii}) \qquad f(\lambda + Re_n) \to \infty \text{ as } R \to \infty, \qquad \forall \lambda \in \Gamma,$$

where $e_n = (0, \ldots, 0, 1)$. Then for every $\phi \in C^\infty(\partial\Omega)$ there exists a unique admissible solution $u \in C^\infty(\overline{\Omega})$ of (3), (4) iff

(iv) $\begin{cases} \partial\Omega \text{ is connected and } \forall x \in \partial\Omega, \text{ if } \mu_1(x), \ldots, \mu_{n-1}(x) \text{ are the} \\ \text{principal curvatures of } \partial\Omega \text{ relative to the interior normal} \\ \text{then for some large } R, \ (\mu_1(x), \ldots, \mu_{n-1}(x), R) \in \Gamma. \end{cases}$

A. Vinacua [25] has extended these results to the complex case.

Let us turn now to problems of type (B). In [6] we construct compact "Weingarten" hypersurfaces which are starshaped about the origin in $\mathbf{R}^{n+1}$ and satisfy (5), i.e., given a positive function $\psi(y)$ in $\mathbf{R}^{n+1}$ satisfying

(6) $$(\rho\psi(y))_\rho \leq 0$$

$(\rho = |y|)$, we find a compact starshaped hypersurface lying in $r_1 \leq |y| \leq r_i$ of

$$f(-\kappa) = \psi(y).$$

Here κ represents the principal curvatures of the hypersurface passing through the point y. f is assumed to satisfy (i) and, in addition, $f(1, \ldots, 1) = 1$, and

$$\sum f_{\lambda_i}, \ \sum \lambda_i f_{\lambda_i} \geq \phi(f)$$

with ϕ a positive increasing function on $\mathbf{R}^+$. We also assume

$$\psi(y) \geq f\left(\frac{1}{r_1}, \ldots, \frac{1}{r_1}\right) \quad \text{for } |y| = r_1,$$

$$\psi(y) \leq f\left(\frac{1}{r_2}, \ldots, \frac{1}{r_2}\right) \quad \text{for } |y| = r_2.$$

The theorem is proved via a priori estimates for the C^2 norm of the hypersurface (in suitable coordinates). In particular we use the equation satisfied by $u(x)$ where $x \in S^n$, and $xu(x) = y$ represents a point on the hypersurface. (In applying A$'$ we have to restrict ourselves to a single equation for a scalar function.) It is interesting that our estimates for the C^2 norm of u involve estimating the maximum principal curvature of the hypersurface, much in the same way as the estimate for the mean curvature carried out by Weyl in [27].

An interesting case is

$$f(\lambda) = \left[\frac{\sigma^{(k)}(\lambda)}{\binom{n}{k}}\right]^{1/k},$$

where $\sigma^{(k)}$ is the kth elementary symmetric function

$$\sigma^{(k)}(\lambda) = \sum_{i_1 < i_2 < \cdots < i_j} \lambda_{i_1} \lambda_{i_2} \cdots \lambda_{i_j}.$$

The case $k = 1$ was treated by A. Treibergs [22] and Treibergs and S. Wei [23], and the case $k = n$ by V. I. Oliker [20].

K. S. Tso [24] has treated the case $k = n$ and, in fact, has obtained a more general result in that case (he does not require (6)) by an interesting variational approach. The problem is to find a compact convex hypersurface X in $\mathbf{R}^{n+1}$ containing the origin inside the convex set Σ it encloses, and satisfying

$$(7) \qquad\qquad K(y) = \psi(y), \qquad y \in X,$$

where K is the Gauss curvature of X, and ψ is a prescribed positive function in $\mathbf{R}^{n+1}$. He considers the variational expression (C is a suitable constant)

$$J = C \int_X \sigma^{(n-1)}(\kappa)\, dS - \int_\Sigma \psi\, dV,$$

where $\sigma^{(n+1)}$ is the $(n-1)$st elementary symmetric function of the principal curvatures of X.

A convex hypersurface X at which J is stationary has (7) as its Euler-Lagrange equation. Assuming ψ is integrable in $\mathbf{R}^{n+1}$ and $F(y) > 1/R^n$ for $|y| = R$, for some R, Tso proves that J has a smooth minimum $\Leftrightarrow J$ is nonpositive for some X. He uses a heat equation approach.

Finally we mention a result in [**7**] for the Dirichlet problem for (5). We have only been able to treat the case that Ω is strictly convex and $u = \phi = \mathrm{const.}$ on $\partial\Omega$. Here we assume f satisfies (i) and $\sum \lambda_i f_{\lambda_i} > 0$ in Γ, and also (iii), as well as the following conditions:

For some constant $c_0 > 0$

$$\sum f_i(\kappa) \geq c_0 \quad \text{whenever } f \geq \min_{\overline{\Omega}} \psi = \psi_0,$$

$$\sum \kappa_i f_i(\kappa) \geq c_0 \quad \text{whenever } \psi_0 = \min \psi \leq f \leq \max \psi = \psi_1$$

and, for some constant $c_1 > 0$, on the set

$$\{\kappa \in \Gamma | \psi_0 \leq f(\kappa) \leq \psi_1, \text{and } \kappa_1 < 0\},$$

we have $f_1 \geq c_1$.

THEOREM. *Suppose f satisfies the preceding conditions. Assume also that there is an admissible subsolution $\underline{u}$ of (5), i.e., a function $\underline{u}$, with $\underline{u} = 0$ on $\partial\Omega$, such that the principal curvatures $\underline{\kappa}$ of the graph satisfy*

$$f(\underline{\kappa}(x)) \geq \psi(x) \quad \text{in } \overline{\Omega}.$$

Then there exists a unique admissible solution u of (5) with $u = 0$ on $\partial\Omega$.

The function

$$f(\kappa) = \left[\sigma^{(k)}\right]^{1/k}, \qquad 1 \leq k \leq n,$$

satisfies all the conditions of the theorem.

The derivation of the a priori estimates in the proof is very tricky.

ACKNOWLEDGMENT. This work was partially supported by ARO-DAA-G29-84-K-0150 and ONR N00014-85-K-0195.

REFERENCES

1. A. Bahri and J. M. Coron, *On a nonlinear elliptic equation involving the critical Sobolev exponent: the effect of the topology of the domain*, Comm. Pure Appl. Math., to appear.

2. R. L. Bryant, P. A. Griffiths and D. Yang, *Characteristics and existence of isometric embeddings*, Duke Math. J. **50** (1983), 893–994.

3. L. Caffarelli, *A priori estimates for second order elliptic equations*, Lectures at Accademia dei Lincei, Cambridge Univ. Press, to appear.

4. L. Caffarelli, J. J. Kohn, L. Nirenberg and J. Spruck, *The Dirichlet problem for nonlinear second order elliptic equations. II: Complex Monge-Ampère and uniformly elliptic equations*, Comm. Pure Appl. Math. **38** (1985), 209–252.

5. L. Caffarelli, L. Nirenberg and J. Spruck, *The Dirichlet problem for nonlinear second order elliptic equations. III: Functions of eigenvalues of the Hessian*, Acta Math. **155** (1985), 261–301.

6. ____, *Nonlinear second order elliptic equations. IV: Starshaped compact Weingarten surfaces*, Current Topics in Partial Differential Equations (Y. Ohya, K. Kasahara and N. Shimakura, eds.), Kinokunize Co., Tokyo, 1986, pp. 1–26.

7. ____, *Nonlinear second order elliptic equations. V: The Dirichlet problem for Weingarten hypersurfaces*, Comm. Pure Appl. Math. **41** (1988), 47–70.

8. L. C. Evans, *Classical solutions of fully nonlinear convex second order elliptic equations*, Comm. Pure Appl. Math. **35** (1982), 333–363.

9. D. Gilbarg and N. S. Trudinger, *Elliptic partial differential equations of second order*, Springer-Verlag, Berlin/Heidelberg/New York, second ed., 1983.

10. J. B. Goodman and D. Yang, *Local solvability of nonlinear partial differential equations of real principal type*, preprint.

11. J. Hong and C. Zuily, in preparation.

12. N. V. Krylov, *Boundedly nonhomogeneous elliptic and parabolic equations in a domain*, Izv. Akad. Nauk SSSR Ser. Mat. **47** (1983), 75–108.

13. ____, *Nonlinear elliptic and parabolic equations of second order*, Moscow Nauk Glavnaya Red. Fisico-Mat. Lit., 1983.

14. N. V. Krylov and M. V. Safonov, *Certain properties of solutions of parabolic equations with measurable coefficient*, Izv. Akad. Nauk SSSR **40** (1980), 161–175; Engl. transl. in Math. USSR-Izv. **16** (1981), 151–164.

15. C. S. Lin, *The local isometric embedding in R^3 of two-dimensional Riemannian manifolds with nonnegative curvature*, J. Differential Geom. **21** (1985), 213–230.

16. ____, *The local isometric embedding in R^3 of two-dimensional Riemannian manifolds with Gaussian curvature changing sign cleanly*, Comm. Pure Appl. Math. **31** (1986), 867–887.

17. L. Nirenberg, *On nonlinear elliptic partial differential equations and Hölder continuity*, Comm. Pure Appl. Math. **6** (1953), 103–156.

18. ____, *The Weyl and Minkowski problems in differential geometry in the large*, Comm. Pure Appl. Math. **6** (1953), 337–394.

19. ____, *Fully nonlinear second order elliptic equations*, Proc. Conf. on Calculus of Variations and Partial Differential Equations dedicated to Hans Lewy, to appear.

20. V. I. Oliker, *Hypersurfaces in R^{n+1} with prescribed Gaussian curvature and related equations of Monge-Ampère type*, Comm. Partial Differential Equations **9** (1984), 807–838.

21. R. Schoen, *Conformal deformation of a Riemannian metric to constant scalar curvature*, J. Differential Geom. **20** (1984), 479–495.

22. A. E. Treibergs, *Existence and convexity for hypersurfaces of prescribed mean curvature*, preprint.

23. A. E. Treibergs and S. W. Wei, *Embedded hyperspheres with prescribed mean curvature*, J. Differential Geom. **18** (1983), 513–521.

24. K. S. Tso, *A variational approach to the existence of convex hypersurfaces with prescribed Gauss-Kronecker curvature*, preprint.

25. A. Vinacua, *Nonlinear elliptic equations written in terms of functions of eigenvalues of the complex Hessian*, submitted to Comm. Partial Differential Equations.

26. H. C. Wente, *Counterexample to a conjecture of H. Hopf*, Pacific J. Math. **121** (1986), 193–243.

27. H. Weyl, *Über die Bestimmung einer geschlossenen konvexen Fläche durch ihr Linienelement*, Vierteljahrsschrift der naturforschenden Gesellschaft. Zürich **61** (1916), 40–72.

28. S. T. Yau, *On the Ricci curvature of a compact Kähler manifold and the complex Monge-Ampère equation. I*, Comm. Pure Appl. Math. **31** (1978), 339–411.

COURANT INSTITUTE OF MATHEMATICAL SCIENCES

Proceedings of Symposia in Pure Mathematics
Volume 48 (1988)

Surfaces in Conformal Geometry

ROBERT L. BRYANT

0. Introduction. Hermann Weyl's interest in conformal geometry is well known. While the subject of conformal invariants of metrics was of direct importance to Weyl and has since received much attention, the subject of conformal invariants of submanifolds has been less studied. The subject was considered by the French school in the latter part of the 19th century, as is evidenced by Darboux's treatment of conformal geometry of surfaces in his monumental *Théorie générale des surfaces*, but almost none of this survived into modern differential geometry. For example, his fundamental mode of calculation in conformal geometry, "pentaspherical coordinates," remains obscure. Later researchers, such as Blaschke and Thomsen [T], took up conformal geometry of surfaces again in the 1920s and made considerable progress, but this work, too, lapsed into obscurity. It was Willmore [W] in 1965 who proposed the first global problem in conformal geometry of surfaces in space, that of minimizing the so-called Willmore functional $\mathscr{W}(\varphi) = \int_M H^2 \, dA$ where $\varphi \colon M^2 \to \mathbf{R}^3$ is an immersion of a compact surface and H and dA are the induced mean curvature and area form respectively. While one easily gets $\mathscr{W}(\varphi) \geq 4\pi$ when $M = S^2$, the minimum value of $\mathscr{W}(\varphi)$ when $M = T^2$ was conjectured by Willmore [W] and remains open. In recent years, new ideas, such as the conformal volume introduced by Li and Yau [LY], have become available and have led to a rapid development of conformal geometry of submanifolds. Quite recently, Leon Simon [Si] has managed to show the existence of an immersion $\varphi \colon T^2 \to \mathbf{R}^3$ realizing the infimum of $\mathscr{W}(\varphi)$ over all immersions of T^2 into $\mathbf{R}^3$. §1 of this paper is a report on two conformal invariants and their relationships with other geometric quantities. Some of these results, such as Proposition 1, are new, while others are reformulations of known results. In §2, we take up the very interesting structure of the moduli space of $\mathscr{W}$-critical immersions $\varphi \colon S^2 \to S^3$. In addition to

1980 *Mathematics Subject Classification* (1985 *Revision*). Primary 53A10, 53A30, 53C42, 53C45; Secondary 58E12, 14H99.

Work done while the author was partially supported by NSF grant DMS-86-01853 and while the author was a Presidential Young Investigator.

describing previous work by the author [**Br**] and by C. K. Peng [**P**], this section recounts a new 'twistor' formulation of the problem, replacing $\mathscr{W}$-critical immersions $\varphi\colon S^2 \to S^3$ by holomorphic rational null curves $\gamma\colon \mathbf{P}^1 \to Q^3 = $ the complex 3-quadric. This reformulation allows us to complete the determination of the critical values of $\mathscr{W}$ on $\mathrm{Imm}(S^2, S^3)$ (begun by the author and Peng) and to describe the moduli space of $\mathscr{W}$-minimizing immersions $\varphi\colon \mathbf{RP}^2 \to S^3$ completely. The fact that this moduli space is noncompact even after reducing modulo the conformal symmetries shows that any direct approach to constructing a minimizer by 'renormalizing' a minimizing sequence will fail. It is hoped that the explicit description of the minimizers for $\mathbf{RP}^2$ will give some clue as to the type of singularities which can develop in the more general case.

1. Basic Conformal Concepts. Let G denote the conformal group acting in $\mathbf{R}^n$. Thus, G is generated by rigid motions, dilations, and inversions. Strictly speaking G is actually well defined on the one-point compactification $S^n = \mathbf{R}^n \cup \{\infty\}$, but, at least initially, we shall not worry about this. G acts transitively on the set of k-spheres and k-planes in $\mathbf{R}^n$, so these are the natural "flat" objects in conformal geometry.

Given a k-dimensional submanifold, $M^k \subseteq \mathbf{R}^n$, it is natural to attempt to describe local conformal invariants on M^k. The induced Euclidean volume form (or density, if M^k is unoriented) Ω_0 is clearly not a conformal invariant, nor is the Euclidean second fundamental form II, though the conformal class of the first fundamental form I is certainly a conformal invariant. Let B_0 denote the function on M which is the square norm of the trace-free part of II with respect to I. Thus, for a hypersurface $M^{n-1} \subseteq \mathbf{R}^n$ with principal curvatures $\kappa_1, \ldots, \kappa_{n-1}$, we have $B_0 = \kappa_1^2 + \cdots + \kappa_{n-1}^2 - (n-1)H^2$ where $H = (1/(n-1))(\kappa_1 + \cdots + \kappa_{n-1})$ is the mean curvature. Note that, in general, $B_0 \geq 0$, with $B_0(p) = 0$ iff M is totally umbilic at $p \in M$. Thus $B_0 \equiv 0$ only if $M^k \subseteq \mathbf{R}^n$ is a k-plane or k-sphere when $k > 1$. (When $k = 1$, $B_0 \equiv 0$ anyway, by definition.) The quantity

$$(1) \qquad\qquad \mathscr{B} = (1/k)B_0^{k/2}\Omega_0$$

is clearly invariant under rigid motions and dilations. A short calculation shows that $\mathscr{B}$ is also invariant under inversions. This conformally invariant volume form (or density) was known to Blaschke and Thomsen [**T**] for surfaces ($k = 2$). It was later rediscovered by Willmore [**W**], although its conformal invariance was not rediscovered until later, by White. For a given immersion $\varphi\colon M^k \to \mathbf{R}^n$ with $k \geq 2$, we may define

$$(2) \qquad\qquad \tilde{\mathscr{W}}(\varphi) = \int_M \mathscr{B}.$$

Clearly $\tilde{\mathscr{W}}(\varphi) \geq 0$ with equality iff $\varphi(M^k)$ is totally umbilic. A fundamental problem in conformal geometry is to compute $\tilde{\mathscr{W}}_n(M)$ where

$$(3) \qquad \tilde{\mathscr{W}}_n(M) = \inf\{\tilde{\mathscr{W}}(\varphi) \mid \varphi\colon M^k \to \mathbf{R}^n \text{ is an immersion}\}.$$

Since $\tilde{\mathscr{W}}_n(M) \geq \tilde{\mathscr{W}}_{n+1}(M)$ for all n, we may set $\tilde{\mathscr{W}}(M) = \lim_{n\to\infty} \tilde{\mathscr{W}}_n(M)$ and try to compute $\tilde{\mathscr{W}}(M)$. It is not at all clear that $\tilde{\mathscr{W}}(M)$ is positive for any M. It is known, however, that $\tilde{\mathscr{W}}(\mathbf{RP}^2) = 4\pi$ and that $\tilde{\mathscr{W}}_3(\mathbf{RP}^2) = 10\pi$ (see §2). The famous Willmore conjecture is that $\tilde{\mathscr{W}}(T^2) \geq 2\pi^2$, where T^2 is the 2-torus. As of this writing it remains unsolved. (Actually, Willmore's original conjecture concerned only $\tilde{\mathscr{W}}_3(T^2)$, but more recent versions in the literature frequently allow the ambient dimension n to be arbitrary.) Leon Simon [**Si**] has shown that, for all n, there exists an imbedding $\varphi_n: T^2 \to \mathbf{R}^n$ so that $\tilde{\mathscr{W}}(\varphi_n) = \tilde{\mathscr{W}}_n(T^2)$. Since an immersion $\varphi_3: T^2 \to \mathbf{R}^3$ is known with $\tilde{\mathscr{W}}(\varphi_3) = 2\pi^2$, it follows that $\tilde{\mathscr{W}}_n(T^2) \leq 2\pi^2$ for all $n \geq 3$. Essentially, his method is to construct a minimizing sequence $\{\varphi_{n,k}: T^2 \to \mathbf{R}^n\}$ and then to show that there exists a sequence of conformal transformations $A_k: \mathbf{R}^n \to \mathbf{R}^n$ so that the sequence $\{A_k \circ \varphi_{n,k}\}$ can be reparametrized so as to have a subsequence converging to a smooth immersion $\varphi_n: T^2 \to \mathbf{R}^n$ which realizes the minimum. This method relies on the compactness of the moduli space of minimizers modulo conformal transformations and reparametrization. As we will see below in §2, this method could not apply to immersions $\varphi: \mathbf{RP}^2 \to \mathbf{R}^3$ since, there, the space of minimizers modulo conformal transformations and reparametrization is noncompact.

In dimension $k = 2$, it is more customary to use the functional $\mathscr{W}(\varphi) = \int_M |H|^2 \Omega_0$, where H is the mean curvature vector. Since $|H|^2 = B_0 + K$ where K is the Gauss curvature of the induced metric, we see that for compact surfaces without boundary we have $\mathscr{W}(\varphi) = \tilde{\mathscr{W}}(\varphi) + 2\pi\chi(M)$ where $\chi(M)$ is the Euler characteristic. Thus, the two functionals differ by a constant and hence have the same Euler-Lagrange equations. A direct calculation [**Br**] shows that an immersion $\varphi: M^2 \to \mathbf{R}^3$ is $\mathscr{W}$-critical iff it satisfies the fourth-order elliptic equation $\Delta H + 2(H^2 - K)H = 0$. In particular, any immersion which is conformal to a minimal immersion is $\mathscr{W}$-critical. When $M^2 = S^2$, the converse is true and can be used to study the moduli space of $\mathscr{W}$-critical immersions of S^2 and $\mathbf{RP}^2$ into $\mathbf{R}^3$; see §2.

A more recent conformal invariant which is first order and global has been introduced independently by M. Gromov [**G**] and P. Li and S. T. Yau [**LY**]. We may describe it as follows. First, we compactify $\mathbf{R}^n$ as $S^n = \mathbf{R}^n \cup \{\infty\}$. Then G acts smoothly on S^n and is isomorphic to $SO(n+1,1)$. It is useful to regard S^n as the 'ideal boundary' of hyperbolic $(n+1)$-space, $H^{n+1} = \{w \in \mathbf{R}^{n+1} \mid |w| < 1\}$. Then G extends to an isometric action of H^{n+1} with the hyperbolic metric

$$(4) \qquad ds_0^2 = \frac{4|dw|^2}{(1-|w|^2)^2}.$$

For each $u \in H^{n+1}$, we let $S_u \subseteq T_u H^{n+1}$ denote the unit sphere. Then the exponential map $\exp_u: T_u H^{n+1} \to H^{n+1}$ gives rise to a diffeomorphism $E_u: S_u \to S^n = \partial_\infty H^{n+1}$ by letting

$$(5) \qquad E_u(v) = \lim_{t\to\infty} \exp_u(tv).$$

The map $E_u\colon S_u \to S^n$ is conformal and hence there exists a conformal metric g_u on S^n for which E_u is an isometry. It is well known that the set $\{g_u \mid u \in H^{n+1}\}$ is the full set of conformal metrics on S^n with sectional curvature $+1$. Explicitly, the metric is

$$(6) \qquad g_u = \frac{(1 - u \cdot u)}{(1 - u \cdot x)^2}|dx|^2.$$

Now, for any immersion $\varphi\colon M^k \to S^n = \mathbf{R}^n \cup \{\infty\}$, we can define a function A_φ on H^{n+1} by the formula

$$(7) \qquad A_\varphi(u) = \int_M d\,\mathrm{vol}\,g_u(M) = \int_M \frac{(1 - u \cdot u)^{k/2}}{(1 - u \cdot \varphi)^k}\Omega_0,$$

where Ω_0 is the Euclidean volume form on M^k induced by $\varphi\colon M^k \to S^n \subseteq \mathbf{R}^{n+1}$.

Gromov [G] calls $A_\varphi(u)/V(S^k)$ the *visual volume* of $\varphi(M^k)$ at $u \in H^{n+1}$. Li and Yau define the *conformal volume* of $\varphi(M^k)$ to be

$$(8) \qquad V_c(\varphi) = \sup\{A_\varphi(u) \mid u \in H^{n+1}\}.$$

It is not difficult to show that $\Delta A_\varphi = k(k - n)A_\varphi$ where Δ is the Laplace-Beltrami operator on H^{n+1}. Thus, A_φ is, in particular, real analytic in H^{n+1}. We have the following elementary formula for radial limits

$$(9) \qquad \lim_{t \to 1^-} A_\varphi(tu) = |\varphi^{-1}(u)|V(S^k)$$

for all $u \in S^n$. In particular, $V_c(\varphi) \geq m_\varphi V(S^k)$, where m_φ is the maximum number of preimages of single points in S^n under $\varphi\colon M^k \to S^n$. Note that this implies that any immersion φ with $V_c(\varphi) < 2V(S^k)$ must be an imbedding. Actually, a more careful estimate shows that, if $u \in S^n$ has only one preimage $p_0 \in M$ under φ, then

$$(10)$$
$$A_\varphi((1-t)u) = \begin{cases} 4\pi + (\pi/2)B_0(p_0)t^2\log(1/t) + O(t^2) & \text{if } k = 2, \\[2mm] V(S^k) + \dfrac{2^{k-1}V(S^{k-1})}{k^2 - 4}B_0(p_0)t^2 + O(t^3\log(1/t)) & \text{if } k \geq 3, \end{cases}$$

where B_0 is the function defined earlier for $\varphi\colon M^k \to S^n \subseteq \mathbf{R}^{n+1}$. This allows us to prove the following proposition.

PROPOSITION 1. *For any $\varphi\colon M^k \to S^n$ we have $V_c(\varphi) > V(S^k)$ except possibly in the following cases:*

(i) *$k = 1$ and M^1 has no compact component.*

(ii) *$\varphi(M^k)$ is totally umbilic.*

PROOF. When $k \geq 2$, (10) shows that $V_c(\varphi) = V(S^k)$ implies that $B_0 \equiv 0$, so $\varphi(M^k)$ is totally umbilic. When $k = 1$, if M has a compact component M_0, then, after a conformal transformation, we may assume that $\varphi(M_0)$ passes through a pair of antipodal points on S^n. It easily follows (since $M_0 \simeq S^1$) that the length of $\varphi(M_0) \geq 2(\pi) = 2\pi = V(S^1)$ with equality iff $\varphi(M_0)$ is a closed geodesic. Q.E.D.

Proposition 1 has the following corollary.

COROLLARY 1. *If M^k is compact and $\varphi\colon M^k \to S^n$ is an imbedding, then there exists a $u_0 \in H^{n+1}$ so that $V_c(\varphi) = A_\varphi(u_0)$. (I.e., A_φ attains its maximum on H^{n+1}.)*

PROOF. If $V_c(\varphi) > V(S^k)$, then any sequence $u_1, u_2, \ldots$ in H^{n+1} satisfying $\lim_j A_\varphi(u_j) = V_c(\varphi)$ cannot accumulate near $S^n = \partial_\infty H^{n+1}$ since we can use (10) to prove that $A_\varphi < V(S^k) + \varepsilon$ outside some compact set $K_\varepsilon \subset H^{n+1}$ for every $\varepsilon > 0$. Of course, this implies that A_φ must attain its maximum on H^{n+1}. If $V_c(\varphi) = V(S^k)$, then Proposition 1 implies that $\varphi(M^k)$ is a totally umbilic k-sphere in S^n. But then $A_\varphi(u) = V(S^k)$ for all u lying on the totally umbilic H^{k+1} in H^{n+1} whose boundary is $\varphi(M^k)$. Q.E.D.

After applying a conformal transformation $g \in G$, we may assume that $V_c(\varphi) = A_\varphi(0)$ whenever $\varphi\colon M^k \to S^n$ is an imbedding of a compact manifold. By (7), we have

$$A_\varphi(u) = A_\varphi(0) + \int_M m(u \cdot \varphi)\Omega_0 + O(|u|^2).$$

Since A_φ has a maximum at $u = 0$, this implies that $\int_M \varphi\Omega_0 = 0$, i.e., that $\varphi(M) \subseteq S^n$ is 'balanced.' In particular, it is easy now to see that the assumption of imbeddedness is necessary in Corollary 1. For example, a pair of tangent circles in S^2 clearly cannot be balanced by any conformal transformation of S^2.

Gromov [G] has shown that there exists an $\varepsilon > 0$ depending on k and n so that $V_c(\varphi) < V(S^k)(1 + \varepsilon)$ for an immersion $\varphi\colon M^k \to S^n$ of a compact M implies that M is diffeomorphic to S^k. Thus, for a given compact manifold M^k, the quantity

$$(11) \qquad \mathscr{A}_n(M) = \inf\{V_c(\varphi) \mid \varphi\colon M^k \to S^n \text{ is an immersion}\}$$

satisfies $\mathscr{A}_n(M) \geq V(S^k)$ with equality only for $M = S^k$. We presently have no computation of $\mathscr{A}_n(M)$ for any $M \neq S^k$ or $\mathbf{RP}^2$ (when $k = 2$). It is known that $\mathscr{A}_3(\mathbf{RP}^2) = 12\pi$ and $\mathscr{A}_n(\mathbf{RP}^2) = 6\pi$ for all $n \geq 4$ (see below).

We can relate the quantities $V_c(\varphi)$ and $\widetilde{\mathscr{W}}(\varphi)$ when $k = 2$ as follows: Let $\varphi\colon M^2 \to S^n \subseteq \mathbf{R}^{n+1}$ denote an immersion of a compact surface and let H be its mean curvature vector in $\mathbf{R}^{n+1}$ while $\overline{H}$ is its mean curvature vector in S^n. We have $|H|^2 = |\overline{H}|^2 + 1$.

Now $B_0 = |H|^2 - K = |\overline{H}|^2 - K + 1$. Also $\Omega_0 = dA$, the induced area form in either S^n or $\mathbf{R}^{n+1}$. Then

$$\begin{aligned}
(12) \qquad \mathscr{W}(\varphi) - 2\pi\chi(M) = \widetilde{\mathscr{W}}(\varphi) &= \int_M (|H|^2 - K)\,dA \\
&= \int_M (|\overline{H}|^2 - K + 1)\,dA \\
&\geq V(\varphi) - 2\pi\chi(M)
\end{aligned}$$

with equality iff $\varphi\colon M^2 \to S^n$ is a minimal immersion (i.e., $\overline{H} = 0$). Thus $\mathscr{W}(\varphi) \geq V(\varphi)$. Now $V(\varphi) = A_\varphi(0)$. Since $\mathscr{W}(\varphi)$ is a conformal invariant, we see that $\mathscr{W}(\varphi) = \mathscr{W}(g \circ \varphi)$ where $g \in G$ is arbitrary. On the other hand,

$V(g \circ \varphi) = A_{g \circ \varphi}(0) = A_{\varphi}(g^{-1}(0))$. Since $g \in G$ acts transitively on H^{n+1}, it follows that $\mathscr{W}(\varphi) \geq V_c(\varphi) = \sup A_{\varphi}$. This fundamental inequality $\mathscr{W}(\varphi) \geq V_c(\varphi)$ is due to Li and Yau.

As an application, note that any immersion $\varphi \colon \mathbf{RP}^2 \to \mathbf{R}^3 \subseteq S^3$ has a triple point [Ba] and hence that $V_c(\varphi) \geq 12\pi$. In §2, we construct an immersion $\varphi \colon \mathbf{RP}^2 \to S^3$ with $\mathscr{W}(\varphi) = 12\pi$. Of course, the above inequality now implies that $V_c(\varphi) = 12\pi$. This establishes the value $\mathscr{A}_3(\mathbf{RP}^2) = 12\pi$. For a different argument involving the Veronese immersion $\mathbf{RP}^2 \hookrightarrow S^4$ and calculating $\mathscr{A}_n(\mathbf{RP}^2)$ for all $n \geq 4$, see [LY].

PROPOSITION 2. *Let $\varphi \colon M^2 \to S^n$ be an immersion of a compact surface which satisfies $\mathscr{W}(\varphi) = V_c(\varphi)$. Then either $g \circ \varphi \colon M \to S^n$ is a minimal immersion for some $g \in G$ or else there is a point $u_0 \in S^n$ so that if $\rho_0 \colon S^n - \{u_0\} \to \mathbf{R}^n$ is stereographic projection, then $\rho_0 \circ \varphi \colon M^2 - \varphi^{-1}(u_0) \to \mathbf{R}^n$ is a minimal immersion with imbedded ends of zero logarithmic growth.*

PROOF. If $V_c(\varphi) = A_{\varphi}(u)$ for some $u \in H^{n+1}$, by applying an element $g \in G$ which satisfies $g(u) = 0$, we may assume that $u = 0$, so that $V_c(\varphi) = A_{\varphi}(0) = V(\varphi)$. Now we have $\mathscr{W}(\varphi) = V(\varphi)$ and (12) above shows that this can only happen if $\overline{H} \equiv 0$. (This part of the argument is due to Li and Yau, see [LY], Lemma 1 of §5. However, they implicitly assume that A_{φ} assumes its maximum in the interior of H^{n+1}, which need not happen. For example, if $\varphi(M^2)$ consists of a pair of distinct totally umbilic 2-spheres in S^n which are tangent at one point $u_0 \in S^n$, then $V_c(\varphi) = 8\pi$, but $A_{\varphi} < 8\pi$ on H^{n+1}.) On the other hand, if $A_{\varphi}(u) < V_c(\varphi)$ for all $u \in H^{n+1}$, then there exists a point $u_0 \in S^n$ and a sequence $u_1, u_2, \ldots$ in H^{n+1} which converges to u_0 and for which $V_c(\varphi) = \lim A_{\varphi}(u_i)$. Using a simple expansion similar to that used to prove (10), we compute that $\lim A_{\varphi}(u_i) = m_0 V(S^2) = 4\pi m_0$, where m_0 is the number of preimages of u_0 in M under $\varphi \colon M \to S^n$. Thus $\mathscr{W}(\varphi) = 4\pi m_0$. Now by an argument of Kusner [K], we see that $\rho_0 \circ \varphi \colon M^2 \to \mathbf{R}^n$ is a minimal immersion with m_0 embedded ends of zero logarithmic growth. Q.E.D.

Li and Yau define a refined notion of conformal volume in [LY] as follows: Fix a conformal structure on M and define

$$(13) \qquad V_c(n, M) = \inf\{V_c(\varphi) \mid \varphi \colon M \to S^n \text{ is branched conformal}\}.$$

Note that $V_c(n, M)$ is defined for n sufficiently large and that $V_c(n, M) \geq \mathscr{A}_n(M)$ (this latter invariant is independent of the conformal structure). This quantity can be used to estimate the first eigenvalue $\lambda_1 > 0$ of the Laplacian of a surface M^2 (assumed compact without boundary). In fact Li and Yau show that

$$(14) \qquad \lambda_1 V(M) \leq 2V_c(n, M)$$

for all n, where M^2 is assumed to be endowed with a Riemannian metric and we use the associated conformal structure to compute the right-hand side of (14). When M is orientable and of genus g, then one can use the Brill-Noether theorem [GH] to produce a branched conformal map $f \colon M \to S^2$ of degree

$d = [(g + 1)/2] + 1$. This leads to the estimate

$$V_c(n, M) \leq 4\pi d \leq (2g + 5 - (-1)^g)\pi$$

for $n \geq 2$. When M is nonorientable and of genus g ($= 2 - \chi(M)$ when M is nonorientable), then one can always find a branched conformal map $f: M \to \mathbf{RP}^2$ of degree g [Sc]. Since $V(4, \mathbf{RP}^2) = 6\pi$, we get the estimate $V_c(n, M) \leq 6\pi g$ for an unorientable surface M of genus g. These estimates somewhat improve those found in [LY].

Further calculations of the conformal volume together with applications may be found in [G], [LY], [MR], and [SI].

2. Algebraic geometry and immersions of S^2 into S^3. For a general surface M^2, the moduli space of $\mathscr{W}$-critical immersions of M^2 into S^3 is unknown. However, for $M = S^2$, the problem of studying this moduli space can be 'reduced' to a problem in algebraic geometry. The essential result is as follows (see Theorem F of [Br]):

THEOREM. *Let $X: S^2 \to S^3$ be a $\mathscr{W}$-critical immersion. Endow S^2 with the induced conformal structure. Then there exists a point $y_0 \in X(S^2) \subseteq S^3$ (unique if X is not totally umbilic) so that $D = X^{-1}(y_0)$ is a divisor on S^2 with $d > 0$ distinct points, a stereographic projection $\rho: S^3 - \{y_0\} \to \mathbf{E}^3$, and a meromorphic curve $f: S^2 \to \mathbf{C}^3$ whose polar divisor is D so that $\rho \circ X = \mathrm{Re}(f)$. Moreover, f is an immersion with null tangents (i.e., $(df, df) = 0$). Conversely, if $f: S^2 \to \mathbf{C}^3$ is a meromorphic immersion with simple poles along D and null tangents, then, regarding $\mathbf{E}^3 = S^3 - \{y_0\}$, we have that $\mathrm{Re}(f): S^2 - D \to \mathbf{E}^3$ completes as a smooth immersion across D to be a conformal $\mathscr{W}$-critical immersion $X: S^2 \to S^3$.*

In the above theorem, we identify S^2 with $\mathbf{CP}^1$, the unique Riemann surface of genus 0. If we choose a rationalizing parameter z on S^2 with its pole at $\infty \in S^2$, then a meromorphic f with simple poles at $z = \lambda_i$ ($i = 1, \ldots, d$) can be written in the form

$$(1) \qquad f(z) = v_0 + \frac{v_1}{z - \lambda_1} + \cdots + \frac{v_d}{z - \lambda_d},$$

where $v_0, \ldots, v_d \in \mathbf{C}^3$. Since changing v_0 only changes the resulting X by a conformal transformation in S^3, we may set $v_0 = 0$. In order to satisfy the condition $(df, df) = 0$, we see that we must impose a large number of algebraic conditions on the complex parameters $\{\lambda_1, \ldots, \lambda_d; v_1, \ldots, v_d\}$. However, we must also require f to be an immersion, which imposes open conditions on these parameters. Direct algebraic methods are not particularly helpful. It is easy to see that the case $d = 1$ corresponds to the umbilic 2-spheres in S^3. One can easily show that $d = 2$ or 3 has no solutions. In [Br] the space of solutions for $d = 4$ is completely described and the author asserts that the moduli space is nonempty for all *even* degrees except 2. C. K. Peng [P] has explicitly computed examples for all even degrees $d \geq 4$ and all odd degrees $d = 2k + 1 \geq 9$. This leaves unsettled the existence of meromorphic null curves with degree 5 or 7. We address this below (Proposition 2).

A more geometric approach to the study of these meromorphic null immersions is as follows: We define Q^3, the complex conformal compactification of $\mathbf{C}^3$, in the usual way. Let $\mathbf{C}^5$ have coordinates z^0, z^1, z^2, z^3, z^4 and give $\mathbf{C}^5$ the complex inner product $\langle \, , \, \rangle$ which corresponds to the quadratic form

$$q = 2z^0 z^4 - (z^1)^2 - (z^2)^2 - (z^3)^2.$$

We define $Q^3 \subseteq \mathbf{CP}^4$ to be the 3-quadric which is the space of null lines in $\mathbf{C}^5$ with respect to q. We can embed $\iota \colon \mathbf{C}^3 \hookrightarrow Q^3$ by the map $\iota(z^1, z^2, z^3) = [1, z^1, z^2, z^3, \frac{1}{2}((z^1)^2 + (z^2)^2 + (z^3)^2)]$. Then $\mathbf{C}^3$ is actually an affine open set in Q^3. If f is a meromorphic null immersion with simple poles along a divisor D in a Riemann surface M, then $\iota \circ f = F \colon M \to Q^3$ is a holomorphic null immersion of degree $d = \deg D$. (We endow Q^3 with its standard holomorphic conformal structure.) Of course, this process is reversible. If $\xi \in \mathbf{C}^5$ is any null vector, then $[\xi] \in Q^3$. If we let $\xi^\perp \subseteq \mathbf{C}^5$ denote the annihilator of $[\xi]$, then $\mathbf{P}\xi^\perp \subseteq \mathbf{CP}^4$ is the tangent hyperplane to Q^3 at $[\xi]$. The complement $Q^3 \backslash (\mathbf{P}\xi^\perp \cap Q^3)$ is naturally isomorphic to $\mathbf{C}^3$ with a nondegenerate complex inner product. If $F \colon M \to Q^3$ is a holomorphic null immersion which meets $\mathbf{P}\xi^\perp \cap Q^3$ transversely in a divisor D, then $f \colon M - D \to Q^3 \backslash (\mathbf{P}\xi^\perp \cap Q^3) \cong \mathbf{C}^3$ is a meromorphic null immersion with simple poles along D (where f is the restriction of F to $M - D$).

Thus, our problem is reduced to studying the holomorphic null immersions $F \colon \mathbf{P}^1 \to Q^3$ or, more generally, $F \colon M^2 \to Q^3$ where M is an arbitrary compact Riemann surface. This is much easier than studying meromorphic null curves $f \colon \mathbf{P}^1 \to \mathbf{C}^3$, mainly because the conformal automorphism group of Q^3, $\mathrm{SO}(5, \mathbf{C})$, is quite large. As a first application of this point of view, note that there cannot be any holomorphic null immersions $F \colon M^2 \to Q^3$ of degree $d = 2$ or 3. To see this, it suffices to observe first that any holomorphic curve in $Q^3 \subseteq \mathbf{CP}^4$ of degree $d < 4$ must be degenerate (see [**GH**]) and hence must lie in a hyperplane $H \subseteq \mathbf{CP}^4$. Now, the intersection $H \cap Q^3$ is either a singular quadric or a smooth quadric $Q^2 \subseteq Q^3$. In either case, the null curves of the conformal structure on Q^3 restricted to $Q^3 \cap H$ are precisely the rulings of $Q^3 \cap H$ and hence are lines in Q^3. Since the image of F lies on a line $L \subseteq Q^3$ and F is assumed to be an immersion, it follows that $\deg F = 1$ and $M = \mathbf{P}^1$.

At this point, it will be advantageous to use the classical Klein correspondence to simplify our problem. We may describe this as follows: Let V be a complex vector space of dimension 4 and let $\Omega \in \Lambda^2(V^*)$ be a nondegenerate 2-form on V. Thus, $\Omega^2 \neq 0$. We let $W \subseteq \Lambda^2(V)$ denote the subspace (of dimension 5) consisting of those $w \in \Lambda^2(V)$ satisfying $\Omega(w) = 0$. (Here, we are using the canonical identification of $\Lambda^2(V^*)$ with $(\Lambda^2(V))^*$.) Now W has a canonical quadratic form defined by $q(w) = \frac{1}{2}\Omega^2(w \wedge w)$. It is easily seen that q is nondegenerate on W. If we denote by $\mathrm{Sp}(2, \mathbf{C})$ the subgroup of $\mathrm{GL}(V, \mathbf{C})$ which fixes Ω, then $\mathrm{Sp}(2, \mathbf{C})$ obviously preserves W and the quadratic form q on W. This gives rise to the classical homomorphism $\mathrm{Sp}(2, \mathbf{C}) \to \mathrm{SO}(5, \mathbf{C})$ which is well known to be a double cover.

There is a distinguished family of 2-planes in V, the space of Ω-null 2-planes. Thus, a 2-plane $A \subseteq V$ is said to be Ω-null if $\Omega(a_1, a_2) = 0$ where $a_1, a_2 \in A$ form a basis. If we let A correspond to $[A] = [a_1 \wedge a_2] \in \mathbf{P}W = \mathbf{C}\mathbf{P}^4$, we see that $a_1 \wedge a_2$ spans a q-null line and hence that $[A] \in Q^3$, the space of q-null lines in W. Conversely, if $B \in \mathbf{P}W$ is represented by $B = [b]$ for some nonzero q-null $b \in W$, then $b \wedge b = 0$ so b must be decomposable, i.e., of the form $b = a_1 \wedge a_2$ for $a_1, a_2 \in V$. Since $b \in W$, it follows that the plane $A \subseteq V$ spanned by a_1 and a_2 is Ω-null. Thus, we have established one of the Klein correspondences:

$$\{\text{points of } Q^3\} = \{q\text{-null lines in } W\}$$

$$\overset{1\text{-}1}{\leftrightarrow} \{\Omega\text{-null planes in } V\}$$

$$\overset{1\text{-}1}{\leftrightarrow} \{\text{contact lines in } \mathbf{P}^3\}.$$

Here we simply write $\mathbf{P}^3$ for $\mathbf{P}V$. By definition, the *contact lines* in $\mathbf{P}^3$ are the lines of the form $\mathbf{P}A$, where $A \subseteq V$ is an Ω-null 2-plane. The other Klein correspondence may be described as follows: Let $a \in V$ be nonzero and let $a \wedge V \subseteq \Lambda^2 V$ denote the 3-plane consisting of multiples of a. It is easy to see that $a \wedge V \not\subseteq W$ since Ω is nondegenerate. Since W has codimension 1 in $\Lambda^2 V$, it follows that $(a \wedge V) \cap W$ has dimension 2 and hence that $L_a = \mathbf{P}[(a \wedge V) \cap W]$ is a projective line in $\mathbf{P}^4 = \mathbf{P}W$, which is the projectivization of a 2-plane consisting of decomposable 2-vectors. Thus $L_a \subseteq Q^3$ and is a null line of the standard holomorphic conformal structure on Q^3. Conversely, if $L \subseteq Q^3$ is a line in $\mathbf{P}^4$, then L is the projectivization of a 2-plane $\tilde{L} \subseteq W$ which consists entirely of q-null (i.e., decomposable) 2-vectors. If $a_1 \wedge a_2$, $b_1 \wedge b_2$ form a basis of $\tilde{L}$, then $a_1 \wedge a_2 \wedge b_1 \wedge b_2 = 0$ since $\tilde{L}$ is null. It follows that there exists a unique $a \in V$ (up to multiples) which "divides" both $a_1 \wedge a_2$ and $b_1 \wedge b_2$ and hence all the elements of $\tilde{L}$. It follows easily that $L = L_a$. Thus

$$\{\text{points of } \mathbf{P}^3\} \overset{1\text{-}1}{\leftrightarrow} \{\text{lines in } Q^3\}$$

$$= \{\text{null lines in } Q^3 \text{ with its standard conformal structure}\}.$$

For more details on the Klein correspondence, see [**GH**]. Our interest in this correspondence is in the *Lie transform*, defined as follows: If $\lambda \colon M \to \mathbf{P}^3$ is a holomorphic curve, we say that λ is a *contact curve* (more explicitly, an Ω-*contact curve*) if, at every point, the tangent line to $\lambda(M)$ is a contact line. For each contact curve $\lambda \colon M \to \mathbf{P}^3$, we define the map $\gamma_\lambda \colon M \to Q^3$ by letting $\gamma_\lambda(p)$ be the tangent line to $\lambda(M)$ at p. We have the following proposition, due to Lie:

PROPOSITION 1. *For any nonlinear contact curve* $\lambda \colon M \to \mathbf{P}^3$, *the Lie transform* $\gamma_\lambda \colon M \to Q^3$ *is a holomorphic nonlinear null curve. Conversely, every holomorphic nonlinear null curve* $\gamma \colon M \to Q^3$ *is of the form* $\gamma = \gamma_\lambda$ *for a unique nonlinear contact curve* $\lambda \colon M \to \mathbf{P}^3$.

SKETCH OF PROOF. If $\lambda \colon M \to \mathbf{P}^3$ is a holomorphic nonlinear contact curve, let $\tilde{\lambda} \colon M \to V$ be a meromorphic lift of λ. Then, in terms of a local holomorphic coordinate z near $p \in M$, we may write $\tilde{\lambda} = \tilde{\lambda}(z)$. Then $\gamma_\lambda = [\tilde{\lambda} \wedge d\tilde{\lambda}/dz] = [\tilde{\lambda} \wedge \tilde{\lambda}']$. By hypothesis, the 2-plane spanned by $\{\tilde{\lambda}(z), \tilde{\lambda}'(z)\}$ is Ω-null. Thus $\gamma_\lambda : M \to Q^3$ since $\tilde{\lambda} \wedge \tilde{\lambda}' : M \to W$ and $\tilde{\gamma}_\lambda = \tilde{\lambda}(z) \wedge \tilde{\lambda}'(z)$ is decomposable for all z. To show that γ_λ is a null curve in Q_3, we note that $\tilde{\gamma}'_\lambda(z) = \tilde{\lambda}(z) \wedge \tilde{\lambda}''(z)$. Thus, the 2-plane spanned by $\{\tilde{\gamma}_\lambda(z), \tilde{\gamma}'_\lambda(z)\} = \{\tilde{\lambda}(z) \wedge \tilde{\lambda}'(z), \tilde{\lambda}(z) \wedge \tilde{\lambda}''(z)\}$ is a q-null 2-plane in W since, by hypothesis, $\{\tilde{\lambda}(z), \tilde{\lambda}'(z), \tilde{\lambda}''(z)\}$ are linearly independent except for at most a discrete set of values of z. Conversely, if $\gamma : M \to Q^3$ is a holomorphic nonlinear null curve, then we may choose a holomorphic lift $\tilde{\gamma} : M \to W$ so that $\{\tilde{\gamma}(z), \tilde{\gamma}'(z)\}$ span a q-null 2-plane at all but a discrete set of values of z.

Algebraically, we know that there must exist a unique holomorphic curve $\tilde{\lambda} : M \to V$ so that $\tilde{\lambda}(z) \wedge \tilde{\gamma}(z) \equiv \tilde{\lambda}(z) \wedge \tilde{\gamma}'(z) \equiv 0$. Differentiating the first of these equations, we see that $\tilde{\lambda}'(z) \wedge \tilde{\gamma}(z) \equiv (\tilde{\lambda}(z) \wedge \tilde{\gamma}(z))' - \tilde{\lambda}(z) \wedge \tilde{\gamma}'(z) \equiv 0$. If $\tilde{\lambda}(z) \wedge \tilde{\lambda}'(z) \equiv 0$, then by scaling, we could take $\tilde{\lambda}(z) \equiv \lambda_0$ so that $\lambda_0 \wedge \tilde{\gamma}(z) \equiv 0$ for all z. However the space of multiples of λ_0 which lie in W is known to be of dimension 2 (see above), so this would force $\tilde{\gamma}(M)$ to lie in a single 2-plane. Thus $\gamma(M) \subseteq Q^3$ would be linear. It follows that $\tilde{\lambda}(z) \wedge \tilde{\lambda}'(z) \not\equiv 0$. Thus, by rescaling $\tilde{\gamma}$ we may write $\tilde{\gamma}(z) = \tilde{\lambda}(z) \wedge \tilde{\lambda}'(z)$. It immediately follows that $\lambda = [\tilde{\lambda}] : M \to \mathbf{P}^3$ is a contact curve and that $\gamma = \gamma_\lambda$. □

Our strategy in studying holomorphic null curves in Q^3 will be to study, instead, the corresponding contact curve in $\mathbf{P}^3$. The reason this is simpler is that the differential equation $\Omega(\tilde{\lambda}, \tilde{\lambda}') = 0$ for contact curves $\lambda = [\tilde{\lambda}] : M \to \mathbf{P}^3$ is linear in the first derivatives of the lift $\tilde{\lambda} : M \to V$ while the differential equation $q(\tilde{\gamma}') = 0$ is quadratic in the first derivatives of the lift $\tilde{\gamma} : M \to W$ and moreover, the condition that $\gamma = [\tilde{\gamma}]$ map M to $Q^3 \subseteq \mathbf{CP}^4$ is a quadratic condition on $\tilde{\gamma}$. If we were only interested in holomorphic null curves $\gamma : M \to Q^3$, then the Weierstrass formula given in [**Br**] would suffice. However, we are mainly interested in holomorphic null *immersions*, i.e., unramified curves, and the Weierstrass formula does not allow one to specify the degree of ramification.

In the terminology of algebraic geometry, the Lie transform, γ_λ, of $\lambda : M \to \mathbf{P}^3$ is the first associated curve of λ. The classical Plücker relations [**GH**] then give the following formulae: Set

R_λ = the ramification divisor of $\lambda \geq 0$,

R_γ = the ramification divisor of $\gamma \geq 0$,

$[D_\lambda]$ = the linear equivalence class of the hyperplane divisor of λ,

$[D_\gamma]$ = the linear equivalence class of the hyperplane divisor of γ,

K = the linear equivalence class of the canonical bundle of M.

Then

$$[D_\gamma] = -[R_\lambda] + 2[D_\lambda] + K, \qquad [R_\gamma] = -2[R_\lambda] + 2[D_\lambda] + 3K.$$

In the second formula, we have used the fact that λ is a contact curve and that γ is its Lie transform. Since we are interested in the case where γ is unramified, i.e., $R_\gamma = 0$, we see that we may eliminate $[R_\lambda]$ up to elements of order 2 in the divisor class group. Taking degrees, we get the formulae

$$r_\lambda = \deg \lambda + 3(g - 1), \qquad \deg \gamma = \deg \lambda - (g - 1),$$

where r_λ is the degree of ramification of λ and g is the genus of M. If we now specialize to the case where $g = 0$, i.e., $M = \mathbf{P}^1$, we get the relations

$$r_\lambda = \deg \lambda - 3, \qquad \deg \gamma = \deg \lambda + 1.$$

These formulae allow us to prove

PROPOSITION 2. *There is no holomorphic null immersion* $\gamma\colon \mathbf{P}^1 \to Q^3$ *of degree 5 or 7.*

SKETCH OF PROOF. Suppose such a γ existed of degree 5. Then the corresponding contact curve $\lambda\colon \mathbf{P}^1 \to \mathbf{P}^3$ would be of degree 4 and would ramify at precisely one point. Let z be a holomorphic coordinate on $\mathbf{P}^1$ so that $z = \infty$ corresponds to the ramification point. Let $\tilde{\lambda}\colon \mathbf{P}^1 - \{\infty\} \to V$ be a meromorphic lift of λ with polar divisor $4(\infty)$. Thus $\tilde{\lambda}$ can be written in the form $\tilde{\lambda} = v_0 + v_1 z + v_2 z^2 + v_3 z^3 + v_4 z^4$ where $v_0, \ldots, v_4 \in V$. Moreover, $v_4 \neq 0$, $v_0 \neq 0$, and $v_3 \wedge v_4 = 0$ since λ ramifies at $z = \infty$. Replacing z by $z + c$ for some complex number c, we may suppose $v_3 = 0$. Thus, we may write $\tilde{\lambda} = v_0 + v_1 z + v_2 z^2 + v_4 z^4$. Since λ cannot be degenerate, we must therefore have $\{v_0, v_1, v_2, v_4\}$ linearly independent in V. Finally, we must consider the contact condition $\Omega(\tilde{\lambda}, \tilde{\lambda}') \equiv 0$. A simple computation shows that this is equivalent to the condition that the set $\{v_0 \wedge v_1, v_0 \wedge v_2, v_1 \wedge v_2, v_0 \wedge v_4, v_1 \wedge v_4, v_2 \wedge v_4\}$ lie in W. In particular, the set of multiples of v_0 lies in W, which is impossible for nondegenerate Ω. Thus $d = 5$ cannot occur. For the case where $\deg \gamma = 7$, there are three possible ramification divisors for λ, namely, a triple point, a double point and a single point, and three single points. A case-by-case analysis as above for degree $\gamma = 5$ shows that none of these ramification divisors can occur. $\square$

It is now easy to give examples of holomorphic null immersions of all even degrees: Let e_0, e_1, e_2, e_3 be a basis of V so that $\Omega = e_0^* \wedge e_3^* + e_1^* \wedge e_2^*$, where e_0^*, e_1^*, e_2^*, e_3^* is the dual basis of V^*. Then the curve

$$\lambda_p = \left[-\left(\frac{1}{2p-1}\right) e_0 + e_1 z^{p-1} + e_2 z^p + e_3 z^{2p-1} \right]$$

is a nonlinear contact curve of degree $2p-1$ and ramification degree $2p-4$ for all $p \geq 2$. The corresponding curve $\gamma_p\colon \mathbf{P}^1 \to Q^3$ is a holomorphic null immersion of degree $2p$ for all $p \geq 2$.

Moreover, we can completely describe the moduli space of holomorphic null immersions $\gamma\colon \mathbf{P}^1 \to Q^3$ of degrees 4 and 6 as follows:

PROPOSITION 3. *If* $\gamma\colon \mathbf{P}^1 \to Q^3$ *is a holomorphic null immersion of degree 4 (resp. 6) then* γ *is equivalent to* γ_2 *(resp.* γ_3*) up to reparametrization in* $\mathbf{P}^1$ *and the action of the holomorphic automorphisms of* Q^3, $SO(5, \mathbf{C})$. $\square$

We can now make a table of what is known so far about $\mathcal{M}_d$, the moduli space of unramified rational null curves of degree d in Q^3:

$$\mathcal{M}_1 = \mathbf{P}^3,$$
$$\mathcal{M}_2 = \mathcal{M}_3 = \varnothing,$$
$$\mathcal{M}_4 = \mathrm{SO}(5, \mathbf{C})/\rho_5(\mathrm{SL}(2, \mathbf{C})),$$
$$\mathcal{M}_5 = \varnothing,$$
$$\mathcal{M}_6 = \mathrm{SO}(5, \mathbf{C})/\mathbf{C}^*,$$
$$\mathcal{M}_7 = \varnothing,$$
$$\mathcal{M}_d \neq \varnothing \quad \text{for all } d \geq 8.$$

In the above table, $\rho_5 \colon \mathrm{SL}(2, \mathbf{C}) \to \mathrm{SO}(5, \mathbf{C})$ is the homomorphism induced by the irreducible representation of degree 5 of $\mathrm{SL}(2, \mathbf{C})$ and $\mathbf{C}^* \subseteq \mathrm{SO}(5, \mathbf{C})$ is a certain (holomorphic) 1-parameter subgroup. The nontriviality of $\mathcal{M}_{2k+1}$ for $k \geq 4$ is due to C. K. Peng [**P**].

Several unanswered questions remain. For example, we do not know whether $\mathcal{M}_d$ is connected or smooth for $d \geq 8$. If $\mathcal{M}_d(g)$ denotes the moduli space of unramified null curves of degree d and genus g in Q^3, we do not know which pairs (d, g) have $\mathcal{M}_d(g) \neq \varnothing$. More specifically, we do not even know whether every compact Riemann surface M has an unramified null curve $\gamma \colon M \to Q^3$, much less do we know an estimate for the least degree of such a curve. Nevertheless, some work has been done, see [**V**] or [**BR**].

In order to return to our original setting in $\mathbf{C}^5$, we must choose an isomorphism $\mathbf{C}^5 \simeq W$ which identifies the quadratic forms on each complex vector space. Since all nondegenerate quadratic forms on $\mathbf{C}^5$ are equivalent, this is always possible. In order to return to immersions $X \colon S^2 \to S^3$, we need a sort of 'real projection' $Q^3 \to S^3$. We do this as follows: We let $\mathbf{R}^5 \subseteq \mathbf{C}^5$ denote the subspace on which each of the coordinates $z^0, \ldots, z^4$ restricts to be real-valued. We define conjugation in $\mathbf{C}^5$ relative to this $\mathbf{R}^5$ in the usual way. If $v = v_0 + iv_1$, where $v_0, v_1 \in \mathbf{R}^5$, then $\bar{v} = v_0 - iv_1$. We then have $Q^3 = \{[p] \in \mathbf{CP}^4 \mid q(p) = 0\}$ and $S^3 = Q^3 \cap \mathbf{RP}^4$, where $\mathbf{RP}^4$ is the projectivization of $\mathbf{R}^5$. Let $\xi \in \mathbf{R}^5 \backslash \{0\}$ satisfy $\langle \xi, \xi \rangle = 0$. Then $[\xi] \in S^3$. We set $Q_\xi^2 = \{[p] \in Q^3 \mid \langle \xi, \rho \rangle = 0\}$. If $[p] \in Q^3 \backslash Q_\xi^2$, then there exists a unique $p \in [p]$ so that $\langle \xi, p \rangle = 1$. We use this to define a 'real projection' $r_\xi \colon Q^3 \to S^3$ by

$$r_\xi([p]) = \begin{cases} [\xi] & \text{if } [p] \in Q_\xi^2, \\ [p + \bar{p} - \tfrac{1}{2}\langle p, \bar{p} \rangle \xi] & \text{if } [p] \in Q^3 \backslash Q_\xi^2. \end{cases}$$

It is easy to see that r_ξ is the identity on $S^3 \subseteq Q^3$ and is smooth on $Q^3 - \{[\xi]\}$. Moreover, r_ξ is a submersion on $Q^3 - Q_\xi^2$. In fact, under the natural identifications $Q^3 - Q_\xi^2 \simeq \mathbf{C}^3$ and $S^3 - \{[\xi]\} \simeq \mathbf{R}^3$, r_ξ simply becomes $\mathrm{Re} \colon \mathbf{C}^3 \to \mathbf{R}^3$. (We warn the reader that r_ξ is not continuous at $[\xi] \in Q^3$.) At points $[p] \in Q_\xi^2 - \{[\xi]\}$, the real rank of dr_ξ is 2 with the kernel being given by the tangent space to Q_ξ^2 at $[p]$.

We have now arrived at the following proposition:

PROPOSITION 4. *Let M be a compact Riemann surface, let $\gamma: M \to Q^3$ be a holomorphic null immersion, and let $[\xi] \in S^3$ be any point so that $\gamma(M)$ intersects Q_ξ^2 transversely. Then $r_\xi \circ \gamma: M \to S^3$ is a conformal immersion which is $\mathscr{W}$-critical and which satisfies $\mathscr{W}(r_\xi \circ \gamma) = 4\pi \deg(\gamma)$.* ☐

Note that, as ξ varies, the immersions $r_\xi \circ \gamma$ are not in general conformally equivalent. (Just consider the case of $\gamma_2: \mathbf{P}^1 \to Q^3$.)

We now turn to another application of this holomorphic representation of $\mathscr{W}$-critical immersions. Recall the classical fact that any immersion of $\mathbf{RP}^2$ into $\mathbf{R}^3$ or S^3 must have at least one triple point [Ba]. By the lower bound in §1, it follows that $\mathscr{W}(X) \geq 12\pi$ for any immersion $X: \mathbf{RP}^2 \to \mathbf{R}^3$. If any immersion can be found satisfying $\mathscr{W}(X) = 12\pi$, then it is certainly $\mathscr{W}$-critical. If we let $\tau: S^2 \to \mathbf{RP}^2$ denote the usual double cover, then $\tilde{X} = X \circ \tau: S^2 \to \mathbf{R}^3$ is a $\mathscr{W}$-critical immersion satisfying $\mathscr{W}(\tilde{X}) = 24\pi$ and must hence come from a holomorphic null immersion $\gamma: \mathbf{P}^1 \to Q^3$ of degree 6. We have already seen in Proposition 3 that, up to the action of $SO(5, \mathbf{C})$ on Q^3, there is a unique holomorphic null immersion of degree 6. Thus $\gamma = A \circ \gamma_3$ where γ_3 is as described above. Of course, $SO(5, \mathbf{C})$ does not preserve the real locus $S^3 \subseteq Q^3$, so some care must be exercised in making algebraic normalizations.

Nevertheless, using the fact that $\gamma = A \circ \gamma_3$, we can arrive at the following formula: There exists a holomorphic parameter z on $\mathbf{P}^1$ so that the antipodal map $\sigma: \mathbf{P}^1 \to \mathbf{P}^1$ which serves as the deck transformation of $\tau: \mathbf{P}^1 \to \mathbf{RP}^2$ is expressed in the form $\sigma(z) = -1/\bar{z}$. Let e_0, e_1, and e_3 denote vectors in $\mathbf{C}^5$ which satisfy the conditions that $e_3 = \bar{e}_3$; e_0 and e_1 span a null plane in $\mathbf{C}^5$ which is perpendicular to e_3 and is transverse to its conjugate plane; and the remaining unspecified products are $e_0 \cdot \bar{e}_1 = 0$, $e_0 \cdot \bar{e}_0 = -4$, $e_1 \cdot \bar{e}_1 = -9$, and $e_3 \cdot e_3 = 10$. Then it is easy to see that the curve

$$(2) \qquad \gamma(z) = [e_0 + e_1 z + i e_3 z^3 - \bar{e}_1 z^5 + \bar{e}_0 z^6]$$

is a holomorphic null immersion $\gamma: \mathbf{P}^1 \to Q^3$ (with simple flexes at $z = 0$ and $z = \infty$) which satisfies $\gamma \circ \sigma = \bar{\gamma}$. Moreover γ has degree 6. A little algebraic argument involving the real form $SO(4, 1)$ of $SO(5, \mathbf{C})$ shows that any holomorphic null immersion $\gamma: \mathbf{P}^1 \to Q^3$ which is of degree 6 and satisfies $\gamma \circ \sigma = \bar{\gamma}$ is, up to reparametrization and a transformation by an element of $SO(4, 1)$, the γ described above. For the generic $[\xi] \in S^3$, it is easy to see that $r_\xi \circ \gamma: \mathbf{RP}^2 \to S^3$ is a $\mathscr{W}$-critical immersion with $\mathscr{W}(r_\xi \circ \gamma) = 12\pi$, the absolute minimum. However, when $\xi = 3e_3 + \sqrt{5}(e_1 + \bar{e}_1)$, the singular quadric $Q_\xi^2 \subseteq Q^3$ is tangent to γ at the points $\gamma(0)$, $\gamma(\infty)$. For this value of ξ, $r_\xi \circ \gamma$ is a branched conformal immersion of $\mathbf{RP}^2$ into S^3. We have now shown all of the pieces of the following proposition.

PROPOSITION 5. *Up to conformal equivalence, all of the conformal immersions $X: \mathbf{RP}^2 \to S^3$ for which $\mathscr{W}(X)$ assumes its minimum value of 12π are of the form $r_\xi \circ \gamma$ where γ is as in (2) above and $[\xi] \in S^3$. However, not all $[\xi] \in S^3$*

give an immersion. It follows that, even after conformal reparametrization, the moduli space of $\mathscr{W}$-minimizing immersions of $\mathbf{RP}^2$ into S^3 is not compact. $\square$

We should remark that R. Kusner has independently demonstrated the existence of an immersion $X: \mathbf{RP}^2 \to S^3$ with $\mathscr{W}(X) = 12\pi$, see [**K**]. His proof uses the holomorphic representation also. The above argument constructs the complete moduli space. The fact that this space is *not* compact, even after reducing modulo the obvious conformal symmetries, shows that the usual variational approach of constructing a solution by passing to the limit of a 'suitably renormalized' minimizing sequence would fail to construct solutions for $\mathbf{RP}^2$. Nevertheless, it is intriguing that the associated holomorphic object, γ, *is* unique modulo the conformal symmetry group $SO(4, 1)$.

Perhaps some information about the failure of compactness can be gained by examining just how singularities develop as one approaches 'nontransverse' $[\xi] \in S^3$. This is the subject of a computer graphics project currently being carried out by the author and G. Yates Fletcher. Our results will be reported on elsewhere.

BIBLIOGRAPHY

[**Ba**] T. Banchoff, *Triple points and surgery of immersed surfaces*, Proc. Amer. Math. Soc. **46** (1974), 407–413.

[**BR**] L. Berzolari and K. Rohn, *Algebraische Raumkurven und abwickelbare Fläschen*, Enzyklopädie der Math. Wissenschaften, vol. III, C9, Teubner, Leipzig, pp. 1359–1387.

[**Br**] R. Bryant, *A duality theorem for Willmore surfaces*, J. Differential Geometry **20** (1984), 23–53.

[**G**] M. Gromov, *Filling Riemannian manifolds*, J. Differential Geometry **18** (1983), 1–147.

[**GH**] P. Griffiths and J. Harris, *Principles of algebraic geometry*, Wiley-Interscience, New York, 1978.

[**K**] R. Kusner, *Conformal geometry and complete minimal surfaces*, Bull. Amer. Math. Soc. (to appear)

[**LY**] P. Li and S.-T. Yau, *A new conformal invariant and its applications to the Willmore conjecture and the first eigenvalue of compact surfaces*, Invent. Math. **69** (1982), 269–291.

[**MR**] S. Montiel and A. Ros, *Minimal immersions of surfaces by the first eigenfunctions and conformal area*, Invent. Math. **83** (1986), 153–166.

[**P**] C. K. Peng, *Some new examples of minimal surfaces in $\mathbf{R}^3$ and its applications*, preprint, MSRI, 1986.

[**PS**] U. Pinkall and I. Sterling, *Willmore surfaces*, preprint, Max-Planck-Institut für Mathematik, Bonn, 1987.

[**Sc**] C. Schoen, private communication, fall 1984.

[**Si**] L. Simon, *Existence of Willmore surfaces*, preprint.

[**SI**] A. E. Soufi and S. Ilias, *Immersions minimales, première valeur propre du laplacien, et volume conforme*, Institut Fourier, preprint, no. 42, 1985.

[**T**] G. Thomsen, *Über konforme Geometrie I: Grundlagen der konformen Flächentheorie*, Abh. Math. Sem. Hamburg (1923), 31–56.

[**V**] J.-L. Verdier, *Two dimensional σ-models and harmonic maps from S^2 to S^{2n}*, Lecture Notes in Phys., vol. 180, Springer-Verlag, 1982, pp. 136–141.

[**W**] T. Willmore, *A note on imbedded surfaces*, An. Ştiinţ. Univ. "Al. I. Cuza" Iaşi Secţ. I a Mat. **11** (1965), 493–496.

DUKE UNIVERSITY

Proceedings of Symposia in Pure Mathematics
Volume 48 (1988)

Algebraic Cycles, Bott Periodicity, and the Chern Characteristic Map

H. BLAINE LAWSON, JR. AND MARIE-LOUISE MICHELSOHN

1. Introduction. It is a fundamental fact that there exists a direct relationship between K-theory and algebraic cycles on an algebraic variety. The correspondence comes essentially by associating to each subvariety the resolution of its structure sheaf. Our purpose here is to show that in a certain universal setting there exists a completely different, but also direct, connection between K-theory and cycles. Our point of departure is the following fact, established in [11]. Let $\mathscr{C}_d^q(\mathbf{P}^n)$ denote the Chow variety of positive algebraic cycles of dimension $n - q$ and of degree d in $\mathbf{P}^n$, and define

$$\mathscr{C}^q(\mathbf{P}^n) = \lim_{d \to \infty} \mathscr{C}_d^q(\mathbf{P}^n),$$

where we consider $\mathscr{C}_d^q(\mathbf{P}^n) \subset \mathscr{C}_{d+1}^q(\mathbf{P}^n)$ by associating $c \mapsto c + l_0$ for some fixed linear subspace l_0 of codimension q.

THEOREM 1.1 [**10, 11**]. *For all $n \geq q$ there is a homotopy equivalence*

$$\mathscr{C}^q(\mathbf{P}^n) \sim K(\mathbf{Z}, 2) \times K(\mathbf{Z}, 4) \times \cdots \times K(\mathbf{Z}, 2q).$$

The inclusion of the degree-one cycles into this limit constitutes a natural embedding

$$\mathscr{G}^q(\mathbf{P}^n) \subset \mathscr{C}^q(\mathbf{P}^n)$$

of the Grassmannian of codimension-q planes. Letting $n \to \infty$ and applying Theorem 1.1 yields a map

(1.2) $$BU_q \xrightarrow{c} K(\mathbf{Z}, 2) \times K(\mathbf{Z}, 4) \times \cdots \times K(\mathbf{Z}, 2q)$$

where $BU_q = \lim_{n \to \infty} \mathscr{G}^q(\mathbf{P}^n)$ is the classifying space for the unitary group U_q. The Eilenberg–Mac Lane spaces $K(\mathbf{Z}, m)$ are classifying spaces for integral cohomology. In particular the map c represents an element in $H^{\mathrm{even}}(BU_q; \mathbf{Z})$. Our first result is the following.

1980 *Mathematics Subject Classification* (1985 *Revision*). Primary 14C05; Secondary 55P15, 55Q52.

Research partially supported by NSF Grant No. DMS 8602645.

THEOREM 1.3. *The map c represents the total Chern class of the universal complex q-plane bundle over BU_q.*

Letting $q \to \infty$ yields a map

(1.4) $$BU \xrightarrow{c} K(\mathbf{Z}, \text{ev})$$

where BU is the classifying space for reduced K-theory and $K(\mathbf{Z}, \text{ev}) = \prod_k K(\mathbf{Z}, 2k)$. The fundamental results of Bott concerning the homotopy of the unitary group [1, 2] can be recast in this setting as follows.

THEOREM 1.5. *The map c is injective on homotopy groups. For each $q \geq 1$, the map*

$$\mathbf{Z} \cong \pi_{2q}(BU) \xrightarrow{c_*} \pi_{2q}(K(\mathbf{Z}, \text{ev})) \cong \mathbf{Z}$$

is multiplication by $(q-1)!$.

There is a natural biadditive operation on algebraic cycles, called the *complex join*

(1.6) $$\divideontimes_{\mathbf{C}} : \mathscr{C}_d^q(\mathbf{P}^n) \times \mathscr{C}_{d'}^{q'}(\mathbf{P}^{n'}) \to \mathscr{C}_{dd'}^{q+q'}(\mathbf{P}^{n+n'+1}),$$

which was introduced in [11] and played a role in generalizing Theorem 1.1 to arbitrary projective varieties. (See Theorem 2.15 below.) This generalization suggests a certain "homology theory" for algebraic varieties which has been developed in a very general setting by Eric Friedlander [7, 8]. Furthermore, Friedlander and B. Mazur have shown that the complex join construction above yields a fundamental operation in this theory which is a reflection of the intersection pairing on cycles. Rationally this pairing can be computed from the universal formula for the rational Chern classes of the Whitney sum of two vector bundles. This was observed by Friedlander who suggested that perhaps our methods could lead to a complete calculation of this operation in the universal case. This was so. In fact we shall present here a "cycle-theoretic" proof of the Whitney sum formula for integral Chern classes.

The definition of the Friedlander-Mazur operations requires passing to some form of group completion. For this reason we consider in §5 the group $\widetilde{\mathscr{C}}^q(\mathbf{P}^n)$ of all codimension-q cycles of degree 0 in $\mathbf{P}^n$. When endowed with the weak topology for the "Chow filtration", this group is shown to be homotopy equivalent to $\mathscr{C}^q(\mathbf{P}^n)$ under a natural embedding $\mathscr{C}^q(\mathbf{P}^n) \hookrightarrow \widetilde{\mathscr{C}}^q(\mathbf{P}^n)$. The complex join naturally extends to a biadditive continuous map

(1.7) $$\divideontimes_{\mathbf{C}} : \widetilde{\mathscr{C}}^q(\mathbf{P}^n) \times \widetilde{\mathscr{C}}^{q'}(\mathbf{P}^{n'}) \to \widetilde{\mathscr{C}}^{q+q'}(\mathbf{P}^{n+n'+1})$$

which represents, at the homotopy level, the intersection pairing on cycles. For all $n \geq q$ we have a homotopy equivalence

$$\widetilde{\mathscr{C}}^q(\mathbf{P}^n) \sim \prod{}^q \overset{\text{def}}{=} K(\mathbf{Z}, 2) \times K(\mathbf{Z}, 4) \times \cdots \times K(\mathbf{Z}, 2q),$$

and the pairing (1.7) gives a map

(1.8) $$\divideontimes_{\mathbf{C}} : \prod{}^q \wedge \prod{}^{q'} \to \prod{}^{q+q'}$$

which is independent of n and n' (up to homotopy). To determine the map completely it suffices to compute the cohomology class $(\divideontimes_{\mathbf{C}})^*(\iota)$ where $\iota = \iota_2 + \iota_4 + \cdots + \iota_{2(q+q')}$ and where ι_{2k} is the generator of $H^{2k}(\prod^{q+q'}; \mathbf{Z}) \cong \mathbf{Z}$. In §6 we prove the following.

THEOREM 1.9.

$$(\divideontimes_{\mathbf{C}})^*(\iota_{2k}) = \sum_{\substack{r+s=k \\ r>0 \\ s>0}} \iota_{2r} \otimes \iota_{2s}.$$

In particular, there are no exotic torsion classes hidden in the join pairing.

The formula in (1.9) is directly related to the Chern characteristic map $BU_q \to \prod^q$ obtained from the inclusion $\mathscr{G}^q(\mathbf{P}^n) \hookrightarrow \widetilde{\mathscr{C}}^q(\mathbf{P}^n)$ by letting $n \to \infty$. We define $\mu : \prod^q \times \prod^{q'} \to \prod^{q+q'}$ by $\mu(c, c') = c \divideontimes_{\mathbf{C}} c' + j(c) + j'(c')$, where $j : \prod^q \to \prod^{q+q'}$, $j' : \prod^{q'} \to \prod^{q+q'}$ are the natural inclusions as factors in the product.

THEOREM 1.10. *There is a commutative diagram*

$$
\begin{array}{ccc}
BU_q \times BU_{q'} & \xrightarrow{\oplus} & BU_{q+q'} \\
\downarrow & & \downarrow \\
\prod^q \times \prod^{q'} & \xrightarrow{\mu} & \prod^{q+q'}
\end{array}
$$

where $\oplus$ denotes the Whitney sum operation and μ is defined as above. The map μ satisfies the formula

$$\mu^*(\iota_{2k}) = \sum_{\substack{r+s=k \\ r\geq 0 \\ s\geq 0}} \iota_{2r} \otimes \iota_{2s}.$$

The existence of this commutative diagram was first observed and shown to us by Eric Friedlander. Friedlander also pointed out that the rational form of the map μ is a direct consequence of this diagram.

In §7 we consider the spaces

$$\mathscr{D}(d) = \lim_{p,q \to \infty} \mathscr{C}^q_d(\mathbf{P}^{q+p})$$

of cycles of a fixed degree. These spaces determine a natural filtration

(1.11) $BU = \mathscr{D}(1) \subset \mathscr{D}(2) \subset \cdots \subset \mathscr{D}(\infty) = K(\mathbf{Z}, \text{ev}).$

We show that the complex join gives a pairing

$$\mathscr{D}(d) \times \mathscr{D}(d') \xrightarrow{\divideontimes_{\mathbf{C}}} \mathscr{D}(dd')$$

for all $d, d' \geq 1$. We also define and compute pairings $\mu_{d,d'}$ on $\mathscr{D}(\infty)$ such that the diagrams

$$
\begin{array}{ccc}
\mathscr{D}(d) \times \mathscr{D}(d') & \xrightarrow{\divideontimes_{\mathbf{C}}} & \mathscr{D}(dd') \\
\cap & & \cap \\
\mathscr{D}(\infty) \times \mathscr{D}(\infty) & \xrightarrow{\mu_{d,d'}} & \mathscr{D}(\infty)
\end{array}
$$

commute. These generalize directly our cycle-theoretic proof of the product formula for Chern classes given in Theorem 1.10.

The authors would like to thank H. Hironaka, N. Shimada, B. Mazur and B. Mann for helpful remarks. We would like to express particular gratitude to Eric Friedlander for his continuing interest in this work and for many valuable and stimulating conversations.

2. Cycle spaces, stability theorems, and fundamental operations. In this section we present a summary of the main ideas and results of [10] and [11]. The discussion will be focused on algebraic varieties defined in complex projective n-space $\mathbf{P}^n$. Recall that $\mathbf{P}^n$ is the manifold of all 1-dimensional linear subspaces of $\mathbf{C}^{n+1}$. There is a natural map

$$(2.1) \qquad\qquad \pi : \mathbf{C}^{n+1} - \{0\} \to \mathbf{P}^n$$

which assigns to a vector $z \neq 0$ the complex line: $\pi(z) = [z] \equiv \mathrm{span}_{\mathbf{C}}(z)$. The points of $\mathbf{C}^{n+1} - \{0\}$ are called *homogeneous coordinates* for $\mathbf{P}^n$.

DEFINITION 2.2. An *algebraic subvariety* of $\mathbf{P}^n$ is a closed subset $V \subset \mathbf{P}^n$ with the property that there is a finite set of homogeneous polynomials $p_1, \ldots, p_r \in \mathbf{C}[Z_0, \ldots, Z_n]$ so that

$$\pi^{-1}(V) = \{z \in \mathbf{C}^{n+1} - \{0\} : p_1(z) = \cdots = p_r(z) = 0\}.$$

A classical theorem of Chow states that any complex analytic subvariety of $\mathbf{P}^n$ is algebraic. It is a basic fact (cf. [14]) that any such subvariety $V \subset \mathbf{P}^n$ has a canonical decomposition

$$V = \mathrm{Reg}(V) \amalg \mathrm{Sing}(V),$$

where $\mathrm{Reg}(V)$ consists of those points of V for which there is a neighborhood $\mathscr{U}$ in $\mathbf{P}^n$ such that $V \cap \mathscr{U}$ is a smooth submanifold of $\mathscr{U}$. The set $\mathrm{Reg}(V)$ is relatively open in V, and the set $\mathrm{Sing}(V)$ is itself an algebraic subvariety of $\mathbf{P}^n$. If $\mathrm{Reg}(V)$ is connected, then V is said to be *irreducible*. Each irreducible subvariety V has a canonical *dimension* p $(= \dim_{\mathbf{C}}(\mathrm{Reg}(V)))$ and determines a canonical homology class $[V] \in H_{2p}(\mathbf{P}^n; \mathbf{Z})$. This class can be written as $[V] = d[\mathbf{P}^p]$, where $\mathbf{P}^p \subset \mathbf{P}^n$ is any p-dimensional linear subspace. The integer d is called the *degree* of V. Of course the concepts of irreducibility, dimension, and degree all have equivalent, purely algebraic definitions.

Suppose we want to put a topology on the space of algebraic varieties. Then we must take into account that the algebraically defined family of irreducible varieties can "degenerate", or "specialize", to a reducible one. This leads quickly to the consideration of algebraic cycles. By definition an (effective) *algebraic cycle of dimension p* in $\mathbf{P}^n$ is a formal finite sum $c = \sum n_\alpha V_\alpha$, where for each α, n_α is a *positive* integer and $V_\alpha \subset \mathbf{P}^n$ is an irreducible p-dimensional algebraic subvariety. The *degree* of such a cycle $c = \sum n_\alpha V_\alpha$ is the integer

$$\deg(c) = \sum n_\alpha \deg(V_\alpha).$$

For fixed p and d the set of all algebraic cycles of dimension p and degree d in $\mathbf{P}^n$ will be denoted by $\mathscr{C}_{p,d}(\mathbf{P}^n)$.

Each of the spaces $\mathscr{C}_{p,d}(\mathbf{P}^n)$ can itself be given the structure of a projective algebraic variety (cf. [12]), and is called a *Chow variety*. Some of these can be identified with well-known spaces. For example, $\mathscr{C}_{n-1,d}(\mathbf{P}^n) \cong \mathbf{P}^{\binom{n+d}{d}-1}$ and $\mathscr{C}_{p,1}(\mathbf{P}^n)$ = the Grassmannian of $(p+1)$-planes in $\mathbf{C}^{n+1}$. Furthermore, one easily sees that $\mathscr{C}_{0,d}(\mathbf{P}^n) = SP^d(\mathbf{P}^n)$ = the d-fold symmetric product of $\mathbf{P}^n$. In general however, $\mathscr{C}_{p,d}(\mathbf{P}^n)$ is both singular and reducible, and its structure as a variety, or even as a topological space, has been a subject of much interest over the years.

We shall be concerned here with the structure of $\mathscr{C}_{p,d}(\mathbf{P}^n)$ in the limit as certain of the integers become large. The first case is the following. Fix a p-dimensional linear subspace $l_0 \subset \mathbf{P}^n$ and consider the sequence of embeddings

$$(2.3) \qquad \cdots \subset \mathscr{C}_{p,d}(\mathbf{P}^n) \subset \mathscr{C}_{p,d+1}(\mathbf{P}^n) \subset \cdots$$

given by mapping c to $c + l_0$. We then define

$$\mathscr{C}_p(\mathbf{P}^n) = \lim_{\substack{\rightarrow \\ d}} \mathscr{C}_{p,d}(\mathbf{P}^n)$$

and endow $\mathscr{C}_p(\mathbf{P}^n)$ with the *weak topology*. This is the topology in which a set F is closed iff $F \cap \mathscr{C}_{p,d}(\mathbf{P}^n)$ is closed for all d. It has the property that any compact subset $K \subset \mathscr{C}_p(\mathbf{P}^n)$ must be contained in $\mathscr{C}_{p,d}(\mathbf{P}^n)$ for some d. (In general all direct limits of spaces here will be given the weak topology.) The space $\mathscr{C}_p(\mathbf{P}^n)$ is an abelian topological monoid under the formal sum $+$. Our first main result is the following.

THEOREM 2.4 [10, 11]. *For each p with $0 \leq p < n$, there exists a homotopy equivalence*

$$\mathscr{C}_p(\mathbf{P}^n) \sim K(\mathbf{Z}, 2) \times K(\mathbf{Z}, 4) \times \cdots \times K(\mathbf{Z}, 2(n-p))$$

where $K(\mathbf{Z}, r)$ denotes the standard Eilenberg–Mac Lane space.

Recall that for a positive integer r and a finitely generated abelian group G, the space $K(G, r)$ is a countable CW-complex uniquely defined up to homotopy type by the requirement that: $\pi_r K(G, r) = G$ and $\pi_s K(G, r) = 0$ if $s \neq r$. These spaces have the important universal property that for any connected countable CW-complex X there is a natural one-to-one correspondence

$$(2.5) \qquad H^r(X; G) \cong [X, K(G, r)],$$

where $[X, Y]$ denotes homotopy classes of base-point preserving maps from X to Y. The cohomology of $K(\mathbf{Z}, r)$ thereby corresponds to certain universal cohomology operations. In general it is quite complicated but well understood. It is shown in [11] that for each d the map

$$(2.6) \qquad H^k(\mathscr{C}_p(\mathbf{P}^n); \mathbf{Z}) \rightarrow H^k(\mathscr{C}_{p,d}(\mathbf{P}^n); \mathbf{Z})$$

is injective for $k < 2d$. Hence, Theorem 2.4 yields concrete information about the topological complexity of the classical Chow varieties.

In the case that $p = 0$, Theorem 2.4 reduces to a special case of the following classical result of Dold and Thom [5, 6]. Given any topological space X with base point x_0, one defines the *infinite symmetric product* of X to be the limit

$$\mathrm{SP}(X) = \varinjlim \mathrm{SP}^d(X),$$

where as above the inclusions $j : \mathrm{SP}^d(X) \subset \mathrm{SP}^{d+1}(X)$ are defined by $j(\sum n_d x_d) = x_0 + \sum n_d x_d$.

THE DOLD-THOM THEOREM. *Let X be a connected, countable CW-complex with base point. Then there is a natural isomorphism*

$$\pi_*(\mathrm{SP}(X)) \xrightarrow{\approx} H_*(X; \mathbf{Z}).$$

Moreover, if X is a finite complex, then there is a homotopy equivalence $\mathrm{SP}(X) \sim \prod_k K(H_k(X; \mathbf{Z}), k)$.

We shall now examine certain synthetic operations on projective algebraic cycles which were introduced in [11]. Fix a pair of disjoint linear subspaces $\mathbf{P}^n$ and $\mathbf{P}^{n'}$ in $\mathbf{P}^{n+n'+1}$ and consider the set of all projective lines in $\mathbf{P}^{n+n'+1}$ joining $\mathbf{P}^n$ to $\mathbf{P}^{n'}$. Any two distinct lines of this type are either disjoint or they meet in exactly one point, which must belong to either $\mathbf{P}^n$ or $\mathbf{P}^{n'}$. In this sense $\mathbf{P}^{n+n'+1}$ can be considered to be the "complex join" of $\mathbf{P}^n$ and $\mathbf{P}^{n'}$. Note that linear projection in homogeneous coordinates yields a pair of vector bundles

$$\mathbf{P}^{n+n'+1} - \mathbf{P}^{n'} \xrightarrow{\sigma} \mathbf{P}^n \quad \text{and} \quad \mathbf{P}^{n+n'+1} - \mathbf{P}^n \xrightarrow{\sigma'} \mathbf{P}^{n'},$$

where for $x \in \mathbf{P}^n$ and $x' \in \mathbf{P}^{n'}$

$$\begin{cases} \overline{\sigma^{-1}(x)} = \text{the union of all lines from } x \text{ to } \mathbf{P}^{n'}, \\ \overline{(\sigma')^{-1}(x')} = \text{the union of all lines from } x' \text{ to } \mathbf{P}^n. \end{cases}$$

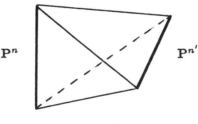

FIGURE

DEFINITION 2.7. Given subsets $V \subset \mathbf{P}^n$ and $V' \subset \mathbf{P}^{n'}$, the *complex join* of V and V' is the subset $V \divideontimes_{\mathbf{C}} V' \subset \mathbf{P}^{n+n'+1}$ consisting of the union of all lines joining V to V'.

Note that if V and V' are closed subsets, then

$$V \divideontimes_{\mathbf{C}} V' = \overline{(\sigma^{-1}V)} \cap \overline{((\sigma')^{-1}V')}.$$

In homogeneous coordinates the complex join can be described simply as follows. Given $V \subset \mathbf{P}^n$, define the *homogeneous cone* of V to be the set $C(V) = \pi^{-1}(V) \cup \{0\}$, where π is the projection (2.1). Then

$$(2.8) \qquad C(V \divideontimes_{\mathbf{C}} V') = C(V) \times C(V').$$

LEMMA 2.9. *If $V \subset \mathbf{P}^n$ and $V' \subset \mathbf{P}^{n'}$ are irreducible algebraic subvarieties, then $V \divideontimes_{\mathbf{C}} V' \subset \mathbf{P}^{n+n'+1}$ is also an irreducible algebraic subvariety with*

$$\dim(V \divideontimes_{\mathbf{C}} V') = \dim V + \dim V' + 1$$

and

$$\deg(V \divideontimes_{\mathbf{C}} V') = (\deg V)(\deg V').$$

The obvious extension of the complex join construction gives a continuous, in fact algebraic, map

$$\mathscr{C}_{p,d}(\mathbf{P}^n) \times \mathscr{C}_{p',d'}(\mathbf{P}^{n'}) \to \mathscr{C}_{p+p'+1,dd'}(\mathbf{P}^{n+n'+1}).$$

DEFINITION 2.10. Fix a pair of disjoint linear subspaces $\mathbf{P}^n$, $\mathbf{P}^{k-1} \subset \mathbf{P}^{n+k}$. Then for any subset $V \subset \mathbf{P}^n$, we define the *k-fold complex suspension of V* to be the subset

$$\Sigma^k(V) = V \divideontimes_{\mathbf{C}} \mathbf{P}^{k-1}.$$

Note that when V is closed and $k = 1$, $\Sigma(V)$ is just the Thom space of the complex line bundle $\mathscr{O}(1)|_V$. Note also that for $V \subset \mathbf{P}^n$ and $V' \subset \mathbf{P}^{n'}$ as above, the definitions imply directly that

$$(2.11) \qquad V \divideontimes_{\mathbf{C}} V' = (\Sigma^{n'+1}V) \cap (\Sigma^{n+1}V').$$

We now move on to examine spaces of cycles contained in a given algebraic subvariety $X \subset \mathbf{P}^n$. The definition presented here was suggested by Eric Friedlander. It differs somewhat from the one given in [11] and is in many ways a vast improvement over it.

Let us fix an algebraic subvariety $X \subset \mathbf{P}^n$. An algebraic cycle $c = \sum n_\alpha V_\alpha$ in $\mathbf{P}^n$ is said to have *support in X* if $\bigcup_\alpha V_\alpha \subset X$. The set of all p-dimensional cycles with support in X will be denoted by $\mathscr{C}_{p,\cdot}(X)$. It is a closed subspace of $\mathscr{C}_{p,\cdot}(\mathbf{P}^n) = \amalg_d \mathscr{C}_{p,d}(\mathbf{P}^n)$ under the natural embedding. In fact each connected component of $\mathscr{C}_{p,\cdot}(X)$ is an *algebraic* subvariety of $\mathscr{C}_{p,d}(\mathbf{P}^n)$ for some d. Thus, under addition of cycles, $\mathscr{C}_{p,\cdot}(X)$ becomes an abelian topological monoid which can be written as a countable disjoint union of algebraic varieties: $\mathscr{C}_{p,\cdot}(X) = \amalg_{\alpha \in A} \mathscr{C}_{p,\alpha}(X)$, where $A = \pi_0(\mathscr{C}_{p,\cdot}(X))$ denotes the set of connected components.

For any such creature $C. = \amalg_\alpha C_\alpha$ there is a notion due to Bousfield and Kan [3] of the *homotopy direct limit*, which is constructed roughly as follows. For each $\alpha \in A$ choose an element $c_\alpha \in C_\alpha$ and consider the continuous map $t_\alpha : C. \times A \to C. \times A$ given by setting $t_\alpha(c, \alpha') = (c + c_\alpha, \alpha' + \alpha)$. Since each C_α is connected, these maps have the property that for each α and β the map $t_{\alpha+\beta}$ is homotopic to $t_\alpha \circ t_\beta$. Thus for any homotopy functor h we can define the direct limit

$$\varinjlim_\alpha h(C. \times \{\alpha\}).$$

The Bousfield-Kan construction gives a space $\underrightarrow{\mathrm{holim}}\,C.$, well defined up to homotopy type, such that

$$h(\underrightarrow{\mathrm{holim}}\,C.) = \underrightarrow{\lim}\,h(C. \times \{\alpha\}).$$

If the $\{c_\alpha\}$ can be chosen so that $c_{\alpha+\beta} = c_\alpha + c_\beta$ for all α, β (e.g., if A is finitely generated and free), then we can define

$$\underrightarrow{\lim}\,C. = \underrightarrow{\lim}_{\alpha}\,C. \times \{\alpha\}$$

directly, and it agrees up to homotopy with $\underrightarrow{\mathrm{holim}}\,C.$.

When applying this construction to cycles it is important to include the component $0 = \{\varnothing\}$ consisting of the unique p-cycle of degree 0. With this in mind we have the following.

DEFINITION 2.13. Given an algebraic subvariety $X \subset \mathbf{P}^n$ and an integer $p \geq 0$, the set $\mathscr{C}_p(X)$ is defined to be the connected component of $(0,0)$ in the homotopy limit $\underrightarrow{\mathrm{holim}}\,\mathscr{C}_{p,\cdot}(X)$.

It is straightforward to see that this agrees with our previous definition when $X = \mathbf{P}^n$.

Eric Friedlander has pointed out that there is a homotopy equivalence

$$\underrightarrow{\mathrm{holim}}\,\mathscr{C}_{p,\cdot}(X) \sim \Omega B\mathscr{C}_{p,\cdot}(X),$$

where for any topological monoid M, $\Omega BM = M^+$ is the group completion obtained by taking loops on the geometric realization of the simplicial space BM given by the standard bar construction. Friedlander has used this second approach to give a generalization of $\mathscr{C}_p(X)$ and the theorem below to varieties defined over any algebraically closed field [7, 8].

Given varieties $X \subset \mathbf{P}^n$ and $X' \subset \mathbf{P}^{n'}$, the complex join gives a map of monoids $\mathscr{C}_{p,\cdot}(X) \times \mathscr{C}_{p',\cdot}(X') \to \mathscr{C}_{p+p'+1,\cdot}(X \ast_{\mathbf{C}} X')$ which extends to a continuous map

(2.14) $\mathscr{C}_p(X) \times \mathscr{C}_{p'}(X') \to \mathscr{C}_{p+p'+1}(X \ast_{\mathbf{C}} X').$

In particular, taking $n' = 0$, we get a map

$$\mathscr{C}_p(X) \xrightarrow{\Sigma} \mathscr{C}_{p+1}(\Sigma X)$$

given by complex suspension of cycles. Note that $\Sigma^k X = \Sigma(\Sigma(\dots(\Sigma X)\dots))$. The central result of [11] asserts the following.

THEOREM 2.15. *For any algebraic subvariety $X \subset \mathbf{P}^n$ and any $k > 0$ the map*

$$\mathscr{C}_p(X) \xrightarrow{\Sigma^k} \mathscr{C}_{p+k}(\Sigma^k X)$$

is a homotopy equivalence.

NOTE. This was originally stated for varieties X containing a p-dimensional linear subspace; cf. [10]. However, as Friedlander points out, the proofs carry over directly to the general situation.

NOTE. Setting $X = \mathbf{P}^q$ (and thus $\Sigma^k X = \mathbf{P}^{k+q}$), we recover Theorem 2.4 from Theorem 2.15 and the Dold-Thom Theorem.

COROLLARY 2.16. *Let $X \subset \mathbf{P}^n$ be any connected algebraic variety. Then for each $k \geq 0$ there is a natural isomorphism*

$$\pi_*(\mathscr{C}_k(\Sigma^k X)) \cong H_*(X; \mathbf{Z}).$$

All the above suggests that for an irreducible subvariety X of dimension N we define

$$\mathscr{C}^q(X) \equiv \mathscr{C}_p(X),$$

where $q = N - p$ is the codimension of the cycles in X. The "Stability Theorem" 2.16 then states that the complex suspension map

$$(2.17) \qquad\qquad \mathscr{C}^q(X) \xrightarrow[\sim]{\Sigma} \mathscr{C}^q(\Sigma X)$$

is a homotopy equivalence. This allows one to define *stable* cycle spaces

$$(2.18) \qquad\qquad \mathfrak{C}^q(X) = \lim_{k \to \infty} \mathscr{C}^q(\Sigma^k X)$$

for any $q \geq 0$. (This definition was suggested by E. Friedlander and B. Mazur.) For example, by Theorem 2.4 we have that $\mathfrak{C}^q(\mathbf{P}^0) \sim K(\mathbf{Z}, 2) \times K(\mathbf{Z}, 4) \times \cdots \times K(\mathbf{Z}, 2q)$. Note also that the pairing (2.14) can be rewritten as

$$\mathfrak{C}^q(X) \times \mathfrak{C}^{q'}(X') \to \mathfrak{C}^{q+q'}(X \bowtie_{\mathbf{C}} X').$$

It has been observed by E. Friedlander and B. Mazur [9] that via the Stability Theorem, this yields a pairing

$$(2.19) \qquad\qquad \mathfrak{C}^q(\mathbf{P}^0) \times \mathfrak{C}^{q'}(X) \to \mathfrak{C}^{q+q'}(X)$$

which factors through the smash product

$$\mathfrak{C}^q(\mathbf{P}^0) \wedge \mathfrak{C}^{q'}(X) \to \mathfrak{C}^{q+q'}(X)$$

and therefore induces a map on homotopy groups. Friedlander and Mazur have computed the bigraded ring $\pi_*\mathfrak{C}^*(\mathbf{P}^0)$ and progress has been made in understanding $\pi_*\mathfrak{C}^*(X)$ as a bigraded $\pi_*\mathfrak{C}^*(\mathbf{P}^0)$-module. Note that via (2.11) and the Stability Theorem, this module multiplication corresponds essentially to intersection of cycles.

3. The Chern characteristic map. Recall that the cycles of degree one in $\mathbf{P}^n$ are exactly the linear subspaces. Hence, there is a natural identification of $\mathscr{C}_1^q(\mathbf{P}^n) \stackrel{\mathrm{def}}{=} \mathscr{C}_{n-q,1}(\mathbf{P}^n)$ with the Grassmannian

$$\mathscr{G}^q(\mathbf{P}^n) = U(n+1)/U(n+1-q) \times U(q)$$

of linear subspaces of codimension q in $\mathbf{C}^{n+1}$. The inclusion of $\mathscr{C}_1^q(\mathbf{P}^n)$ into the monoid of all codimension-q cycles gives a canonical map

$$(3.1) \qquad \mathscr{G}^q(\mathbf{P}^n) \hookrightarrow \mathscr{C}^q(\mathbf{P}^n) \sim K(\mathbf{Z}, 2) \times K(\mathbf{Z}, 4) \times \cdots \times K(\mathbf{Z}, 2q).$$

Now the homotopy equivalence on the right is independent of n, so we are naturally led to take a limit as $n \to \infty$. To do this we fix an infinite flag

$$(3.2) \qquad\qquad \mathbf{C}^{q+1} \subset \mathbf{C}^{q+2} \subset \mathbf{C}^{q+3} \subset \cdots$$

and a family of compatible splittings: $\mathbf{C}^{n+1} = \mathbf{C}^n \oplus \mathbf{C}$ for each n. Sending a linear subspace $V \subset \mathbf{C}^n$ to $V \oplus \mathbf{C} \subset \mathbf{C}^n \oplus \mathbf{C}$ gives an inclusion

$$(3.3) \qquad \mathscr{G}^q(\mathbf{P}^{n-1}) \hookrightarrow \mathscr{G}^q(\mathbf{P}^n)$$

which is exactly the complex suspension map to the point $\mathbf{P}^0 = [0 \oplus \mathbf{C}]$ in $\mathbf{P}^n$. This yields a commutative diagram

$$(3.4) \qquad
\begin{array}{ccc}
\mathbf{P}^q & \to & \mathscr{C}^q(\mathbf{P}^q) \\
\not\cong \uparrow & & \updownarrow \not\cong \\
\vdots & & \vdots \\
\not\cong \uparrow & & \updownarrow \not\cong \\
\mathscr{G}^q(\mathbf{P}^{n-1}) & \to & \mathscr{C}^q(\mathbf{P}^{n-1}) \\
\not\cong \uparrow & & \updownarrow \not\cong \\
\mathscr{G}^q(\mathbf{P}^n) & \to & \mathscr{C}^q(\mathbf{P}^n) \\
\not\cong \uparrow & & \updownarrow \not\cong \\
\vdots & & \vdots
\end{array}$$

where each natural map on the right is a homotopy equivalence. Passing to the limit in (3.4) gives a canonical map

$$(3.5) \qquad BU_q \to K(\mathbf{Z}, 2) \times K(\mathbf{Z}, 4) \times \cdots \times K(\mathbf{Z}, 2q)$$

where $BU_q = \varinjlim_n \mathscr{G}^q(\mathbf{P}^n)$, the Grassmannian of q-planes in $\mathbf{C}^\infty$, is the classifying space for the unitary group U_q. Recall that over BU_q there is a universal complex q-plane bundle $\xi_q \to BU_q$ whose fibre at a plane P consists of all vectors in P, i.e., is tautologically P itself. Recall also that the homotopy class of the map (3.5) corresponds canonically to an element in $H^2(BU_q; \mathbf{Z}) \oplus \cdots \oplus H^{2q}(BU_q; \mathbf{Z})$. (See (2.5).) The main result of this section is the following.

THEOREM 3.6. *The integral cohomology class determined by the canonical cycle map (3.5) is exactly the total Chern class of the universal q-plane bundle ξ_q over BU_q.*

PROOF. Essentially the argument proceeds by induction on q. Consider the infinite subflag: $\mathbf{C}^q \subset \mathbf{C}^{q+1} \subset \mathbf{C}^{q+2} \subset \cdots$ of the flag (3.2) defined by taking the first coordinate in each space to be zero. Considering cycles of codimension $(q-1)$ in these spaces gives us a commutative diagram

$$(3.7) \qquad
\begin{array}{ccc}
\mathbf{P}^{q-1} & \to & \mathscr{C}^{q-1}(\mathbf{P}^{q-1}) \\
\not\cong \uparrow & & \updownarrow \not\cong \\
\vdots & & \vdots \\
\not\cong \uparrow & & \updownarrow \not\cong \\
\mathscr{G}^{q-1}(\mathbf{P}^{n-2}) & \to & \mathscr{C}^{q-1}(\mathbf{P}^{n-2}) \\
\not\cong \uparrow & & \updownarrow \not\cong \\
\mathscr{G}^{q-1}(\mathbf{P}^{n-1}) & \to & \mathscr{C}^{q-1}(\mathbf{P}^{n-1}) \\
\not\cong \uparrow & & \updownarrow \not\cong \\
\vdots & & \vdots
\end{array}$$

which maps into the diagram (3.4) above. Passing to the limit gives us the commutative diagram

$$
(3.8) \quad
\begin{array}{ccccc}
\mathbf{P}^{q-1} & \subset & BU_{q-1} & \to & K(\mathbf{Z},2) \times \cdots \times K(\mathbf{Z},2q-2) \\
\cap & & \cap & & \cap j \\
\mathbf{P}^q & \subset & BU_q & \to & K(\mathbf{Z},2) \times \cdots \times K(\mathbf{Z},2q-2) \times K(\mathbf{Z},2q)
\end{array}
$$

where the vertical inclusion j on the right is the limit as $n \to \infty$ of the natural maps

$$(3.9) \qquad \mathscr{C}^{q-1}(\mathbf{P}^{n-1}) \subset \mathscr{C}^q(\mathbf{P}^n)$$

defined by including the $(n-q)$-dimensional cycles on $\mathbf{P}^{n-1}$ into the set of all $(n-q)$-cycles on $\mathbf{P}^n$.

LEMMA 3.10. *The vertical inclusion j in (3.8) is homotopic to the standard inclusion as a factor: $j_0(x) = (x, x_0)$ for a fixed point $x_0 \in K(\mathbf{Z}, 2q)$. Hence, after projecting $K(\mathbf{Z},2) \times \cdots \times K(\mathbf{Z},2q)$ onto $K(\mathbf{Z},2q)$, the maps in (3.8) descend to maps of the quotients*

$$(3.11) \qquad S^{2q} = \mathbf{P}^q/\mathbf{P}^{q-1} \subset BU_q/BU_{q-1} \to K(\mathbf{Z},2q).$$

The composition in (3.11) represents the generator of $\pi_{2q}K(\mathbf{Z},2q) = \mathbf{Z}$.

PROOF. Since the vertical inclusions in (3.4) and (3.7) are homotopy equivalences, it suffices to check our statements at the first level. Note that $\mathscr{C}^q(\mathbf{P}^q) = SP(\mathbf{P}^q)$ and that the inclusion $\mathscr{C}^{q-1}(\mathbf{P}^{q-1}) \subset \mathscr{C}^q(\mathbf{P}^q)$ is exactly the homomorphism of monoids $SP(\mathbf{P}^{q-1}) \subset SP(\mathbf{P}^q)$ induced by the inclusion $\mathbf{P}^{q-1} \subset \mathbf{P}^q$. One of the fundamental results of Dold and Thom [6] is that the map

$$SP(\mathbf{P}^q) \to SP(\mathbf{P}^q)/\!/SP(\mathbf{P}^{q-1})$$

to the quotient monoid is a quasifibration with fibre $SP(\mathbf{P}^{q-1})$, and that

$$SP(\mathbf{P}^q)/\!/SP(\mathbf{P}^{q-1}) \cong SP(\mathbf{P}^q/\mathbf{P}^{q-1}) = SP(S^{2q}) \cong K(\mathbf{Z},2q).$$

(We henceforth use $/\!/$ for monoid or group quotients.) The lemma now follows. $\square$

REMARK 3.12. It is in fact shown in [11] that for all n the map

$$\mathscr{C}^q(\mathbf{P}^n) \to \mathscr{C}^q(\mathbf{P}^n)/\!/\mathscr{C}^{q-1}(\mathbf{P}^{n-1})$$

to the quotient monoid is a quasifibration with fibre $\mathscr{C}^{q-1}(\mathbf{P}^{n-1})$ and quotient $K(\mathbf{Z},2q)$.

We now invoke the fundamental fact that the integral cohomology of BU_q is a polynomial algebra

$$H^*(BU_q; \mathbf{Z}) \cong \mathbf{Z}[c_1, \ldots, c_q]$$

generated by the Chern classes $c_k = c_k(\xi_q) \in H^{2k}(BU_q; \mathbf{Z})$ of the universal bundle ξ_q, and that the kernel of the homomorphism

$$(3.13) \qquad H^*(BU_q; \mathbf{Z}) \to H^*(BU_{q-1}; \mathbf{Z})$$

induced by the inclusion $BU_{q-1} \subset BU_q$ is the ideal generated by c_q. Let us express the map (3.5)

$$BU_q \xrightarrow{a_1 \times \cdots \times a_q} K(\mathbf{Z}, 2) \times \cdots \times K(\mathbf{Z}, 2q)$$

as a set of cohomology classes $a_k \in H^{2k}(BU_q; \mathbf{Z})$, $k = 1, \ldots, q$.

Now by Lemma 3.10 we know that the class a_q, when restricted to BU_{q-1}, is zero. Hence, $a_q = n_q c_q$ for some $n_q \in \mathbf{Z}$. Furthermore, the class a_q, when restricted to $\mathbf{P}^q$, factors through the quotient $\mathbf{P}^q/\mathbf{P}^{q-1}$ and represents a generator of $\pi_{2q} K(\mathbf{Z}, 2q)$. Hence, we have that $(a_q, [\mathbf{P}^q]) = 1$ and $n_q = \pm 1$. To determine the sign we examine the map $i: \mathbf{P}^q \to BU_q$ defined above. One sees directly from the definition that

$$i * \xi_q = \lambda^{\perp},$$

where $\lambda \to \mathbf{P}^q$ is the tautological line bundle. Hence, we have that

$$a_q \circ i = n_q c_q \circ i \cong n_q i * c_q(\xi_q) = n_q c_q(\lambda^{\perp}) = n_q (-1)^q \omega^q$$

where $\omega = c_1(\lambda)$ is the "negative" generator of $H^2(\mathbf{P}^q; \mathbf{Z})$, i.e., the one which satisfies $(\omega, [\mathbf{P}^1]) = -1$. Pairing with $\mathbf{P}^q$ we find that

$$1 = (a_q, [\mathbf{P}^q]) = n_q (-1)^q (\omega^q, [\mathbf{P}^q]) = n_q (-1)^q (-1)^q,$$

and so $n_q = 1$ as claimed. To compute a_{q-1} we can restrict to BU_{q-1} and repeat the argument. This completes the proof. $\square$

4. Bott periodicity. In the last section we considered the cycle inclusion $\mathscr{C}_1^q(\mathbf{P}^n) \subset \mathscr{C}^q(\mathbf{P}^n)$ and took the limit as $n \to \infty$. We now examine what happens as $q \to \infty$. To this end we consider the sequence of infinite flags:

$$
\begin{array}{ccccccc}
\vdots & & \vdots & & \vdots & & \\
\cap & & \cap & & \cap & & \\
\mathbf{C}^{q-1} & \subset & \mathbf{C}^q & \subset & \mathbf{C}^{q+1} & \subset & \cdots \\
\cap & & \cap & & \cap & & \\
\mathbf{C}^q & \subset & \mathbf{C}^{q+1} & \subset & \mathbf{C}^{q+2} & \subset & \cdots \\
\cap & & \cap & & \cap & & \\
\mathbf{C}^{q+1} & \subset & \mathbf{C}^{q+2} & \subset & \mathbf{C}^{q+3} & \subset & \cdots \\
\cap & & \cap & & \cap & & \\
\vdots & & \vdots & & \vdots & &
\end{array}
\tag{4.1}
$$

where each horizontal inclusion is given by adding a new variable on the right, and each vertical inclusion adds a new variable on the left. We define an associated grid of maps on algebraic cycles by complex suspension in horizontal directions and by inclusion in vertical directions. Taking horizontal limits as in

§3, we get a diagram

$$
(4.2) \quad
\begin{array}{ccccc}
\vdots & & \vdots & & \vdots \\
\cap & & \cap & & \cap \\
\mathbf{P}^{q-2} & \subset & BU_{q-2} & \subset & \mathbb{C}^{q-2} \\
\cap & & \cap & & \cap \\
\mathbf{P}^{q-1} & \subset & BU_{q-1} & \subset & \mathbb{C}^{q-1} \\
\cap & & \cap & & \cap \\
\mathbf{P}^{q} & \subset & BU_{q} & \subset & \mathbb{C}^{q} \\
\cap & & \cap & & \cap \\
\vdots & & \vdots & & \vdots
\end{array}
$$

Taking vertical limits gives us maps

$$(4.3) \qquad \mathbf{P}^{\infty} \subset BU \xrightarrow{c} \mathbb{C} = K(\mathbf{Z}, \mathrm{ev}),$$

where BU is the classifying space for reduced K-theory and where $K(\mathbf{Z}, \mathrm{ev}) = \prod_q K(\mathbf{Z}, 2q)$ (the weak product).

Observe now that by the Bott Periodicity Theorem there is an isomorphism $\pi_* BU \cong \pi_* K(\mathbf{Z}, \mathrm{ev})$. On the other hand there can be no map between these spaces which realizes this isomorphism, since the homotopy types of the two spaces are drastically different. (The cohomology of BU is a polynomial ring, while that of $K(\mathbf{Z}, \mathrm{ev})$ is quite a bit more complicated.) Nevertheless our cycle map c comes as close as possible to inducing such an isomorphism.

THEOREM 4.4. *The mapping $c : BU \to K(\mathbf{Z}, \mathrm{ev})$ defined above induces an injective map on homotopy groups. In fact, for each $q > 0$, the map*

$$c_* : \pi_{2q} BU \to \pi_{2q} K(\mathbf{Z}, \mathrm{ev})$$

is multiplication by $(q-1)!$.

This theorem is essentially a reformulation of the classical results of Bott [1].

PROOF. Let $f : S^{2q} \to BU$ be a map representing the generator of $\pi_{2q} BU$, and let ι_{2q} denote the generator of $H^{2q}(K(\mathbf{Z}, 2q); \mathbf{Z})$. We want to compute the integer $k = ((c \circ f)^* \iota_{2q}, S^{2q})$. By Theorem 3.6 we know that $(c \circ f)^* \iota_{2q} = c_q(E)$ where E is the (stable) vector bundle over S^{2q} induced by the map f. Hence, by Bott's Theorem [1] we know that $(q-1)! | k$.

On the other hand let $f_1 : S^{2q} \to BU$ be the map classifying the complex bundle of positive spinors $\mathbb{S}^+$ on S^{2q}. Then direct computation using the character of the Spin representation shows that $(c_{2q}(\mathbb{S}^+), S^{2q}) = (q-1)!$. Hence, $k | (q-1)!$. $\square$

This gives yet another perspective on the classical Bott theorems. It seems not unlikely that a proof of these theorems could be given purely in terms of algebraic cycles.

5. Groups of cycles. One might naturally wonder what happens in Theorem 2.4 if the topological monoid $\mathscr{C}_p(\mathbf{P}^n)$ is replaced by its naive group completion. Interestingly, the conclusion remains unchanged. In fact much of the theory for cycles in $\mathbf{P}^n$ carries over to this context, and occasionally this proves to be quite useful.

Let $\mathscr{GC}_p(\mathbf{P}^n)$ denote the free abelian group generated by the irreducible algebraic subvarieties of dimension p in $\mathbf{P}^n$, and let

$$(5.1) \qquad \widetilde{\mathscr{C}}_p(\mathbf{P}^n) = \left\{ \sum_\alpha n_\alpha V_\alpha \in \mathscr{GC}_p(\mathbf{P}^n) : \sum_\alpha n_\alpha \deg V_\alpha = 0 \right\}$$

be the subgroup of cycles of degree zero. For each $d > 0$ we define the subset $F_d = \{c \in \widetilde{\mathscr{C}}_p(\mathbf{P}^n) : c$ can be written $c = \sum_\alpha n_\alpha V_\alpha$, where $\sum_\alpha |n_\alpha| \deg V_\alpha \leq 2d\}$. This set can be expressed as a quotient of the variety $\mathscr{C}_{p,d}(\mathbf{P}^n) \times \mathscr{C}_{p,d}(\mathbf{P}^n)$. As such it inherits the topology of a compact Hausdorff space. We give $\widetilde{\mathscr{C}}_p(\mathbf{P}^n)$ the weak limit topology for the filtration $F_1 \subset F_2 \subset F_3 \subset \cdots$. This makes $\widetilde{\mathscr{C}}_p(\mathbf{P}^n)$ a connected abelian topological group.

The complex suspension defined in 2.10 extends by linearity to these groups, and we have the following.

THEOREM 5.2. *For each p, $0 \leq p < n$, the complex suspension homomorphism*

$$\widetilde{\mathscr{C}}_p(\mathbf{P}^n) \xrightarrow{\ \Sigma\ } \widetilde{\mathscr{C}}_{p+1}(\mathbf{P}^{n+1})$$

is a homotopy equivalence.

PROOF. The argument follows exactly the lines of the one given in [11] to prove the homotopy equivalence $\mathscr{C}_p(\mathbf{P}^n) \to \mathscr{C}_{p+1}(\mathbf{P}^{n+1})$. Here the family of sets $\{\mathscr{C}_{p,d}(\mathbf{P}^n)\}_{d=1}^\infty$ is replaced by the family $\{F_d\}_{d=1}^\infty$. The only point requiring a little care is the verification that the lifting of cycles via divisors remains continuous in this case. $\square$

If we fix now a p-dimensional linear subspace $l_0 \subset \mathbf{P}^n$, we can define an embedding of our monoid

$$(5.3) \qquad \mathscr{C}_p(\mathbf{P}^n) \xhookrightarrow{(\tilde{\cdot})} \widetilde{\mathscr{C}}_p(\mathbf{P}^n)$$

by setting $\tilde{c} = c - dl_0$ for each $c \in \mathscr{C}_{p,d}(\mathbf{P}^n)$. (Note that this map respects the sequence of inclusions (2.3) and therefore extends to the limit.) The embedding (5.3) is a continuous monoid homomorphism which realizes $\widetilde{\mathscr{C}}_p(\mathbf{P}^n)$ as a topological group completion of $\mathscr{C}_p(\mathbf{P}^n)$.

THEOREM 5.4. *The embedding* (5.3) *is a homotopy equivalence.*

PROOF. For $p = 0$ this is a result of Dold and Thom [6]. For $p > 0$ it follows from the fact that the complex suspension maps $\Sigma : \mathscr{C}_p(\mathbf{P}^n) \to \mathscr{C}_{p+1}(\mathbf{P}^{n+1})$ and $\Sigma : \widetilde{\mathscr{C}}_p(\mathbf{P}^n) \to \widetilde{\mathscr{C}}_{p+1}(\mathbf{P}^{n+1})$ are homotopy equivalences which are compatible with the embeddings (5.3). $\square$

We now fix a linear subspace $\mathbf{P}^m \subset \mathbf{P}^n$ and consider the closed submonoid $\mathscr{C}_p(\mathbf{P}^m) \hookrightarrow \mathscr{C}_p(\mathbf{P}^n)$. From [11] we know that the map to the quotient $\mathscr{C}_p(\mathbf{P}^n) \to \mathscr{C}_p(\mathbf{P}^n)/\!\!/\mathscr{C}_p(\mathbf{P}^m)$ is a quasifibration with fibre $\mathscr{C}_p(\mathbf{P}^m)$ (cf. 3.12). There exist an analogous subgroup and quotient for the cycle groups, and the maps (5.3) give a commutative diagram

(5.5)
$$\begin{array}{ccccc}
\mathscr{C}_p(\mathbf{P}^m) & \hookrightarrow & \mathscr{C}_p(\mathbf{P}^n) & \overset{\beta}{\to} & \mathscr{C}_p(\mathbf{P}^n)/\!/\mathscr{C}_p(\mathbf{P}^m) \\
\uparrow & & \uparrow & & \downarrow \\
\widetilde{\mathscr{C}}_p(\mathbf{P}^m) & \hookrightarrow & \widetilde{\mathscr{C}}_p(\mathbf{P}^n) & \overset{\tilde{\beta}}{\to} & \widetilde{\mathscr{C}}_p(\mathbf{P}^n)/\!/\widetilde{\mathscr{C}}_p(\mathbf{P}^m)
\end{array}$$

THEOREM 5.6. *The quotient map $\tilde{\beta}$ is a principal fibration with group $\widetilde{\mathscr{C}}_p(\mathbf{P}^m)$. Furthermore, the diagram (5.5) is a morphism of quasifibrations which is a homotopy equivalence at each level.*

PROOF. The argument that $\tilde{\beta}$ is a quasifibration follows exactly the argument given in Dold and Thom [6] for the case where $p = 0$. That (5.5) is a homotopy equivalence of quasifibrations follows from Theorem 5.4 and the 5-lemma. $\square$

The remainder of this section is devoted to proving certain generalizations of the results above which will be technically useful later on. Essentially they amount to carrying the theorems over to "thickenings" of linear subspaces.

For convenience we introduce on $\mathbf{P}^n$ the standard Fubini-Study metric with holomorphic curvature 1. For any fixed linear subspace $\mathbf{P}^m \subset \mathbf{P}^n$ and any number t, $0 \leq t < \pi$, we define the closed tubular neighborhood

$$\mathbf{P}_t^m \equiv \{x \in \mathbf{P}^n : \mathrm{dist}(X, \mathbf{P}^m) \leq t\}.$$

Let $\mathbf{P}^{n-m-1} = \mathbf{P}^n - \mathrm{int}(\mathbf{P}_\pi^m)$ be the *polar linear subspace*, and let $\varphi_s : \mathbf{P}^n \to \mathbf{P}^n$, $s \in \mathbf{C}$, denote the family of analytic automorphisms which fixes $\mathbf{P}^m \cup \mathbf{P}^{n-m-1}$ and represents scalar multiplication by s in the vector bundle $\mathbf{P}^n - \mathbf{P}^{n-m-1} \to \mathbf{P}^m$. Note that for each s the map φ_s gives an analytic automorphism

$$\varphi_s : \mathbf{P}_t^m \overset{\approx}{\longrightarrow} \mathbf{P}_{t(s)}^m,$$

where $t(s) \searrow 0$ as $|s| \searrow 0$.

We now consider the closed subgroups

(5.7)
$$\widetilde{\mathscr{C}}_p(\mathbf{P}_t^m) = \{c \in \widetilde{\mathscr{C}}_p(\mathbf{P}^n) : \mathrm{supp}(c) \subset \mathbf{P}_t^m\}$$

where "$\mathrm{supp}(c) \subset X$" means that c can be written as $\sum_\alpha n_\alpha V_\alpha$ with $\bigcup_\alpha V_\alpha \subset X$. These groups have a nice structure. For each d, the set $\mathscr{C}_{p,d}(\mathbf{P}_t^m) = \mathscr{C}_{p,d}(\mathbf{P}^n) \cap \widetilde{\mathscr{C}}_p(\mathbf{P}_t^m)$ is semianaltyic. In fact it is a regular neighborhood of $\mathscr{C}_{p,d}(\mathbf{P}^m)$ in $\mathscr{C}_{p,d}(\mathbf{P}^n)$. Furthermore we have the following.

PROPOSITION 5.8. *For each $t < \pi$, the subset $\widetilde{\mathscr{C}}_p(\mathbf{P}^m) \subset \widetilde{\mathscr{C}}_p(\mathbf{P}_t^m)$ is a deformation retract.*

PROOF. The flow φ_s, $0 \leq s \leq 1$, induces a deformation retraction of $\widetilde{\mathscr{C}}_p(\mathbf{P}_t^m)$ down to $\widetilde{\mathscr{C}}_p(\mathbf{P}^m)$. $\square$

Consider now a linear inclusion $\mathbf{P}^n \subset \mathbf{P}^{n+1}$ and the complex suspension Σ to the polar point $\mathbf{P}^0 \subset \mathbf{P}^{n+1}$.

THEOREM 5.9. *Set* $\mathbf{P}^{m+1} = \oint \mathbf{P}^m$. *Then for all* $t < \pi$, *the complex suspension homomorphism*

$$\oint : \widetilde{\mathscr{C}}_p(\mathbf{P}_t^m) \to \widetilde{\mathscr{C}}_{p+1}(\mathbf{P}_t^{m+1})$$

is a homotopy equivalence.

PROOF. The arguments of [11] apply directly to prove that $\oint : \widetilde{\mathscr{C}}_p(\mathbf{P}_t^m) \to \widetilde{\mathscr{C}}_{p+1}(\oint(\mathbf{P}_t^m))$ is a homotopy equivalence. Applying the flow φ_s (which can be chosen to commute with the "suspension flow") shows that

$$\widetilde{\mathscr{C}}_{p+1}(\mathbf{P}^{m+1}) \subset \widetilde{\mathscr{C}}_{p+1}(\oint(\mathbf{P}_t^m))$$

is a deformation retract. Hence, by 5.8 the inclusion $\widetilde{\mathscr{C}}_{p+1}(\oint(\mathbf{P}_t^m)) \subset \widetilde{\mathscr{C}}_{p+1}(\mathbf{P}_t^{m+1})$ is a homotopy equivalence. $\square$

Consider now a linear embedding $\mathbf{P}^n \subset \mathbf{P}^N$ and fix $t < \pi$. Then we have a commutative diagram of continuous group homomorphisms

$$
\begin{array}{ccccc}
\widetilde{\mathscr{C}}_p(\mathbf{P}^m) & \hookrightarrow & \widetilde{\mathscr{C}}_p(\mathbf{P}^n) & \overset{\tilde{\beta}}{\to} & \widetilde{\mathscr{C}}_p(\mathbf{P}^n)/\!\!/\widetilde{\mathscr{C}}_p(\mathbf{P}^m) \\
\uparrow & & \uparrow & & \downarrow \\
\widetilde{\mathscr{C}}_p(\mathbf{P}_t^m) & \hookrightarrow & \widetilde{\mathscr{C}}_p(\mathbf{P}^n) & \overset{\tilde{\beta}}{\to} & \widetilde{\mathscr{C}}_p(\mathbf{P}^n)/\!\!/\widetilde{\mathscr{C}}_p(\mathbf{P}_t^m) \\
\uparrow & & \uparrow & & \downarrow \\
\widetilde{\mathscr{C}}_p(\mathbf{P}_t^m) & \hookrightarrow & \widetilde{\mathscr{C}}_p(\mathbf{P}_t^n) & \overset{\tilde{\beta}}{\to} & \widetilde{\mathscr{C}}_p(\mathbf{P}_t^n)/\!\!/\widetilde{\mathscr{C}}_p(\mathbf{P}_t^m)
\end{array}
$$

where in each case $\tilde{\beta}$ is the canonical projection to the quotient group. In the middle line of (5.10) $\mathbf{P}_t^m$ is the t-tubular neighborhood of $\mathbf{P}^m$ in $\mathbf{P}^n$. In the bottom line it denotes the t-tubular neighborhood of $\mathbf{P}^m$ in $\mathbf{P}^N$.

THEOREM 5.11. *In the diagram* (5.10) *each of the maps* $\tilde{\beta}$ *is a principal fibration and each of the vertical maps is a homotopy equivalence.*

PROOF. The proof of the first assertion follows exactly the lines of that given by Dold and Thom in [6] for the case $p = 0$. (It is important here that the sets $\mathscr{C}_{p,d}(\mathbf{P}_t^m)$ are triangulable.) The second assertion follows by applying the deformation used in 5.8. $\square$

REMARK 5.12. The above discussion from (5.7) onward remains true if $\widetilde{\mathscr{C}}_p$ is everywhere replaced by $\mathscr{C}_p$ and "principal fibration" is replaced by "quasifibration".

6. Whitney sum and the complex join construction. Recall that the complex join defined in §2 gives a family of maps

(6.1) $\mathscr{C}_{p,d}(\mathbf{P}^n) \times \mathscr{C}_{p',d'}(\mathbf{P}^{n'}) \overset{*_{\mathbf{C}}}{\longrightarrow} \mathscr{C}_{p+p'+1,dd'}(\mathbf{P}^{n+n'+1}).$

In particular, it induces a pairing on the cycles of degree one $\mathscr{C}_{p,1}(\mathbf{P}^n) \equiv \mathscr{G}^q(\mathbf{P}^n)$, $p + q = n$. From (2.8) we see that in homogeneous coordinates this pairing is given exactly by the direct sum of linear subspaces.

(6.2) $\mathscr{G}^q(\mathbf{P}^n) \times \mathscr{G}^{q'}(\mathbf{P}^{n'}) \overset{\oplus}{\to} \mathscr{G}^{q+q'}(\mathbf{P}^{n+n'+1}).$

(Here the direct sum applies both to the subspaces and to their orthogonal complements.) Stabilizing with respect to n and n' gives a map

$$(6.3) \qquad\qquad BU_q \times BU_{q'} \overset{\oplus}{\to} BU_{q+q'}$$

which classifies the Whitney sum operation on vector bundles. It satisfies $\oplus^*(\xi_{q+q'}) = \xi_q \oplus \xi_{q'}$.

We proved in §3 that the natural cycle inclusion $\mathscr{G}^q(\mathbf{P}^n) \hookrightarrow \mathscr{C}^q(\mathbf{P}^n)$ represents the total Chern class of ξ_q. This leads us to consider defining the complex join on the stabilized spaces $\mathscr{C}^q(\mathbf{P}^n)$. Unfortunately, the pairing (6.1) does not extend naively to a map $\mathscr{C}^q(\mathbf{P}^n) \times \mathscr{C}^{q'}(\mathbf{P}^{n'}) \to \mathscr{C}^{q+q'}(\mathbf{P}^{n+n'+1})$. For this purpose it is better to use the groups of degree-zero cycles $\widetilde{\mathscr{C}}^q(\mathbf{P}^n) \equiv \widetilde{\mathscr{C}}_{n-q}(\mathbf{P}^n)$ defined in §5. Here the complex join does extend to a continuous bilinear map $\widetilde{\mathscr{C}}^q(\mathbf{P}^n) \times \widetilde{\mathscr{C}}^{q'}(\mathbf{P}^{n'}) \to \widetilde{\mathscr{C}}^{q+q'}(\mathbf{P}^{n+n'+1})$ which, since $c \underset{\mathbf{C}}{\times} 0 = 0 \underset{\mathbf{C}}{\times} c' = 0$, descends to the smash product

$$(6.4) \qquad\qquad \widetilde{\mathscr{C}}^q(\mathbf{P}^n) \wedge \widetilde{\mathscr{C}}^{q'}(\mathbf{P}^{n'}) \overset{\times_{\mathbf{C}}}{\longrightarrow} \widetilde{\mathscr{C}}^{q+q'}(\mathbf{P}^{n+n'+1}).$$

For convenience let us set $K_{2k} = K(\mathbf{Z}, 2k)$ and $\prod^q = K_2 \times K_4 \times \cdots \times K_{2q}$. Then under the fundamental homotopy equivalence $\widetilde{\mathscr{C}}^q(\mathbf{P}^n) \sim \prod^q$ we see that the complex join (6.4) defines a cohomology pairing. The first main result of this section will compute this pairing. Let ι_{2k} denote the generator of

$$H^{2k}(K_{2k} : \mathbf{Z}) \cong \mathbf{Z}.$$

THEOREM 6.5. *The complex join* $\times_{\mathbf{C}} : \prod^q \wedge \prod^{q'} \to \prod^{q+q'}$ *has the property that*

$$(\times_{\mathbf{C}})^*(\iota_{2k}) = \sum_{\substack{r+s=k \\ r>0 \\ s>0}} \iota_{2r} \otimes \iota_{2s}$$

in integral cohomology.

Observe now that via the embedding (5.3) the join determines a map

$$(6.6) \qquad \mathscr{G}^q(\mathbf{P}^n) \wedge \mathscr{G}^{q'}(\mathbf{P}^{n'}) \subset \mathscr{C}^q(\mathbf{P}^n) \wedge \mathscr{C}^{q'}(\mathbf{P}^{n'}) \overset{\times_{\mathbf{C}}}{\longrightarrow} \widetilde{\mathscr{C}}^{q+q'}(\mathbf{P}^{n+n'+1}).$$

Since (5.3) is a homotopy equivalence, this map is essentially the same as (6.4). However, *this map does not restrict to give the Whitney sum on* $\mathscr{G}^q(\mathbf{P}^n) \times \mathscr{G}^{q'}(\mathbf{P}^{n'})$. For positive cycles c, c' of degree d and d' respectively, the map (6.6) is given by

$$(6.7) \qquad\qquad (c, c') \mapsto (c - dl_0) \underset{\mathbf{C}}{\times} (c' - d'l'_0).$$

This leads us to consider a second pairing. Given cycles $c \in \widetilde{\mathscr{C}}(\mathbf{P}^n)$ and $c' \in \widetilde{\mathscr{C}}^{q'}(\mathbf{P}^{n'})$, we set $\Sigma^{n'+1}(c) = c \underset{\mathbf{C}}{\times} l'_0$ and $\Sigma^{n+1}(c') = l_0 \underset{\mathbf{C}}{\times} c'$, where l_0, l'_0 are the distinguished linear subspaces in $\mathbf{P}^n$ and $\mathbf{P}^{n'}$ respectively. We then define

$$(6.8a) \qquad\qquad \widetilde{\mathscr{C}}^q(\mathbf{P}^n) \times \widetilde{\mathscr{C}}^{q'}(\mathbf{P}^{n'}) \overset{\mu}{\to} \widetilde{\mathscr{C}}^{q+q'}(\mathbf{P}^{n+n'+1})$$

by

$$(6.8b) \qquad\qquad \mu(c, c') = c \underset{\mathbf{C}}{\times} c' + \Sigma^{n+1}(c') + \Sigma^{n'+1}(c).$$

It is straightforward to check that the diagram

(6.9)
$$\begin{array}{ccc} \mathscr{G}^q(\mathbf{P}^n) \times \mathscr{G}^{q'}(\mathbf{P}^{n'}) & \xrightarrow{\oplus} & \mathscr{G}^{q+q'}(\mathbf{P}^{n+n'+1}) \\ \downarrow & & \downarrow \\ \widetilde{\mathscr{C}}^q(\mathbf{P}^n) \times \widetilde{\mathscr{C}}^{q'}(\mathbf{P}^{n'}) & \xrightarrow{\mu} & \widetilde{\mathscr{C}}^{q+q'}(\mathbf{P}^{n+n'+1}) \end{array}$$

commutes, where the vertical maps are given as above via (5.3)

THEOREM 6.10. *The map* $\mu : \prod^q \times \prod^{q'} \to \prod^{q+q'}$ *has the property that*

$$\mu^*(\iota_{2k}) = \sum_{\substack{r+s=k \\ r \geq 0 \\ s \geq 0}} \iota_{2r} \otimes \iota_{2s}.$$

By stabilizing (6.9) to the diagram

(6.11)
$$\begin{array}{ccc} BU_q \times BU_{q'} & \xrightarrow{\oplus} & BU_{q+q'} \\ \downarrow & & \downarrow \\ \mathfrak{C}^q \times \mathfrak{C}^{q'} & \xrightarrow{\mu} & \mathfrak{C}^{q+q'} \end{array}$$

and applying 6.10, we obtain a "cycle-theoretic" proof of the classical formula

$$c(E \oplus E') = c(E) \cup c(E')$$

for the integral Chern classes of complex vector bundles.

REMARK 6.12. The diagram (6.11) can be stabilized with respect to q and q'. This can be done formally by passing to the group completions of the monoids $(BU\cdot, \oplus)$ and $(\widetilde{\mathfrak{C}}^\cdot, \mu)$. It can also be done by direct construction. This will be discussed in §7.

The remainder of this section will be devoted to the proof of Theorems 6.5 and 6.10.

PROOF OF THEOREM 6.5. To begin we fix flags

(6.13) $\quad l_0 = \mathbf{P}^p \subset \mathbf{P}^{p+1} \subset \cdots \subset \mathbf{P}^n$ and $l_0' = \mathbf{P}^{p'} \subset \mathbf{P}^{p'+1} \subset \cdots \subset \mathbf{P}^{n'}$

and consider the images

$$\mathbf{P}^r \ast_{\mathbf{C}} \mathbf{P}^{r'} \subset \mathbf{P}^{n+n'+1}$$

of these subspaces under the complex join map. It is possible to choose a corresponding flag

(6.14)
$$\mathbf{P}^{p+p'+1} \subset \mathbf{P}^{p+p'+2} \subset \cdots \subset \mathbf{P}^{n+n'+1}$$

and a number $\varepsilon < \pi$ with the property that

(6.15)
$$\mathbf{P}^r \ast_{\mathbf{C}} \mathbf{P}^{r'} \subset \mathbf{P}^k_\varepsilon \quad \text{for all } r + r' < k,$$

where $\mathbf{P}^k_\varepsilon$ is the ε-tubular neighborhood of $\mathbf{P}^k$ in $\mathbf{P}^{n+n'+1}$. To see this one simply chooses a polar flag

$$\widetilde{\mathbf{P}}^{n+n'-p-p'-1} \supset \widetilde{\mathbf{P}}^{n+n'-p-p'} \supset \cdots \supset \widetilde{\mathbf{P}}^0$$

with the property that $\widetilde{\mathbf{P}}^l_{\pi-\varepsilon} \cap (\mathbf{P}^r \ast_{\mathbf{C}} \mathbf{P}^{r'}) = \varnothing$ for all (r, r') with $r + r' + l \leq n + n'$. (Such a flag exists by general position arguments.) One then defines $\mathbf{P}^k_\varepsilon = \mathrm{int}(\mathbf{P}^{n+n'+1} - \widetilde{\mathbf{P}}^{n+n'-k}_{\pi-\varepsilon})$.

Let us set $U(k) = \mathbf{P}^k_\varepsilon$. Then (6.14) gives rise to a "thickened flag"

$$(6.16) \qquad U(p + p' + 1) \subset U(p + p' + 2) \subset \cdots \subset U(n + n')$$

and under the restriction of the join map we have by (6.15) that

$$(6.17) \qquad \widetilde{\mathscr{C}}_p(\mathbf{P}^r) \wedge \widetilde{\mathscr{C}}_{p'}(\mathbf{P}^{r'}) \xrightarrow{*_{\mathrm{C}}} \widetilde{\mathscr{C}}_{p+p'+1}(U(k))$$

for all $r + r' < k$.

Using 2.13 and the homotopy equivalences discussed in §5, we now identify the projection $\prod^q \to K_{2q}$ (up to homotopy) with the corresponding monoid quotients (cf. (5.10)). In particular we have a homotopy equivalence of fibrations

$$(6.18)$$
$$\begin{array}{ccccc}
\widetilde{\mathscr{C}}^{q+q'-1}(U(n+n')) & \hookrightarrow & \widetilde{\mathscr{C}}^{q+q'}(\mathbf{P}^{n+n'+1}) & \to & \widetilde{\mathscr{C}}^{q+q'}(\mathbf{P}^{n+n'+1})/\!\!/\widetilde{\mathscr{C}}^{q+q'-1}(U(n+n')) \\
\downarrow \wr & & \downarrow \wr & & \downarrow \wr \\
\prod^{q+q'-1} & \hookrightarrow & \prod^{q+q'} & \xrightarrow{\mathrm{proj}} & K_{2(q+q')}
\end{array}$$

where we have chosen the convention that $\widetilde{\mathscr{C}}^q(U(k)) \equiv \widetilde{\mathscr{C}}_{k-q}(U(k))$. Similarly, there are homotopy equivalences of fibrations

$$(6.19)$$
$$\begin{array}{ccccc}
\widetilde{\mathscr{C}}^{q-1}(\mathbf{P}^{n-1}) & \hookrightarrow & \widetilde{\mathscr{C}}^q(\mathbf{P}^n) & \to & \widetilde{\mathscr{C}}^q(\mathbf{P}^n)/\!\!/\widetilde{\mathscr{C}}^{q-1}(\mathbf{P}^{n-1}) \\
\downarrow & & \downarrow & & \downarrow \\
\prod^{q-1} & \hookrightarrow & \prod^q & \xrightarrow{\mathrm{proj}} & K_{2q}
\end{array}$$

for all $q \leq n$.

REMARK 6.20. These diagrams generalize inductively to other splittings $\prod^r \hookrightarrow \prod^q \to K_{2r+2} \times \cdots \times K_{2q}$.

PROPOSITION 6.21. *Under the identifications above, the complex join map factors into the commutative diagram*

$$\begin{array}{ccc}
\prod^q \wedge \prod^{q'} & \to & \prod^{q+q'} \\
{\scriptstyle (\mathrm{proj}) \wedge (\mathrm{proj})} \downarrow & & \downarrow {\scriptstyle \mathrm{proj}} \\
K_{2q} \wedge K_{2q'} & \to & K_{2(q+q')}
\end{array}$$

PROOF. Apply 6.17 with $r = n - 1$, $r' = n' - 1$, and $k = n + n'$, and use bilinearity to see that the map

$$\widetilde{\mathscr{C}}^q(\mathbf{P}^n) \times \widetilde{\mathscr{C}}^{q'}(\mathbf{P}^{n'}) \to \widetilde{\mathscr{C}}^{q+q'}(\mathbf{P}^{n+n'+1})/\!\!/\widetilde{\mathscr{C}}^{q+q'-1}(U(n+n'))$$

factors through the quotient

$$(\widetilde{\mathscr{C}}^q(\mathbf{P}^n)/\!\!/\widetilde{\mathscr{C}}^{q-1}(\mathbf{P}^{n-1})) \times (\widetilde{\mathscr{C}}^{q'}(\mathbf{P}^{n'})/\!\!/\widetilde{\mathscr{C}}^{q'-1}(\mathbf{P}^{n'-1})). \quad \square$$

PROPOSITION 6.22. *The map* $J : K_{2q} \wedge K_{2q'} \to K_{2(q+q')}$ *given in 6.21 induces an isomorphism on integral cohomology in dimension* $2(q + q')$, *i.e.*,

$$J^*(\iota_{2(q+q')}) = \iota_{2q} \otimes \iota_{2q'}.$$

PROOF. Let $\mathbf{P}^q \hookrightarrow K_{2q}$ be the natural embedding given in (3.8). Then there is a commutative diagram of maps

$$\begin{array}{ccc}
K_{2q} \wedge K_{2q'} & \to & K_{2(q+q')} \\
\cup & & \cup \\
(\mathbf{P}^q/\mathbf{P}^{q-1}) \wedge (\mathbf{P}^{q'}/\mathbf{P}^{q'-1}) & \xrightarrow{\approx} & \mathbf{P}^{q+q'}/\mathbf{P}^{q+q'-1}
\end{array}$$

which carries the generator $S^{2q} \wedge S^{2q'} = S^{2(q+q')}$ of $\pi_{2(q+q')}(K_{2q} \wedge K_{2q'})$ to the generator $S^{2(q+q')}$ of $\pi_{2(q+q')}(K_{2(q+q')})$ (cf. (3.11)). $\square$

The remainder of the proof of 6.5 proceeds essentially by induction. Let Sk_{2k+1} denote the $(2k+1)$-skeleton of $\prod^q \wedge \prod^{q'}$. Then since each K_{2l} is $(2l-1)$-connected, we have

$$(6.23) \qquad Sk_{2k+1} \subset \bigcup_{r+r' \leq k} \left(\prod^r \wedge \prod^{r'} \right).$$

The main observation now is that essentially $*_C(\prod^r \wedge \prod^{r'}) \subset \prod^{r+r'}$. To be specific, we identify the factor $\prod^r \times \prod^{r'} \subset \prod^q \times \prod^{q'}$ with the subgroup $\widetilde{\mathscr{C}}_p(\mathbf{P}^{p+r}) \times \widetilde{\mathscr{C}}_{p'}(\mathbf{P}^{p'+r'})$ as in (6.19) and 6.20. Then by 6.17 we get a commutative diagram of maps

$$
\begin{array}{ccccccc}
Sk_{2k+1} & \subset & \bigcup_{r+r' \leq k} \left(\prod^r \wedge \prod^{r'} \right) & \xrightarrow{*_C} & \widetilde{\mathscr{C}}(U(k)) & \sim & K_2 \times \cdots \times K_{2k} \\
\cup & & \cup & & \cup & & \cup \\
Sk_{2k-1} & \subset & \bigcup_{s+s' \leq k-1} \left(\prod^s \wedge \prod^{s'} \right) & \xrightarrow{*_C} & \widetilde{\mathscr{C}}(U(k-1)) & \sim & K_2 \times \cdots \times K_{2k-2}
\end{array}
$$

(for simplicity we drop the dimension of cycles). Projecting to the group quotients gives maps

$$(6.24)$$
$$Sk_{2k+1}/Sk_{2k-1} \to \bigcup \left(\prod^r \wedge \prod^{r'} \right) \Big/ \bigcup \left(\prod^s \wedge \prod^{s'} \right) \to \widetilde{\mathscr{C}}(U(k))/\!/\widetilde{\mathscr{C}}(U(k-1)) \sim K_{2k}.$$

This composition defines our $2k$-dimensional cohomology class $(*_C)^*(\iota_{2k})$. It factors through

$$\bigcup_{r+r' \leq k} \left(\prod^r \wedge \prod^{r'} \right) \Big/ \bigcup_{s+s' \leq k-1} \left(\prod^s \wedge \prod^{s'} \right) \to \bigvee_{r+r'=k} \left(\prod^r \wedge \prod^{r'} \Big/ \left(\prod^r \wedge \prod^{r'-1} \cup \prod^{r-1} \wedge \prod^{r'} \right) \right)$$

$$= \bigvee_{r+r'=k} \left\{ \left(\prod^r / \prod^{r-1} \right) \wedge \left(\prod^{r'} / \prod^{r'-1} \right) \right\}.$$

The symbol $\prod^r / \prod^{r-1}$ represents "collapsing", not "group quotient". However, since we have a bouquet we can consider the factor individually. Recalling that $\prod^r = \widetilde{\mathscr{C}}_p(\mathbf{P}^{p+r})$, one can check directly, as in the proof of 6.21, that the map $\prod^r \wedge \prod^{r'} \to \widetilde{\mathscr{C}}(U(k))/\!/\widetilde{\mathscr{C}}(U(k-1))$ factors through

$$\left(\prod^r /\!/ \prod^{r-1} \right) \wedge \left(\prod^{r'} /\!/ \prod^{r'-1} \right) = K_{2r} \wedge K_{2r'}$$

and is exactly the map discussed in the original case. Hence, the composition (6.24) factors through a bouquet of maps

$$Sk_{2k+1}/Sk_{2k-1} \to \bigvee_{r+r'=k} (K_{2r} \wedge K_{2r'}) \to K_{2k}$$

each of which is of the type discussed above. $\square$

PROOF OF THEOREM 6.10. The map $\widetilde{\mathscr{C}}^q(\mathbf{P}^n) \to \widetilde{\mathscr{C}}^{q+q'}(\mathbf{P}^{n+n'+1})$ given by $c \mapsto \Sigma^{n'+1}(c) = c *_C l'_0$ is homotopic to the inclusion $\prod^q \hookrightarrow \prod^{q+q'}$ as a factor. Therefore, Theorem 6.10 is a direct consequence of Theorem 6.5 and the following.

LEMMA 6.25. *Let $f, g : X \to \prod^q = \widetilde{\mathscr{C}}^q(\mathbf{P}^n)$ be two basepointed maps. Then for $\iota = \iota_2 + \iota_4 + \cdots + \iota_{2q}$, we have*

$$(f + g)^*(\iota) = f^*(\iota) + g^*(\iota).$$

PROOF. The homotopy equivalence $K_2 \times \cdots \times K_{2q} = \mathrm{SP}(S^2) \times \cdots \times \mathrm{SP}(S^{2q}) \to \widetilde{\mathscr{C}}^q(\mathbf{P}^n)$ can be chosen to be a homomorphism of monoids. (Map $\bigvee_k S^{2k}$ onto generators of π_* and extend.) Hence, the lemma reduces to the corresponding statement for $K_{2k} = \mathrm{SP}(S^{2k})$. For this consider the composition

$$X \xrightarrow{\Delta} X \times X \xrightarrow{f \times g} K_{2k} \times K_{2k} \xrightarrow{+} K_{2k}$$

and note that $(+)^* \iota_{2k} = \iota_{2k} \otimes 1 + 1 \otimes \iota_{2k}$. $\square$

7. The cycles of a fixed degree. Let us fix an integer $d \geq 1$. Then using the grid of suspension and inclusion maps associated to (4.1), we can define the space

$$(7.1) \qquad \mathscr{D}(d) = \varinjlim_{p,q} \mathscr{C}_{p,d}(\mathbf{P}^{p+q})$$

consisting essentially of all projective cycles of degree d (with arbitrary dimension and codimension). Note that there is a filtration

$$(7.2) \qquad BU = \mathscr{D}(1) \subset \mathscr{D}(2) \subset \cdots \subset \mathscr{D}(\infty) = K(\mathbf{Z}, \mathrm{ev})$$

which refines the Chern characteristic map $\mathscr{D}(1) \subset \mathscr{D}(\infty)$. These inclusions have certain basic connectivity properties.

THEOREM 7.3. *The inclusion map $\mathscr{D}(d) \subset \mathscr{D}(\infty)$ has a right homotopy inverse up to dimension d; that is, there exists a finite subcomplex $\mathscr{D}_0(d) \subset \mathscr{D}(d)$ such that the inclusion $\mathscr{D}_0(d) \subset \mathscr{D}(\infty)$ is $(2d - 1)$-connected.*

PROOF. The inclusion $\mathrm{SP}^d(\mathbf{P}^q) \subset \mathrm{SP}(\mathbf{P}^q)$ is $(2d - 1)$-connected for any q, and the complex suspension

$$\mathrm{SP}(\mathbf{P}^q) = \mathscr{C}^q(\mathbf{P}^q) \xrightarrow{\Sigma^p} \mathscr{C}^q(\mathbf{P}^{q+p})$$

is a homotopy equivalence for all p. Hence, we can choose $\mathscr{D}_0(d) = \mathrm{SP}^d(\mathbf{P}^d)$. $\square$

The complex join construction defined in §2 can be extended to these spaces. To see this we first note that $\mathscr{D}(d) = \lim_{n \to \infty} \mathscr{C}_{n-1,d}(\mathbf{P}^{2n-1})$. We then consider the "shuffle" isomorphism $\sigma : \mathbf{C}^{2n} \times \mathbf{C}^{2n} \to \mathbf{C}^{4n}$ defined by $\sigma(z, w) = (z_1, w_1, z_2, w_2, \ldots, z_{2n}, w_{2n})$. Composing the standard complex join (which takes place "in blocks") with the isomorphism induced by σ, gives us an "enmeshed" join

$$\mathscr{C}_{n-1,d}(\mathbf{P}^{2n-1}) \times \mathscr{C}_{n-1,d'}(\mathbf{P}^{2n-1}) \xrightarrow{\times_\mathbf{C}} \mathscr{C}_{2n-1,dd'}(\mathbf{P}^{4n-1})$$

which is compatible with our family of inclusion-suspension maps. Passing to the limit, we obtain the following.

PROPOSITION 7.4. *The complex join determines continuous mappings*

$$\mathscr{D}(d) \times \mathscr{D}(d') \to \mathscr{D}(dd')$$

for all (finite) integers $d, d' \geq 1$.

When $d = d' = 1$, this is exactly the Whitney sum map

$$BU \times BU \overset{\oplus}{\to} BU.$$

It is important to note that the join maps are *not* compatible with the inclusions (7.2). Nevertheless, by using the shuffle as above we can extend the "reduced" join (6.4) to a map

$$\widetilde{\mathfrak{C}} \wedge \widetilde{\mathfrak{C}} \overset{\times_{\mathbf{C}}}{\to} \widetilde{\mathfrak{C}}$$

where $\widetilde{\mathfrak{C}} = \lim_{n \to \infty} \widetilde{\mathscr{C}}_{n-1}(\mathbf{P}^{2n-1})$. The assignment $c \mapsto c - dl_0$ gives an embedding $\mathscr{C}_{n-1,d}(\mathbf{P}^{2n-1}) \subset \widetilde{\mathscr{C}}_{n-1}(\mathbf{P}^{2n-1})$ which extends to an embedding

$$(7.5) \qquad\qquad \mathscr{D}(d) \subset \widetilde{\mathfrak{C}}.$$

These are compatible with the inclusions (7.2), and the final embedding $\mathscr{D}(\infty) \subset \widetilde{\mathfrak{C}}$ is a homotopy equivalence. Now for each pair of integers $d, d' \geq 1$ we define a map

$$(7.6) \qquad \begin{aligned} \mu_{d,d'} &: \widetilde{\mathfrak{C}} \times \widetilde{\mathfrak{C}} \to \widetilde{\mathfrak{C}} \\ \mu_{d,d'}(c, c') &= c \times_{\mathbf{C}} c' + d' j_1(c) + d j_2(c') \end{aligned}$$

where $j_1 : \widetilde{\mathfrak{C}} \to \widetilde{\mathfrak{C}}$ is the homotopy equivalence defined as the limit of the maps $\widetilde{\mathscr{C}}_{n-1}(\mathbf{P}^{2n-1}) \to \widetilde{\mathscr{C}}_{2n-1}(\mathbf{P}^{4n-1})$ given by $c \mapsto c \times_{\mathbf{C}} l_0'$, and j_2 is determined similarly by $c' \mapsto l_0 \times_{\mathbf{C}} c'$. Invoking the homotopy equivalence $\widetilde{\mathfrak{C}} \sim K(\mathbf{Z}, \mathrm{ev})$ gives the following.

THEOREM 7.7. *For each pair of integers $d, d' \geq 1$ there is a commutative diagram*

$$\begin{array}{ccc} \mathscr{D}(d) \times \mathscr{D}(d') & \overset{\times_{\mathbf{C}}}{\longrightarrow} & \mathscr{D}(dd') \\ \downarrow & & \downarrow \\ K(\mathbf{Z}, \mathrm{ev}) \times K(\mathbf{Z}, \mathrm{ev}) & \overset{\mu_{d,d'}}{\longrightarrow} & K(\mathbf{Z}, \mathrm{ev}) \end{array}$$

where the vertical maps are given by the inclusion (7.5) (the "extended Chern characteristic map"), and where $\mu_{d,d'}$ is the map uniquely determined up to homotopy by the property that

$$(7.8) \qquad (\mu_{d,d'})^*(\iota_{2k}) = d'(\iota_{2k} \otimes 1) + d(1 \otimes \iota_{2k}) + \sum_{\substack{r+s=k \\ r>0 \\ s>0}} \iota_{2r} \otimes \iota_{2s}$$

for all k.

PROOF. It is straightforward to verify the commutativity of the diagram with $\widetilde{\mathfrak{C}} \equiv K(\mathbf{Z}, \mathrm{ev})$ and $\mu_{d,d'}$ given by (7.6). The formula (7.7) follows from 6.5 and 6.25. $\square$

There remain several interesting speculations concerning the spaces $\mathscr{D}(d)$:

I. Is $\pi_*\mathscr{D}(d) \to \pi_*\mathscr{D}(\infty)$ injective? If so, the sequence of maps

$$\pi_{2q}\mathscr{D}(1) \longrightarrow \pi_{2q}\mathscr{D}(2) \longrightarrow \cdots \longrightarrow \pi_{2q}\mathscr{D}(2q) \xrightarrow{\approx} \pi_{2q}\mathscr{D}(2q+1) \xrightarrow{\approx} \cdots$$
$$\parallel \qquad\qquad \parallel \qquad\qquad\qquad\qquad \parallel \qquad\qquad\quad \parallel$$
$$\mathbf{Z} \xrightarrow{n_1} \mathbf{Z} \xrightarrow{n_2} \cdots \xrightarrow{n_{2q-1}} \mathbf{Z} \xrightarrow{n_{2q}} \mathbf{Z} \xrightarrow{n_{2q+1}} \cdots$$

corresponds to a sequence of integers $\{n_j\}_{j=1}^{\infty}$ with $n_j = 1$ for $j \geq 2q$ and with $\prod_j n_j = (q-1)!$.

II. Are the spaces $\mathscr{D}(d)$ universal for some natural geometric construction?

III. Are the spaces $\mathscr{D}(d)$ infinite loop spaces?

IV. Is $H_*(\mathscr{D}(d), \mathscr{D}(d-1))$ identifiable? (computable?)

There seems to be a generalization of the Schubert calculus to these spaces, the details of which should prove useful and interesting.

REFERENCES

1. R. Bott, *The space of loops on a Lie group*, Michigan Math. J. **5** (1958), 35–61.

2. ――, *The stable homotopy of the classical groups*, Ann. of Math. **70** (1959), 179–203.

3. A. K. Bousfield and D. M. Kan, *Homotopy limits, completions and localizations*, Lecture Notes in Math., vol. 304, Springer-Verlag, New York, 1972.

4. A. Dold, *Homology of symmetric products and other functors of complexes*, Ann. of Math. (2) **68** (1958), 54–80.

5. A. Dold and R. Thom, *Une généralisation de la notion d'espaces fibré. Applications aux produits symétriques infinis*, C. R. Acad. Sci. Paris **242** (1956), 1680–1682.

6. ――, *Quasifaserungen und unendliche symmetrische Produckte*, Ann. of Math. (2) **67** (1958), 230–281.

7. E. M. Friedlander, *Etale homotopy of stable Chow varieties* (to appear).

8. ――, *Homotopy theory for algebraic cycles using Chow varieties* (to appear).

9. E. M. Friedlander and B. Mazur, personal communication.

10. H. B. Lawson, Jr., *The topological structure of the space of algebraic varieties*, Bull. Amer. Math. Soc. **17** (1987), 326–330.

11. ――, *Algebraic cycles and homotopy theory*, Ann. of Math. (2) (to appear).

12. I. R. Shafarevitch, *Basic algebraic geometry*, Springer-Verlag, New York, 1978.

13. G. Whitehead, *Elements of homotopy theory*, Springer-Verlag, New York, 1974.

14. H. Whitney, *Complex analytic varieties*, Addison-Wesley, Reading, Mass., 1972.

STATE UNIVERSITY OF NEW YORK AT STONY BROOK

Proceedings of Symposia in Pure Mathematics
Volume 48 (1988)

Uniformization of Geometric Structures

S.-T. YAU

In geometry, a lot of efforts have been devoted to provide a "normal form" to certain geometric objects. A famous example is the Weyl's embedding theorem which says that any abstractly defined closed surface with positive curvature can be isometrically embedded in R^3 as a convex surface.

In the process of finding a normal form, there are local problems and global problems. In this article, we shall concentrate only on the problem of uniformization of geometric structures on smooth manifolds. Classically we have the famous uniformization theorem that any conformal structure on a surface can be normalized in such a way that it is the quotient of S^2, R^2 or D^2 by a group of conformal transformations. While the program of "uniformization" originated from function theoretic consideration, we shall take a geometric viewpoint. (It should, however, be stressed that the function theoretic method is actually an important tool in a major part of the program that we are going to discuss.)

Our goal would be to generalize these two-dimensional theorems to higher-dimensional manifolds. In our proposal of uniformization, we should have special models. A typical model has the following form. Let G be a Lie group acting transitively on a manifold $\widetilde{M}$. Let Ω be a domain in $\widetilde{M}$ and Γ be a subgroup of G so that Ω/Γ is compact. (In many cases, compactness can be replaced by the concept of completeness.) Then Ω/Γ is our model space.

The question of uniformization can be phrased as follows: Given a manifold M with certain structure (related to G) can we find a model space Ω/Γ so that M is isomorphic to Ω/Γ? Most of the known uniformization theorems are existent in nature; it is normally very difficult and interesting to give an explicit description of the uniformization. An interesting example is the following: Let M be an algebraic surface defined by an ideal of homogeneous polynomials in the complex projective space. Suppose that we know that M is uniformized by the ball in $\mathbb{C}^n$. Can we calculate the covering map mapping the ball onto M? This is clearly a difficult question and is largely unknown. In this article, we shall only discuss the existence problem.

1980 *Mathematics Subject Classification* (1985 *Revision*). Primary 53C55; Secondary 35J60.
Supported by NSF Grant # AMS-8711394.

There are several levels of questions we can ask. The fundamental question is to give an intrinsic topological characterization of manifolds which can be uniformized. An example is the famous theorem of Thurston that, in three dimensions, a compact manifold which is "atoroidal" and "sufficiently large" admits a hyperbolic structure and can be uniformized. For dimension greater than three, no theorem of this type exists. It is certainly a challenging problem to find an analogue of Thurston's theorem in higher dimension and in a wider category.

The question of uniformization becomes more tractable when the manifold has certain special structures. The most notable structures are complex structures, conformally flat structures, affine structures and projective structures.

From the point of view of Riemannian geometry, the most homogeneous and isotropic manifolds are manifolds with constant curvature. They are isometrically covered by the sphere, the euclidean space or the hyperbolic space. From the point of view of complex geometry, the corresponding spaces will be complex projective space, the complex euclidean space or the complex hyperbolic space.

In Riemannian geometry, there has been a great effort to classify complete manifolds with positive sectional curvature or negative sectional curvature. While there has been much progress on the subject, it is still far from giving a complete list. If we assume that the curvature operator is positive definite, the most recent work of Richard Hamilton shows that the manifold may in fact be diffeomorphic to the quotient of the sphere by a finite group of isometries. He has demonstrated this for four-dimensional manifolds by using his famous method of deforming the metric along the Ricci direction. B. Chow and D. Yang [6] have recently demonstrated that if the curvature operator is only positive semidefinite, then the manifold is covered by the product of a finite number of manifolds which are either symmetric or admit a metric with positive definite curvature operator. It is interesting to know whether the negative analogue holds or not. It should be noted that the proof of Hamilton can be considered as part of the uniformization idea because he deforms the metric by an explicitly defined equation to one with constant sectional curvature which can then by uniformized. Can we prove Thurston's theorem by such a procedure? B. Chow discovered a new equation which preserves metrics with negative curvature. This is encouraging. Perhaps Chow's flow can be used to study metrics with negative curvature.

For manifolds which are equipped with affine structures, projective structures or conformally flat structures, it is by definition that we can locally embed the structure into R^n, RP^n or S^n. Since the transition functions can be realized by global transformations on these model spaces, it is not difficult to see that any simply connected manifolds with the above structures admit a structural preserving map into the corresponding model spaces. This is done by the well-known monodromy argument and the map is called the developing map. The major problem in this subject is that the developing map is not injective and therefore the manifold is in general not the quotient of a domain by a discrete group which preserves the corresponding structures. At present there is no good criterion for deciding the embedness of the developing map.

However, for conformally flat manifolds, R. Schoen and the author [30] have come up with some criterion. First of all, we should note that, by the outstanding solution of the Yamabe problem by R. Schoen, every compact conformally flat manifold admits a conformally flat metric with constant scalar curvature. The sign of this scalar curvature is a conformal invariant. We demonstrated that when the scalar curvature is nonnegative, the conformally flat manifold is indeed the quotient of a domain by a discrete group of conformal transformations. There are certain constructions that we can make to construct manifolds with positive scalar curvatures from the quotient of manifolds of type $S^{n-k} \times H^k$ where H^k is either empty or S^1 or hyperbolic manifolds. It is not clear whether this construction gives the full class of compact conformally flat manifolds with positive scalar curvature or not. It is also not clear whether every complete conformally flat manifold with positive scalar curvature is an open set of some compact conformally flat manifold with positive scalar curvature or not. According to the work of R. Schoen and the author, the complement set should have small Hausdorff dimension.

For three-dimensional conformally flat manifolds Gusevskii and Kapovich [11] proved that if the image of the developing map is not onto S^3, then it is a covering map. There is also a paper by Kapovich [13] which gave more detailed investigations.

For affine flat structure, projective flat structure, Lorentz structure, and various structures modelled after some Lie group, one should consult works of Bogomolov, Kulkarni, Goldman, Gunning, Sullivan, Thurston and others. It has been a temptation for a long time to put some of these structures on a compact simply-connected three-dimensional manifold to prove the Poincaré conjecture.

When the manifold has a complex structure, the question of uniformization becomes more classical. Of course when the dimension of the manifold is greater than one, there are many types of questions we can ask. The fundamental question is to find a good sufficient condition for an almost complex manifold to be deformable to a complex manifold. This is a difficult question and we restrict ourselves to algebraic manifolds in the following.

We want to know how close M resembles Ω/Γ. (Note that Ω can be compact and Γ can be trivial.) There are three levels of question 1. Is Ω/Γ isomorphic (or biregular) to M? 2. Is Ω/Γ birational to M? 3. Is there a nonzero degree map from Ω/Γ to M? We shall discuss what is known in the following. Roughly speaking, we divide the discussion into three categories.

I. **The compact case.** The model spaces would be compact Hermitian symmetric spaces. It is natural to ask whether the complex structure over these spaces is "rigid" or not. If the Hermitian symmetric space is reducible, the structures are not necessarily rigid. For example, there are an infinite number of Hirzebruch structures on $CP^1 \times CP^1$. For irreducible Hermitian symmetric spaces, it is expected that they admit only one Kählerian structure. For the complex projective space, this was demonstrated for odd dimension by Hirzebruch-Kodaira [12] and by the author for even dimension. A similar statement should hold for other irreducible symmetric spaces. (J. Morrow [26] did demonstrate the rigidity for hyperquadrics with dimension greater than two.) It is interesting to know whether "rigidity" still holds if we only fix the homotopic type of the model space. For dimension not greater than six, rigidity in this generality was demonstrated by A. Libgober and J. Wood. It is difficult to decide whether these manifolds have non-Kählerian complex structure or not. If such structure exists, they should have low algebraic dimension. The question is already not known even when the algebraic dimension is top dimensional, i.e., when the manifold is Moishezon. For reducible Hermitian symmetric space, rigidity for Kählerian structure may still hold if we allow fiber space structures. (Note that Hirzebruch structures have these fiber structures.)

Besides topological data, we can also use some geometric data to characterize Hermitian symmetric spaces. The famous Frankel-Hartshorne conjecture says that M is biregular to CP^n iff the tangent bundle of M is positive. (The Frankel conjecture assumes the bisectional curvature of M to be positive.) It was solved by Mori [24] (for the stronger Hartshorne conjecture) and Siu-Yau (for the weaker Frankel conjecture). The method of Mori is so strong that we expect to have better understanding of manifolds with $C_1(M) > 0$. In fact, Miyaoka and Mori [21] demonstrated that such manifolds are uniruled (and have a lot of rational curves). There may in fact be nonzero degree maps from CP^n onto M.

As a generalization of the Frankel conjecture, the author asked whether compact algebraic manifolds with nonnegative bisectional curvature are covered by finite products of Hermitian symmetric manifolds. Under the stronger assumption on the curvature operator, Cao and Chow [3] demonstrated this statement. The question was finally solved by Mok [22], making use of Mori's result and improving an argument started by Bando who solved the case of three-dimensional manifolds. A more natural question is still unsolved, i.e., One would like to prove the same statement if we merely assume that the tangent bundle is semipositive. Perhaps one wants to add the assumption that $C_1(M)$ is positive also. It is an interesting question to know: What happens when we weaken the curvature assumption? What are manifolds with positive holomorphic sectional curvature? It is not hard to prove that two-dimensional manifolds with positive holomorphic sectional curvature are precisely rational manifolds. What is the algebraic geometric meaning of holomorphic sectional curvature? When does a projective manifold admit Kähler metrics with negative holomorphic sectional curvature? The only known obstructions for the later class of metrics come from the existence of rational and elliptic curves. Will that be the only obstruction?

II. Higher dimensional generalization of the complex torus and the complex Euclidean space. A beautiful theorem in this direction is given by H. C. Wang [36] which says that if the tangent bundle of a compact complex

manifold M is holomorphically trivial, it is covered by a complex Lie group. Conversely, what are manifolds that are covered holomorphically by a complex Lie group? Would they have a finite cover whose tangent bundle is trivial? It is likely that they must be the quotient of the complex Lie group by a discrete group of affine transformations. An important step would be to prove these manifolds have zero first Chern class. Note that if M is Kähler, $C_1(M) = 0$ and $\pi_2(M) = 0$, then M is covered by the torus. This is a corollary of Calabi's conjecture as was proved by the author. If M is noncompact and is covered by a complex Lie group, is M a Zariski open set of some compact complex manifold? Can we classify these compact manifolds? Their Kodaira dimension should be nonpositive.

In general, it seems difficult for a compact Kähler manifold M to be covered by a Zariski open set of another compact complex manifold unless this Zariski open set contains a pencil of complex lines. It is not difficult to verify such a statement for two-dimensional manifolds. If furthermore we assume $\pi_2(M) = 0$, one should expect M to be covered by a complex torus. This is consistent with a conjecture of Iitaka which says that compact Kähler manifolds covered by $\mathbf{C}^n$ must be covered by a complex torus. There is also a beautiful characterization [28] of $\mathbf{C}^2$ as an affine manifold which is homeomorphic to $\mathbf{C}^2$. On the geometric side, the author has long ago conjectured that a complete noncompact Kähler manifold with positive bisectional curvature is biholomorphic to $\mathbf{C}^n$. Unfortunately we do not even know such a manifold is Stein. If we assume the sectional curvature to be positive, then the theorem of Gromoll-Meyer and Greene-Wu says that it is a Stein manifold which is diffeomorphic to $\mathbf{C}^n$. It would be an important step if one can prove that such a manifold is a Zariski open set of a compact complex manifold. As a matter of fact, we expect that a complete Kähler manifold with nonnegative Ricci curvature is a Zariski open set of a compact complex manifold. It is also important to know whether the compact complex manifold is algebraic or not. The compact manifold clearly has nonpositive Kodaira dimension and is most likely a unirational manifold if the Ricci curvature is positive at some point. Clearly one can ask a similar question for complete manifolds with positive holomorphic sectional curvature.

There is another class of manifold which resembles euclidean space. This is the class of complete Kähler manifolds with zero Ricci curvature. Tian and the author will publish a more detailed paper on this subject.

III. **Algebraic manifolds of noncompact type.** As a step toward understanding the topology of compact algebraic manifolds with infinite fundamental group, we propose to show that the universal cover of such a manifold is always birational to a subdomain of another compact algebraic manifold. It may even be biholomorphic to the domain if M has no rational curves. When we have already decided that a complete Kähler manifold is a subdomain of a compact complex manifold, can we find a canonical representation of such a domain?

Hence we have to study discrete groups that act on the domain. In many cases, each element in the group can be extended by a birational transformation of the ambient manifold. If there is a canonical way of embedding the domain into an algebraic manifold so that the function field of the algebraic manifold is generated by sections of certain line bundles arising from the quotient manifold, then most likely the discrete group acts on the function field and can be embedded as a subgroup of the group of birational transformations of the manifold. On the other hand, most algebraic manifolds do not contain a subdomain Ω where an infinite subgroup of birational transformations acts properly discontinuously. Can we classify these manifolds? If such a domain exists, are there infinitely many of them in the given algebraic manifold?

There are potentially many manifolds which are the quotient of a domain in a compact Hermitian symmetric space. An important example is the famous theorem of Griffith's that every point in an algebraic manifold has a Zariski neighborhood which is the quotient of a bounded domain in $\mathbf{C}^n$. By using theorems of Cheng-Yau [4] and Mok-Yau [23], we then conclude that the Zariski neighborhood admits a unique complete Kähler Einstein metric with finite volume. Hence every algebraic manifold is a compactification of such manifolds. It is therefore of significance to recognize those manifolds which are covered by a domain.

Compactification of a noncompact complex manifold is closely related to the program of uniformization. Besides the above compactification questions related to uniformization, it will also be useful to know the following questions. Many years ago, the author conjectured that every complete Kähler manifold with Ricci curvature bounded by two negative constants and with finite volume can be compactified as a Zariski open set of an algebraic manifold. It can be considered as a generalization of the famous works of Borel-Baily, Satake, Mumford and other people. Siu and the author [33] managed to prove the conjecture if the curvature is strongly negative. There is some progress recently made by Nadel-Tsuji [27]. From the point of view of Kähler geometry the idea of compactification is to find a geometric characterization of Kähler manifolds that admit a quasiprojective structure. In this regard, it is very interesting to know the uniqueness question: if two quasiprojective manifolds are biholomorphic to each other, can the biholomorphic transformation be realized by an algebraic map? (An algebraic map is a holomorphic map that can be extended to certain compactification of the manifold.) This is indeed the case when the manifold is of a general type. The uniqueness question is difficult when the manifold is rational.

The basic tools that we propose to study the uniformization theorem will be analytic in nature. It will be nice to construct a universal cover of an algebraic manifold by algebraic means so that the fundamental group acts algebraically. A lot of the above-mentioned problems will be more transparent if this is possible. If the fundamental group is residually finite, there may be a good chance of doing algebraic constructions by taking the inverse limit of the finite covers.

(Perhaps fundamental groups of algebraic manifolds are residually finite in any case.) Since the algebraic method has not borne much fruit in our problem so far, we shall concentrate on analytic tools.

There are two basic types of theorems that we can use for the purpose of the uniformization theorem. The first one is the existence of Kähler-Einstein metrics and the second one is the existence of Hermitian Yang-Mills connections on stable vector bundles.

Nonsingular Kähler Einstein metrics exist on a compact Kähler manifold only if the first Chern class of the manifold has a definite sign. When the sign is either negative or zero, the existence and uniqueness of Kähler Einstein metric is well understood. When the sign is positive, only the uniqueness of the metric is understood. While the existence is still not well understood, a fundamental breakthrough was made recently by Tian [**34**].

The part that is relevant to the uniformization program is the observation that for a compact Kähler Einstein manifold M of dimension n, $(-1)^n C_2(M) C_1^{n-1}(M) \geq \frac{(-1)^n 2n}{n+1} C_1^n(M)$. When $C_1(M) \neq 0$, equality holds only if the holomorphic sectional curvature of M is constant. Therefore when $C_1(M) < 0$, M is covered by the ball holomorphically iff $(-1)^n C_2(M) C_1^{n-2}(M) = \frac{(-1)^n 2n}{n+1} C_1^n(M)$. When $C_1(M) = 0$, we prove that M is covered by C^n iff $C_2(M) \cup \omega^{n-2} = 0$ for some Kähler class ω. A similar statement is presumably true if $C_1(M) > 0$.

It turns out that $C_1(M) < 0$ is not necessary if we keep in mind that in higher dimensional uniformization we probably should allow manifolds to have singularities. Let Γ be a discrete group acting properly discontinuously on a domain Ω. Then in most cases, Γ has a subgroup of finite index Γ' so that Γ' acts on Ω without fixed point. (This is a lemma of Selberg when Γ is a discrete subgroup of a semisimple Lie group.) The orbit space Ω/Γ' is then a manifold. In any case, this means that Ω/Γ is a variety whose singularities are quotient singularities. We can then look for uniformization of algebraic manifolds which are obtained by desingularizing Ω/Γ. Naturally when Ω/Γ is a ball quotient with singularities, the equality $C_2 C_1^{n-2} = \frac{n}{2(n+1)} C_1^n$ no more holds. There is a correction term due to the resolution. This point of view led Chang and the author [**5**] to express a lower bound of $3C_2(M) - C_1^2(M)$ in terms of contribution from rational curves where $C_1 = 0$. (This was obtained by Miyaoka independently.) Furthermore, equality holds only if the manifold is the orbit space of the ball. In particular, we demonstrated that for an algebraic surface of general type, $3C_2(M) = C_1^2(M)$ is the condition for M to be covered by a ball. The higher dimensional analogue is not well understood, partially because our understanding of manifolds of general type is far less complete than the understanding of two-dimensional surfaces of general type. However, by the recent works of Mori and Kawamata, we have a better understanding of the minimal model of manifolds of general type. Moduli the question of singularities it is likely that the canonical bundle is semiample and all the above discussions are still valid.

If M is a quotient manifold of symmetric domain with finite volume, then the famous work of Baily, Borel, Satake and Mumford gave a compactification of M so that $M = \overline{M} \backslash D$, where D is a divisor with normal crossing. As was observed by Tian and Wang, $K_{\overline{M}} + D$ is semiample and $(K_{\overline{M}} + D)^n \neq 0$ for the Mumford compactification. Conversely, Tian and the author proved that for any compact Kähler manifold $\overline{M}$ so that $K_{\overline{M}} + D$ is numerically effective and $(K_{\overline{M}} + D)^n \neq 0$, then

$$(-1)^n c_1(T(D))^n \leq \frac{(-1)^n 2(n+1)}{n} c_2(T(D)) c_1(T(D))^{n-2}$$

and equality holds iff M is the quotient of a ball with finite volume. This is obtained by demonstrating the existence of canonical Kähler Einstein metric on M.

Therefore we have a rather good uniformization theorem for rank one Hermitian symmetric spaces. For quotients of higher rank Hermitian symmetric domain, we can also find algebraic geometric characterizations. A compact algebraic manifold M is covered by Hermitian symmetric domain iff the following statement holds. The canonical bundle of M is ample and T_M can be decomposed as a direct sum of irreducible subbundles $E_1, \ldots, E_k$ so that when $r > 1$, either $c_1^2(E_i) c_1^{n-2}(M) = \frac{2(r+1)}{r} c_2(E_i) c_1^{n-2}(M)$ or $S^m(E_i) \otimes (\bigwedge^r E_i^*)^m$ has nontrivial sections. Here r is the fiber dimension of E_i and $S^m(E_i)$ denotes the m-fold symmetric product of E_i. If $M = \overline{M} \backslash D$, we have a corresponding characterization of quotients of Hermitian symmetric domains with finite volume. (It may be interesting to note that Kazhdan's theorem on arithmetic variety is a consequence of this characterization.)

The argument can be roughly divided into two parts. If the canonical line bundle of an algebraic manifold M is ample and if the tangent bundle T can be decomposed into a direct sum of E_i's, then M is covered by a product manifold $N_1 \times \cdots \times N_k$ where the tangent bundle of N_i projects to E_i for all i. This is obtained by studying the splitting of the Kähler Einstein metric on M. (Each E_i has a unique Hermitian Yang-Mills connection which adds up to the metric connection induced by the metric.) The deRham theorem then says that the universal cover of M must split accordingly. Each factor N_i defines a totally geodesic foliation on M. The Chern class calculation that was used to prove the Chern number inequality can still be applied to the metric on N_i to show that $(-1)^r c_2(E_i) - \frac{r}{2(r+1)} c_1^2(E_i)$ is a positive semidefinite $(2, 2)$ form. Its product with $(-1)^n c_2^{n-2}(M)$ is always nonnegative and is zero iff N_i has constant negative holomorphic sectional curvature. This settles the first case when the Chern numbers of E_i have an equality. In the second case, one observes that $S^m(E_i) \otimes (\bigwedge^r E_i^*)^m$ has a natural Hermitian Yang-Mills connection induced from E_i. On the other hand, by the well-known Bochner argument, any holomorphic section of a holomorphic bundle with Hermitian Yang-Mills connection must be parallel. This means that the holonomy group of N_i must be reduced to a smaller subgroup. As N_i is irreducible, N_i must be Hermitian symmetric and therefore M is covered by a Hermitian symmetric space.

It is also possible to discuss uniformization theorems of homogeneous complex manifold using a theorem of Li-Yau [17], which is a generalization of the theorem of Donaldson [7] and Uhlenbeck-Yau [35]. (It should be mentioned that Buchdahl also found the same generalization when dimension is equal to two.) We will report about this work in a separate paper with Li.

In the above theorems, we made the assumption that the canonical line bundle is semiample. It is most likely that most of the theorems are still valid if the tangent bundle is stable. A very interesting development was made by Simpson [31], using a generalization of the theorem of Donaldson and Uhlenbeck-Yau. Both the Kähler Einstein metric and the Hermitian Yang Mills connection have a great deal to do with the uniformization. In the above discussions, we only used the special form of Chern numbers and the Bochner vanishing theorem. There are many other applications related to uniformization which we have not included here. We hope to discuss them elsewhere.

REFERENCES

1. W. Baily and A. Borel, *Compactification of arithmetic quotients of bounded symmetric domains*, Ann. of Math. **84** (1966), 442–528.

2. N. Buchdahl, *Stable vector bundle on compact complex surfaces*, to appear.

3. H. D. Cao and B. Chow, *Compact Kähler manifolds with nonnegative curvature operator*, Invent. Math. **83**(3) (1986), 553–556.

4. S. Y. Cheng and S. T. Yau, *On the existence of a complete Kähler metric on noncompact complex manifolds and the regularity of Fifferman's equation*, Comm. Pure Appl. Math. **33** (1980), 507–544.

5. ____, *Inequality between Chern numbers of singular Kähler surfaces and characterization of orbit space of discrete group of* SU(2, 1), Contemp. Math. **49** (1986), 31–43.

6. B. Chow and D. Yang, *Rigidity of nonnegatively curved compact quaternionic Kähler manifolds*, J. Differential Geom. (to appear).

7. S. K. Donaldson, *Anti-self-dual Yang-Mills connections on complex algebraic surfaces and stable vector bundles*, Proc. London Math. Soc. (3) **50** (1985), 1–26.

8. R. Greene and H. Wu, *On Kähler manifolds of positive bisectional curvature and a theorem of Hartogs*, Abh. Math. Sem. Univ. Hamburg **47** (1978), 171–185.

9. D. Gromoll and W. Meyer, *On complete open manifolds of positive curvature*, Ann. of Math. **90**(10) (1969), 75–90.

10. R. Gunning, *On uniformization of complex manifolds: The role of connections*, Math. Notes, no. 22, Princeton Univ. Press, Princeton, N.J., 1978.

11. N. A. Gusevskii and M. E. Kapovich, Dokl. Akad. Nauk SSSR **290** (1986), 537–541; *Conformal structures on three-dimensional manifolds*, English transl. in Soviet Math. Dokl. **34**(2) (1987), pp. 314–318.

12. F. Hirzebruch and K. Kodaira, *On the complex projective spaces*, J. Math. Pures Appl. **36** (1957), 201–216.

13. M. E. Kapovich, *Some properties of developments of conformal structures on three-dimensional manifolds*, Soviet Math. Dokl. **35** (1987), 146–149.

14. Y. Kawamata, *On the plurigenera of minimal algebraic 3-folds with* $K \approx 0$, Functional Anal. Appl. **12** (1978), 51–61.

15. ____, *On the classification of non-complete algebraic surfaces*, Algebraic Geometry (Proc. Copenhagen 1978), Lecture Notes in Math., Vol. 732, Springer-Verlag, 1979, pp. 215–232.

16. D. Kazhdan, *On arithmetic varieties*, Lie groups and their representations, Halsted Press, New York, 1975.

17. J. Li and S.-T. Yau, *Hermitian Yang-Mills connection on non-Kähler manifolds*, Mathematical aspects of string theory (S.-T. Yau, ed.) World Scientific, 1986.

18. M. Miyanishi and T. Sugie, *Affine surfaces containing cylinder-like open sets*, J. Math. Kyoto Univ. **20** (1980), 133–176.

19. M. Miyanishi, *An algebro-topological characterization of the affine space of dimension three*, Amer. J. Math. **106** (1984), 1469–1486.

20. Y. Miyaoka, *On the Chern numbers of surfaces of general type*, Invent. Math. **42** (1977), 225–237.

21. Y. Miyaoka and S. Mori, *A numerical criterion for uniruledness*, Ann. of Math. **124** (1986), 65–69.

22. N. Mok, *The Uniformization Theorem for compact Kähler manifolds of nonnegative holomorphic bisectional curvature*, Invent. Math. **27** 2(1988), 179–214.

23. N. Mok and S.-T. Yau, *Completeness of the Kähler-Einstein metric on bounded domains and the characterization of domains of holomorphy by curvature conditions*, Proc. Sympos. in Pure Math., vol. 39, Amer. Math. Soc., Providence, R.I., 1983, pp. 41–59.

24. S. Mori, *Projective manifolds with ample tangent bundles*, Ann. of Math. **110**(2) (1979), 593–606.

25. ____, *Threefolds whose canonical bundles are not numerically effective*, Ann. of Math. **116**(2) (1982), 133–176.

26. J. Morrow, *A survey of some results on complex Kähler manifolds*, Global Analysis: Papers in honor of K. Kodaira (D. C. Spencer/S. Iyanaga, eds.), Princeton Univ. Press, Princeton, N.J., 1969, pp. 315–324.

27. A. Nadel and H. Tsuji, *Compactification of complete Kähler manifolds of negative Ricci curvature*, J. Differential Geom. (to appear).

28. C. P. Ramanujan, *A topological characterization of the affine plane as an algebraic variety*, Ann. of Math. **94**(2) (1971), 69–88.

29. I. Satake, *On compactification of the quotient spaces for arithmetically defined discontinuous groups*, Ann. of Math. **72** (1960), 555–580.

30. R. Schoen and S.-T. Yau, *Conformally flat manifolds, Kleinian groups and scalar curvature*, Invent. Math. (to appear).

31. C. Simpson, *Constructing variations of Hodge structure using Yang-Mills theory and applications to uniformization*, to appear.

32. Y. T. Siu and S.-T. Yau, *Compact Kähler manifolds of positive bisectional curvature*, Invent. Math. **59** (1980), 189–204.

33. ____, *Compactification of negatively curved complete Kähler manifolds of finite volume*, Seminar on differential geometry (S.-T. Yau, ed.), Ann. of Math. **102** (1982), 363–380.

34. G. Tian, *On Kähler Einstein metrics on certain Kähler manifolds with $C_1(M) > 0$*, Invent. Math. **89**(2) (1987), 225–246.

35. K. Uhlenbeck and S.-T. Yau, *On the existence of Hermitian Yang-Mills connections on stable bundles*, Comm. Pure and Appl. Math. **39** (1986), 257–293.

36. H. C. Wang, *Complex parallelizable manifolds*, Proc. Amer. Math. Soc. **5** (1954), 771–776.

37. S.-T. Yau, *Calabi's conjecture of some new results in algebraic geometry*, Proc. Nat. Acad. Sci. U.S.A. **74** (1977), 1798–1799.

38. ____, *On the Ricci curvature of compact Kähler manifolds and the complex Monge-Ampère equation. I*, Comm. Pure Appl. Math. **31** (1978), 339–411.

HARVARD UNIVERSITY

Proceedings of Symposia in Pure Mathematics
Volume 48 (1988)

Elliptic Invariants for Differential Operators

R. G. DOUGLAS

To honor the memory of Hermann Weyl and his contributions to mathematics, I will discuss a topic which his work has strongly influenced. Since I started in functional analysis and abstract operator theory, my early knowledge of Weyl was based mainly on his books. Hence I knew him as one of the towering figures at the center of this century's revolution in geometry and physics. Although I knew the theorem of Weyl and von Neumann on compact perturbations, this was seen as an interesting result, not especially important, and certainly not the beginning of anything fundamental. However, the development of index theory and of K-theory and cyclic cohomology for operator algebra, and, more generally, of the interaction between geometry and operator algebras changed that. Moreover, Weyl's seminal role in all of this is now clear. His results and ideas paved the way for much of the recent development of "noncommutative topology and geometry."

The work I shall report on relates to several of Weyl's themes: the already mentioned theorem on the spectral behavior of selfadjoint operators under compact perturbation, the distribution of eigenvalues of differential operators, the equidistribution of subsets of a group, the representation theory for compact groups, and probably others.

The point of view I will use in the following is based on my joint work with many collaborators. However, the new results I will report on have been obtained recently in joint work with Steve Hurder and Jerry Kaminker [13, 14, 15].

We begin by discussing index theory, but exactly what does that mean? Since a review of the literature will not reveal one clearcut answer, let me explain what I mean by index theory. Index theory can be described using an analogy to the vertices and sides of a triangle. The first vertex consists of geometrical/topological data, while the second consists of a Hilbert space operator, possibly unbounded, along with additional structure such as an associated operator algebra. The corresponding side is some concrete method for constructing the Hilbert space

---

1980 *Mathematics Subject Classification* (1985 *Revision*). Primary 58G10; Secondary 58G12, 53C12, 46L99, 35P05.

operator from the given data. The third vertex is an appropriate analytical invariant defined for some class of Hilbert space operators to which the constructed operator belongs and this gives the second side of the triangle. Finally, the third side is the index formula which relates the analytical operator-theoretic invariant to the geometrical/topological data with which one began. Such a theory is usually referred to in terms of the index formula.

The most familiar example of an index theory is that of Atiyah and Singer. In this case the data consists of a smooth, compact, closed manifold M along with a class $[\xi]$ in $K^1(\mathrm{Th}\, M)$, the K-theory of the Thom space $\mathrm{Th}\, M$ of M. In [3] Atiyah and Singer showed how to construct a zeroth-order elliptic pseudodifferential operator D acting between the spaces of sections of two bundles E_0 and E_1 on M with the class of the principal symbol $\sigma(D)$ equal to $[\xi]$ in $K^1(\mathrm{Th}\, M)$. After specifying metrics on M, E_0, and E_1, a bounded operator, also denoted D, is defined between $L^2(E_0)$ and $L^2(E_1)$. This operator is Fredholm and the Atiyah-Singer index theorem states that

$$\mathrm{index}\, D = \int_{\tilde{M}} \mathrm{ch}(E_\sigma) \cup \mathrm{Td}(T\tilde{M}),$$

where $\tilde{M}$ is the compact spin^c-manifold obtained from M by glueing together two copies of the unit ball bundle in the tangent bundle TM, $\mathrm{Td}(T\tilde{M})$ is its Todd class, and E_σ is the vector bundle obtained by pulling E_0 and E_1 back to the two unit ball bundles over TM with $\sigma(D)$ serving as a clutching function.

Although this formula is given in terms of ordinary homology and cohomology, Atiyah and Singer realized that it could be profitably expressed in terms of K-theory. This led Atiyah to introduce a realization of K-homology in terms of generalized elliptic operators [1] and hence one has an analytical cycle $[D]$ in $K_0(M)$. This notion was fully developed by Kasparov [16] and Brown, Douglas, and Fillmore [5]. We do not plan to discuss this case in detail but we do need to discuss the pairing which results between elliptic operators as K-homology and vector bundles as K-cohomology.

Continuing the above notation with D on M, we let F be a smooth subbundle of the trivial bundle $M \times \mathbf{C}^n$. If $P(x)$ denotes the projection of $x \times \mathbf{C}^n$ onto F_x for x in M, then $D_F = P(D \otimes I_n)P$ defines an elliptic pseudodifferential operator on M and we define the pairing

$$\langle [D], [F] \rangle = \mathrm{index}\, D_F.$$

Since the class of the principal symbol of D_F in $K^1(\mathrm{Th}\, M)$ can be readily calculated in terms of that for D in $K^1(T^*M)$ and F in $K^0(M)$, the Atiyah-Singer index formula can be extended to D_F. The consequence of this pairing determines the element $[D]$ in $K_0(M)$ up to torsion [11].

A closer examination of the definition of D_F reveals the implicit role played by the differential geometric notion of a connection. Although every bundle F can be realized as a subbundle of a trivial bundle over M, this realization is not unique and the definition of D_F as a Hilbert space operator will depend on which realization is chosen. But the principal symbol of D_F does not! If D is actually a differential operator and ∇_F is a connection on F, then D_F can be defined canonically (cf. [17]) using ∇_F and is a differential operator which will be denoted by $D \otimes_{\nabla_F} I_F$. For $F = M \times \mathbf{C}^n$ and ∇ the trivial product connection, we have $D \otimes_{\nabla_F} I_n = D \otimes I_n$. However, if F is the trivial bundle but ∇_F is not the product connection, then the difference $D \otimes_{\nabla_F} I_n - D \otimes I_n$ is not zero but is a differential operator of lower order than D. A rather different kind of index theory applies in this case which culminates in a theorem of Atiyah, Patodi and Singer [2].

Let D be an elliptic selfadjoint differential operator defined on the smooth vector bundle E over M. Thus we are assuming that metrics have been chosen on M and E such that the unbounded Hilbert space operator D is selfadjoint on $L^2(E)$. Let $\tilde{M}$ denote the universal covering space of M, $\tilde{E}$ the pullback of E to $\tilde{M}$, and $\tilde{D}$ the corresponding differential operator defined on $\tilde{E}$. (We will suppress the roles of E and $\tilde{E}$ in what follows.) Let $\alpha \colon \pi_1(M) \to G$ be a homomorphism of the fundamental group $\pi_1(M)$ to the compact Lie group G such that $\operatorname{im} \alpha$ is dense in G. If $G = U_n(\mathbf{C})$, then we can use α to construct the Hermitian vector bundle $F = \tilde{M} \times_{\pi_1(M)} \mathbf{C}^n$ and this yields the differential operator $\tilde{D} \times_{\pi_1(M)} I_n$ induced on F by $\tilde{D} \otimes I_n$. Suppose F is a trivial bundle and that $\theta \colon M \times \mathbf{C} \to F$ is a trivialization of F. Then $\theta^*(\tilde{D} \times_{\pi_1(M)} I_n)\theta$ defines a differential operator D_α on $M \times \mathbf{C}^n$ equal to $D \otimes_{\nabla_\alpha} I_n$, where ∇_α is the flat connection induced on $M \times \mathbf{C}^n$ by θ.

Since both D_α and $D \otimes I_n$ have the same principal symbol, their primary analytical index invariants are equal. As we shall discuss shortly, this means that $[D_\alpha]$ and $[D \otimes I_n]$ are equal in $K_1(M)$. However, Atiyah, Patodi, and Singer showed in [2] how to define a secondary invariant for the pair D_α and $D \otimes I_n$ in terms of the eta invariant and went on to obtain an index formula. We will describe this now.

Recall that an elliptic selfadjoint differential operator D on a closed compact manifold has a complete set of real eigenvalues $\{\lambda_k\}$. Moreover, the eigenvalues grow such that $\eta(D, s) = \sum_k \operatorname{sign} \lambda_k \cdot |\lambda_k|^{-s}$ defines a holomorphic function for $\operatorname{Re} s$ sufficiently large such that η can be extended to a meromorphic function with finite value at 0, called the eta invariant for D. The study of the distribution of the eigenvalues of a positive differential operator using the zeta function which coincides with the eta function in that case was begun by Weyl [21]. We modify the eta invariant by setting

$$\xi(D) = \tfrac{1}{2}(\eta(D,0) + \dim \ker D).$$

Although $\xi(D_t)$ may not be continuous for a smooth path of operators $\{D_t\}$, the derivative $\dot{\xi}(D_t)$ of $\xi(D_t)$ with respect to t is continuous, and we set

$$\operatorname{Eta}(D_t) = \int_0^1 \dot{\xi}(D_t)\, dt + \xi(D_0).$$

For an arbitrary path D_t, $\operatorname{Eta}(D_t)$ depends on the specific path of operators. However, if the principal symbol of D_t is unchanged, then $\operatorname{Eta}(D_1, D_0) = \operatorname{Eta}(D_t)$

depends only on the endpoints D_0 and D_1. If we apply this definition to the pair $(D_\alpha, D \otimes I_n)$, then the result of Atiyah, Patodi, and Singer states that

$$\text{Eta}(D_\alpha, D \otimes I_n) = \int_{S(T\mathscr{F}_\alpha)} \text{Tch}(\rho \circ \alpha) \cup \text{Td}(T\mathscr{F}_\alpha) \cup \text{ch}(\sigma_1(D))\, d\mu$$

where Tch stands for the transgression of the Chern class of $\rho \circ \alpha$ to the sphere bundle $S(T\mathscr{F}_\alpha)$ and $d\mu$ is the Ruelle-Sullivan current for the foliation $\mathscr{F}_\alpha$ determined by θ. (We will discuss $\mathscr{F}_\alpha$ in more detail later.)

One of the starting points for my joint work with Hurder and Kaminker was our realization that the topological expression in this index formula also represents the analytical index for a very different problem. To explain that, we must discuss the primary invariants for *selfadjoint* elliptic pseudodifferential operators.

Let D be an elliptic selfadjoint pseudodifferential operator defined on the Hermitian vector bundle E over the closed compact manifold M. It is not difficult to show that $[D] = 0$ in $K_0(M)$. However, the relevant index problem for D concerns the generalized Toeplitz operators defined for the positive spectral space $\mathscr{H}_+$ for D. If P denotes the orthogonal projection of $L^2(E)$ onto $\mathscr{H}_+$, then for φ in $C(M)$ we set $T_\varphi = PM_\varphi P$, where M_φ denotes the multiplication operator on $L^2(E)$. If we let $\mathscr{T}$ denote the C^*-subalgebra of $\mathscr{L}(\mathscr{H}_+)$ generated by the compact operators $\mathscr{K}(\mathscr{H}_+)$ and $\{T_\varphi : \varphi \in C(M)\}$, then we obtain a short exact sequence $0 \to \mathscr{K} \to \mathscr{T} \to C(M) \to 0$, which defines an element $[D]$ in $K_1(M)$ by the work of Brown, Douglas, and Fillmore [5]. The Weyl–von Neumann Theorem [20, 19] is used in this case to establish the existence of the identity element in $K_1(M)$.

Now for an invertible φ in $C(M)$, it follows that T_φ is Fredholm and index T_φ is a primary invariant for D. In fact, the pairing between $[D]$ in $K_1(M)$ and $[\varphi]$ in $K^1(M)$ can be defined by

$$\langle [D], [\varphi] \rangle = \text{index}\, T_\varphi.$$

Moreover, by replacing D by $D \otimes I_n$, we can define T_φ for φ in $\text{GL}_n(M)$ and hence we obtain the pairing of $[D]$ with all of $K^1(M)$.

This index theory is often referred to as the "odd index theorem of Atiyah and Singer" since the index formula can be obtained from that for the even case. Although implicit in the work of Atiyah [1] and Kasparov [16], this was first made explicit in [4]. The index formula in this case is

$$\text{index}\, T_\varphi = \int_{\hat{M}} \pi^* \text{ch}(\varphi) \cup \text{ch}(E_+) \cup \pi^* \text{Td}(TM \otimes_{\mathbf{R}} \mathbf{C}),$$

where E_+ is the subbundle of the pullback $\pi^* E$ of E to the cotangent bundle $T^* M$ representing the positive subspaces of $\sigma(D)$, $TM \otimes_{\mathbf{R}} \mathbf{C}$ is the complexified tangent bundle of M, and $\hat{M}$ is the unit sphere bundle in $T^* M$. Taken in toto the index of the generalized Toeplitz operators defined for the positive spectral subspace is a complete primary invariant for the selfadjoint elliptic operator up to torsion [11].

Although the index formula we need to describe is for the odd selfadjoint case, it is for nonelliptic operators. There is no index theory, at present, for such operators without the imposition of additional structure. From one point of view index theory attempts to measure or quantify the failure of the basic operator D and the multiplication operators M_φ to commute. When D is elliptic, the commutators $[D, M_\varphi]$ are of strictly lower order than D and, hence, are compact relative to D. The latter statement can be made precise using the calculus for pseudodifferential operators, and we are able to reduce the problem to an algebra of bounded operators modulo the compact operators. And this quotient algebra can be identified with the algebra of principal symbols. This procedure could be described as "inverse quantization" since operators are being replaced by functions. When D fails to be elliptic, the commutator $[D, M_\varphi]$ is not compact but is only "compact-like" in the directions in the cotangent bundle T^*M in which D is elliptic. In order to exploit this we need some geometrical regularity to the partial ellipticity of D. One way to proceed is based on the notion of a foliation.

A foliation $\mathcal{F}$ is a smooth manifold V together with a decomposition of V into leaves so that locally V looks like $\mathbf{R}^p \times \mathbf{R}^q$, where $\mathbf{R}^p \times x$ is contained in a leaf of $\mathcal{F}$. It is important to note that different "local leaves" $\mathbf{R}^q \times x_1$ and $\mathbf{R}^q \times x_2$ can be part of the same leaf in V. The totality of tangent vectors to V which lie in the leaf direction forms the tangent bundle $T\mathcal{F}$ to $\mathcal{F}$ and is a subbundle of TV.

Returning to the case of $\alpha\colon \pi_1(M) \to G$ with a fixed trivialization

$$\theta\colon \tilde{M} \times_\alpha G \to M \times G$$

of the principal bundle $\tilde{M} \times_\alpha G$, we obtain the natural foliation $\mathcal{F}_\alpha$ for $V = \tilde{M} \times_\alpha G$, where the leaves are just the images of $\tilde{M}$ in V. For example, let $M = S^1$, $G = S^1$, $\tilde{M} = \mathbf{R}$, and let α be defined from $\pi_1(S^1) = \mathbf{Z}$ to $S^1 = \mathbf{R}/2\pi\mathbf{Z}$ by sending 1 to $\overline{2\pi\alpha}$. For the trivialization $\theta((\varphi, \overline{\psi})) = (\overline{\varphi}, \overline{\psi + \alpha\varphi})$ of $\mathbf{R} \times_\alpha S^1$, we obtain the Kronecker foliation of the two-torus with leaves the images of $\mathbf{R} \times \{\overline{\psi}\}$ in $\mathbf{R} \times_\alpha S^1$.

If D is an elliptic selfadjoint differential operator on M, then one can define $\tilde{D}$ on $\tilde{M}$ and then $\tilde{D} \otimes I$ on $\tilde{M} \times G$. Finally, $\tilde{D} \otimes I$ descends to a differential operator $\tilde{D} \otimes_\alpha I$ on the principal bundle $\tilde{M} \times_\alpha G$. The trivialization $\theta\colon \tilde{M} \times_\alpha G \to M \times G$ defines a unitary operator, also denoted θ, from $L^2(\tilde{M} \times_\alpha G)$ to $L^2(M \times G)$. We set $D_\alpha = \theta(\tilde{D} \otimes_\alpha I)\theta^*$, which will be a selfadjoint differential operator on $M \times G$ which acts elliptically along the leaves of $\mathcal{F}_\alpha$ and hence is longitudinally elliptic. We seek to study the Toeplitz operators defined for the positive spectral subspace $\mathcal{P}_\alpha$ of D_α.

To proceed as in the elliptic case depends on our being able to show that the commutators $[P_\alpha, \varphi]$, where φ is in $C(M \times G)$ and P_α is the orthogonal projection onto $\mathcal{P}_\alpha$, lie in some special C^*-algebra which plays the role of the compact operators for $\mathcal{F}_\alpha$. The most natural algebra to use is the foliation C^*-algebra $C^*(\mathcal{F}_\alpha)$ of Connes [7] although that causes technical problems since the commutators will not be in it. This problem is circumvented by using $f(D_\alpha)$ instead of P_α, where f is an approximation to the characteristic function for $\mathbf{R}^+$ in $C^\infty(\mathbf{R})$. If we now let $\mathscr{S}_\alpha$ be the C^*-subalgebra of $\mathscr{L}(L^2(M \times G))$ generated by $C^*(\mathscr{F}_\alpha)$, $f(D_\alpha)$, and $C(M \times G)$, then we have the short exact sequence

$$0 \to C^*(\mathscr{F}_\alpha) \to \mathscr{S}_\alpha \to C(M \times G) \oplus C(M \times G) \to 0,$$

where $f(D_\alpha)$ corresponds to $1 \oplus 0$. This extension yields an element $[D_\alpha]$ in

$$KK^1(C(M \times G) \oplus C(M \times G), C^*(\mathscr{F}_\alpha)),$$

the odd Kasparov group, which can be used to define the index homomorphism[1]

$$K^1(M \times G) \to K_0(C^*(\mathscr{F}_\alpha)).$$

Haar measure on G is an invariant transverse measure for $\mathscr{F}_\alpha$ and hence defines a trace Tr_α on $C^*(\mathscr{F}_\alpha)$.

We use the trace on $C^*(\mathscr{F}_\alpha)$ to define a homomorphism from $K_0(C^*(\mathscr{F}_\alpha))$ to $\mathbf{R}$ and hence obtain the real-valued index

$$K^1(M \times G) \to \mathbf{R}$$

which generalizes that defined for Toeplitz operators with almost periodic symbol by Coburn, Douglas, Schaeffer, and Singer in [6]. One can establish the odd analogue of the index theorem of Connes and Skandalis [10] to obtain the formula[2]

$$\mathrm{Index}_{\mathbf{R}} T_\varphi = \int_{S(T\mathscr{F}_\alpha)} \mathrm{Tch}(\varphi) \cup \mathrm{Td}(T\mathscr{F}_\alpha) \cup \mathrm{ch}(\sigma_1(D)) \, d\mu,$$

where the terms are as defined before. As is apparent, the expression for the topological index here is identical to that obtained by Atiyah, Patodi, and Singer for the relative eta invariant. One important starting point for the joint work with Hurder and Kaminker [13, 14, 15] was the question of whether this was a coincidence or whether reason lay behind it. Although we succeeded in establishing the connection, the path was not an obvious one and will require some further preliminaries to describe.

We have described the primary elliptic invariants for differential operators in terms of Fredholm index and K-theory, especially K-homology. Connes introduced another kind of invariance group for operator algebras, cyclic cohomology, and showed how to define the Chern character from K-homology to cyclic cohomology directly in terms of the operators [8]. Also, he showed in [9] and in [10] with Skandalis how to define such invariants for partially elliptic operators in the context of foliations, at least in the even case. We need, however, to discuss the Chern character in the odd case for selfadjoint operators on compact manifolds and in the context of foliations.

Let D be a selfadjoint elliptic pseudodifferential operator defined on the compact, closed manifold M of dimension n. Let F be a symmetric unitary for

---

[1] We are using the injection $K^1(M \times G) \to K^1(M \times G \vee M \times G)$ via the first copy of $M \times G$ followed by the product with $[D_\alpha]$.

[2] We use the notation T_φ for an operator in $\mathscr{S}_\alpha$ which maps to $\varphi \oplus 1$.

the polar decomposition of D. For φ in $C^\infty(M)$, the commutator $[F, \varphi]$ is a pseudodifferential operator of order -1 and hence belongs to the Schatten–von Neumann class $\mathcal{L}^k(L^2(M))$ for $k > n$. For $\varphi_0, \ldots, \varphi_k$ in $C^\infty(M)$ with k odd and $k > n$, we define the cyclic cocycle

$$c_D(\varphi_0, \ldots, \varphi_k) = C_k \operatorname{Tr}\{\varphi_0 [F, \varphi_1] \cdots [F, \varphi_k]\},$$

where C_k is a constant depending only on k. The cyclic cocycle is well-defined by the preceding remarks. The class $[c_D]$ in $HC^k(C^\infty(M))$ is the Chern character of $[D]$ in $K_1(M)$ [8]. Moreover, for an invertible φ in $C^\infty(M)$, we have

$$\operatorname{index} T_\varphi = \langle [D], [\varphi] \rangle = c_D(\varphi, \varphi^{-1}, \ldots, \varphi, \varphi^{-1}).$$

Thus the cyclic cocycle enables one to proceed directly from "small" commutators to the elliptic invariant.

If D_α is the longitudinally elliptic selfadjoint differential operator on $L^2(M \times G)$ defined before, then after some slight technicalities which we won't discuss (we must further modify $f(D_\alpha)$ by forming a two-by-two matrix to obtain a projection; cf. [13, 14]), the formula

$$c_{D_\alpha}^L(\varphi_0, \ldots, \varphi_k) = C_k \operatorname{Tr}_\alpha\{\varphi_0 [f(D_\alpha), \varphi_1] \cdots [f(D_\alpha), \varphi_k]\}$$

defines a cyclic cocycle on $C^\infty(M \times G)$, where Tr_α denotes the trace on the smooth foliation algebra $C^\infty(\mathscr{F}_\alpha)$. Moreover, the class $[c_{D_\alpha}^L]$ of this cyclic cocycle in $HC^k(C^\infty(M \times G))$ can also be regarded as the Chern character of $[D_\alpha]$ (although there is more to the story in this case; cf. [12]). Moreover, we have the pairing

$$\operatorname{index}_{\mathbf{R}} T_\varphi = c_{D_\alpha}^L(\varphi, \varphi^{-1}, \ldots, \varphi, \varphi^{-1})$$

for an invertible φ in $C^\infty(M \times G)$.

There is an additional way to define a cyclic cocycle using $f(D_\alpha)$ and basically the same formula. For φ in $C^\infty(M \times G)$, the commutator $[f(D_\alpha), \varphi]$ is not compact but is in $C^*(\mathscr{F}_\alpha)$. Although it is impossible to use the ordinary trace to define the cyclic cocycle, we can use the foliation trace. If instead of using φ in $C^\infty(M \times G)$, we use an operator k in $\mathscr{L}(L^2(M \times G))$ that is already "compact-like" in the direction transverse to $\mathscr{F}_\alpha$, then the $[f(D_\alpha), k]$ would be a compact operator and we would attempt to define a cyclic cocycle on these operators using the ordinary trace. Again we need some additional structure to accomplish this.

Let $\mathscr{F}_G$ be the trivial product foliation of $M \times G$, that is, the foliation in which the leaves are $x \times G$. Then the operator D_α is transversally elliptic relative to $\mathscr{F}_G$. Therefore, the above situation holds and we can define the transverse cyclic cocycle $c_{D_\alpha}^{\mathscr{A}}$ on $C^\infty(\mathscr{F}_G)$ by

$$c_{D_\alpha}^{\mathscr{A}}(k_0, \ldots, k_k) = C_k \operatorname{Tr}\{k_0 [f(D_\alpha), k_1], \ldots, [f(D_\alpha), k_k]\}.$$

Although the class of this cyclic cocycle expresses the index of the Toeplitz operators $P_\alpha k P_\alpha$ for k in $C^*(\mathscr{F}_G)$ to which an identity has been adjoined, our interest in $c_{D_\alpha}^{\mathscr{A}}$ concerns its connection with the eta invariant and spectral flow.

Suppose that $\rho: G \to U_n(\mathbf{C})$ is a unitary representation of G. Then the composition $\rho \circ \alpha: \pi_1(M) \to U_n(\mathbf{C})$ enables us to construct the vector bundle $\tilde{M} \times_{\rho\circ\alpha} \mathbf{C}^n$ over M which $\rho \circ \theta$ identifies with $M \times \mathbf{C}^n$. Since $M \times \mathbf{C}^n$ has the product Hermitian structure, the algebra $\mathrm{Hom}(M \times \mathbf{C}^n)$ is the C^*-algebra $C(M) \otimes M_n(\mathbf{C})$. If we consider the regular representation of G on $L^2(G)$, then by the Peter-Weyl theorem [18], we have $L^2(G) = \bigoplus_\rho \mathscr{V}_\rho$, where the $\mathscr{V}_\rho$ are finite-dimensional and the representation of G on $L^2(G)$ decomposes into irreducible unitary representation on the $\mathscr{V}_\rho$. Corresponding to any finite-dimensional invariant subspace $\mathscr{V}$ of $L^2(G)$, we have the natural inclusion of $\mathrm{Hom}(M \times \mathscr{V})$ in $C^*(\mathscr{F}_G)$. Moreover, $C^*(F_G)$ is just the completion of $C^\infty(M \times G \times G)$ acting on $L^2(M \times G)$ by "matrix multiplication" on the G factors and pointwise multiplication on the M.

Since $\mathrm{Hom}(M \times \mathscr{V}_\rho) \subseteq C^*(\mathscr{F}_G)$, then we can restrict the cyclic cocycle $c_{D_\alpha}^{\mathscr{A}}$ to $\mathrm{Hom}(M \times \mathscr{V}_\rho)$. It is not difficult to see that we obtain the cyclic cocycle that is defined by $D \times_{\nabla_{\rho\alpha}} I_n$. Evaluating $c_{D_\alpha}^{\mathscr{A}}$ at the function $\rho \circ \theta$ and its inverse determines not only the index of the associated Toeplitz operator but also the spectral flow between the two elliptic selfadjoint differential operators $D \otimes I_n$ and $D \times_{\nabla_\alpha} I_n$ both defined on $L^2(M) \otimes \mathbf{C}^n$ and having the same principal symbol. Since the relative eta invariant $\mathrm{Eta}(D \otimes I_n, D \times_{\nabla_{\rho\alpha}} I_n)$ can be shown to equal an average of spectral flow, it is possible to calculate the relative eta invariant using $c_{D_\alpha}^{\mathscr{A}}$. Moreover, the calculation is not ad hoc but "natural", although it involves new ideas. We consider this and return later to the connection with the eta invariant.

We have used the partially elliptic operator D_α on $M \times G$ to define two odd cyclic cocycles $c_{D_\alpha}^L$ on $C^\infty(M \times G)$ and $c_{D_\alpha}^{\mathscr{A}}$ on $C^\infty(\mathscr{F}_G)$ using the same formula. It is reasonable to ask whether the two are related, and, in particular, whether $c_{D_\alpha}^L$ can be obtained from $c_{D_\alpha}^{\mathscr{A}}$. Since $HC^k(C^\infty(\mathscr{F}_G)) = HC^k(C^\infty(M))$ and the class of $c_{D_\alpha}^{\mathscr{A}}$ equals $[c_D]$ in $HC^k(C^\infty(M))$, then $c_{D_\alpha}^{\mathscr{A}}$ contains less cyclic cohomological data than $c_{D_\alpha}^L$. Therefore, the method of going from $c_{D_\alpha}^{\mathscr{A}}$ to $c_{D_\alpha}^L$ will have to be "transcendental", not "topological".

Our technique could be described as "asymptotic quantization". Since there is no homomorphism $\Psi: C^\infty(M \times G) \to C^\infty(\mathscr{F}_G)$ intertwining the actions of $C^\infty(M)$, we define an asymptotic family of homomorphisms $\{\Psi_t\}_{t\in(0,1]}$ which does this as follows:

$$\Psi_t(\varphi) = \frac{e^{-t\Delta}}{\mathrm{Tr}\, e^{-t\Delta_1}} \cdot \varphi \quad \text{for } \varphi \text{ in } C^\infty(M \times G),$$

where Δ_1 is the invariant Laplacian on G, $\mathrm{Tr}\, e^{-t\Delta_1}$ is calculated on $L^2(G)$ and $\Delta = I \times \Delta_1$ on $M \times G$. It is not difficult to see that $\Psi_t(\varphi)$ is in $C^\infty(\mathscr{F}_G)$. The part that is less obvious is that

$$T c_{D_\alpha}^{\mathscr{A}}(\varphi_0, \varphi, \dots, \varphi_k) = \lim_{t \to 0^+} c_{D_\alpha}^{\mathscr{A}}(\Psi_t(\varphi_0), \dots, \Psi_t(\varphi_k))$$

exists for $\varphi_0, \varphi_1, \dots, \varphi_k$ in $C^\infty(M \times G)$ and that

$$T c_{D_\alpha}^{\mathscr{A}} = c_{D_\alpha}^L.$$

(Actually, this is slightly more complicated because one must also take a second limit allowing the function f to approach the characteristic function of $\mathbf{R}^+$ in the formula for both $c^{\varnothing}_{D_\alpha}$ and $c^L_{D_\alpha}$, but we omit the details; cf. [13, 15].) Thus we are able to renormalize the cyclic cocycle $c^{\varnothing}_{D_\alpha}$ to obtain $c^L_{D_\alpha}$. Moreover, the proof [15] is based on a generalization of the equidistribution results of Weyl.

Although the above identity may seem to be unsymmetric since the Laplacian Δ appears to be used only in defining $Tc^{\varnothing}_{D_\alpha}$, that is not the case. One can show that for k in $C^\infty(\mathscr{F}_\alpha)$, we have $\Psi_t(k)$ in $\mathscr{L}^1(L^2(M \times G))$ and

$$\mathrm{Tr}_\alpha k = \lim_{t\to 0^+} \mathrm{Tr}\{\Psi_t(k)\}.$$

Again, this involves a generalization of the equidistribution results of Weyl. Therefore, the cyclic cocycle $c^L_{D_\alpha}$ can also be obtained using the ordinary trace via an analogous renormalization. This is an example of a Fubini-type theorem for cyclic cocycles described in [12].

Now although the motivation for the renormalization process was to obtain $c^{\varnothing}_{D_\alpha}$ from $c^L_{D_\alpha}$, it also happens that the resulting averaging process expresses the relative eta invariant in terms of spectral flow. Hence one has that

$$\mathrm{Eta}(D \times I_n, D \times_{\nabla_{\rho\circ\alpha}} I_n) = Tc^{\varnothing}_{D_\alpha}(\rho \circ \alpha, (\rho \circ \alpha)^{-1}, \ldots, (\rho \circ \alpha)^{-1}),$$

where $Tc^{\varnothing}_{D_\alpha}$ has been extended to $M_n^\infty(M \times G)$ using the normalized trace on the fiber algebras. Since $[c^L_{D_\alpha}]$ yields the index class for the Toeplitz operators defined for D_α, while $[Tc^{\varnothing}_{D_\alpha}]$ yields the class for the relative eta invariant, we see that the equality of the topological formulas is not an accident and that the preceding development leads to a new and different proof of the index theorem of Atiyah, Patodi, and Singer [2]. Full details will appear in [14] and [15].

One advantage of the procedure just described is that topological formulas for the index of longitudinally elliptic operators, obtained via the heat equation method, can be used to understand the dependence or independence of the index on the basic data. On the other hand, such an understanding for the invariants of transversally elliptic operators is difficult to obtain but important since the invariants or the "renormalized" invariants often relate to secondary and spectral data. Thus the above bridge should be useful for this reason.

REFERENCES

1. M. F. Atiyah, *Global theory of elliptic operators*, Proc. Internat. Congr. on Functional Analysis and Related Topics (Tokyo, 1969), Univ. of Tokyo Press, 1970, pp. 21–29.

2. M. F. Atiyah, V. K. Patodi, and I. M. Singer, *Spectral asymmetry and Riemannian geometry*. I, II, III, Math. Proc. Cambridge Philos. Soc. **77** (1975), 43–69, **78** (1975), 405–432, **79** (1976), 71–99.

3. M. F. Atiyah and I. M. Singer, *The index of elliptic operators*. I, Ann. of Math. (2) **87** (1968), 484–530.

4. P. Baum and R. G. Douglas, *Toeplitz operators and Poincaré duality*, Proc. Toeplitz Memorial Conference (Tel Aviv, 1981) (ed. I. C. Gohberg), Birkhäuser, Basel, 1982, pp. 137–166.

5. L. G. Brown, R. G. Douglas, and P. A. Fillmore, *Extensions of C*-algebras and K-homology*, Ann. of Math. (2) **105** (1977), 265–324.

6. L. A. Coburn, R. G. Douglas, D. G. Schaeffer and I. M. Singer, *On C^*-algebras of operators on a half-space. II. Index theory*, Inst. Hautes Études Sci. Publ. Math. **40** (1971), 69–79.

7. A. Connes, *A survey of foliations and operator algebras*, Operator Algebras and Applications (ed. R. V. Kadison), Proc. Sympos. Pure Math., vol. 38, Amer. Math. Soc., Providence, R.I., 1981, pp. 521–628.

8. ____, *Non-commutative differential geometry*, Publ. Inst. Hautes Études Sci. **62** (1985), 257–360.

9. ____, *Cyclic cohomology and the transverse fundamental class of a foliation*, Geometric Methods in Operator Algebras (H. Araki and E. G. Effros, eds.), Pitman Research Notes, vol. 123, Longman, Harlow, 1986.

10. A. Connes and G. Skandalis, *The longitudinal index theorem for foliations*, Publ. Res. Inst. Math. Sci. Kyoto Univ. **20** (1984), no. 6, 1139–1183.

11. R. G. Douglas, *Invariant theory for elliptic operators*, Proc. Roy. Irish Acad. **86A** (1986), 161–174.

12. ____, *Elliptic invariants and operator algebras: Toroidal examples*, to appear.

13. R. G. Douglas, S. Hurder, and J. Kaminker, *Toeplitz operators and the eta invariant: The case of S^1*, Index Theory of Elliptic Operators, Foliations, and Operator Algebras (C. Schochet, ed.), Contemp. Math., vol. 70, Amer. Math. Soc., Providence, R.I., 1988.

14. ____, *The longitudinal cyclic cocycle and the index of Toeplitz operators*, in preparation.

15. ____, *Cyclic cocycles, renormalization and von Neumann eta invariants*, in preparation.

16. G. G. Kasparov, *Topological invariants of elliptic operators. I: K-homology*, Math. USSR-Izv. **9** (1975), 751–792.

17. R. Palais, *Seminar on the Atiyah-Singer index theorem*, Ann. of Math. Studies, no. 57, Princeton Univ. Press, Princeton, N.J., 1965.

18. F. Peter and H. Weyl, *Die Vollständigkeit der Primitiven Darstellungen einer Geschlossenen Kontinuierlichen Gruppe*, Math. Ann. **97** (1927), 737–755.

19. J. von Neumann, *Charakteristerung des Spectrums einer Integral Operator*, Hermann, Paris, 1935.

20. H. Weyl, *Über beschrankte quadratische Formen deren Differenz vollstetig ist*, Rend. Circ. Mat. Palermo **27** (1909), 373–392.

21. ____, *Ramifications, old and new, of the eigenvalue problem*, Bull. Amer. Math. Soc. **56** (1950), 115–139.

STATE UNIVERSITY OF NEW YORK AT STONY BROOK

Proceedings of Symposia in Pure Mathematics
Volume **48** (1988)

New Invariants of 3- and 4-Dimensional Manifolds

MICHAEL ATIYAH

1. Introduction. Hermann Weyl was probably the most influential mathematician of the twentieth century. The topics he chose to study, the lines he initiated and his general outlook have proved remarkably fruitful and have underpinned much of the development of the past fifty years. Weyl saw mathematics, and to some extent theoretical physics, as an organic whole and not as a collection of special subjects. The relation between geometry and physics was perhaps his central interest, but this led him deeply into the theory of Lie groups and differential equations. He would, I am sure, have been delighted in the resurgence of interest in this whole area which we are now witnessing. The geometry and physics of gauge theories, in its modern form, is one of the most exciting developments of our time and it rests ultimately on Weyl's pioneering ideas.

In this talk I am going to describe some of the most recent, and still incomplete, developments involving the application of ideas from the physics of gauge theories to the study of manifolds in 3 and 4 dimensions. The main results are due to S. K. Donaldson who initiated the whole programme a few years ago, but important contributions are being made by C. Taubes and A. Floer. Moreover mathematicians have learnt a great deal about the geometrical interpretation of physical ideas from E. Witten. His paper [**14**] on supersymmetry and Morse theory has been very influential in a number of ways.

I should emphasize that many of the results I shall describe have not yet been fully written up, but the general picture seems to be fairly clear and the ideas are so beautiful and simple that they deserve a nontechnical presentation. Detailed treatments will hopefully be provided in due course by Donaldson, Floer, and Taubes. I am grateful to all of them for explaining their ideas to me at this early stage.

1980 *Mathematics Subject Classification* (1985 *Revision*). Primary 81E13, 57R55 ;Secondary 58E05, 58F05.

Let me now outline in broad terms the picture I want to describe. It may be illustrated by the following scheme.

Dimension 2		Dimension 3		Dimension 4
π_1		π_1		$(\pi_1 = 0)$
$\operatorname{Aut} \pi_1$		Homology spheres		Differentiable
		$(H_1 = 0)$		structures
	$\leftarrow$	Casson invariant	$\leftarrow$	Donaldson
		Floer homology		invariants
X	$\leftarrow$	Y	$\leftarrow$	Z

In each dimension I have indicated the object of interest and the invariants used to study them. Thus in dimensions 2 and 3 the fundamental group π_1 is our main concern (and homology spheres concentrate on the nonabelian part), while in dimension 4, even when we restrict to simply connected manifolds, there are many different differentiable structures on the same topological manifold. These are detected by Donaldson's invariants [5], while in 3 dimensions there is the integer invariant recently introduced by Casson [3] and, as I shall explain, refined by Floer to give a homology theory. The horizontal arrows and the diagrams at the bottom are meant to indicate that the invariants in n dimensions (for $n = 4, 3$) can be related to those of an $(n-1)$-dimensional submanifold when we 'cut it in half'.

2. The Casson invariant. Let Y be an oriented homology 3-sphere, i.e., with $H_1(Y) = 0$. Then the Casson invariant $\lambda(Y)$ is roughly defined by

$$\lambda(Y) = \tfrac{1}{2}\{\text{number of irreducible representations } \pi_1(Y) \to \mathrm{SU}(2)\}.$$

Here, of course, we identify conjugate representations, and since $H_1(Y)$, the abelianization of $\pi_1(Y)$, is zero the only reducible representation in $\mathrm{SU}(2)$ is the trivial one, so irreducible = nontrivial.

The main problem is to give a proper way of counting the number of representations, so as to get a well-defined (and finite) integer. The most natural way has been developed by Taubes and involves identifying representations $\pi_1(Y) \to \mathrm{SU}(2)$ with *flat connections* on Y. We proceed as follows. Let

$\mathscr{A}$ = {space of $\mathrm{SU}(2)$-connections for the trivial bundle over Y},

$\mathscr{G}$ = {group of gauge transformations, i.e. maps $Y \to \mathrm{SU}(2)$},

$\mathscr{C} = \mathscr{A}/\mathscr{G}$.

Then, except for reducible connections, $\mathscr{C}$ is an infinite-dimensional manifold. Moreover $A \to F_A$ (the curvature of A) defines a natural 1-form F on $\mathscr{C}$. To see this note that tangents to $\mathscr{A}$ are 1-forms on Y with values in the Lie algebra of

SU(2), which pair naturally (since dim $Y = 3$) with Lie algebra–valued 2-forms such as the curvature. The Bianchi identity asserts that the $\mathscr{G}$-invariant 1-form on $\mathscr{A}$ defined by the curvature descends to a 1-form on $\mathscr{C}$.

The zeros of F, i.e. flat connections, correspond naturally to representations $\pi_1(Y) \to$ SU(2). Thus the "number of irreducible representations" appears as the "number of zeros of F" on the nonsingular part of $\mathscr{C}$. To make sense of this number one must, as in finite dimensions, consider perturbations of the 1-form to get simple zeros and then count them up with appropriate signs determined by orientations. Here, in an infinite-dimensional setting one must use appropriate Fredholm perturbations, and this has been carried out and justified by Taubes. The determination of signs is a subtle story and I shall return to this later in §3.

REMARKS. (1) With this definition it is not clear why $\lambda(Y)$ (rather than $2\lambda(Y)$) is an integer. In fact no very satisfactory explanation appears to be known at present.

(2) In principle SU(2) could by replaced here by SU(n), but then more care would need to be taken with reducible representations. So far this has not been fully investigated.

(3) The Casson invariant $\lambda(Y)$ has a more computational definition (due to Casson) which will be described in §5.

(4) The Casson invariant is quite a powerful invariant and was used to settle an outstanding problem on 3-manifolds.

3. Morse theory. The 1-form F on $\mathscr{C}$ given by the curvature turns out to be *closed*. It should therefore locally be the differential of a function on $\mathscr{C}$. In fact there is a well-known function $f: \mathscr{C} \to \mathbf{R}/\mathbf{Z}$ introduced by Chern and Simons and an elementary calculation shows that

$$F = 4\pi^2 \, df.$$

For convenience I recall the definition of f. Given a connection A for the trivial bundle $Y \times$ SU(2), let A_0 be the trivial or product connection and put $A_t = (1-t)A + tA_0$ for $0 \le t \le 1$. This is a path of connections on Y or equivalently a connection on $Y \times I$, where I is the unit interval. Now define

$$f(A) = \frac{1}{8\pi^2} \int_{Y \times I} \mathrm{Tr}\, F^2$$

where F is the connection on $Y \times I$. Note that this integral on a closed 4-manifold gives the second Chern class and so is an integer. For similar reasons $f(A)$ is invariant under the connected component $\mathscr{G}_0$ of the gauge group $\mathscr{G}$, but changes by integers under the full group (note that $\mathscr{G}/\mathscr{G}_0 \cong \mathbf{Z}$). Thus, as a function on $\mathscr{C}$, f is well-defined provided we take its values in $\mathbf{R}/\mathbf{Z}$.

Now in finite dimensions if a 1-form is the differential of a real-valued function then the zeros of the 1-form become critical values of the function. The number of zeros is the Euler characteristic of the manifold, but the Morse theory of the function gives us more precise information related to the homology of the manifold. Traditionally this relation is given by the *Morse inequalities* between

Betti numbers and numbers of critical points of different types. However, as Witten suggested in [15], one can do better. It is possible to use the critical point information to *construct* the homology of the manifold. Witten's idea is so beautiful and so important that I will now review it, before passing on to explain how Floer adapts it to the infinite-dimensional case of our manifold $\mathscr{C}$ of connections.

Assume therefore that we have a compact manifold M and a real-valued function f with only nondegenerate critical points. At each such critical point P the Hessian $H_P(f)$ is a nondegenerate quadratic form on the tangent space at P. It has a "type" (n_P^+, n_P^-), the integers n^+ and n^- being the number of $+$ and $-$ entries in a diagonalization. Of course $n_P^+ + n_P^- = \dim M$ is independent of P. As a first approximation to the homology of M Witten forms the chain groups C_q having one generator for each critical point P with $n_P^- = q$. The next and crucial step is to introduce a boundary operator $\partial: C_q \to C_{q-1}$. This will be given by a matrix with one entry for each pair of critical points (P, Q) with $n_P^- = q$, $n_Q^- = q - 1$. To define ∂ we first choose a generic metric on M and introduce the corresponding (descending) gradient flow of f. We then look at trajectories of this flow that start at P and end at Q. The number of such trajectories is *finite* and, counted with an appropriate sign, this gives the (P, Q) entry of ∂. One then verifies that $\partial^2 = 0$, so that we can define the *homology groups* of the complex C_*. This can be shown to be independent of the metric and finally identified with the homology of M.

REMARKS. (1) For Witten the homology of M is the Hodge–de Rham homology, represented by harmonic forms or, using f, as the zero eigenforms of the modified Laplacian Δ_{tf} where d is replaced by $e^{-tf} d e^{tf}$. In Δ_{tf} we have a potential term $t^2 |\text{grad } f|^2$ and so, for $t \to \infty$, the zero eigenforms are concentrated near the critical points of f. These are the classical ground states. However, the eigenform belonging to P has an exponentially small correction due to Q which is approximately computed by using the trajectories of grad f from P to Q. This is *quantum mechanical tunnelling*, which describes the probability of the transition $P \to Q$, which is forbidden classically, but can occur quantum mechanically. Thus the boundary operator ∂ of Witten's chain complex C_* is to be interpreted in terms of such tunnelling. This remark will acquire even more significance in the Floer theory we shall discuss next.

(2) For any pair of critical points (P, Q) with $n_P^- - n_Q^- = r$ say, the trajectories of grad f from P to Q form (for generic metrics) an $(r - 1)$-dimensional family. They are the intersections of the *unstable* manifold of P (with dimension n_P^-) and the *stable* manifold of Q (with dimension $n_Q^+ = \dim M - n_Q^- = \dim M - n_P^- + r$). This intersection has dimension r and is made up of an $(r-1)$-dimensional family of curves.

4. Floer homology. After our finite-dimensional digression in §3 on Witten's version of Morse theory we return to the space $\mathscr{C}$ of (classes of) connections on our homology 3-sphere Y and the Chern-Simons function $f: \mathscr{C} \to \mathbf{R}/\mathbf{Z}$.

To develop the corresponding Morse theory for this situation we have two main problems:

(1) f takes values in $\mathbf{R}/\mathbf{Z}$, not in $\mathbf{R}$.

(2) The Hessian of f at a critical point has both Morse indices n^+ and n^- *infinite*.

The first problem is not too serious and can be dealt with by passing to the infinite cyclic covering $\mathscr{C}_0 = \mathscr{A}/\mathscr{G}_0$. The second problem is more fundamental and presents essentially new features. To understand this let us recall that for the classical Morse theory of geodesics in Riemannian manifolds the Hessian (viewed as operator, rather than quadratic form, by using the metric) is of *Laplace type*; i.e., it is a second-order elliptic operator and hence is bounded below, so that n^- is always finite. In our present situation the Hessian is of *Dirac type*, i.e., it is of first order. In fact, this Hessian is essentially the operator $*d$ on $\Omega^1/d\Omega^0$, suitably extended to Lie algebra–valued forms.

The way around our difficulty is to observe that the important quantity in the Morse theory is not the Morse index n_P^-, but the *relative* Morse indices $n_{P,Q}^- = n_P^- - n_Q^-$ for pairs of critical points P, Q. Although $n_P^- = \infty$ for all P we can make sense of the differences $n_{P,Q}^-$, i.e, we can define relative Morse indices. This is done as follows.

First, using a fixed metric on Y we can extend the Hessians at the critical points P to a continuous family of self-adjoint (Dirac type) operators H_C for all $C \in \mathscr{C}$. In particular, for any continuous path $\alpha(t)$ from P to Q we get a 1-parameter family H_t of self-adjoint operators connecting H_P to H_Q. In this situation there is a standard integer invariant, the *spectral flow*, that can be defined [2]. It describes the net number of negative eigenvalues of H_P that cross over and end up as positive eigenvalues of H_Q. This is clearly a regularization of the formal quantity $n_P^- - n_Q^-$. It is a topological invariant and so depends only on the homotopy class of the path α from P to Q. If $\mathscr{C}$ (or better the irreducible part of $\mathscr{C}$) were simply connected α would be unique up to homotopy and so the relative Morse index $n_{P,Q}^-$ would be well-defined. Since $\mathscr{C}$ is not simply connected, but has the infinite cyclic covering $\mathscr{C}_0$, it follows that $n_{P,Q}^-$ is defined modulo the spectral flow round a generating closed loop in $\mathscr{C}$. But such a spectral flow can be computed from the index theorem on $Y \times S^1$ [2] and one finds the answer 8. This is essentially the same index calculation which determines the dimension of the instanton moduli space, and which figures prominently in Donaldson's work.

It is now clear how one should proceed, at least formally. Assume first that all nontrivial critical points of f (i.e., all irreducible representations $\pi_1(Y) \to \mathrm{SU}(2)$) are nondegenerate: if not we would have to make a Fredholm perturbation as in Taubes' approach to the Casson invariant. Now we follow Witten and define chain groups C_* indexed modulo 8. Finally, using trajectories of grad f from P to Q we define a boundary operator ∂, prove $\partial^2 = 0$ and derive homology groups H_*. Finally one should prove these homology groups are independent of the various choices (perturbations) made.

The groups obtained in this way have a mode 8 grading but no obvious start of the grading, i.e., no obvious H_0. In fact with more care, comparing with the trivial representation, one can fix the grading by identifying H_0. Thus we get groups labelled by the integers modulo 8.

REMARK. The problem of signs involved in the definition of ∂ is essentially the same as the problem of giving signs to the zeros of grad f in Taubes' approach. Also the labelling of the dimensions modulo 8 is a refinement of the modulo 2 labelling that determines the sign of the Casson invariant.

The most important ingredient in this whole procedure is the definition of the boundary ∂, using trajectories of grad f. Now, such a trajectory is explicitly a solution of the differential equation

$$(4.1) \qquad \frac{dA}{dt} = - * F_A$$

(since grad $f = *F_A$). This equation, interpreted on the infinite cylinder $Y \times R$, is just the anti-self-duality equation which defines instantons. The boundary conditions we impose are that for $t \to -\infty$ the connection converges to the flat connection corresponding to P, while $t \to +\infty$ corresponds similarly to Q. Recalling Witten's interpretation of ∂ as a tunnelling effect, we see that we are using instantons to tunnel from the ground state, or vacuum, of one flat connection on Y to the ground state of another flat connection. This is precisely the way physicists use instantons and was their original motivation. For this reason, Witten even used the word 'instanton' in the classical Morse theory picture for the trajectories connecting consecutive critical points.

Of course the analytical justification of the formal procedure just outlined requires much careful analysis. However, the essential ingredients centre round the analytical properties of instantons, in particular compactness questions and Fredholm perturbation theory. These properties are by now well understood as a result of the basic work of Uhlenbeck, Taubes, and Donaldson. The conclusion is that we have new invariants defined for homology 3-spheres Y in the form of homology groups, denoted $HF_q(Y)$, indexed by $q \in \mathbf{Z}_8$.

Actually, just as in finite dimensions, reversing the sign of f interchanges n^+ and n^- and corresponds to Poincaré duality. Therefore we should really distinguish between two dual homology groups $HF^+(Y)$ and $HF^-(Y)$. They depend on the orientation of Y and switch if we reverse the orientation.

Of course, from the way we derived these homology groups, the Casson invariant (up to a factor 2) is just the corresponding Euler characteristic:

$$2\lambda(Y) = \sum_{q=0}^{7} (-1)^q \dim HF_q^+(Y).$$

In this sense the groups HF represent a refinement of the Casson invariant, and therefore should be very interesting invariants of Y.

REMARKS. (1) In finite dimensions the Witten-Morse approach leads to the usual homology of the underlying manifold. In the infinite-dimensional case of our space $\mathscr{C}$ the homology groups HF that we obtained are not related to the ordinary homology groups of $\mathscr{C}$. I shall explain later that they should be viewed as "middle-dimensional" homology groups.

(2) The letter F in HF can stand for Floer who introduced these groups [8]. It can also appropriately stand for Fredholm, Fermi or Fock. This will become clear in a later section.

5. Relation with 2 dimensions.

I will now indicate how the Casson invariant, and, more generally, the Floer homology groups, can be calculated when our homology 3-sphere Y is presented in terms of a *Heegaard splitting*.

A Heegaard splitting of Y is a decomposition into 2 halves $Y^\pm$, where

$$\partial Y^+ = X = -\partial Y^-$$

with X a compact Riemann surface of genus g say, and $Y^\pm$ are handlebodies of genus g (connected sums of g solid tori).

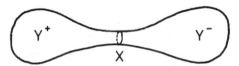

In a suitable basis $\pi_1(X)$ is generated by $A_1, \ldots, A_g$, $B_1, \ldots, B_g$ with the single relation

$$\Pi[A_i, B_i] = 1,$$

$\pi_1(X) \to \pi_1(Y^+)$ sends each $B_i \to 1$ and the images of the A_i freely generate $\pi_1(Y^+)$. For $\pi_1(Y^-)$ a similar description holds relative to a different basis—i.e., after applying an automorphism of $\pi_1(X)$.

Casson's procedure [3] to define his invariant $\lambda(Y)$ is to count the "number of representations of $\pi_1(Y)$" using the diagram

$$
\begin{array}{ccc}
\pi_1(X) & \longrightarrow & \pi_1(Y^+) \\
\downarrow & & \downarrow \\
\pi_1(Y^-) & \longrightarrow & \pi_1(Y)
\end{array}
$$

This shows that a representation of $\pi_1(Y)$ is the same as a pair of representations of $\pi_1(Y^+)$, $\pi_1(Y^-)$ which agree when pulled back to $\pi_1(X)$.

Now the (classes of) representations $\pi_1(X) \to \mathrm{SU}(2)$ form a moduli space M which has been extensively studied in algebraic geometry [1, 12, 13]. In particular, after removing the reducible representations, it is a manifold of dimension $6g - 6$. If X is given a complex structure then M becomes naturally a complex (algebraic) variety and a metric on X induces a Kähler metric on M (outside its singularities). The representations of $Y^\pm$ give subspaces $L^\pm \subset M$ of dimension $3g - 3$. We can therefore define the Casson invariant $\lambda(Y)$ by

$$2\lambda(Y) = L^+ \cap L^-,$$

where we count the intersections in M (away from the singularities of M). We are now dealing with ordinary homology and usual intersection theory (except that care has to be taken over the singular points of M).

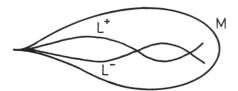

REMARKS. (1) If this approach is taken as a definition of $\lambda(Y)$ then one must show independence of the Heegaard splitting.

(2) Casson uses this approach to produce an effective algorithm for computing $\lambda(M)$, when M is described in terms of surgery on links in S^3 (another way of presenting homology 3-spheres).

Since the Floer homology groups are a refinement of the Casson invariant, it is now reasonable to ask if there is a way of computing $HF(Y)$ using the Heegaard splitting and the Riemann surface X. I shall outline an approach to this problem, which has yet to be fully worked out.

Since M is a Kähler manifold (with singularities) it is in particular symplectic. In fact, as shown in [1], the symplectic structure is canonical and independent of the metric on X. Moreover $L^\pm$ are *Lagrangian* submanifolds, i.e. submanifolds of middle dimension on which the symplectic 2-form ω of M is identically zero. Now Floer [6, 7] has studied, in general, the problem of intersections of Lagrangian submanifolds of compact symplectic manifolds and, for this purpose, has developed a homology theory. From an analytical point of view this is very similar to the theory leading to the HF groups described in §4. When applied to the particular case of $L^\pm$ in M above it is highly plausible that it should coincide with the theory of §4, as I shall indicate later. So first let me outline Floer's "symplectic Morse theory".

We start from any compact symplectic manifold M and two (connected) Lagrangian submanifolds L^+ and L^-. Consider the space Q of paths in M starting on L^- and ending on L^+. Assume for simplicity that $L^+ \cap L^-$ is not empty (otherwise the theory will be trivial) and choose a base point $m_0 \in L^+ \cap L^-$. Define a function $f(p)$ on Q as the area (integral of the symplectic 2-form ω) of a strip obtained by deforming the path p to the constant path m_0.

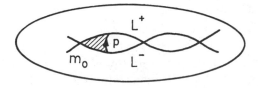

Since L^+ and L^- are Lagrangian, and ω is closed, this area is unchanged under continuous variations of the strip (with p fixed). However topologically inequivalent strips will differ in area by a "period" of ω. If for simplicity we assume L^+ and L^- simply connected these periods are just the usual periods of ω over 2-cycles in M. If $b_2(M) = 1$ there is just one period which by rescaling ω we can take to be 1. This means f becomes a function

$$f\colon Q \to \mathbf{R}/\mathbf{Z}.$$

The critical points of f are easily seen to be the constant paths corresponding to the points of intersection of L^+ and L^-. The Hessian is again of Dirac type and one can define a relative Morse index as in §4. This turns out to be well-defined modulo $2N$, where $c_1(M) = N[\omega]$, $c_1(M)$ being the first Chern class of M (note that symplectic manifolds have Chern classes) and $[\omega]$ is the class of ω in $H^2(M)$.

The trajectories of $\operatorname{grad} f$ correspond to *holomorphic* strips (with boundaries in $L^\pm$) in the sense of Gromov [10]. If M is actually complex Kähler then these are just holomorphic strips in the usual sense.

In this way, following Witten as in §4, Floer defines homology groups graded by Z_{2N} as intrinsic invariants of (M, L^+, L^-).

If now we take (M, L^+, L^-) to be the moduli spaces arising from a Heegaard splitting of a homology 3-sphere Y it is then reasonable to conjecture that the groups defined in the symplectic context (with care taken of the singularities of M) coincide with the groups $HF(Y)$ of §4.

Note that in both cases the representations $\pi_1(Y) \to \mathrm{SU}(2)$ give the generators of the chain group (provided these representations are nondegenerate). One has then to compare the relative Morse indices and the boundary operator ∂.

Geometrically, a path on M, i.e. a 1-parameter family of flat connections on the Riemann surface X, can be viewed as a connection on the cylinder $X \times R$. Moreover the boundary conditions (corresponding to L^+ and L^-) imply that, asymptotically as $t \to \pm\infty$, the connection extends (as a flat connection) over $Y^\pm$, thus giving essentially a connection on Y. In this way the symplectic theory for paths in M should be related to a limiting case of the Floer theory for the space $\mathscr{C}$ of connections on Y. Note that the limit is one in which Y is stretched out along its "neck", so that the two ends get further and further apart.

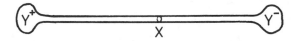

6. Donaldson invariants. Donaldson [5] has introduced certain invariants for smooth 4-manifolds which appear to be extremely powerful in distinguishing different differentiable structures. These invariants are defined in the following context. Let Z be an oriented simply connected differentiable 4-manifold and let b_2^+ and b_2^- be the number of $+$ and $-$ terms in a diagonalization of the quadratic (intersection) form on $H_2(Z)$. We assume b_2^+ odd and > 1. Note that, for a complex algebraic surface, we have the theorem of Hodge:

$$b_2^+ = 1 + 2p_g,$$

where p_g is the geometric genus (number of independent holomorphic 2-forms). Thus b_2^+ is odd and > 1 when $p_g \neq 0$.

The Donaldson invariants are a sequence of integer polynomials ϕ_k on $H_2(Z)$. Here $k > k_0$ and the degree of ϕ_k is

$$(6.1) \qquad\qquad d(k) = 4k - 3(b_2^+ + 1)/2.$$

Two key results of Donaldson [5] indicate the power of these invariants:

THEOREM 1. *If $Z = Z_1 \ast Z_2$ is a connected sum with $b_2^+(Z_i) \neq 0$ for $i = 1, 2$ then $\phi_k(Z) \equiv 0$ for all k.*

THEOREM 2. *If Z is algebraic, then for $k > k_1(Z)$, $\phi_k(Z) \not\equiv 0$.*

These theorems together show that, viewed as smooth manifolds, algebraic surfaces are essentially indecomposable. Note however that blowing-up points always leads to a decomposition in which one factor has $b_2^+ = 0$, $b_2^- = 1$.

Donaldson's invariants are defined using instantons and, in general, are impossible to compute directly. However, for algebraic surfaces another theorem of Donaldson [4] implies that his invariants can be calculated algebraically and this in particular leads to Theorem 2.

For nonalgebraic indecomposable 4-manifolds, the question of computing the Donaldson invariants is therefore an interesting and important one. In particular suppose the quadratic form A of Z is a direct sum $A = A_1 \oplus A_2$ with $b_2^+(A_i) \neq 0$. If $\phi_k(Z) \neq 0$ for all k then from Theorem 1 we know that we cannot decompose Z as a connected sum with Z_i having quadratic form A_i. However, it is known [9] that one can always decompose Z along a *homology 3-sphere Y*, inducing the algebraic decomposition $A = A_1 \oplus A_2$ on homology.

(6.2)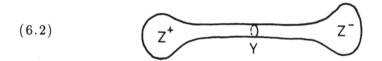

This shows that the indecomposability of Z as a usual connected sum (along a genuine 3-sphere), measured by the nonvanishing of the Donaldson invariants, is somehow reflected in the nontriviality of the homology 3-sphere Y.

My aim is now to explain (in the next section) how Donaldson relates his invariants for Z to the Floer homology groups of Y. For this we shall first have to recall briefly the definition of the Donaldson invariants ϕ_k.

We fix a Riemannian metric on Z, a positive integer k and look at the moduli space $M_k(Z)$ of k-instantons on Z, i.e., solutions of the anti-self-duality equations $*F = -F$ with $c_2 = k$. For generic metrics this is a manifold of dimension $2d(k)$, where $d(k)$ is defined by (6.1). If $d(k) = 0$ then M_k is a finite set of points and, when suitably counted, this is Donaldson's invariant. For $d(k) > 0$ we fix $d(k)$ spherical cycles $\alpha_i \colon S^2 \to Z$ with homology classes $[\alpha_1], \ldots, [\alpha_d]$. Each such cycle defines a codimension 2 submanifold A_i of M_k, consisting of connections on Z which pull back via α_i to *special* connections on S^2. A special connection is one which defines a *nontrivial* holomorphic bundle (note: in dimension 2 a unitary connection defines a holomorphic structure). We now consider the intersection number

$$A_1 \cap A_2 \cap \cdots \cap A_d$$

as a function of $\alpha_1, \ldots, \alpha_d$. It depends only on the classes $[\alpha_1], \ldots, [\alpha_d]$ and its values defines $\phi_k([\alpha_1], \ldots, [\alpha_d])$ as a symmetric d-linear function on $H_2(Z)$.

Because M_k is not compact, care has to be taken "at ∞" and this is where the restriction $k > k_0$ enters. Essentially M_k can be compactified and $k > k_0$ ensures that ∂M_k has codimension ≥ 2, which is enough to define intersection numbers (via the fundamental class of M_k).

7. Relation to Floer homology. I now consider a decomposition of the 4-manifold Z along a homology 3-sphere Y as in Figure (6.2). The idea is to study the instanton equations on Z by considering them separately on Z^+ and Z^- and then matching boundary values along Y.

A much simpler prototype problem may help to illustrate the ideas. Consider the 2-sphere S^2 cut in half along the equator. To construct holomorphic functions on S^2 we consider holomorphic functions in each hemisphere and compare their boundary values. This gives the usual Hardy spaces of Fourier series

$$H^+ = \left\{ \sum_{n \geq 0} a_n z^n \right\}, \qquad H^- = \left\{ \sum_{n \leq 0} a_n z^n \right\},$$

and the intersection $H^+ \cap H^-$ gives of course just the constant functions (the only global holomorphic functions on S^2).

This example was both lower-dimensional and linear. We can keep to dimension 2 but make the problem *nonlinear* by looking at the construction of holomorphic maps

$$S^2 \to P,$$

where P is some complex manifold (e.g. projective space). Again, cutting S^2 into half along the equator, a holomorphic map in either hemisphere is determined by its restriction to S^1 which is a point of the free loop space LP. We thus get H^+ and H^- subspaces of LP, but this time these are not linear. Nevertheless the global holomorphic maps are still given by $H^+ \cap H^-$. If we linearize this problem we recover our earlier one, so that $H^\pm$ should be seen as infinite-dimensional manifolds of approximately "half" the dimension of LP.

After this 2-dimensional digression let us return to the 4-dimensional situation. Since the instanton equations (4.1) are (like Cauchy-Riemann) of first order, a solution is determined by the appropriate boundary data, which in this case is just a connection on Y (up to equivalence). Thus we should look in the space $\mathscr{C}(Y)$ at the two spaces $\Sigma^\pm$ consisting of boundary values of solutions of the instanton equations in $Z^\pm$ respectively. Their intersection gives global solutions on Z.

Suppose for simplicity that $d(k) = 0$, so that the Donaldson invariant is just an integer and describes simply the algebraic number of k-instantons on Z. This number should then be computed as an intersection number of Σ^+ and Σ^- in

$\mathscr{C}(Y)$. For this we need to have an appropriate homology theory for $\mathscr{C}(Y)$ in which Σ^+ and Σ^- represent cycles. The Floer homology groups $HF^+(Y)$ and $HF^-(Y)$ provide just such a framework, as I shall try to explain.

Let us revert briefly to the finite-dimensional Morse theory. There if we have a geometric cycle α and we want to associate to it a cycle in Witten's complex, we push it along the gradient flow and see which critical points it "hangs" on. If β is a cycle of complementary dimension and we want to compute the intersection number $\alpha \cdot \beta$ we should deform β along the *ascending* gradient flow of f and find which critical points it "hangs" on. The intersection number $\alpha \cdot \beta$ is now reduced to local calculations near the critical points.

For our infinite-dimensional manifold $\mathscr{C}$ the gradient flows are only defined for appropriately restricted initial data, since we have to solve the heat equation for an operator which is bounded neither above nor below. However, the cycles Σ^+ and Σ^- provide suitable data for the two opposite flows. In this way Donaldson assigns classes $[\Sigma^+] \in HF^+(Y)$ and $[\Sigma^-] \in HF^-(Y)$ whose pairing, under the natural (Poincaré duality) map $HF^+(Y) \otimes HF^-(Y) \to \mathbf{Z}$, gives the Donaldson invariant $\phi_k(Z)$.

The general case when $d(k) > 0$ is essentially similar. Given classes $[\alpha_i] \in H_2(Z)$ for $i = 1, 2, \ldots, d$, we choose an integer r with $0 \le r \le d$, and representative spherical cycles

$$\alpha_i \colon S^2 \to Z^+, \qquad i = 1, \ldots, r,$$
$$\alpha_j \colon S^2 \to Z^-, \qquad j = r+1, \ldots, d.$$

The boundary values of instantons on Z^+ which are special on $\alpha_1, \ldots, \alpha_r$ define a "cycle" $\Sigma^+(\alpha_1, \ldots, \alpha_r) \subset \mathscr{C}$ and hence a Floer homology class $[\Sigma^+(\alpha_1, \ldots, \alpha_r)] \in HF^+(Y)$. Similarly we define $[\Sigma^-(\alpha_{r+1}, \ldots, \alpha_d)] \in HF^-(Y)$. Pairing these together then defines a homomorphism of symmetric products:

$$(7.1) \qquad S^r(H_2(Z^+)) \otimes S^{d-r}(H_2(Z^-)) \to \mathbf{Z}.$$

Since

$$H_2(Z) = H_2(Z^+) \oplus H_2(Z^-),$$
$$S^d(Z) = \sum_{r=0}^{d} S^r(H_2(Z^+)) \otimes S^{d-r}(H_2(Z^-)),$$

summing (7.1) over r gives a homomorphism

$$S^d(H_2(Z)) \to \mathbf{Z}$$

and Donaldson proves that this is his invariant ϕ_k.

REMARKS. (1) Formalizing the construction described above, we see that, when $Y = \partial Z^+$ with Y a homology 3-sphere, there is a sequence of polynomials ϕ_r on $H_2(Z^+)$ with values in $HF^+(Y)$. These may be viewed as a generalization, or a relative version, of the Donaldson polynomials for closed manifolds.

(2) When Y is the standard 3-sphere so that Z is just the usual connected sum, then $HF^+(Y) = 0$ and Donaldson's Theorem 1 follows as immediate corollary

of the general procedure for computing $\phi_k(Z)$ just outlined. Conversely, if we know that Z is indecomposable, e.g. if Z is an algebraic surface, it follows that $HF^+(Y) \neq 0$. Thus Donaldson's Theorem 2 implies the nontriviality of the Floer homology groups for many homology 3-spheres Σ, namely those occurring in a pseudodecomposition of algebraic surfaces.

8. Concluding remarks. Let me return now to the Floer homology groups $HF^\pm(Y)$ of a homology 3-sphere Y, and make some general heuristic comments.

For a finite-dimensional manifold there are various ways of defining the homology groups depending on the choice of chain complex. In particular we have:

(1) the Witten complex of a Morse function,

(2) the classical chain complex of geometrically defined cycles (of various kinds),

(3) the de Rham complex,

(4) the Hodge theory.

Witten's argument using Δ_{tf} with $t \to \infty$ relates (1) and (4), while pushing along the gradient flow of f relates (1) and (2). The relation of (3) to (4) is by standard elliptic theory. Definition (1) is the most "finite", using least analysis, while (4) uses the most.

For our infinite-dimensional manifold $\mathscr{C}$ of connections on Y (or better, for the infinite cyclic covering $\mathscr{C}_0$) the Floer definition follows (1), using the Chern-Simons function. Donaldson's work essentially uses (2), since the "cycles" $\Sigma^\pm \subset \mathscr{C}$ occur as boundary values of instantons.

Both the Floer and Donaldson approaches show clearly that the homology in question occurs in the "middle dimensions" of the infinite-dimensional manifold $\mathscr{C}$. We could attempt to give a rigorous definition of such middle-dimensional cycles (at least if they are smooth) on the following lines. As already pointed out, a metric on Y enables us to define a continuous family of self-adjoint operators H_C acting on the tangent space to $\mathscr{C}$ at the point C. These are of Dirac type and their spectral decomposition defines subspaces T^+ and T^- of the tangent bundle of $\mathscr{C}$. These are not quite continuous because there are finite jumps in dimension whenever C is in the exceptional set $\mathscr{C}'$ where H_C has a zero eigenvalue. Now given a submanifold $K \subset \mathscr{C}$ we can define it to be a *positive cycle* if, for each $C \in \mathscr{C}$, the tangent space TK is close, in an appropriate sense, to T^+. Close should mean that the projection $TK \to T^+$ is *Fredholm* while the projection $TK \to T^-$ is *compact*. Modulo an appropriate equivalence relation, these should define the *positive* homology groups. Similarly, replacing T^+ by T^-, we would have negative cycles leading to *negative* homology. The index of the Fredholm operator $TK \to T^+$ of a positive cycle would define its (renormalized) dimension, an integer ($+$ or $-$). The jumps in T^+ mean that this dimension would depend on the component of $\mathscr{C} - \mathscr{C}'$ in which one was working. The Donaldson cycles $\Sigma^\pm$ should fit into this framework.

A more elementary and obvious case when such "middle-dimensional" homology groups enter is for a product $\mathscr{S}^+ \times \mathscr{S}^-$. Each factor has ordinary homology (represented by finite-dimensional cycles) and dually ordinary cohomology represented by finite-codimensional cycles. In the product we then have four types of cycle:

$$\text{finite} \times \text{finite}, \qquad \text{cofinite} \times \text{cofinite},$$
$$\text{cofinite} \times \text{finite}, \qquad \text{finite} \times \text{cofinite}.$$

The first two give ordinary homology and cohomology respectively. The other two are quite different from these and give in an obvious sense "middle-dimensional" homology, one "positive" and one "negative".

Our space $\mathscr{C}$ is not globally a product but only infinitesimally (because its tangent bundle decomposes). Thus its middle-dimensional homology cannot be reduced to ordinary homology and cohomology by a factorization. Thus the Floer homology groups are, from a topological point of view, something essentially new.

Let us now go on to consider the appropriate de Rham theory for $\mathscr{C}$. We clearly want a de Rham complex Ω^+ on the following lines. If e_n ($n \in \mathbf{Z}$) is an orthonormal base for T_C given by the spectrum of H_C, so that T_C^+ is spanned by $n \geq 0$, consider the "volume element" of T_C^+:

$$\omega = e_0 \wedge e_1 \wedge \cdots$$

and also those infinite wedge products which differ from ω by only finitely many terms. Dualizing and taking linear combinations should define the "positive" differential forms at C. Similarly there would be Ω^- and these should lead to the correct de Rham theories.

The semi-infinite volume element ω is of course familiar to all physicists as the vacuum vector of a Fermionic Fock space. Differential forms in Ω^+ are therefore fields of a Fermionic quantum field theory. If we go further and introduce the Laplacian Δ_f^+ (where f is the Chern-Simons function) then this should be the Hamiltonian of the quantum field theory, and the harmonic forms are therefore the ground states of the QFT. Thus, purely formally (and ignoring the trivial ground state), we conclude that the *Floer homology groups $HF^+(Y)$ are the ground states of the QFT with Hamiltonian Δ_f^+.*

All of this discussion is essentially implicit in Witten's original paper [**14**]. Certainly Witten was well aware that QFT was concerned with middle-dimensional homology. Of course our discussion here is completely formal since QFT in dimension $3 + 1$ does not yet exist as a proper mathematical theory and in particular there is no rigorous definition of the Hamiltonian Δ_f^+. However, if we considered instead Floer's symplectic theory for paths in a symplectic manifold then QFT in $1 + 1$ dimensions is in much better shape and is intensively investigated in connection with strings. The connection between Floer homology and QFT should therefore be taken more seriously in this context.

Finally let me list a few of the major problems that are still outstanding in this area.

(1) Prove that the two definitions of $HF(Y)$ agree (one using $\mathscr{C}(Y)$ and the other via a Heegaard splitting). NOTE. In view of the physical observations

above, this problem suggests a comparison between two quantum field theories, one in $1 + 1$ dimensions and the other in $3 + 1$ dimensions.

(2) Find an algorithm to compute $HF(Y)$ which generalizes Casson's algorithm for his invariant.

(3) Find a method to compute the Donaldson invariants $S^*(H_2(Z^+)) \to HF^+(Y)$ when $Y = \partial Z^+$.

More speculatively, I would like to end with

(4) Find a connection with the link invariants of Vaughan Jones [11].

As circumstantial evidence that this is reasonable I will list some properties shared by Floer homology and the Jones polynomial.

(i) Both are subtle 3-dimensional invariants.

(ii) They are sensitive to orientation of 3-space (unlike the Alexander polynomial).

(iii) They depend on Lie groups: $SU(2)$ in the first instance but capable of generalization.

(iv) There are 2-dimensional schemes for computing these 3-dimensional invariants.

(v) Whereas the variable in the Alexander polynomial corresponds to $\pi_1(S^1)$, the variable in the Jones polynomial appears to be related to $\pi_3(S^3)$, the origin of "instanton numbers".

(vi) Both have deep connections with physics, specifically quantum field theory (and statistical mechanics).

REFERENCES

1. M. F. Atiyah and R. Bott, *The Yang-Mills equations on Riemann surfaces*, Philos. Trans. Roy. Soc. London A **308** (1982), 523–615.

2. M. F. Atiyah, V. K. Patodi and I. M. Singer, *Spectral asymmetry and Riemannian geometry*. III, Math. Proc. Cambridge Philos. Soc. **79** (1976), 71–99.

3. A. Casson, *An invariant for homology 3-spheres*, Lectures at MSRI, Berkely, 1985.

4. S. K. Donaldson, *Anti-self-dual Yang-Mills connections over complex algebraic surfaces and stable vector bundles*, Proc. London Math. Soc. **50** (1985), 1–26.

5. ____, *Geometry of 4-manifolds*, Proc. Internat. Congress of Mathematicians (Berkeley, 1986), Amer. Math. Soc., Providence, R.I., 1987.

6. A. Floer, *A relative Morse index for the symplectic action*, Courant Institute preprint, 1987.

7. ____, *Morse theory for Lagrangian intersections*, Courant Institute preprint, 1987.

8. ____, in preparation.

9. M. Freedman and L. Taylor, Λ-*splitting 4-manifolds*, Topology **16** (1977), 181–184.

10. M. Gromov, *Pseudoholomorphic curves in symplectic manifolds*, Invent. Math. **82** (1985), 307–347.

11. V. F. R. Jones, *A polynomial invariant for knots via von Neumann algebras*, Bull. Amer. Soc. (N.S.) **12** (1985), 103–111.

12. M. S. Narasimhan and C. S. Seshadri, *Stable and unitary vector bundles on a compact Riemann surface*, Ann. of Math. (2) **82** (1965), 540–567.

13. P. E. Newstead, *Topological properties of some spaces of stable bundles*, Topology **6** (1967), 241–262.

14. E. Witten, *Supersymmetry and Morse theory*, J. Differential Geom. **17** (1982), 661–692.

MATHEMATICAL INSTITUTE, OXFORD UNIVERSITY, ENGLAND

Proceedings of Symposia in Pure Mathematics
Volume 48 (1988)

Moduli Spaces and Homotopy Theory

CLIFFORD HENRY TAUBES

Topology and Geometry name mathematical subjects with robust interaction, often along the basic theme: Study geometric representatives for topological invariants. And, the search for geometry to represent topological invariants has proved immensely stimulating to both disciplines. This lecture provides three examples; one classical, and two modern.

The classical example is Hodge theory, where the rational cohomology groups of a compact manifold (a topological invariant) can be represented (after the choice of a Riemannian metric) by the vector space of harmonic differential forms. Hermann Weyl completed the proof the Hodge theorem.

The first modern example is Graeme Segal's theorem concerning the moduli space of holomorphic maps from S^2 to itself. The second modern example is a theorem of the author concerning the moduli space of self-dual connections on a compact 4-dimensinal manifold.

The basic theme is also manifest in the work of Blaine Lawson [L] which he described in his lecture here in this symposium.

I. Hodge Theory.. A compact, oriented, smooth manifold, M, is the arena. The specification of the differential structure on M provides sufficient data to construct the de Rham complex on M

$$(1.1) \qquad 0 \to \Omega^0 \xrightarrow{d} \Omega^1 \xrightarrow{d} \Omega^2 \xrightarrow{d} \cdots \xrightarrow{d} \Omega^n \to 0.$$

Here, Ω^p is the vector space of smooth p-forms on M (sections of $\Lambda^p T^* M$), and $d\colon \Omega^p \to \Omega^{p+1}$ is the exterior derivative. Also, $n \equiv \text{dimension}(M)$.

Since $d^2 \equiv 0$, the pth de Rham cohomology group

$$(1.2) \qquad H^p \equiv (\ker d \subset \Omega^p)/\operatorname{Im} d$$

is well defined. De Rham proved that H^p is a homotopy invariant of M, and that it is isomorphic to the singular cohomology of M with rational coefficients.

PROBLEM. Find a geometric representation for H^p.

1980 *Mathematics Subject Classification* (1985 *Revision*). Primary 53C05, 58B05, 58E99.
Research supported in part by the NSF.

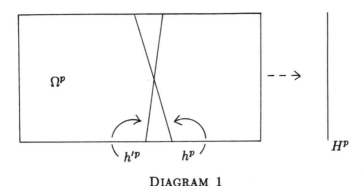

DIAGRAM 1

Hodge, with help at the end from Weyl, described such a geometric representation. (See, e.g., F. Warner's book [**W**] for a full discussion of the Hodge theorem.) To specify the *geometry*, Hodge requires that a Riemannian metric for T^*M be chosen. Such a metric defines a volume form for integration on M, and it defines metrics on all the exterior powers of T^*M. Then, L^2-metrics can be defined on each vector space Ω^p. Introduce $d^*: \Omega^p \to \Omega^{p-1}$, the (formal) L^2-adjoint of d.

HODGE THEOREM. $H^p \approx h^p \equiv \{\omega \in \Omega^p : d\omega = 0 \text{ and } d^*\omega = 0\}$.

A new choice of Riemannian metric produces a different adjoint for the exterior derivative, and so a different vector space, h'^p, to represent the cohomology H^p. See Diagram 1.

It is amusing to see a variational approach to the Hodge theorem: Introduce the "energy" functional on Ω^p which assigns to a p-form ω the number

$$(1.3) \qquad E(\omega) \equiv \langle d\omega, d\omega \rangle_{L^2} + \langle d^*\omega, d^*\omega \rangle_{L^2}.$$

This functional is nonnegative on Ω^p and the Hodge theorem is summarized by

$$(1.4) \qquad H^p \approx E^{-1}(0).$$

II. G. Segal's Theorem.. Let Maps* denote the space of smooth maps from S^2 (the standard 2-dimensional sphere) to S^2 which take the north pole to the north pole. Every map from S^2 to S^2 has a well-defined degree (winding number), and the specification of the degree defines the decomposition of Maps* into its connected components,

$$(2.1) \qquad \text{Maps}^* = \bigcup_{k \in \mathbf{Z}} \text{Maps}_k^*.$$

A classical theorem in topology asserts that a weak homotopy equivalence exists between any pair Maps_k^*, $\text{Maps}_{k'}^*$. Indeed, the homotopy groups of Maps_k^* can be computed from the homotopy groups of S^2 using the formula

$$(2.2) \qquad \pi_m(\text{Maps}_k^2) \approx \pi_{m+2}(S^2).$$

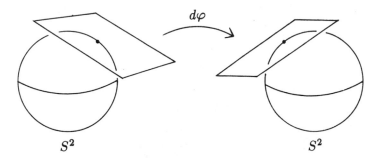

$$d\varphi$$

$$S^2 \qquad\qquad S^2$$

Graeme Segal [**S1**] gave an answer to the following:

PROBLEM. Find a geometric subspace of Maps_k^* whose inclusion into Maps_k^* induces $\pi_m(\mathrm{Maps}_k^*)$.

Segal introduced the following geometry: Think of S^2 as the *complex* projective space $\mathbf{CP}^1$. The complex structure allows the notion of a holomorphic map from S^2 to S^2 to be defined. And, sitting in Maps_k^* is the subspace $\mathscr{h}_k$ of based, holomorphic maps of degree k.

If one thinks of $\mathbf{CP}^1$ as $\mathbf{C} \cup \infty$, then a map $\varphi \in \mathscr{h}_k$ is just a rational function on $\mathbf{C}$—a function which sends the complex coordinate z to

$$(2.3) \qquad \varphi(z) = (z^k + a_1 \cdot z^{k-1} + \cdots + a_k)/(z^k + b_1 \cdot z^{k-1} + \cdots + b_k).$$

Here, the numerator and denominator are constrained only by the requirement that the set of roots of the numerator be disjoint from the set of roots of the denominator. By inspection, $\mathscr{h}_k$ is smooth of real dimension $4 \cdot |k|$.

There is a variational definition of $\mathscr{h}_k$ which comes from an energy functional. To specify this functional, use the defining embedding of S^2 in $\mathbf{R}^3$ to build a metric on TS^2. A map $\varphi \in \mathrm{Maps}^*$ defines the vector bundle $\varphi^* TS^2 \otimes T^* S^2$, and the metric on TS^2 induces a metric $(\,,\,)$ on $\varphi^* TS^2 \otimes T^* S^2$.

The map φ has a differential, $d\varphi$, which is a section over S^2 of $\varphi^* TS^2 \otimes T^* S^2$. The differential describes the first-order behavior of the map φ about each point in S^2, as in Diagram 2.

Now, define the energy of the map φ to be

$$(2.4) \qquad\qquad E(\varphi) \equiv \int_{S^2} (d\varphi, d\varphi) \cdot d\mathrm{vol}.$$

This energy functional on Maps_k^* is bounded from below by $4\pi \cdot |k|$, and, in analogy with (1.4),

$$(2.5) \qquad\qquad E^{-1}(4\pi \cdot |k|) \cap \mathrm{Maps}_k^* \equiv \mathscr{h}_k.$$

As illustrated in Diagram 3, the spaces $\mathrm{Maps}_0^*, \mathrm{Maps}_1^*, \ldots$ are all homotopically the same, while the geometric spaces $\mathscr{h}_0, \mathscr{h}_1, \ldots$ are growing in dimension.

It is natural to ask whether the contribution from $\mathscr{h}_k$ to the topology of Maps_k^* is also growing with growing k.

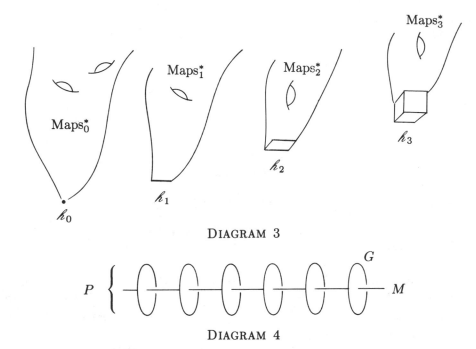

DIAGRAM 3

DIAGRAM 4

SEGAL'S THEOREM. *The inclusion* $i: \mathscr{k}_k \rightarrow \mathrm{Maps}_k^*$ *induces* $i_*: \pi_m(\mathscr{k}_k) \rightarrow \pi_m(\mathrm{Maps}_k^*)$ *which is an isomorphism for* $m < |k|$.

III. Yang-Mills Theory.. In this section, M will be a compact, oriented, 4-dimensional manifold, and G will be a compact, simple Lie group. For example, $G = \mathrm{SU}(2)$, the group of 2-by-2 unitary matrices with determinant equal to 1. The discussion begins with some basic geometric constructions. See [**KN**] for a basic reference.

IIIa. *Principal bundles.* A principal G-bundle, P, over M generalizes the product manifold $M \times G$. Indeed, by definition, a principal G-bundle over M is a compact manifold, P, on which the group G acts freely together with a surjection,

$$(3.1) \qquad\qquad P \xrightarrow{\pi} M.$$

The following properties are obeyed by π:

(1) $\pi^{-1}(\text{point}) = G$.
(2) The G action commutes with π so that $\pi(p \cdot g) = \pi(p)$ for all $(p, g) \in P \times G$.

(3.2)
(3) The surjection π is locally trivial in a G-equivariant way. That is, every point in M has a neighborhood U, and a G-equivariant map from $P|_U$ to $U \times G$ which covers the identity map on U.

A schematic rendition of P is provided in Diagram 4.

There is a natural notion of isomorphism between two principal G-bundles P, P' over M: The bundle P is isomorphic to P' when there exists a G-equivariant map $\varphi \colon P \to P'$ which covers the identity map of M. For simply connected G, the set of isomorphism classes of principal G-bundles over M is

$$(3.3) \qquad\qquad \operatorname{Iso}(M; G) \approx \mathbf{Z}.$$

The preceding isomorphism is provided by the first Pontrjagin class (in $H^4(M; \mathbf{Z}) \approx H_0(M; \mathbf{Z}) \approx \mathbf{Z}$) of the associated vector bundle

$$(3.4) \qquad\qquad \operatorname{Ad} P \equiv P \times_G \quad \text{(Lie algebra } G\text{)}.$$

The Pontrjagin class actually maps $\operatorname{Iso}(M; G)$ into $c(G) \cdot \mathbf{Z}$ with $c(G)$ a positive integer which depends on the group G. For notational simplicity, it is convenient to divide by the constant $c(G)$. The resulting integer is called the "instanton number" of the bundle P.

The product bundle, $P_0 \equiv M \times G$, has instanton number 0. One can choose a set of bundles $\{P_k\}_{k \in \mathbf{Z}}$ with k labeling the instanton number, and any principal G-bundle over M is isomorphic to a unique bundle in this set.

An isomorphism of a principal G-bundle P with itself is called an automorphism of P. It is convenient to fix a base point $x_0 \in M$ and introduce the group of automorphisms

$$(3.5) \qquad \mathscr{G}(P) \equiv \{\text{Automorphisms } \varphi \colon P \to P \colon \varphi(x_0) \equiv \text{identity}\}.$$

The set $\mathscr{G}(P)$ is naturally a topological group; indeed there is a weak homotopy equivalence

$$(3.6) \qquad\qquad \mathscr{G}(P) \approx \operatorname{Maps}^*(M; G),$$

where $\operatorname{Maps}^*(M; G)$ is the group of based maps from M to G. Note that the homotopy type of $\mathscr{G}(P)$ depends only on the homotopy type of the manifold M.

To see the equivalence in (3.6), start with the following observations: An automorphism φ must commute with the identity map of M, so it acts as an automorphism of each fiber of the surjection π. That is, φ rotates $\pi^{-1}(x) \approx G$ for each $x \in M$. Since φ is G-equivariant, it must rotate each point in $\pi^{-1}(x)$ by the same amount. Thus, φ acts on $\pi^{-1}(x)$ as multiplication by some $g(x) \in G$. Then, for those automorphisms in $\mathscr{G}(P)$, the assignment of $g(x)$ to x defines a map from M into G which uniquely determines the automorphism.

Since $\mathscr{G}(P)$ is a topological group, abstract arguments produce a contractible topological space on which $\mathscr{G}(P)$ acts freely, and the quotient is, by definition, a classifying space $B\mathscr{G}(P)$. The space $B\mathscr{G}(P)$ is unique up to weak homotopy equivalence. From (3.6)

$$(3.7) \qquad\qquad B\mathscr{G}(P) \approx \operatorname{Maps}^*_P(M; BG).$$

Here, BG is the classifying space for the Lie group G, and $\operatorname{Maps}^*_P(M; BG)$ is the space of continuous, based maps from M into BG which pull back the bundle P (see [A-B] and [D2]).

Due to (3.6), the homotopy type of $\operatorname{Maps}^*_P(M; BG)$ is independent of P and depends only on the homotopy type of M. Thus, for any two principal G-bundles $P, P' \to M$, there is a weak homotopy equivalence

$$(3.8) \qquad\qquad B\mathscr{G}(P) \approx B\mathscr{G}(P').$$

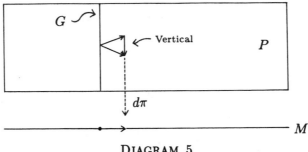

DIAGRAM 5

For example, when $M = S^4$, then

(3.9) $\mathrm{Maps}_P^*(S^4; BG) \approx \mathrm{Maps}^*(S^3; G),$

and

(3.10) $\pi_m(\mathrm{Maps}_P^*(S^4; BG)) \approx \pi_{m+3}(G).$

IIIb. *Connections.* For any M, remember that $B\mathscr{G}(P)$ is constructed from a contractible topological space on which $\mathscr{G}(P)$ acts freely. The space $\mathscr{A}(P)$ of connections on the principal bundle P is such a space. Thus,

(3.11) $\mathscr{B}(P) \equiv \mathscr{A}(P)/\mathscr{G}(P)$

provides a geometric representation for $B\mathscr{G}(P)$.

A connection on a principal G-bundle is not such a mysterious object. The short digression that follows summarizes the basic notions: To motivate the definition of connection, introduce the tautological exact sequence of vector bundles over P:

(3.12) $0 \to \mathrm{Vert}\, P \to TP \to \pi^*TM \to 0.$

Here, $\mathrm{Vert}\, P$ is the subbundle of TP which is annihilated by the differential, $d\pi$, of the surjection π. The preceding exact sequence summarizes the fact that two vectors in $TP|_p$ which project to the same vector in $TM|_{\pi(p)}$ differ by a vector in $\mathrm{Vert}\, P$; examine Diagram 5.

Since G acts on P, the action lifts naturally to TP. And, since G preserves the fibers of π, G acts on $\mathrm{Vert}\, P$. Indeed, since the fiber of π is a copy of G, the exponential map on G provides a canonical, G-equivariant trivialization of $\mathrm{Vert}\, P \to P$,

(3.13) $\mathrm{Vert}\, P \approx P \times \mathrm{Lie\ Algebra}\ G.$

The group G acts on π^*TM trivially, which makes the sequence in (3.12) G-equivariant.

An exact sequence, as in (3.12), need not have a natural splitting. And, in general, (3.12) has none. A connection on P is precisely the choice of a G-equivariant splitting of (3.12). That is, a connection A is a G-equivariant, linear map

(3.14) $A: TP \to \mathrm{Vert}\, P$

with the property that the composition

(3.15) $$\text{Vert}\,P \to TP \overset{A}{\to} \text{Vert}\,P$$

is the identity.

The kernel of A in TP defines the horizontal subbundle, $H_A \subset TP$. This G-invariant vector bundle is isomorphic to $\pi^* TM$. The diagram below summarizes:

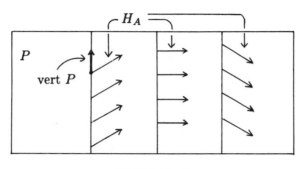

DIAGRAM 6

A partition of unity construction produces connections on any principal bundle.

Equation (3.15) asserts that the difference between two connections on P is a geometric object which is pulled up from M since said difference is G-equivariant and annihilates the vertical subspace of TP. Two connections differ by the pullback of a 1-form on M with values in the vector bundle $\text{Ad}\,P$ of (3.4). Conversely, a connection on P plus a section of $\text{Ad}\,P \otimes T^*M$ defines a second connection on P.

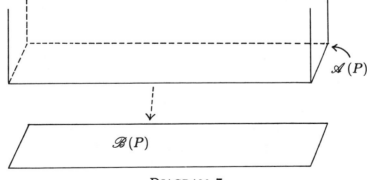

DIAGRAM 7

The preceding paragraph is summarized by the assertion that the space of connections on P is an affine space which obeys the noncanonical isomorphism

(3.16) $$\mathscr{A}(P) \approx \Gamma(\text{Ad}\,P \otimes T^*M).$$

A connection is, among other things, a G-equivariant section over P of $\text{Vert}\,P \otimes T^*P$. An automorphism of P pulls back forms over P, and preserves their equivariance. So, $\mathscr{G}(P)$ acts naturally on $\mathscr{A}(P)$. This action is seen to be a

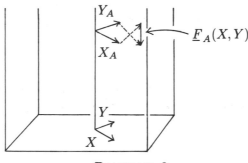

DIAGRAM 8

free action. And, since $\mathscr{A}(P)$ is an affine space, the quotient $\mathscr{B}(P)$ from (3.11) provides a model for the classifying space $B\mathscr{G}(P)$.

When P is isomorphic to P', there are *natural* identifications between $\mathscr{B}(P)$ and $\mathscr{B}(P')$. Since each P is isomorphic to some P_k with instanton number k, it is convenient to consider $\mathscr{B}(k) \equiv \mathscr{B}(P_k)$. According to (3.8), there is a weak homotopy equivalence between $\mathscr{B}(k)$ and $\mathscr{B}(k')$ when $k \neq k'$.

IIIc. *Curvature.* The topology of $\mathscr{B}(k)$ will be represented geometrically using an energy function on $\mathscr{B}(k)$, the Yang-Mills function.

The Yang-Mills function is defined using the curvature of a connection, and so a digression on curvature is appropriate.

Fix a connection, A, on P. Think of A as in (3.14). The kernel of A is the subbundle $H_A \subset TP$, a G-invariant subbundle which is isomorphic to $\pi^* TM$. Think of H_A as a distribution on P and measure its failure to be integrable in the sense of Frobenius.

The failure of integrability is measured as follows: At $p \in P$, pick two vectors $(X, Y) \subset TM_{\pi(p)}$. Extend these vectors locally as vector fields in a neighborhood of $\pi(p)$. Since $H_A|_p \approx TM_{\pi(p)}$, there is a unique lift of (X, Y) to G-invariant vector fields (X_A, Y_A) on a neighborhood of p which lie in H_A. The subbundle H_A is integrable at p if and only if the commutator $[X_A, Y_A]$ defines a vector field in H_A for all possible vectors X, Y. So, the failure of integrability is measured by

$$(3.17) \qquad \underline{F}_A(X, Y) \equiv A \circ [X_A, Y_A] \in \operatorname{Vert} P.$$

Equation (3.17) summarizes the geometric picture of Diagram 8.

The right-hand side of (3.17) is G-equivariant, depends only on $(X, Y)|_{\pi(p)}$ and changes sign when X and Y are switched. Therefore, the right-hand side of (3.17) defines a section,

$$(3.18) \qquad F_A \in \Gamma(\operatorname{Ad} P \otimes \Lambda^2 T^* M),$$

whose pull-back to P is $\underline{F}_A$ of (3.17). This F_A is called the curvature of the connection A.

IIId. *The Yang-Mills functional.* The Yang-Mills function on $\mathscr{B}(P)$ requires a Riemannian metric on TM for its definition. The choice of such a metric inputs *geometry* into what has been, up to this point, a differential topology discourse.

Fix the Riemannian metric. The Yang-Mills functional assigns to the $\mathscr{G}(P)$ orbit of a connection A the number

$$(3.19) \qquad YM(A) \equiv \int_M |F_A|^2 \cdot d\,\mathrm{vol}.$$

Here, the norm is defined using the chosen Riemannian metric on TM and the unique (up to scale) Killing metric on Lie Algebra G. The volume form is defined by the Riemannian metric on TM.

The use of the Killing metric from Lie Algebra G to measure norms on the vector bundle $\mathrm{Ad}\,P$ insures that the right-hand side of (3.19) is $\mathscr{G}(P)$-invariant. The Killing metric is normalized so that

$$(3.20) \qquad YM\colon \mathscr{B}(k) \to [\,|k|, \infty).$$

It is known [**Ta1**] that YM is surjective onto the open interval $(|k|, \infty)$. Furthermore, given the Riemannian metric, there exists $k(M)$ (see [**Ta2**], [**Ta3**]) such that for all k with absolute value greater than $k(M)$,

$$(3.21) \qquad \mathscr{M}(k) \equiv YM^{-1}(k) \neq \varnothing.$$

The space $\mathscr{M}(k)$ is called the pointed moduli space of (anti-) self-dual connections on P; the prefix "anti" is used for $k < 0$. An orbit $[A]$ in $\mathscr{M}(k)$ is called self-dual (or antiself-dual) because the curvature of A must obey the algebraic constraint

$$(3.22) \qquad F_A = (\pm) * F_A,$$

with $*\colon \Lambda^2 T^*M \to \Lambda^2 T^*M$ being the Riemannian metric's Hodge star operator. These spaces play a major role in the work of Simon Donaldson [**D1**–**D4**]; see Michael Atiyah's lecture in these proceedings.

Karen Uhlenbeck (see [**FU**]) has proved that when the chosen metric on TM is suitably generic, then $\mathscr{M}(k)$ is a smooth manifold, a manifold which is naturally embedded in $\mathscr{B}(k)$.

Atiyah, with N. Hitchin and I. M. Singer [**AHS**] have used the Atiyah-Singer index theorem to prove that the dimension of $\mathscr{M}(k)$ ($k \geq 0$) is

$$(3.23) \qquad \begin{aligned} \dim \mathscr{M}(k) &= 2 \cdot p_1(\mathrm{Ad}\,P) - (\dim G) \cdot (\chi(M) - \tau(M))/2 \\ &\equiv 2 \cdot c(G) \cdot k - c_1(G, M). \end{aligned}$$

Here, χ is the Euler characteristic of M, and τ is the signature of M; both are homeomorphism invariants of the underlying topological manifold.

IIIe. *The inclusion $\mathscr{M}(k) \subset \mathscr{B}(k)$.* Equation (3.23) asserts that the spaces $\mathscr{M}(k)$ get larger as k increases; their dimensions grow linearly with k. On the other hand, the spaces $\mathscr{B}(k)$ are homotopically equivalent for all k. Compare Diagram 9 with Diagram 3.

The schematic in Diagram 9 leads one to conjecture that as $|k| \to \infty$, the inclusion of $\mathscr{M}(k)$ into $\mathscr{B}(k)$ becomes topologically interesting. The validity of

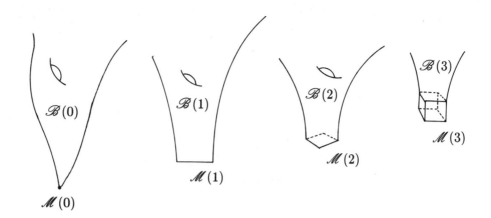

DIAGRAM 9

this conjecture has been verified by the author in [**Ta3**]:

THEOREM 1. *Let M be a compact, oriented, 4-dimensional manifold. Fix a Riemannian metric on M. Let G be a simple and simply connected Lie group. Given a positive integer m_0, there exists a positive integer k_0 such that for all pairs of integers k, m that obey $|k| > k_0$ and $0 \leq m \leq m_0$, the inclusion $i: \mathcal{M}(k) \to \mathcal{B}(k)$ induces epimorphisms*

$$i_*: \pi_m(\mathcal{M}(k)) \to \pi_m(\mathcal{B}(k)) \to 0 \quad and \quad i_*: H_m(\mathcal{M}(k)) \to H_m(\mathcal{B}(k)) \to 0.$$

There is a sense in which a direct limit of the pairs of spaces $(\mathcal{B}(k), \mathcal{M}(k))$ can be taken. For each $k > 0$, there exists an integer $J(k) > 0$ and for all $j \geq k + J(k)$, there exists a continuous map of pairs of spaces

$$(3.24) \qquad T_{j,k}: (\mathcal{B}(k), \mathcal{M}(k)) \to (\mathcal{B}(j), \mathcal{M}(j))$$

which has the following two properties

$$(3.25) \qquad \begin{array}{l} (1)\ T_{j,k} \circ T_{k,h}\ \text{is homotopy equivalent rel}\ \mathcal{M}(j)\ \text{to}\ T_{j,h}. \\ (2)\ T_{j,k}: \mathcal{B}(k) \to \mathcal{B}(j)\ \text{is a homotopy equivalence.} \end{array}$$

Property (1) of the set $\{T_{j,k}\}$ allows for an unambiguous definition of the direct limits,

$$(3.26) \qquad \mathcal{M}(\infty) = \mathrm{dir}\lim_{k \to \infty} \mathcal{M}(k) \quad and \quad \mathcal{B}(\infty) \equiv \mathrm{dir}\lim_{k \to \infty} \mathcal{B}(k).$$

Property (2) of (3.25) implies that $\mathcal{B}(\infty)$ is weakly homotopic to each $\mathcal{B}(k)$.

THEOREM 2. *Let M be a compact, oriented, Riemannian 4-manifold and let G be a simple and simply connected Lie group. Using the set of maps $\{T_{j,k}\}$ in (3.25), construct the direct limits in (3.26). There exists a weak homotopy equivalence between $\mathcal{M}(\infty)$ and $\mathcal{B}(\infty)$.*

There are analogous theorems for principal bundles with non-simply connected structure groups.

The topology of the inclusion $\mathscr{M}(k) \to \mathscr{B}(k)$ was originally studied by Atiyah and Jones [**AJ**]; they considered the case $M \equiv S^4$ with the standard metric. In this context, they proved the homology epimorphism assertion of the first theorem above.

Atiyah and Jones conjectured that both epimorphisms in Theorem 1 should be isomorphism when S^4 has its standard metric. To aficionados, this is the "Atiyah-Jones conjecture".

Progress on the Atiyah-Jones conjecture has been made by a student of Graeme Segal [**S2**], who proved Theorem 2 for S^4 with its standard metric. Frances Kirwan [**Ki**] has also made considerable progress towards proving the Atiyah-Jones conjecture.

Recently, Charles Boyer and Ben Mann [**BM**] have studied the properties of the maps $\{T_{j,k}\}$ in the S^4 case. They used these maps to define homology loop sum operations on $\{\mathscr{M}(k)\}$. These operations gave detailed information about the homology groups of $\{\mathscr{M}(k)\}$.

IIIf. *Morse Theory on* $\bigcup_k \mathscr{B}(k)$. Theorems 1 and 2 are proved in [**Ta3**] by considering classes in the relative homotopy groups of the pair $(\mathscr{B}(k), \mathscr{M}(k))$. Indeed, suppose that $z \in \pi_m(\mathscr{B}(k), \mathscr{M}(k))$ has been given. The class z is the homotopy class (rel $\mathscr{M}(k)$) of a map φ from the standard m-dimensional ball into $\mathscr{B}(k)$ whose boundary, the $(m-1)$-dimensional sphere, is mapped into $\mathscr{M}(k)$.

It is convenient to define, for each positive number $\lambda \geq k$, the set $\mathscr{B}(k, \lambda) \equiv \{b \in \mathscr{B}(k): E(b) \leq \lambda\}$. The inclusion of $\mathscr{M}(k)$ into $\mathscr{B}(k, \lambda)$ induces a group homomorphism from $\pi_*(\mathscr{B}(k), \mathscr{M}(k))$ into $\pi_*(\mathscr{B}(k), \mathscr{B}(k, \lambda))$.

Associate to a class $z \in \pi_m(\mathscr{B}(k), \mathscr{B}(k, \lambda))$ the maximum of λ and the number

$$(3.27) \qquad E_k(z) \equiv \inf_{\varphi \in z} \sup_{y \in B} E(\varphi(y)).$$

Here, B is the m-ball, and $E(\cdot)$ is the Yang-Mills functional.

The specification of a connected family of homotopy equivalences between $\mathscr{B}(k)$ and $\mathscr{B}(k+1)$ which maps $\mathscr{B}(k, \lambda)$ into $\mathscr{B}(k+1, \lambda')$ allows one to compare $\pi_*(\mathscr{B}(k), \mathscr{B}(k, \lambda))$ with its image in $\pi_*(\mathscr{B}(k+1), \mathscr{B}(k+1, \lambda'))$. A generalized connected sum (see [**Ta3**]) with a localized self-dual connection on the $k = 1$ principal bundle over S^4 provides such a connected family of homotopy equivalences. The parameters of this family are given by allowing continuous variations over $\mathscr{B}(k)$ of (1) the position in M of the connect-sum point; (2) the size of the ball in M on which the connect sum is made; (3) the choice of a homomorphism $\rho: SU(2) \to G$ to generate $\pi_3(G)$.

Given positive ε, one can find members of this family which send $\mathscr{B}(k, \lambda)$ into $\mathscr{B}(k+1, \lambda + \varepsilon)$. This is accomplished through judicious choice of the size dependence in the connect-sum construction.

The aforementioned family of homotopy equivalence allows a comparison between $E(z, k)$ and $E(z, k+1)$ for fixed $z \in \pi_m(\mathscr{B}(k), \mathscr{B}(k, \lambda))$. Is $E(z, k+1)$

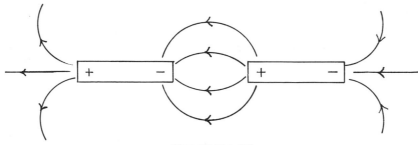

DIAGRAM 10

smaller? Not necessarily, but

$$(3.28) \qquad\qquad \lim_{k \to \infty} (E(z, k) - k) \to 0.$$

There is physics intuition behind (3.28). In physics, the connection is supposed to describe the force field due to certain subnuclear particles. For a physicist, the Yang-Mills functional measures the energy that is contained in the force field. The connect-sum construction can be interpreted in a heuristic sense as the addition of a specific, extra particle to a system of k particles (each particle contributes a unit charge to the instanton number).

If the new particle can be added to make the net forces attractive, then the normalized energy, $E(\cdot) - k$, will decrease upon addition of this particle. Indeed, when the normalized energy is positive, the physics does allow a special, self-dual force field to be added which causes a decrease in the normalized energy. Think here of bar magnets: If the alignment is correct, then the forces are attractive. (If the alignment is reversed, then the forces are repulsive.)

A simple calculation verifies the physics; a connect-sum of the standard self-dual connection on S^4 with a connection with positive normalized energy can be made to produce a new connection on a bundle with larger instanton number and with smaller normalized energy. This is accomplished through the choice of the homomorphism from SU(2) into G. By allowing multiple connect-sums, the normalized energy can be decreased uniformly on any compact set in $\mathscr{B}(k) \backslash \mathscr{M}(k)$. Equation (3.28) becomes an immediate consequence.

IIIg. *The obstruction to self-duality.* Theorems 1 and 2 are obtained from (3.28) with the help of a "tubular neighborhood" theorem. The ideal tubular neighborhood theorem should assert the existence of a positive number ε with the existence, for all large k, of a homotopy of $\mathscr{B}(k, \varepsilon)$ onto $\mathscr{M}(k)(\mathrm{rel}\,\mathscr{M}(k))$. Unfortunately, there is usually no such ideal theorem. The cohomology classes which are constructed in [D2] give obstructions.

One must make due with the following [Ta3]:

PROPOSITION 3. *Let M be a compact, oriented, 4-dimensional Riemannian manifold, and let G be a simple Lie group. There exists $\varepsilon > 0$, an integer $N < \infty$ and, for all large k, a nonempty subvariety $\mathscr{B}(k)^* \subseteq \mathscr{B}(k, \varepsilon)$. These have the following properties:*

(1) *The codimension of any component of $\mathscr{B}(k)^*$ is bounded by N.*

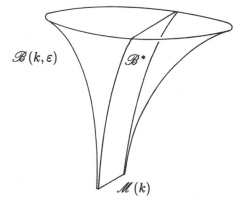

DIAGRAM 11

(2) $\mathscr{M}(k) \subset \mathscr{B}(k)^*$.

(3) *There exists a retraction of $\mathscr{B}(k)^*$ onto $\mathscr{M}(k)(rel\,\mathscr{M}(k))$.*

For example, when S^4 has its standard metric, $\mathscr{B}(k)^* = \mathscr{B}(k, \varepsilon)$. In general, the subvariety $\mathscr{B}(k)^*$ is defined locally as the zero set of a section of an "obstruction bundle" over $\mathscr{B}(k, \varepsilon)$; it is the subvariety on which the obstruction to self-duality vanishes. The schematic picture is given in Diagram 11.

The subvariety $\mathscr{B}(k)^*$ globalizes a local construction which was formulated by Kuranishi [**Ku**]. Consider an orbit $[A] \in \mathscr{B}(k, \varepsilon)$. Suppose that $[A + \omega]$ is self-dual. Then, the $\operatorname{Ad} P$-valued 1-form ω must obey the equation

$$(3.29) \qquad (1 - *) \cdot (d_A \omega + \omega \wedge \omega + F_A) = 0,$$

with $d_A: C^\infty(\operatorname{Ad} P \otimes T^*) \to C^\infty(\operatorname{Ad} P \otimes \Lambda^2 T^*)$ being the exterior covariant derivative. By the Fredholm alternative, (3.29) is solvable if and only if $(\omega \wedge \omega + F_A)$ is orthogonal to the kernel of the adjoint of the differential operator $((1 - *) \cdot d_A)(\cdot)$.

If ε is small, then one might expect that $[A]$ is close to $\mathscr{M}(k)$. Then, the 1-form ω in (3.29) will be small and the term $\omega \wedge \omega$ will be a small perturbation to F_A. (Solvability of (3.29) for small ω is necessary for the construction of a *continuous* deformation (rel $\mathscr{M}(k)$) onto $\mathscr{M}(k)$.) With $\omega \wedge \omega$ small, the Fredholm alternative makes (3.29) solvable for small ω only if F_A is orthogonal to the set of eigenvectors of the nonnegative, differential operator $((1-*) \cdot d_A)^*((1-*) \cdot d_A)(\cdot)$ which have small eigenvalues.

This last orthogonality condition defines an obstruction whose vanishing defines a subvariety in $\mathscr{B}(k, \varepsilon)$ which gives the leading approximation (in ε) to the subvariety $\mathscr{B}(k)^*$. The corrections are higher order in ε and stem from the nonlinear nature of (3.29). As $[A]$ varies in $\mathscr{B}(k, \varepsilon)$, a uniform upper bound (depending on ε and not on k) for the number of eigenvectors with small eigenvalue for the operator $((1 - *) \cdot d_A)^*((1-*) \cdot d_A)(\cdot)$ produces the integer N in the statement of Proposition 3.

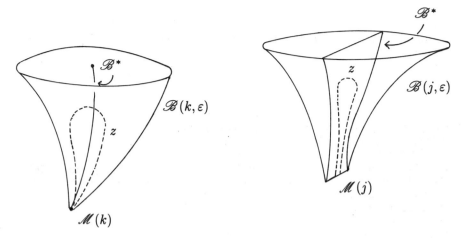

DIAGRAM 12

Proposition 3 and (3.28) raise the following questions: Can a map

$$\varphi\colon (B^n, S^{n-1}) \to (\mathscr{B}(k,\varepsilon), \mathscr{M}(k))$$

be homotoped (rel $\mathscr{M}(k)$) into the variety $\mathscr{B}(k)^*$ on which the obstruction to self-duality vanishes? If not, can a homotopy equivalence $T\colon \mathscr{B}(k) \to \mathscr{B}(j)$ be found which sends $\mathscr{M}(k)$ to $\mathscr{M}(j)$ and sends the image of φ into $\mathscr{B}(j)^*$?

The second question is answered by

PROPOSITION 4. *Let M be a compact, oriented, 4-dimensional Riemannian manifold, and let G be a simple Lie group. Given a positive integer k, there exists an integer $J(k)$, and for all $j > k + J(k)$, there exists a connected family, $\mathscr{T}(j,k)$, of homotopy equivalences from $\mathscr{B}(k)$ to $\mathscr{B}(j)$ which have the following properties*:

(1) *Each $T \in \mathscr{T}(j,k)$ maps $\mathscr{M}(k)$ to $\mathscr{M}(j)$.*

(2) *If $T \in \mathscr{T}(j,k)$ and $T' \in \mathscr{T}(h,j)$, then $T' \circ T \in \mathscr{T}(h,k)$.*

(3) *Given $z \in \pi_m(\mathscr{B}(k), \mathscr{M}(k))$, there exists $J(k,z)$ and, for all $j > J(k,z)$, there exists $T \in \mathscr{T}(j,k)$ and φ representing z such that $T \circ \varphi$ maps the m-dimensional ball into $\mathscr{B}(j)^*$.*

The schematic in Diagram 12 summarizes.

The assertions of Proposition 4 descend ultimately from a simple calculation (see [**Ta2**], [**Ta3**]) which shows how the connect-sum map from $\mathscr{B}(k)$ to $\mathscr{B}(k+1)$ changes the self-duality obstruction. Indeed, when the connect-sum construction is made using small balls, the obstruction on the image in $\mathscr{B}(j)$ of $\mathscr{B}(k)$ approximately decouples into a sum of two contributions.

This decoupling becomes more exact by using the size parameters to decrease the size of the region which is affected in the connect-sum. The first contribution gives the original obstruction on $\mathscr{B}(k)$. The second contribution is localized at the points in M where the connect-sums have been made. And, changing the connect-sum parameters changes the second contribution.

A judicious choice of the parameters which define the connect-sum construction will fine-tune the second contribution to cancel the first.   The result is Proposition 4.

REFERENCES

[**AHS**] M. F. Atiyah, N. J. Hitchin and I. M. Singer, *Self-duality in 4-dimensional Riemannian geometry*, Proc. Roy. Soc. London Ser. A. **362** (1978), 425.

[**AJ**] M. F. Atiyah and J. D. S. Jones, *Topological aspects of Yang-Mills theory*, Comm. Math. Phys. **61** (1978), 97.

[**BM**] C. P. Boyer and B. M. Mann, *Homology operations on instantons*, preprint, 1986.

[**D1**] S. K. Donaldson, *An application of gauge theory to the topology of 4-manifolds*, J. Differential Geom. **18** (1983), 279.

[**D2**] _____, *Connections, cohomology and the intersection forms of 4-manifolds*, J. Differential Geom. **24** (1986), 275.

[**D3**] _____, *Irrationality and the h-cobordism conjecture*, J. Differential Geom. **126** (1987), 141.

[**D4**] _____, *The orientation of Yang-Mills moduli spaces and 4-manifold topology*, J. Differential Geom., to appear.

[**FU**] D. Freed and K. K. Uhlenbeck, *Instantons and 4-manifolds*, Springer-Verlag, New York and Berlin, 1984.

[**Ki**] F. Kirwan, private communication.

[**Ku**] M. Kuranishi, *New proof for existence of locally complete families of complex structures*, Proc. Conf. Complex Analysis (Minneapolis, 1964), Springer-Verlag, Berlin and New York, 1965.

[**KN**] S. Kobayashi and K. Nomizu, *Foundations of differential geometry*, Interscience, New York, 1963.

[**L**] H. B. Lawson, *Algebraic cycles and homotopy theory*, to appear.

[**S1**] G. Segal, *The topology of spaces of rational functions*, Acta Math. **143** (1979), 39.

[**S2**] G. Segal, private communication.

[**Ta1**] C. H. Taubes, *Self-dual connections on nonself-dual 4-manifolds*, J. Differential Geom. **17** (1982), 139.

[**Ta2**] _____, *Self-dual connections on 4-manifolds with indefinite intersection matrix*, J. Differential Geom. **19** (1982), 517.

[**Ta3**] _____, *Stable topology of self-dual moduli spaces*, J. Differential Geom., to appear.

[**W**] F. W. Warner, *Foundations of differentiable manifolds and Lie groups*, Springer-Verlag, New York and Berlin, 1983.

HARVARD UNIVERSITY

Proceedings of Symposia in Pure Mathematics
Volume 48 (1988)

Fundamental Asymmetry in Physical Laws

R. PENROSE

Weyl and asymmetry in physics. The concept of *symmetry* was clearly a central theme in much of Hermann Weyl's work; and, as is well known, he wrote a beautiful semi-popular book, with that title, late in his life (Weyl 1952). It would be clear from this book alone that, along with being a great mathematician, Weyl was a philosopher, and he much appreciated the arts. But in addition to these other interests, he was a physicist of definite distinction, having had deep insights into the workings of the natural world and having introduced several important and highly influential ideas (e.g., in particular, gauge theory). Symmetry was an underlying theme in many of Weyl's contributions to physics. Indeed, one of Weyl's most significant early physical realizations was the importance of symmetry in quantum theory, and his book *The theory of groups and quantum mechanics* (Weyl 1931) was an important landmark.

Perhaps almost as significant for physics as his recognition of the role of symmetry was his suspicion of a possible fundamental role for *a*symmetry in the laws governing physical behaviour. Weyl seems to have been particularly intrigued by the *discrete* symmetries of space-reflection, particle-antiparticle interchange, and time-reversal—these being the now well-known operations denoted by **P**, **C** and **T**, respectively. He pointed out (Weyl 1931) that Dirac's "hole" theory implied that the holes must have the *same* mass as the electrons—and so could not be protons, as Dirac had initially hoped (invariance under **C**). He considered the question of invariance under **P**, **C** and **T**, noting that such invariance was an additional issue to the normal Lorentz invariance requirements. In 1929 Weyl introduced the wave equation that now bears his name,[1]

$$(\partial/\partial t - \partial/\partial z)\nu_0 = (\partial/\partial x - i\partial/\partial y)\nu_1, \qquad (\partial/\partial t + \partial/\partial z)\nu_1 = (\partial/\partial x + i\partial/\partial y)\nu_0,$$

1980 *Mathematics Subject Classification* (1985 *Revision*). Primary 83C45, 81B05.

[1]Ironically, it is the physicists who pay tribute to the mathematician Weyl by referring to this (massless) equation as the "Weyl equation", whereas the mathematicians honour a physicist, calling it the "Dirac equation"! (Strictly, the mathematicians are historically more correct, since Dirac had hit upon this wave equation first—and discarded it for what were good physical reasons at the time—before introducing his more famous *massive* Dirac equation; cf. Dirac 1928, 1982.

and which is now the standard equation used to describe the (massless) neutrino. The lack of space-reflection invariance of this equation had got him into trouble with the established physicists, particularly Pauli, since physical laws had been thought, at that time, to be completely reflection-invariant. The fact that **P** is actually *not* a symmetry of *weak* interactions (and nor are **C**, **CT**, **PT**) was not discovered until 1957 (Wu et al. 1957, cf. Lee and Yang 1956), about a year after Weyl's death.

Almost symmetries in nature. It is a puzzling fact that there are basic symmetries in nature that are only approximately true—or true only for some interactions but not for others. The **P** and **C** symmetries are examples of this, being symmetries of the strong, electromagnetic and gravitational interactions, but not of the weak interactions. There is also a *single* phenomenon known, for which **CP** is not an exact symmetry—and neither is **T**: the decay of the K^0 particle (Christenson et al. 1964)! Such puzzling "almost symmetry" in nature is a feature of other symmetry operations, also. In the modern theories of particle classification, symmetry groups play a fundamental role. The original example of such a group was SU(2), which referred to a concept called *isotopic* (or, more correctly, *isobaric*) *spin*, introduced in the 1930s. Elements of this SU(2) group can "rotate" protons into neutrons. Strong interactions respect this symmetry, whereas weak and electromagnetic interactions do not. This SU(2) group was extended, in 1962, to SU(3) by Gell-Mann, Ne'eman and others (cf. Gell-Mann and Ne'eman 1964). Again the group is (essentially) respected by strong, but not weak or electromagnetic, interactions. Nowadays, theoreticians often use much larger symmetry groups. The idea is now to have a more unified type of theory in which strong, weak and electromagnetic interactions are *all* to be thought of as different aspects of *one* grand scheme. Thus, according to the various "grand unified theories" (abbreviated GUT), all these different interactions should come together as different aspects of a single concept, and transform among themselves under the action of the proposed group. As yet there is no consensus as to which GUT theory would be most appropriate, and none of these theories appears to have significant experimental support so far. However, the more modest *electroweak* theory (using the group SU(2) × U(1)), due to Salam, Ward, Weinberg and Glashow, which just unifies weak interactions with electromagnetic ones, has indeed had some impressive successes, and the theory is now regarded as standard.

Spontaneous symmetry breaking. In all of these unified theories, the entire symmetry group of the theory is *not* an exact symmetry of the relevant observed physical phenomena. A crucial idea is needed, in order to make such a theory work. This is to invoke the procedure of *spontaneous symmetry breaking*. Accordingly, when the ambient temperature is greater than a certain critical value, the physical phenomena do indeed respect the entire symmetry; but below this critical temperature, symmetry breaking sets in, and only some smaller symmetry group remains, and is respected by the observed phenomena. An example taken from standard everyday physics is often cited as a particular instance of this type of phenomenon: the example of *ferromagnetism*. Above the critical temperature, a piece of iron will not spontaneously exhibit magnetic properties. Though the various atoms behave individually like little magnets, and though the atoms have a tendency to align themselves with the magnets in the neighbouring atoms, the random motions induced by the ambient temperature are too great to allow the atoms to align themselves collectively. But when the temperature drops below this value, these random motions are not sufficient to overcome the tendency for alignment, and a lower energy configuration is achieved if the atoms *are* all aligned with one another. Now imagine a solid spherical ball of this iron which is initially above the critical temperature, but is then allowed to cool. As the iron cools below the critical temperature, it becomes energetically favoured that the ball should, as a whole, become a magnet, pointing in some particular direction. What direction does it choose? Assuming that there is no background magnetic field, the direction must be chosen *at random*. The original rotational symmetry is now broken. The physics as a whole does have rotational symmetry, and above the critical temperature the equilibrium state is indeed spherically symmetric. But below this temperature, the equilibrium state does *not* have full rotational symmetry, but only axial symmetry. The symmetry has been *broken*—from SO(3) down to SO(2).

More abstractly, situations of this general kind are frequently described in terms of a "Mexican hat potential", e.g.

$$V = V_0 + (1 - r^2)^2,$$

where r is the distance from the centre in, say, a Euclidean k-space representing the different possible states of a system, depicted horizontally, and the potential energy V is depicted as the height. The lowest energy is achieved around the "rim of the hat", namely the S^{k-1} given by $r = 1$, this energy being *degenerate*. At low temperature, the system settles into this degenerate energy, but no one point on the S^{k-1} is preferred over any other. Which point is actually chosen is a matter of pure chance, and when the choice is made, the symmetry of the system is reduced to that of S^{k-2}, providing us with a typical example of spontaneous symmetry breaking.

This kind of procedure is a very basic ingredient, common to almost all modern unified theories. Consider electroweak theory. The view would be that in the very early stages of the universe when, according to the standard "hot big bang" theory, it was vastly smaller and at a far higher temperature than it is today, the forces of electromagnetism and weak interactions were on an equal footing with each other as part of one comprehensive symmetrical scheme. As the universe expanded, it would have cooled and dropped below the critical temperature. At that stage, a "direction" would be singled out in the space on which the symmetry acts, and the electromagnetic interaction would become distinguished from weak interactions—i.e., photons distinguished from W-bosons, the latter

acquiring masses—and the SU(2) × U(1) symmetry group of the theory would
become reduced to (a diagonal) U(1).

Two possible viewpoints. One may contrast this idea of a symmetry-
breaking process with a completely opposed point of view, according to which
higher symmetry is not taken as fundamental at all, but just an approximation,
the underlying exact theory being not supposed actually to possess this higher
symmetry. Approximate symmetries of this kind are not unfamiliar in situations
in which a good underlying theory happens to be known. An example is given by
the theory of atomic structure, where groups and group representations can play
important roles, even when the group involved provides only an approximate
symmetry of the entire physics (cf. Dyson 1966). In this example, the underlying theory comes from standard quantum mechanics and electromagnetism.
In such a situation, one does *not* expect the approximate theory to become exact when the temperature exceeds a critical value, so the implications of such a
viewpoint are quite different. For some reason, viewpoints of this kind do not
appear to find much favour with modern particle physicists, and the alternative
of symmetry-breaking is the viewpoint almost universally adopted. (One of the
reasons for this has to do with renormalization theory. A good way of producing
a quantum field theory for which troublesome infinities cancel one another out is
to have a gauge theory with a high degree of symmetry, the lack of such symmetry at the *observed* level being attributed to a symmetry-breaking phenomenon
like that discussed above—referred to as a *Higgs mechanism* 1966).

We thus have two basic alternative viewpoints with regard to the deviations
from perfect symmetry in physics:

  S: spontaneous symmetry-breaking,

  N: underlying physics does not have the full symmetry.

Of course, for some physical phenomena, S can be more appropriate, while for
others it can be N. I do not mean to imply that either one of these mechanisms
is universally to be preferred in physics generally—and we have seen above that
there are examples in ordinary physics where each of these holds. However, the
issue seems to become more ideological, if one believes that one is approaching a
fundamental truth underlying physical laws. Does one believe in an underlying
physics in which some large Lie group (or perhaps some large finite group—such
as the "monster group", as has occasionally been suggested) is a fundamental
symmetry of the scheme? Or is a large symmetry group, just in itself, something
of a complication? Might one prefer, instead, some underlying scheme in which
symmetry, as such, has no particular fundamental role to play? My own personal
preference is indeed for the latter possibility—i.e., for N rather than S, at the
fundamental level. I do not see why symmetry, as such, should be fundamentally
important, though many basic mathematical structures do indeed possess important symmetry properties. Symmetry is very important for the description and
understanding of such basic structures, but I do not see why it should be put in
there at the start. This seems to be a minority view among present-day particle

physicists—for symmetry does indeed have a special beauty of its own—but I would like to feel that Weyl might actually have been on my side!

The discrete symmetries P, C and T. Let us examine the symmetries **P**, **C** and **T**, in relation to the two possible viewpoints **S** and **N**. (Here it is not quite so unfashionable to adopt **N**, at the fundamental level, as it is with the continuous symmetries.) It should be remarked, first, that there is an important theorem, known as the **PCT**-theorem, which asserts that the product **PCT** of all three symmetries **P**, **C** and **T** should be an exact symmetry for any quantum field theory subject to certain "physically reasonable" assumptions. These assumptions have to do with causality and positivity of energy, and invariance under the nonreflective transformations of the Poincaré group (which is the symmetry group of Minkowski space-time). Note the necessity of this last assumption. In my view this could be significant, because this assumption is explicitly violated by general relativity. We do not yet know the correct way to combine the basic ideas of quantum field theory with the principles of curved-space geometry—principles which are fundamental to Einstein's theory. I shall be indicating shortly my reasons for believing that, in fact, **PCT** ought to be *false* for the correct quantum gravity theory!

I have remarked that, along with **CP**, the symmetry **T** is false for K^0-decay. In fact the operation of **T** is a puzzle for physics in other ways. Let us disregard **T**-violation in K^0-decay for the moment. That is a very tiny effect, and not directly significant for any known action shaping the universe. We now find that there is a striking discrepancy between the symmetry principles which hold for the large-scale behaviour of macroscopic objects and those which seem to hold at the level of submicroscopic physics and exact physical laws. Invariance under **T** holds for the latter, whereas time-*a*symmetry is manifest for many large-scale phenomena.

Entropy and the second law. The most far-reaching of these time-asymmetries is that provided by the *second law of thermodynamics*. According to the second law, the quantity known as *entropy*, which measures the manifest disorder of a system, increases with time. (The entropy of a state of a system can be taken to be the logarithm of the phase-space volume of all the states which are to be regarded as macroscopically indistinguishable from the given state. This seems somewhat subjective, but this subjectiveness is not the important issue for the points that I wish to make. Reasonable changes in one's viewpoint as to which states are to be regarded as "macroscopically indistinguishable" make virtually no difference to the entropy values in most circumstances.) If a system is given to us in some state having a low value for its entropy, then it is not really surprising that its most probable behaviour in the future will be such as to increase this entropy. (For the point in phase space which represents the system will be likely to move into larger and larger phase-space regions as time progresses.) However, the *puzzle* that the second law presents us with is that if we take the same given low-entropy state and ask for its most likely behaviour in the *past* direction in time, then—in the actuality of our universe—the system gets into more and more manifestly improbable states as time progresses farther and farther back in time. In the absence of some other given constraint on the system in the remote past, we ought to expect that the most likely way to achieve some given low-entropy state would be to approach it with other *more* probable (i.e., higher entropy) states first, and achieve the desired given low-entropy configuration as a local minimum. This is blatantly *not* what happens in our universe. The entropy of the whole universe decreases and decreases in the past direction in time, until it reaches its minimum value at the very beginning of time. (If our universe is infinite, then one needs to be a little careful about talking about its "total entropy". But again, the technical problems concerned with precise definitions are not the essential issue here.)

The "beginning of time" occurred with the big bang. So the constraint on the universe which set the second law on its course must have occurred then. One point should be made clear. This constraint was not simply that the universe started out as very "tiny". There were as many degrees of freedom available at the beginning of time as there are at any other time. The entropy did not *need* to be low simply because the universe was small. Perhaps the easiest way to see this is to consider a universe model which is collapsing. The entropy can continue to go up, but one finds that most of the entropy resides in black holes. This black-hole entropy can be calculated by use of a formula due to Bekenstein (1974) and Hawking (1978) which asserts that (in natural units: the velocity of light, Planck's constant$/2\pi$, Newton's gravitational constant and Boltzmann's constant are all set equal to unity) this entropy is just $A/4$, where A is the surface area of the hole's horizon. In the final collapse, these black holes all come together, and the space-time singularities which lie at their cores (where curvatures diverge to infinity) all congeal into one final singularity— referred to as the big crunch. The entropy in a big crunch singularity would be far larger than that in the big bang. Essentially, this entropy provides a measure of the complication of the singularity. The big bang singularity was indeed highly constrained—and one can more or less understand the nature of this constraint, namely that the Weyl conformal curvature was zero—or at least very small in comparison with the remaining Ricci curvature (Penrose 1979a). Such a constraint cannot apply to the big crunch, or to the singularities inside black holes generally. If such a constraint did apply, then one would have severe violations of the second law of thermodynamics.

The difference between these latter types of singularity and the singularity of the big bang is that they are *future* singularities (i.e., timelike curves can enter them in the future direction) whereas the big bang was a *past* singularity (timelike curves can come out of it in the future direction). The rules for past singularities must be different from those which apply at future singularities. Past singularities (or, rather, the only past singularity that we know about, namely the big bang) seem to be highly constrained in having zero (or else very small) Weyl curvature. I refer to this constraint as the *Weyl curvature hypothesis* (cf. Penrose 1981, 1986a, b).[2]

The common belief is that it should be *quantum gravity* which determines whatever these rules are—and so it should presumably be quantum gravity which provides the Weyl curvature hypothesis (and hence the second law). However, we note that the Weyl curvature hypothesis is not invariant under **T** (nor under **PT**, nor **CT**, nor even under **PCT**). This seems to imply that quantum gravity theory is not invariant under these symmetries either. At least this would be the view consistent with N. Some people might prefer an explanation more in line with S. However, it is difficult to see how the ideas of symmetry breaking can apply here. Symmetry breaking, in the ordinary sense, refers to a system "settling down" into an asymmetrical configuration in the normal future direction in time—and it thereby *makes use of* the second law of thermodynamics rather than being able to provide an *explanation* for that law! It seems that the asymmetry in time is already being assumed in the standard symmetry-breaking mechanisms, and I do not see how S can be invoked to explain time-asymmetry.

Inflationary universes. I should make some comment concerning the fashionable "inflationary models" (cf. Hawking, Gibbons and Siklos 1986) which purport to explain some of the puzzling features of the early universe in terms of mechanisms which depend heavily upon an S-type process. The idea is that in the extremely early stages of the big bang, the universe expanded by an absolutely enormous factor because of some phase change in the vacuum at very early times, using some S mechanism. This enormous expansion is supposed to explain the big bang's spatial uniformity of expansion, which is estimated to be "fine-tuned" to a factor of at least one part in 10^{60}. However, inflation provides no real explanation, since the physics in the inflationary models is taken to be consistent with **T**. Thus, the inflationary scheme (in reverse) should apply equally to a generic collapsing universe—which it blatantly need not, because of the black-hole singularities. A time-asymmetric constraint is still needed at space-time singularities, and we appear to be back with the Weyl curvature hypothesis, as before, for the initial singularity. An explanation for the time-asymmetric second law is *not* provided by the inflationary models. In fact the puzzle of fine tuning to one part in 10^{60} is totally dwarfed by the constraint needed to remove the time-reverses of black holes. For a closed universe with about 10^{80} baryons such a constraint would need fine tuning to about one part in 10^{123}. For a larger universe, this fine tuning would need to be even more precise. It is this that we need the Weyl curvature hypothesis for, whether or not an inflationary phase of the universe actually occurred.

---

[2] Weyl himself seems to have been aware that some important constraint was needed for the early universe. According to "Weyl's postulate" (cf. Bondi 1952) the idealized substratum particles must have world-lines which are orthogonal to spacelike hypersurfaces. This implies some degree of uniformity for the big-bang singularity. There seems to be some loose connection between Weyl's postulate and what I am calling the Weyl curvature hypothesis.

It should be remarked that the inflationary models are also invoked in order to explain other seeming puzzles, namely the absence of a proliferation of such extravagant phenomena as "domain walls", "cosmic strings", "axions" and "monopoles". These are all animals of the menagerie conjured up by a belief in the reality of $, in current unified gauge theories. If we eventually find that we do not need $, then the inflationary models will not be needed either!

T-noninvariance at the fundamental level? I have indicated by belief that a fundamental theory of physics ought to be more in line with N than with $. Can one provide a basic theory of physics which, at the *fundamental* level, is *not* invariant under P, C, T, or under any nontrivial combination of these three? In my opinion this will be ultimately necessary. I have indicated that I believe that the correct theory of quantum gravity, when it is found, should be time-asymmetric. I believe, also (cf. Penrose 1981, 1986a, 1987), that such a theory ought to give some kind of objective picture of what happens when "collapse of the wave function" takes place in ordinary quantum mechanics. I have argued elsewhere that this procedure also must be time-asymmetric (Penrose 1987). According to the standard quantum-mechanical prescription, quantum amplitudes are converted into probabilities—by our taking the squares of their moduli—whenever an "observation" (and hence "wave-function collapse") is deemed to have taken place. This procedure works marvellously, and is fully consistent with observational facts, provided that it is employed in the normal future direction. However, it would give completely the wrong answers if it were to be applied in the past direction!

The procedures of unitary evolution (i.e., Schrödinger equation, or equivalent), on the other hand, are completely time symmetric. Classical general relativity, also, is an entirely time-symmetric theory. It would seem, therefore, that in the absence of any mechanism of the $ type, quantum gravity ought also to be time-symmetric if it is simply some kind of union between these two time-symmetric procedures. It is thus my opinion that the union between quantum theory and gravitational theory will entail an actual change in the structure of quantum theory, where the unified theory will be time-asymmetric at the fundamental level—in accordance with N—and some objective and mathematically well-defined procedure will take over the rather unsatisfactory "collapse of the wave function" that is a feature of present quantum theory.

P- and C-noninvariance: twistor theory. What about the symmetries P and C? Can asymmetry with respect to these operations also be built in at the fundamental level? Some modern string theories attempt to do just this (e.g., the proposal of so-called "heterotic strings", though I have found some difficulty, myself, in understanding just how these theories are to be interpreted. A radically different approach from string theory, albeit with certain common ideas (cf. Shaw 1986, Hughston and Shaw 1987a, 1987b), is that provided by *twistor* theory (see Huggett and Tod 1985, Penrose and Rindler 1986, Ward and Wells 1988). According to the twistor approach, the structures of holomorphic geometry are taken as fundamental. However, the basic geometry is not that of space-time, but of twistor space (or of portions of products of twistor spaces). At the most visualizable level, twistor space can be regarded as describing, in a holomorphic way, the space of (straight) light rays in four-dimensional Minkowski space-time **M**. The light rays constitute a five-dimensional system, but the 5-manifold describing them imbeds naturally as a real hypersurface **PN**, in a complex projective 3-space **PT**. Here, **PT** is the *projective* version of the full twistor space **T**, which is a complex vector space with a pseudo-Hermitian structure of signature $(+ + --)$, the null elements of which determine **PN**. The points of **M** are interpreted as complex projective lines, in **PT**, which lie in **PN**.

There is some immediate relation to string theory in four-dimensional space-time (at least for "string histories" whose induced metric from the space-time is definite rather than Lorentzian), since it turns out that holomorphic curves in **PT** describe general solutions of the "minimal surface equation" in **M**. (This construction has been studied by Shaw 1986 and generalized to other induced-metric signatures and to other space-time dimensions—up to ten—by Hughston and Shaw 1987a, 1987b.) However the underlying philosophy of twistor theory is different from that of string theory. With twistor theory one tries to build up the space-time points themselves as secondary structures, the twistor space being regarded as more fundamental. In standard twistor theory, the space-time is four-dimensional and the signature of the space-time metric is Lorentzian, though many higher-dimensional generalizations and departures from this original scheme have been studied.

One of the underlying thoughts of twistor theory is that symmetry groups are not necessarily to be taken as fundamental, but they may arise as "incidental" structures. The basic example of this sort of thing is provided by the complex projective lines in **PN** which represent points of **M**. The *points* of such a line constitute a Riemann sphere in **PN** representing the light rays in **M** passing through a particular point **x** in **M**. Thus, this Riemann sphere represents the "sky" or celestial sphere of an observer situated at **x**. Now, this celestial sphere has a natural structure as a one-dimensional complex manifold (i.e., it is naturally an oriented conformal real 2-manifold). This structure is preserved as we pass from one observer at **x** to another. The group which preserves this structure is $PL(2, \mathbb{C})$, which is the same as the *restricted* (i.e., nonreflective) *Lorentz group*. This group may be regarded as the most important symmetry group of physics, and it is seen as arising here as the simplest symmetry group of a "structureless" complex manifold: the group of symmetries of a complex 1-manifold of simplest (i.e., S^2) topology. This kind of idea also extends (though perhaps less impressively) to certain higher-dimensional symmetry groups.

Massless fields and cohomology. Linear massless space-time fields—or, rather more appropriately, wave functions—can be represented very neatly in terms of *sheaf cohomology* elements for appropriate regions of **T** (or of **PT**).

There is a relation between the homogeneity degree n of the functions—called *twistor functions*—providing the required representative (Čech) cocycle—and the *helicity s* of the field:

$$n = -2s - 2$$

(where units are chosen for which Planck's constant/2π is taken as unity). This means, in particular, that the *right-handed photon* (described by a wave function which is a self-dual Maxwell field) has a twistor function of homogeneity degree $n = -4$, whereas a *left-handed photon* (anti-self-dual Maxwell field) has a twistor function of degree $n = 0$. For a Weyl *neutrino* (left-handed) we have $n = -1$, while for an *anti-neutrino* (right-handed), $n = -3$. If we take gravitons to be described by linearized solutions of Einstein's vacuum equations (i.e., vanishing Ricci tensor), then we find that the *right-handed graviton* (wave function describing a self-dual linearized Weyl tensor) has a twistor function of homogeneity degree $n = -6$, while a *left-handed graviton* (anti-self-dual linearized Weyl tensor) has a twistor function of degree $n = +2$.

Note that there is a fundamental left-right asymmetry in this description. A space-reflection (operation of **P**) would merely reverse the sign of the helicity,[3] but this does something more complicated to the homogeneity degree of the twistor function, namely effect the interchange

$$n \longleftrightarrow -n - 4.$$

This is the kind of thing that can make a significant difference in a cohomology description, though such a difference does not actually show up for the (twistor) cohomology group elements describing free wave functions.

Left-right asymmetry in nonlinear twistor descriptions. A more serious left-right asymmetry arises with the *nonlinear* versions of these cohomology elements. In close relation to the Kodaira-Spencer (1958) theory of deformations of complex manifolds and the theory of holomorphic vector bundles, such cohomology elements can be "exponentiated", in certain cases, to give nonlinear fields (or "nonlinear wave functions"). In the *nonlinear graviton construction* (Penrose 1976) this is achieved for the case $n = 2$, and provides the general (complex) solution of the Einstein vacuum equations with anti-self-dual curvature, while in the *Ward* (1977) *construction* (cf. also Atiyah and Ward 1977, Atiyah, Hitchin, Manin and Drinfeld 1978), the case $n = 0$ is generalized to provide the general solution of the anti-self-dual Yang-Mills equations. For a more complete physical theory, one would need to be able to code *self*-dual nonlinear fields into deformations of twistor structures in addition to anti-self-dual ones (this is the *googly* problem, cf. Penrose 1979b) and then to combine the two to give the full description of a nonlinear field which need be neither self-dual nor anti-self-dual.

The self-dual nonlinear fields seem to be very much more difficult to handle in terms of the standard twistor space (the googly problem) than do the

---

[3]The different roles of **P** and **C** are a little difficult to disentangle in all this. There is actually a choice, in the twistor description which, via the sign of the given Hermitian form, serves to single out one half of **PT\PN**, denoted by **PN**$^+$, as preferred over the other.

anti-self-dual ones. A description in the gravitational case seems to be possible, but it involves some rather strange ideas, involving the "exponentiation" of *relative* (second) cohomology elements (Penrose 1986b) and leading to peculiar nonlinear deformations up the fibres of $T \rightarrow PT$, providing a type of locally extendible "conical singularity". There is a blatant left-right asymmetry in this that many would no doubt find disturbing—since Einstein's vacuum equations, for example, are perfectly left-right symmetric. The asymmetry arises because of a fundamental choice having been made right at the start: *twistor space has been chosen in preference to dual twistor space.* Had we started with dual twistor space instead, the right-handed (self-dual) cases would be the easy ones and the left-handed (anti-self-dual) cases, the difficult ones. An alternative route to follow is that provided by the *ambitwistor* descriptions, which work with a space representing the complex null geodesics in complexified space-time which can, in a certain sense, be realised as a subspace of the product space of twistor space with its dual. This route has been pioneered by Isenberg, Yasskin and Green (1978), LeBrun (1983), Witten (1978, 1986) and Manin (1982). It does not exhibit left-right asymmetry, and has its own disadvantages as well as advantages. Its connection with twistor wave functions is less clear than is the case for the original asymmetrical twistor approach.

In my own view, despite its technical difficulties, the fundamentally asymmetrical approach is likely to be ultimately the most fruitful. Only time—and a great deal of hard work, insight and good luck—will tell!

REFERENCES

M. F. Atiyah and R. S. Ward (1977), *Instantons and algebraic geometry*, Comm. Math. Phys. **55**, 111–124.

M. F. Atiyah, N. J. Hitchin, V. G. Drinfeld, and Yu. I. Manin (1978), *Construction of instantons*, Phys. Lett. **65A**, 185–187.

J. D. Bekenstein (1974), *Generalized second law of thermodynamics in black-hole physics*, Phys. Rev. **D9**, 3292–3300.

H. Bondi (1952), *Cosmology*, Cambridge Univ. Press, Cambridge.

J. H. Christenson, J. W. Cronin, V. L. Fitch, and R. Turlay (1964), *Evidence for the 2π decay of the K^0 meson*, Phys. Rev. Lett. **13**, 138–140.

P. A. M. Dirac (1928), *The quantum theory of the electron*, Proc. Roy. Soc. London **A117**, 610–624; part II, ibid. **A118**, 351–361.

—— (1982), *Pretty mathematics*, Int. J. Theor. Phys. **21**, 603–605.

F. J. Dyson (1966), *Symmetry groups in nuclear and particle physics*, Benjamin, New York.

M. Gell-Mann and Y. Ne'eman (1964), *The eightfold way*, Benjamin, New York/Amsterdam.

S. W. Hawking (1978), *Particle creation by black holes*, Commun. Math. Phys. **43**, 199–220.

S. W. Hawking, G. Gibbons, and S. Siklos (1986), *The very early universe*, Cambridge Univ. Press, Cambridge.

P. Higgs (1966), *Spontaneous symmetry breakdown without massless bosons*, Phys. Rev. **145**, 1156–1163.

L. P. Hughston and W. Shaw (1987a), Classical Quantum Gravity **4**, 869.

—— (1987b), *Classical strings in ten dimensions*, Proc. Roy. Soc. London **A414**, 423–431.

J. Isenberg, P. B. Yasskin, and P. S. Green (1978), *Non-self-dual gauge fields*, Phys. Lett. **78B**, 462–464.

K. Kodaira and D. C. Spencer (1958), *On deformations of complex analytic structures*. I, II, Ann. Math. **67**, 328–401, 403–466.

S. A. Huggett and K. P. Tod (1985), *An introduction to twistor theory*, London Math. Soc. Student Texts (L.M.S. publ.).

T. D. Lee and C. N. Yang (1956), *Question of parity conservation in weak interactions*, Phys. Rev. **104**, 254–258.

C. R. LeBrun (1983), *Spaces of complex null geodesics in complex Riemannian geometry*, Trans. Amer. Math. Soc. **284**, 601–616.

Yu. I. Manin (1982), *Gauge field and cohomology of analytic sheaves*, Twistor Geometry and Non-Linear Systems (H. D. Doebner and T. D. Palev, eds.), Springer-Verlag, Berlin.

R. Penrose (1976), *Non-linear gravitons and curved twistor theory*, Gen. Rel. Grav. **7**, 31–52.

_____ (1979a), *Singularities and time-asymmetry*, General Relativity: An Einstein Centenary (S. W. Hawking and W. Israel, eds.), Cambridge Univ. Press, Cambridge.

_____ (1979b), *A googly graviton?* Advances in Twistor Theory (L. P. Hughston and R. S. Ward, eds.), Pitman Press, Boston, pp. 168–176.

_____ (1981), *Time-asymmetry and quantum gravity*, Quantum Gravity 2 (C. J. Isham, R. Penrose and D. W. Sciama, eds.), Oxford Univ. Press, Oxford.

_____ (1986a), *Gravity and state-vector reduction*, Quantum Concepts in Space and Time (R. Penrose and C. J. Isham, eds.), Oxford Univ. Press, Oxford.

_____ (1986b), *Hermann Weyl, space-time and conformal geometry*, Hermann Weyl 1885–1985 (K. Chandrasekharan, ed.), I. T. H., Zurich and Springer-Verlag, Berlin.

_____ (1987), *Newton, quantum theory and reality*, 300 Years of Gravity (S. W. Hawking and W. Israel, eds.), Cambridge Univ. Press, Cambridge.

R. Penrose and W. Rindler (1986), *Spinors and space-time.* Vol. 2: *Spinor and twistor methods in space-time geometry*, Cambridge Univ. Press, Cambridge.

W. T. Shaw (1986), Classical Quantum Gravity **3**, 753–761.

R. S. Ward (1977), *On self-dual gauge fields*, Phys. Lett. **61A**, 81–82.

R. S. Ward and R. O. Wells, Jr. (1988), *Twistor geometry and field theory*, Cambridge Univ. Press, Cambridge.

H. Weyl (1918), *Gravitation und Electrizität*, Sitz. Ber. Preuss. Ak. Wiss., 465–480.

_____ (1929), *Elektron und Gravitation*. I, Z. Phys. **56**, 330–352.

_____ (1931), *The theory of groups and quantum mechanics*, Methuen, London.

_____ (1952), *Symmetry*, Princeton Univ. Press, Princeton, N. J.

E. Witten (1978), *An interpretation of classical Yang-Mills theory*, Phys. Lett. **77B**, 394–398.

_____ (1986), Nuclear Phys. **B 266**, 245–264.

C. S. Wu, E. Ambler, R. Hayward, D. Hoppes, and R. Hudson (1957), *Experimental test of parity conservation in beta decay*, Phys. Rev. **105**, 1413–1415; *Further experiments on beta decay of polarized nuclei*, Phys. Rev. **106**, 1361–1363.

MATHEMATICAL INSTITUTE, OXFORD, UNITED KINGDOM

Proceedings of Symposia in Pure Mathematics
Volume 48 (1988)

Free Fermions on an Algebraic Curve

EDWARD WITTEN

In these notes, I would like to describe certain aspects of conformally invariant quantum field theory on Riemann surfaces, especially those aspects that are analogous to the theory of automorphic representations.

Over the years, there have been several sources of inspiration for work on conformal field theory. Many of the important insights came in work whose motivation was to understand phase transitions in two-dimensional surfaces. Other investigators were mainly interested in finding simple, soluble models of quantum field theory in a world of one space, one time dimension. Finally, and increasingly in recent years, work on conformal field theory has been motivated by its applications to string theory. (For instance, see [1].)

String theory is a remarkable framework for physics in which general relativity and Riemannian geometry are replaced even in the classical limit by completely new—and as yet little understood—structures. The really central geometrical ideas in string theory have not yet been properly identified and formulated. To do so is, I think, a challenge that would have pleased Hermann Weyl. I will not attempt a general overview of the subject here; whatever I would have had to say can be found in [2]. Rather, I would like to describe some aspects of what is certainly an important part of the puzzle—though the pieces haven't been properly fitted in place yet—namely quantum field theory on Riemann surfaces.

Riemann surfaces can be viewed as C^∞ objects (two-dimensional real manifolds) or holomorphic objects (one-dimensional complex manifolds). Likewise, quantum field theory on Riemann surfaces has two branches, which we may roughly call C^∞ and holomorphic. These two branches are very different in flavor. C^∞ quantum field theory on Riemann surfaces is related to analysis and index theory on function spaces, to geometry and topology; in its applications in string theory it is related to space-time geometry. Holomorphic quantum field theory on Riemann surfaces is more closely related to algebraic geometry and the theory of automorphic representations. The two subjects are related by an

1980 *Mathematics Subject Classification* (1985 *Revision*). Primary 14H99, 83C47; Secondary 83E99.

Research supported in part by NSF grants PHY 80-19754, 86-16129, 86-20266.

"index" map. Roughly speaking (I am making an optimistic simplification), the index of a (supersymmetric) C^∞ quantum field theory is a holomorphic one. A structure essentially equivalent to this index map was first used by Schellekens and Warner in their work on anomalies in string theory [3]. Later on, it became clear [4] that a certain topological conjecture that was first stated in generality in [5, 6] should have a natural proof in the context of the index map; this point of view has recently been made rigorous [7].

The index map certainly shows that C^∞ and holomorphic quantum field theories on Riemann surfaces are closely related. For a brief introduction to mathematical applications of C^∞ quantum field theory on Riemann surfaces, I refer to [4]; in these lecture notes, we will concentrate on the holomorphic side of the story. Thus, our Riemann surfaces will be viewed as curves over $\mathbf{C}$; and more generally, we will consider algebraic curves over an arbitrary ground field k. For illustrative purposes, we will study in some detail the simplest quantum field theory that can be defined on an algebraic curve, namely the theory of free fermions.

Along the way, we will find a purely mathematical result which I will state here. Let k be a field and X a curve defined over k. Let L be the canonical line bundle of X, and let S be a square root of L (a line bundle with $S \otimes S \approx L$) that is defined over k. Let h be a rational function on X.

Write the divisor of h as

$$(1) \qquad (h) = - \sum_{m_i \text{ even}} m_i(Q_i) - \sum_{n_j \text{ odd}} n_j(P_j),$$

where Q_i and P_j are certain points on X. The P_j, $j = 1, \ldots, r$ are the points at which h is of odd order. (We assume for the moment that the Q_i and P_j are defined over k. The more general situation will be treated in §3). For each P_j, $j = 1, \ldots, r$, pick a rational section ψ_j of S which is of order $(n_j - 1)/2$ at P_j. Thus, the differential form $h\psi_j^2$ has a simple pole at P_j. Let

$$(2) \qquad \alpha_j = \text{Res}_{P_j}(h\psi_j^2).$$

Note that the residue class of α_j in $k^\times/(k^\times)^2$ ($k^\times$ is the multiplicative group of k) is independent of the choice of ψ_j.

I claim

$$(3) \qquad (-1)^{r/2} \prod_{j=1}^{r} \alpha_j = 1 \quad \mod(k^\times)^2.$$

The proof (using the Riemann-Roch theorem and simple facts about quadratic forms) can be found in §3, which is independent of the rest of these notes. However, the theoretical context for the statement, in terms of free fermion quantum field theory, will emerge in §§1 and 2.

To clarify the content of formula (3), let f and g be two rational functions on k. Then a well-known formula of Weil asserts that

$$(4) \qquad \prod_{P \in X} (-1)^{\text{ord}_P f \cdot \text{ord}_P g} \frac{f(P)^{\text{ord}_P g}}{g(P)^{\text{ord}_P f}} = 1.$$

Let us refer to the left-hand side of (4) as $W(f,g)$. Also, let us refer to the left-hand side of (3) as $A(h)$. Then it is easy to see that

$$(5) \qquad A(f)A(g) = W(f,g)A(fg) \quad \mod(k^{\times})^2.$$

Thus (3) and (4) are closely related; the validity of (4) $\mod(k^{\times})^2$ follows from (3). From a physical point of view, (3) is closely related (as we will see) to one-component Majorana-Weyl fermions, while Weil's formula (4) would arise in the theory of two-component fermions with GL(1) symmetry (current algebra).

1. Generalities about holomorphic conformal field theory.

Let k be a field and X a curve over k. In describing a holomorphic field theory on X, the first ingredient is the space of observables. For each point $P \in X$, there is a k vector space V_P of "observables at P." Every, or almost every, V_P contains a distinguished vector 1_P which in physical terminology corresponds to the identity operator or to the "vacuum vector."

The space of observables V is then the restricted or adelic tensor product of the V_P; that is, V consists of tensor products $\bigotimes_P u_P$ with $u_P \in V_P$ for all P and $u_P = 1_P$ for almost all P. Following custom we denote this as

$$(6) \qquad V = \coprod_P V_P.$$

The next ingredient in the story is the action of a certain adelic Lie algebra (or superalgebra) $\mathcal{U}$ on the V_P and their restricted product V. The deeper side of conformal field theory is that V is uniquely determined by $\mathcal{U}$, V_P being in a sense the fiber at P of the global Lie algebra underlying $\mathcal{U}$. This is a key to the richness of conformal field theory, but a proper mathematical language for describing it has not been developed yet, and so I will not try here to formulate the relation between $\mathcal{U}$ and V. However, this should be borne in mind as an important area for future investigation.

Thus (as physicists commonly do in practice), we will work with a subalgebra $\mathcal{G}$ of $\mathcal{U}$ that is of manageable size. There are several manageable examples that might be considered; for simplicity we will describe a situation that in physical terminology corresponds to "current algebra."

Let $\mathcal{R}$ be a simple Lie algebra over k with an invariant bilinear form $(\, , \,)$. Let $\mathcal{G}$ be the Lie algebra of rational maps of X into $\mathcal{R}$. For each $P \in X$, let $\mathcal{G}_P$ be the completion of $\mathcal{G}$ at P. If f, g are rational maps f, $g: X \to R$ then (f, dg) is a rational differential form on X. For f, $g \in \mathcal{G}_P$,

$$(7) \qquad \omega(f,g) = \mathrm{Res}_P(f, dg)$$

is a cocycle for a central extension $\widehat{\mathcal{G}}_P$ of $\mathcal{G}_P$ by k^+ (the additive group of k).

If $k = \mathbf{C}$ (the field of complex numbers), then $\widehat{\mathcal{G}}_P$ is isomorphic to a completion of an affine Lie algebra $\widehat{\mathcal{R}}$. Indeed, if z is a local parameter at P, then an element of $\widehat{\mathcal{G}}_P$ is a formal series

$$(8) \qquad g = \sum_{n \gg -\infty} z^n a_n$$

with $a_n \in \mathcal{R}$, and the Lie bracket

(9) $$[z^n a_n, z^m a_m] = z^{m+n}[a_m, a_n] + m\delta_{m+n}(a_n, a_m).$$

In current algebra on an algebraic curve, the V_P are chosen to be $\widehat{\mathcal{G}}_P$ modules, with the property that the distinguished vector 1_P (which exists for almost all P) obeys

(10) $$g \cdot 1_P = 0$$

whenever $g \in \widehat{\mathcal{G}}_P$ is regular at P. This corresponds to highest weight modules in the theory of affine Lie algebras.[1]

Now, let $\mathcal{G}_A$ be the adelic product of the $\mathcal{G}_P$:

(11) $$\mathcal{G}_A = \coprod_P \mathcal{G}_P.$$

$\mathcal{G}_A$ consists of objects $\bigoplus_{P \in X} g_P$ with $g_P \in \mathcal{G}_P$ for all P, and g_P regular at P for almost all P. Likewise, let $\widehat{\mathcal{G}}_A$ be the adelic product of the $\widehat{\mathcal{G}}_P$:

(12) $$\widehat{\mathcal{G}}_A = \coprod_P \widehat{\mathcal{G}}_P.$$

Thus, $\widehat{\mathcal{G}}_A$ is an extension of $\mathcal{G}_A$ by k^+. Explicitly, $\widehat{\mathcal{G}}_A$ is described by the cocycle $\omega_A(f, g) = \sum_P (f_P, dg_P)$ for $f = \bigotimes_P f_P$, $g = \bigotimes_P g_P$ in $\mathcal{G}_A$. This is a finite sum since f_P, g_P are regular at P for all but finitely many P.

The condition (10) ensures that $V = \coprod_P V_P$ is a module for $\widehat{\mathcal{G}}_A$ in a natural way, with $\bigoplus_{P \in X} g_P \in \widehat{\mathcal{G}}_A$ acting componentwise on $\bigoplus_{P \in X} u_P \in V$.

The global Lie algebra $\mathcal{G}$ has a natural "diagonal" embedding in $\mathcal{G}_A$ (via $g \to \bigoplus_{P \in X} g_P$ with $g_P = g$ for all P; this gives an element of $\mathcal{G}_A$ since g is regular at P for all but finitely many P). The formula

(13) $$\sum_P \operatorname{Res}_P(f, dg) = 0$$

shows that the cocycle (7) splits over $\mathcal{G} \subset \mathcal{G}_A$. Thus, there is in fact a natural embedding of Lie algebras

(14) $$\phi: \mathcal{G} \hookrightarrow \widehat{\mathcal{G}}_A.$$

The last crucial ingredient in describing quantum field theory on an algebraic curve is an operation of "integration," which in favorable cases can be defined by a Feynmann path integral, and otherwise must be defined more abstractly. For our purposes the "integral" is simply a linear functional on V, which we will denote as $\int$:

(15) $$\int : V \to k.$$

---

[1] At finitely many P at which the distinguished vector 1_P may not exist, we require the existence of a nontrivial, finite-dimensional subspace $U \subset V_P$ such that $gu = 0$ for $u \in U$, if $g = 0$ at P. In physical terminology, such points are points with insertions of "spin operators."

If u is an "observable," that is, an element of $V = \coprod_P V_P$, then $\int u$ is called by physicists the "expectation value of u." The basic requirement on $\int$ is that, for $g \in \mathcal{G}$ and $u \in V$,

$$(16) \qquad \int \phi(g) \cdot u = 0.$$

Thus $\int$ is really a linear functional on the k vector space $V^{\mathcal{G}} = V/\mathcal{G}V$, which is one-dimensional in favorable cases and finite-dimensional in general. When $V^{\mathcal{G}}$ is one-dimensional, $\int$ is uniquely determined (up to normalization) by (16); when $V^{\mathcal{G}}$ has dimension bigger than one, to determine $\int$ uniquely requires additional considerations, which depend on the particular physical context. Compatibility with the Clifford algebra introduced in the next section would be a typical example.

The above description of holomorphic quantum field theory on a Riemann surface is not quite the standard formulation usually used by physicists but an adaptation that makes sense over any field. For instance, (16) is a restatement of the "Ward identities," as those were presented by Belavin, Polyakov, and Zamolodchikov [8]; for an account of the relation of (16) (and the preceding formulas) to standard physical discussions, see [9], especially §3.

One might wonder whether in the above, instead of the *Lie algebra* $\mathcal{G}$ of rational maps $X \to \mathcal{R}$, we could consider the *group* $G(X)$ of rational maps $X \to R$, with R being an algebraic group (over k) corresponding to the Lie algebra $\mathcal{R}$. In fact, since the highest weight modules of affine Lie algebras admit the action of the corresponding groups, compatibly with the Lie algebra, it is possible to do so. In this way, we would be led to consider the completion G_P of $G(X)$ at $P \in X$; the same spaces V_P considered earlier are naturally modules for central extensions $\widehat{G}_P$ of the G_P (the kernel of these extensions now being $k^\times$). Taking adelic products, we would form the adele group $G_A = \coprod_P G_P$ and the corresponding central extension $\widehat{G}_A = \coprod_P \widehat{G}_P$. The analogue of (13) would be the assertion that the central extension

$$(17) \qquad 1 \to k^\times \to \widehat{G}_A \to G_A \to 1$$

should split over the image of the diagonal embedding $G(X) \hookrightarrow G_A$. The existence of an extension of the adele group G_A with this property is an old result [10]. In [9], I sketched a proof for $G = O(N)$ using compatibility with the Clifford algebra considered in the next section.

Finally, instead of considering the Lie algebras and groups considered above, it is natural to consider the Lie algebra S of rational vector fields on X, its various completions S_P, $P \in X$, and the adelic Lie algebra $S_A = \coprod_P S_P$. Each S_P then has a central extension $\widehat{S}_P$, with kernel k^+ (a proper geometric account was first given in [11]; see also the appendix to [9] for another approach). One again forms the adelic central extension $\widehat{S}_A = \coprod_P \widehat{S}_A$, which splits over the image of the diagonal embedding $S \hookrightarrow S_A$. One considers adelic modules $V = \coprod_P V_P$, each V_P being a highest weight $\widehat{S}_P$ module, and one defines the "expectation value of a product of observables" via a linear map $\int : V \to k$ such that $\int s \cdot u = 0$ for $s \in S$, $u \in V$. A discussion of these matters, which are very close to the heart of conformal field theory, would be very similar to the case we actually considered but slightly more involved.

2. Free fermions. In this section, I will describe in somewhat more detail what is perhaps the simplest of conformal field theories on a Riemann surface, namely the theory of free fermions. Apart from giving some further illustration for conformal field theory, my motivation for describing free fermions here is to explain the theoretical context for formula (3).

As in the introduction, let k be a field, X a curve over k, L the canonical line bundle of k, and S a square root of L, defined over k. To simplify life, we suppose that the characteristic of k is not 2. For the moment we assume k to be algebraically closed.

Let Y be the space of rational sections of S defined over k; it is an infinite-dimensional k vector space. Let $\bigwedge Y$ be the exterior algebra on Y; as a vector space it is

$$(18) \qquad \bigwedge Y = 1 \oplus Y \oplus \bigwedge^2 Y \oplus \cdots$$

with $\bigwedge^k Y$ the kth exterior power and "1" a one-dimensional vector space. For $P \in X$, let Y_P and $\bigwedge Y_P$ denote the completions of Y and $\bigwedge Y$ at P.

Let h be a rational function on X. For each $P \in X$, we define a quadratic form $(\, , \,)_P$ on Y_P by the formula

$$(19) \qquad (y, z)_P = \mathrm{Res}_P(hyz)$$

for y, $z \in Y_P$. (Note that for $y, z \in Y_P$, yz and hyz are differential forms, or more exactly sections of the completion of L at P.)

Next we form a Clifford algebra on Y_P with respect to the quadratic form $(\, , \,)_P$. This Clifford algebra, which we will call CY_P, is generated by objects $\hat{z}_P$, $z \in Y_P$, with relations

$$(20) \qquad \hat{y}_P \hat{z}_P + \hat{z}_P \hat{y}_P = 2(y, z)_P$$

for all y, $z \in Y_P$.

Let $-n$ be the order of h at P. An element $X \in CY_P$ is said to be "regular at P" with respect to $(\, , \,)_P$ if it is a sum of products, $\hat{y}_{1,P}\hat{y}_{2,P}\cdots\hat{y}_{n,P}$ with $\mathrm{ord}_P(y_i) \geq n/2$ for all i (so that hy_i^2 is regular at P in the usual sense). We then define the "global Clifford algebra"

$$(21) \qquad CY = \coprod_P CY_P$$

to consist of formal sums $\bigoplus_{P \in X} x_P$, with $x_P \in CY_P$ for all P and x_P regular at P for all but finitely many P.

Given $y \in Y$, y is regular at P for almost all P, so there is a diagonal embedding $\phi : Y \hookrightarrow CY$ given by

$$(22) \qquad y \to \hat{y} = \bigoplus_P \hat{y}_P.$$

Since

(23)
$$\sum_P (y, z)_P = 0$$

for all y, $z \in Y$, we have

(24)
$$\hat{y}\hat{z} + \hat{z}\hat{y} = 0$$

and thus ϕ gives an embedding of the global exterior algebra $\bigwedge Y$ in CY:

(25)
$$\phi \colon \bigwedge Y \hookrightarrow CY.$$

Now, we wish to pick for each $P \in X$ an irreducible CY_P module V_P, in such a way that the product $V = \coprod_P V_P$ (the restricted product being taken in an appropriate sense) will have a natural structure of CY module. Once this is done, we will complete the description of free fermion quantum field theory by introducing a linear functional $\int \colon V \to k$ with the property

(26)
$$\int \hat{y} \cdot u = 0$$

for all $y \in Y$, $u \in V$. For the modules V_P described below, it can be shown that the k vector space $V^Y = V/YV$ is one-dimensional (YV is the subspace of V spanned by all $\hat{y} \cdot u$, $y \in Y$, $u \in V$), so $\int$ is determined uniquely, up to normalization, by (26). (For $h = 1$, this was shown in detail in [**9**, §4], and the general case is similar.) The one-dimensionality of V/YV makes very powerful the compatibility of the Clifford algebra with, for instance, the action of an algebraic group.

The one-dimensional vector space $V^Y = V/YV$ has an interesting interpretation in algebraic geometry; I will state the basic facts without an attempt at proof. Consider the case $h = 1$. Consider an algebraic family of curves X_t (with spin bundle S_t) over a base T. (The letter t denotes a point in T; X_t is the corresponding curve, with spin bundle S_t.) One defines the determinant of cohomology $(\det H^0(X_t, S_t)) \otimes (\det H^1(X_t, S_t))^{-1}$, $t \in T$, as a line bundle over the base T. (Although $H^0(X_t, S_t)$ and $H^1(X_t, S_t)$ are in general not locally free over T, one can define a locally free sheaf on T that reduces to $(\det H^0(X_t, S_t)) \otimes (\det H^1(X_t, S_t))^{-1}$ when H^0 and H^1 are locally free.)

By Serre duality, $H^1(X_t, S_t)$ is isomorphic to the dual of $H^0(X_t, S_t)$, so one may expect to be able to give a definition of the square root of the determinant of cohomology. In fact, the one-dimensional vector space $V^Y = V/YV$, canonically defined for any curve X with spin bundle S, can serve as a definition of the square root of the determinant of cohomology. In an algebraic family over a base T, the V^Y are fibers of a line bundle L over T. If S is an "even" spin structure, so that $H^0(X_t, S_t)$ and $H^1(X_t, S_t)$ are trivial for generic $t \in T$, then the identification of L with the square root of the determinant of cohomology can proceed as follows. For each $t \in T$ and $P \in X_t$, there is a canonical "vacuum vector" $1_{P,t}$ in the local module V_P (which we will be defining presently); then $w_t = \bigotimes_{P \in X_t} 1_{P,t}$ is a canonically defined element of the fiber L_t of L at $t \in T$. This gives a section w of L. It can be seen that w vanishes precisely when $H^0(X_t, S_t) \neq 0$; more precisely, w has a zero of order n on the locus on which $H^0(X_t, S_t)$ is n-dimensional. The existence of such a section shows that L can be identified with the square root of the determinant of cohomology.

Returning to our task of defining free fermion quantum field theory, what remains is to describe the relevant Clifford modules V_P. Let us first recall some facts about Clifford algebras in finite dimensions. (For more detail on many points that follow, see [12].)

Let B be an even-dimensional vector space over k (still algebraically closed), say of dimension $n = 2r$ with a nondegenerate quadratic form $(\ ,\ )$. The associated Clifford algebra C_B is a central simple algebra over k so it has up to isomorphism a unique irreducible module V_B. This may be constructed as follows. Pick any decomposition $B = W \oplus \widetilde{W}$ with W and $\widetilde{W}$ isotropic subspaces of dimension r. Let e_i be a basis of W and f_i the dual basis of $\widetilde{W}$. Thus the Clifford algebra is described by generators $\widehat{e}_i$, $\widehat{f}_j$ with relations

$$(27) \qquad \{\widehat{e}_i, \widehat{e}_j\} = \{\widehat{f}_i, \widehat{f}_j\} = 0, \qquad \{\widehat{e}_i, \widehat{f}_j\} = \delta_{ij}.$$

Let us prove that any Clifford module V_B contains a vector u with $\widehat{e}_i u = 0$, $i = 1, \ldots, k$. Indeed, picking any $u \in V$, if $\widehat{e}_1 u \neq 0$ we replace u by $u' = \widehat{e}_1 u$, so that $\widehat{e}_1 u' = 0$; repeating this for $\widehat{e}_2, \ldots, \widehat{e}_r$, we find a vector u with $\widehat{e}_i u = 0$, $i = 1, \ldots, r$. It is now easily seen from (27) that the 2^r vectors

$$(28) \qquad \widehat{f}_1^{u_1} \widehat{f}_2^{u_2} \ldots \widehat{f}_r^{u_r} \cdot u,$$

with $u_i = 0, 1$, for $i = 1, \ldots, r$, are linearly independent and closed under the action of the $\widehat{e}_i$, $\widehat{f}_j$. The $\widehat{f}$, $f \in \widetilde{W}$, act on the vectors (28) via the "wedge product," and the $\widehat{e}$, $e \in W$, act on V_B via "interior multiplication"—they are "annihilation operators." We have thus proved uniqueness of the irreducible Clifford module V_B and determined its structure. For any decomposition $B = W \oplus \widetilde{W}$ in maximal isotropic subspaces W, $\widetilde{W}$, one can write V_B as

$$(29) \qquad V_B = \bigoplus_{j=0}^{r} \wedge^j \widetilde{W}.$$

What happens if B is an infinite-dimensional vector space with a nondegenerate quadratic form $(\ ,\ )$? Then it is no longer true that the Clifford module V_B is unique. The step in the above which fails is the construction of a vector u annihilated by the $\widehat{e}_i$, $i = 1, \ldots, r$; this may not exist if $r = \infty$. However, *if* we are given a decomposition $B = W \oplus \widetilde{W}$ in maximal isotropic subspaces W, $\widetilde{W}$, then we can define a Clifford module V_B by the same formula as in finite dimensions. As a vector space

$$(30) \qquad V_B = \bigoplus_{j=0}^{\infty} \wedge^j \widetilde{W}$$

and the $\widehat{e}_i$, $\widehat{f}_j$ act just as in finite dimensions.

Now, let us go back to our algebraic curve X. For $P \in X$, Y_P was defined as the completion at P of the space of rational sections of the spin bundle S. Y_P is endowed with the nondegenerate quadratic form

$$(31) \qquad (y, z) = \operatorname{Res}_P(hyz),$$

with y, $z \in Y_P$ and h an arbitrarily chosen rational function on X.

Consider first the case $h = 1$. Then for each P one has a natural choice of a maximal isotropic subspace $W \subset V_P$. W consists of sections of S that are regular at P. There isn't a natural isotropic complement $\widetilde{W}$ of W. However, a complement can be picked by choosing a trivialization of S in a neighborhood of P and a uniformizer z at P; then W is spanned by sections $\{1, z, z^2, \dots\}$ of S, and one picks $\widetilde{W}$ to be spanned by $\{z^{-1}, z^{-2}, z^{-3}, \dots\}$. Although the choice of $\widetilde{W}$ is not natural, it can be seen that the Clifford module

$$(32) \qquad V_P = \bigoplus_{j=0}^{\infty} \wedge^j \widetilde{W}$$

is independent of the arbitrary choices. In fact, the module V_P may be described invariantly as an irreducible Clifford module on CY_P which contains a vector 1_P annihilated by $\widehat{w}$ for any $w \in W$. This choice of Clifford module, together with our earlier comments, completes the description of the quantum field theory of free fermions, in the case $h = 1$.

Before generalizing to $h \neq 1$, let us recall a few more facts from finite-dimensions. Given a finite-dimensional vector space B, with a nondegenerate quadratic form, a decomposition $B = W \oplus \widetilde{W}$ in isotropic spaces W, $\widetilde{W}$ will only exist if B is even-dimensional. If B is odd-dimensional, the closest we can come is to write $B = B_0 \oplus B'$, where B_0 is one-dimensional (and not isotropic) and B' is the even-dimensional orthocomplement of B_0. B' will have a decomposition $B' = W \oplus \widetilde{W}$ in isotropic subspaces.

Also, if $B = B_1 \oplus B_2$ where B_1, B_2 are even-dimensional and B_2 is the orthocomplement of B_1, then the Clifford modules V_B, V_{B_1}, and V_{B_2} are related by $V_B = V_{B_1} \widehat{\otimes} V_{B_2}$ ($\widehat{\otimes}$ is a Z_2 graded tensor product). This can be seen, for instance, from our earlier determination of the structure of the Clifford modules.

Now let us go back to (31) with $h \neq 1$. If $n = -\operatorname{ord}_P h$ is even, we can proceed much as before. We pick the maximal isotropic subspace W of Y_P to consist of sections y of S with $\operatorname{ord}_P y \geq n/2$. The complement $\widetilde{W}$ we take to be spanned by sections $\{z^{n/2-1}, z^{n/2-2}, z^{n/2-3}, \dots\}$. This gives a decomposition $Y_P = W \oplus \widetilde{W}$ with W, $\widetilde{W}$ isotropic, and we define $V_P = \bigoplus_{j=0}^{\infty} \wedge^j \widetilde{W}$ as before.

Thus, we have described the Clifford module V_P if $\operatorname{ord}_P(h)$ is even. What shall we do at the finitely many points at which $n = -\operatorname{ord}_P(h)$ is odd? We define the maximal isotropic space W of Y_P consisting of $\{y \in Y_P \mid \operatorname{ord}_P(y) \geq (n+1)/2\}$. And we define $\widetilde{W}$ to be spanned by $\{z^{(n-3)/2}, z^{(n-5)/2}, z^{(n-7)/2}, \dots\}$. W and $\widetilde{W}$ are both maximal isotropic subspaces, but it is not true that $Y_P = W \oplus \widetilde{W}$; rather, $Y_P = W \oplus \widetilde{W} \oplus U_P$, where U_P is the one-dimensional vector space spanned by a section $\{z^{(n-1)/2}\}$ of Y_P. In a more invariant way, the one-dimensional vector space U_P consists of sections of S of order $\geq (n-1)/2$ at P, modulo those of higher order.

Let $\widetilde{Y}_P = W \oplus \widetilde{W}$. The quadratic form on Y_P induces a nondegenerate quadratic form on $\widetilde{Y}_P$. We form the Clifford algebra $C\widetilde{Y}_P$, and define the Clifford module

$$(33) \qquad \widetilde{V}_P = \bigoplus_{j=0}^{\infty} \wedge^j \widetilde{W}.$$

This will do as a module for the Clifford algebra $C\widetilde{Y}_P$. Actually, we want to define a Clifford module for the adelic Clifford algebra $CY = \coprod_P CY_P$ that was introduced earlier (equation (21)).

To do so, let P_i, $i = 1, \ldots, s$ be the points at which h has odd order. (Of course, s is finite and even.) These are the points at which $Y_P \neq \widetilde{Y}_P$ and $CY_P \neq C\widetilde{Y}_P$. For each such P_i, we have defined a one-dimensional vector space U_i consisting of sections of S of order $(n-1)/2$ at P_i, modulo those of higher order. Let

$$(34) \qquad U = \bigoplus_{i=1}^{s} U_i.$$

U is endowed with a natural quadratic form. Indeed, if $Y = (y_1, \ldots, y_s)$ and $Z = (z_1, \ldots, z_s)$ are in U (so y_i, $z_i \in U_i$ for $i = 1, \ldots, s$), we let

$$(35) \qquad (Y, Z) = \sum_{i=1}^{s} \mathrm{Res}_{P_i}(h y_i z_i).$$

This quadratic form is evidently nondegenerate. Let CU be the Clifford algebra on U with this quadratic form.

Now, to define a Clifford module on CY, notice that

$$(36) \qquad CY = C\widetilde{Y} \hat{\otimes} CU$$

where $C\widetilde{Y}$ is the modified adelic Clifford algebra

$$(37) \qquad C\widetilde{Y} = \coprod_p C\widetilde{Y}_p$$

and CU is the Clifford algebra on U. We already defined the $C\widetilde{Y}_p$ module

$$(38) \qquad \widetilde{V}_P = \bigoplus_{j=0}^{\infty} \wedge^j \widetilde{W}$$

in equation (33); and the corresponding adelic product

$$(39) \qquad V = \coprod_P \widetilde{V}_P$$

is a natural $C\widetilde{Y}$ module. The irreducible CY module V that we want is then

$$(40) \qquad V = \widetilde{V} \hat{\otimes} R,$$

where R is an irreducible CU module; as we know, this exists and is unique up to isomorphism, since V is finite-dimensional.

Over an algebraically closed field k, this completes the description of what one means by free fermion quantum field theory, for arbitrary h. (Of course, one must also prove the one-dimensionality of V/YV, so that an integration law obeying (26) exists and is unique up to normalization; for $h = 1$ and an even spin structure this was proved in [**9**].) What if k is not algebraically closed?

To ask the right question we first need some facts from the algebraically closed case. If B is an $n = 2r$-dimensional k vector space with a basis u_i and a quadratic form (,), the Clifford algebra CB is generated by $\widehat{u}_i$ with

$$(41) \qquad \widehat{u}_i\widehat{u}_j + \widehat{u}_j\widehat{u}_i = 2(u_i, u_j).$$

This algebra clearly possesses the automorphism $\tau\colon \widehat{u}_i \to -\widehat{u}_i$. It follows from the uniqueness of the irreducible Clifford module V_B that the automorphism τ acts on V_B. Indeed, in terms of a decomposition $B = W \oplus \widetilde{W}$, with W, $\widetilde{W}$ maximal isotropic subspaces, one can take $V_B = \bigoplus \bigwedge^j \widetilde{W}$, with τ acting on $\bigwedge^j \widetilde{W}$ as multiplication by $(-1)^j$.

If now k is *not* necessarily algebraically closed, by a Clifford module V_B defined over k, we will mean a k vector space V_B such that:

(i) V_B is an absolutely irreducible module for the k-algebra (41); i.e. $V_B \otimes_k \overline{k}$ is irreducible, $\overline{k}$ being the algebraic closure of k.

(ii) V_B is Z_2 graded; there is an operator $\tau\colon V_B \to V_B$ with $\tau^2 = 1$, $\tau\widehat{u}_i = -\widehat{u}_i\tau$. In contrast to the algebraically closed case, if k is not algebraically closed, the Clifford module V_B does not necessarily exist in that sense. Suppose that (,) corresponds to the quadratic form

$$(42) \qquad F(X_1, \dots, X_n) = \sum \alpha_i X_i^2$$

in $n = 2r$ variables. The algebra (41) is then

$$(43) \qquad \widehat{u}_i^2 = \alpha_i, \qquad \widehat{u}_i\widehat{u}_j + \widehat{u}_j\widehat{u}_i = 0, \quad i \neq j.$$

Let $\sigma = \widehat{u}_1\widehat{u}_2 \cdots \widehat{u}_n \cdot \tau$. One sees that

$$(44) \qquad [\sigma, \widehat{u}_i] = 0, \qquad i = 1, \dots, n,$$

so σ must be multiplication by an element of $k^\times$ if V_B is absolutely irreducible. On the other hand, one computes

$$(45) \qquad \sigma^2 = (-1)^{n/2} \prod_{i=1}^n \alpha_i.$$

Thus a necessary (not sufficient) condition for V_B to exist over k is that the right-hand side of (45) should be a square in k.

In our study of free fermions on a curve X over a ground field k that is not necessarily algebraically closed, we would like to know whether the adelic Clifford module V, described in (40), can be defined over k. To this aim, note that in (40) V is given as a graded tensor product. The first factor, $\widetilde{V}$, can be defined over any ground field k, as we see from its definition (38), (39). The issue therefore is entirely whether R can be defined over k; here R is the Clifford module for the Clifford algebra CU on the finite-dimensional vector space $U = \bigoplus_{i=1}^{s} U_i$. A necessary condition for defining R over k is that the right-hand side of (45) should be a square in k. This condition was written out explicitly in equation (3) in the introduction, which we have thus set in its theoretical context. That R can be defined over k will be proved in the next section.

This completes our description of the adelic Clifford module V that appears in fermion free field theory as usually studied by physicists. It would be desirable, and presumably possible, to give a more intrinsic description without the decompositions $Y_P = W \oplus \widetilde{W} \oplus U$, $CY = C\widetilde{Y} \widehat{\otimes} CU$ that were used in the above.

Perhaps I should conclude this section by briefly comparing the treatment of fermion spin operators given here with more usual treatments, described for instance in [1]. Unfortunately, the discussion will not be self-contained.

Physicists usually start with an even number of points $P_i \in X$ at which "spin operators" are to be inserted. This means that the fermi field ψ is to be "double-valued" near the P_i. At least over an algebraically closed field, the double-valuedness can be avoided as follows. The divisor $\sum(P_i)$ is twice a divisor, say

$$(46) \qquad \sum(P_i) - 2\sum m_j(Q_j) = 0$$

modulo principal divisors.

Let h be a rational function with $(h) = \sum(P_i) - 2\sum m_j(Q_j)$. Then the P_i are the points at which h has odd order. Instead of working with ψ, work with $\widetilde{\psi} = \sqrt{h}\psi$. This eliminates the usual double-valuedness from the correlation functions. On the other hand, the short distance expansion becomes

$$\widetilde{\psi}(P)\widetilde{\psi}(Q) \xrightarrow{P \to Q} \frac{1}{P - Q} h(Q),$$

and this corresponds to the use of our quadratic form $(y, z) = \operatorname{Res}_P(hyz)$ on Y_P.

While the usual formalism with a double-valued ψ makes sense over $\mathbf{C}$, the variant with $\widetilde{\psi}$ makes sense over an arbitrary ground field k. In particular, by eliminating the trivial local square roots from more usual descriptions via $\psi \to \widetilde{\psi} = \sqrt{h}\psi$, we have been able to identify the global square root (i.e., the question of solving (45) with $\sigma \in k^{\times}$) that is really relevant to defining free fermions over an arbitrary ground field k. In the next section, we will prove that free fermions, with arbitrary h, can indeed be defined over any ground field k.

3. Proof of the formula. Here we will indicate the proof of the formula which was stated in the introduction as equation (3) and whose significance for free fermion quantum field theory hopefully became clear in the last section.[2] Recall then that k is a field, X is a curve over k, L is the canonical line bundle of X, and S is a square root of L, defined over k. If h is a rational function on

---

[2]The argument benefited from criticism by E. de Shalit.

X, we write its divisor as

$$(47) \qquad (h) = - \sum_{m_i \text{ even}} m_i(Q_i) - \sum_{n_i \text{ odd}} n_j(P_j).$$

We assume for the moment that the Q_i and P_j are defined over k. The more general case is considered at the end of this section. For each of the points P_j, $j = 1, \ldots, r$, at which h is of odd order, define an element $\alpha_j \in k^\times/(k^\times)^2$ by

$$(48) \qquad \alpha_j = \mathrm{Res}_{P_j}(h\psi_j^2) \quad \mathrm{mod}(k^\times)^2$$

with ψ_j any rational section of S that is of order $(n_j - 1)/2$ at P_j. We will study the quadratic form in r variables

$$(49) \qquad F(X_1, \ldots, X_r) = \sum_{j=1}^r \alpha_j X_j^2$$

and show that it is equivalent over k to the form

$$(50) \qquad G(U_1 \cdots U_s, V_1 \cdots V_s) = \sum_{i=1}^s U_i V_i$$

where $s = r/2$ (note that r is necessarily even). The equivalence of F and G implies in particular that their discriminants are equal, or in other words that

$$(51) \qquad (-1)^{r/2} \prod_{j=1}^r \alpha_j = 1 \quad \mathrm{mod}(k^\times)^2,$$

and this is the formula stated in the introduction. To prove that F and G are equivalent we use the following criterion: a quadratic form F in r variables is equivalent over k to G if and only if there is an isotropic subspace of dimension $r/2$, that is, a subspace of dimension $r/2$ on which $F = 0$.

To prove that F has such an isotropic subspace, the crucial step is to interpret F in the fashion in which it arose at the end of the last section. For each $j = 1, \ldots, r$, let W_j be the one-dimensional k vector space consisting of sections of S of order at least $(n_j - 1)/2$ at P_j, modulo those of order at least $(n_j + 1)/2$. Let $W = \bigoplus_{j=1}^r W_j$. An element of W is an r-plet $(\psi_1, \ldots, \psi_r)$, with $\psi_j \in W_j$, $j = 1, \ldots, r$. On W we place the quadratic form

$$(52) \qquad F(\psi_1, \ldots, \psi_r) = \sum_{j=1}^r \mathrm{Res}_{P_j}(h\psi_j^2).$$

Clearly the form F is precisely that considered in equation (49).

Let E be the line bundle represented by the divisor

$$(53) \qquad (E) = - \sum \frac{m_i}{2}(Q_i) - \sum \frac{n_j - 1}{2}(P_j)$$

with m_i, n_j as in equation (47). The degree of E is $r/2$, so by Riemann-Roch,

$$(54) \qquad \dim H^0(E \otimes S) - \dim H^1(E \otimes S) = r/2.$$

A global section ψ of $E \otimes S$ can be identified as a rational section of S such that $(\psi) + (E) \geq 0$; such a ψ is of order at least $(n_j - 1)/2$ at P_j, for $j = 1, \ldots, r$. There is consequently a natural map

$$\text{(55)} \qquad\qquad \phi \colon H^0(E \otimes S) \to W$$

given by $\phi(\psi) = (\psi, \psi, \ldots, \psi)$. The image of $H^0(E \otimes S)$ in W is an isotropic subspace of W, since for $\psi \in H^0(E \otimes S)$,

$$\text{(56)} \qquad\qquad F(\phi(\psi)) = \sum_{j=1}^{r} \text{Res}_{P_j}(h\psi^2) = 0.$$

(Note that the condition $(\psi) + (E) \geq 0$ means that the residues of the differential form $h\psi^2$ are all zero except at the P_j. (56) is zero since it is the sum of all residues of this differential form.) Therefore, if we can prove that $\phi(H^0(E \otimes S))$ has dimension $r/2$ we will have obtained the required maximal isotropic subspace of W and thus proved the equivalence of F and G.

Clearly,

$$\text{(57)} \qquad \dim(\phi(H^0(E \otimes S))) = \dim H^0(E \otimes S) - \dim(\ker \phi).$$

Now, $\psi \in H^0(E \otimes S)$ is in the kernel of ϕ if and only if its order at each P_j is at least $(n_j + 1)/2$. In this case, $(\psi) + (\widetilde{E}) \geq 0$, where $(\widetilde{E})$ is the divisor

$$\text{(58)} \qquad\qquad (\widetilde{E}) = -\sum \frac{m_i}{2}(Q_i) - \sum \frac{n_j + 1}{2}(P_j).$$

Thus, the kernel of ϕ can be regarded as $H^0(\widetilde{E} \otimes S)$, with $\widetilde{E}$ the line bundle represented by the divisor (58). Since $(E) + (\widetilde{E})$ is the principal divisor (h), the line bundle $\widetilde{E}$ is the inverse of the line bundle E, and so by Serre duality $\dim H^0(\widetilde{E} \otimes S) = \dim H^1(E \otimes S)$. Hence (57) becomes

$$\begin{aligned}
\text{(59)} \qquad \dim(\phi(H^0(E \otimes S))) &= \dim H^0(E \otimes S) - \dim H^0(\widetilde{E} \otimes S) \\
&= \dim H^0(E \otimes S) - \dim H^1(E \otimes S) \\
&= r/2,
\end{aligned}$$

where we have used (57), the identification $\ker \phi = H^0(\widetilde{E} \otimes S)$, the duality of $H^0(\widetilde{E} \otimes S)$ and $H^1(E \otimes S)$, and the Riemann-Roch formula (54). (59) says that $\phi(H^0(E \otimes S))$ is an isotropic subspace of W of dimension $r/2$, as we wished to show. This completes the argument.

So far we have assumed that the points Q_i and P_j at which h vanishes are defined over k. In the general case, (47) is replaced by

$$(h) = -\sum_{m_i \text{ even}} m_i(Q_i) - \sum_{n_j \text{ odd}} n_j(P_j),$$

where Q_i and P_j are prime divisors, defined over k, of degrees s_i and t_j respectively. Thus, P_j would decompose under an extension k' of k into the union of t_j points $P_{j,k}$, $k = 1, \ldots, t_j$.

For each j, let W_j be the k vector space of sections of the spin bundle S that are of order $(n_j - 1)/2$ at P_j, modulo those of order at least $(n_j + 1)/2$. W_j is a k vector space of dimension t_j. W_j is endowed with a natural quadratic form defined as follows. For $\psi \in W_j$, let

$$(60) \qquad\qquad F_j(\psi) = \operatorname{Res}_{P_j}(h\psi^2).$$

The residue operation in (60) can, for instance, be regarded as the sum of residues of the differential form $h\psi^2$ at the points $P_{j,k}$; that is, the right-hand side of (60) can be computed over k', but, being Galois invariant, this gives a quadratic form F_j defined over k. F_j is the generalization of the number α_j, or better the quadratic form in the one variable $\alpha_j X^2$, introduced in (48). F_j is nondegenerate since this is obvious if one works over k'.

The generalization of (49) is the quadratic form

$$(61) \qquad\qquad F = \bigoplus_{j=1}^{r} F_j$$

in $r' = \sum_{j=1}^{r} t_j$ variables.

The generalization of the theorem that (49) is equivalent to (50) is the statement that the quadratic form F is equivalent to the form

$$(62) \qquad\qquad G = \sum_{i=1}^{s'} U_i V_i$$

where $s' = r'/2$. The proof is exactly as before; one uses the Riemann-Roch theorem to construct a space of dimension s' on which $F = 0$.

If we define α_j to be the discriminant of F_j, the equivalence of F and G implies that

$$(63) \qquad\qquad (-1)^{s'} \prod_{j=1}^{r} \alpha_j = 1 \quad \mod (k^\times)^2,$$

and this is the general statement that reduces to our previous assertions when the zeros and poles of h are defined over k.

Note added in proof. D. Kazhdan has recently pointed out that a number theoretic analogue of equation (63) can be found in A. Weil, *Sur Certaines Groupes D'Operateurs Unitaires*, Acta. Math. III 143 (1964), Proposition (5).

REFERENCES

1. M. B. Green and D. J. Gross (eds.), *Unified string theories*, World Scientific, 1986 [especially the articles by O. Alvarez, S. Shenker, D. Friedan, and E. Martinec].

2. E. Witten, *Physics and geometry*, Proc. Internat. Congr. Math. (Berkeley, August 1986), to appear.

3. A. Schellekens and N. Warner, Phys. Lett. **177B** (1986), 317.

4. E. Witten, *The index of the Dirac operator in loop space*, Proc. Conf. Elliptic Curves and Modular Forms in Algebraic Topology (P. Landweber, ed.), Springer-Verlag, to appear.

5. P. Landweber and R. Stong, *Circle actions on spin manifolds and characteristic numbers*, Topology, to appear.

6. S. Ochanine, *Sur les genres multiplicatifs definis par des integrals elliptiques*, Topology **36** (1987), 143.

7. C. Taubes, S^1 *actions and elliptic genera*, Harvard preprint, 1987.

8. A. A. Belavin, A. M. Polyakov, and A. B. Zamolodchikov, *Infinite conformal symmetry in two dimensional quantum field theory*, Nuclear Phys. **B247** (1984), 333.

9. E. Witten, *Quantum field theory, Grassmannians, and algebraic curves*, Princeton preprint, 1987.

10. H. Matsumoto, *Sur les sous-groupes arithmétiques des groupes semi-simples déployés*, Ann. Sci. École Norm. Sup. (4) **2** (1969), 1.

11. A. A. Beilinson, Yu. I. Manin, and V. A. Schechtman, *Localization of the Virasoro and Neveu-Schwarz algebras*, preprint, 1986.

12. T. Y. Lam, *Algebraic theory of quadratic forms*, Benjamin, 1973, Chapter 5.

INSTITUTE FOR ADVANCED STUDY

ABCDEFGHIJ — 898